Sustainable Development and Biodiversity

Volume 17

Series editor

Kishan Gopal Ramawat,
Botany Department, M.L. Sukhadia University, Udaipur, Rajasthan, India

This book series provides complete, comprehensive and broad subject based reviews about existing biodiversity of different habitats and conservation strategies in the framework of different technologies, ecosystem diversity, and genetic diversity. The ways by which these resources are used with sustainable management and replenishment are also dealt with. The topics of interest include but are not restricted only to sustainable development of various ecosystems and conservation of hotspots, traditional methods and role of local people, threatened and endangered species, global climate change and effect on biodiversity, invasive species, impact of various activities on biodiversity, biodiversity conservation in sustaining livelihoods and reducing poverty, and technologies available and required. The books in this series will be useful to botanists, environmentalists, marine biologists, policy makers, conservationists, and NGOs working for environment protection.

More information about this series at http://www.springer.com/series/11920

M.R. Ahuja · S. Mohan Jain
Editors

Biodiversity and Conservation of Woody Plants

Springer

Editors
M.R. Ahuja
Formerly Forestry Consultant
Zobel Forestry Associates
New Paltz, NY
USA

S. Mohan Jain
University of Helsinki
Helsinki
Finland

ISSN 2352-474X ISSN 2352-4758 (electronic)
Sustainable Development and Biodiversity
ISBN 978-3-319-66425-5 ISBN 978-3-319-66426-2 (eBook)
DOI 10.1007/978-3-319-66426-2

Library of Congress Control Number: 2017949151

Printed on acid-free paper

This Springer imprint is published by Springer Nature
The registered company is Springer International Publishing AG
The registered company address is: Gewerbestrasse 11, 6330 Cham, Switzerland

Preface

Biodiversity or biological diversity refers to diversity in the genes, species, community of species, and the ecosystem. Biodiversity includes all living forms on earth. Biodiversity encompasses genetic, species, and ecosystem diversity. Genetic diversity represents the heritable variation within and between populations of a species. Biodiversity is essential for the survival of a species in a changing environment. Climate change creates a new challenge for biodiversity conservation. Species ranges and ecological dynamics are already responding to recent climate shifts, and current reserves/parks will not continue to support all plant/tree species.

Forests cover approximately 31% of the world's total land area, of which 93% is natural forest and only 7% are planted trees. Forest decline is progressing at an alarming rate worldwide. In addition to human activities (logging, deforestation, and forest lands for agriculture and industrial use), a number of other diverse factors, including pests and diseases, drought, soil acidity, radiation, ozone, are cumulatively contributing to global forest decline. The present situation forces us to focus on forest conservation strategies for the present and future. Gene conservation and maintaining of genetic diversity in the forest ecosystems are crucial for the preservation of forest genetic resources. This calls for integrated action for in situ (on site) preservation of forest stands and ex situ (away from the original site) strategies for the conservation of woody plant genetic resources. Selected priority areas include: (1) assessing patterns of genetic diversity and threats, (2) understanding biological processes regulating genetic diversity, (3) assessing the impact of human activities and climate change on genetic diversity, and (4) finding methods to prioritize species and populations for the conservation of forest tree genetic resources. Chapters in this volume are written by leading scientists in their fields of specialization that include woody plant diversity, ecology and evolution, assessment of genetic diversity in forest tree populations, conservation planning under climate change, in situ and ex situ strategies, including biotechnological approaches, in a number of diverse and economically important woody plant species for the conservation of their genetic resources.

This volume provides complete, comprehensive, and broad subject-based reviews for students, teachers, researchers, policy makers, conservationists, and NGOs interested in the field of biodiversity and conservation of woody plants.

New Paltz, USA M.R. Ahuja
Helsinki, Finland S. Mohan Jain

Contents

Part I Genetic Diversity and Conservation

1 **Coconut Genetic Diversity, Conservation and Utilization** 3
V. Arunachalam and M.K. Rajesh

2 **Genetic Diversity and Conservation of Mexican Forest Trees** 37
Christian Wehenkel, Samantha del Rocío Mariscal-Lucero, Juan Pablo Jaramillo-Correa, Carlos Antonio López-Sánchez, J. Jesús Vargas-Hernández and Cuauhtémoc Sáenz-Romero

3 **Climate Change, Genetic Diversity, and Conservation of Paleoendemic Redwoods** 69
M.R. Ahuja

4 **Mulberry (*Morus* Spp.) Genetic Diversity, Conservation and Management** 95
Amalendu Tikader and Kunjupillai Vijayan

5 **Conservation of Forest Genetic Resources in Norway in a Climate Change Perspective** 129
Tore Skrøppa and Kjersti Bakkebø Fjellstad

6 **Use of Molecular Markers in the Conservation Management of Tropical Trees** 155
S. Changtragoon, R. Jalonen and A.J. Lowe

7 **Genetic Diversity Utilization and Conservation of Neem (*Azadirachta indica* A. Juss.)** 197
A.V.C. Silva, A.S. Jesus, A.N.R. Soares and A.S. Ledo

Part II Biodiversity and Conservation

8 **Biotechnological Approaches for Conservation of the Genus *Pistacia*** 221
Hülya Akdemir and Ahmet Onay

9 Biodiversity and Conservation of *Elaeis* Species 245
Maizura Ithnin, Norhalida Mohamed Serdari, Norziha Abdullah, Ahmad Kushairi and Rajinder Singh

10 Conservation of Woody Plant Diversity in Kenya 273
Kimberly E. Medley, John K. Maingi and Christine M. Mutiti

11 Woody Plant Diversity and Density in Selected Eucalyptus and Other Plantation Forest Species in Ethiopia 291
Shiferaw Alem and Muhammad Nakhooda

12 Biodiversity and Conservation of Date Palm 313
Adel A. Abul-Soad, S. Mohan Jain and Mushtaque A. Jatoi

13 Approaches for In Vitro Conservation of Woody Plants Germplasm 355
María Teresa González-Arnao, Natalia Dolce, María Elena González-Benito, Carlos R. Castillo Martínez and Carlos A. Cruz-Cruz

14 Conservation of Hardwood Forest Species 421
E. Corredoira, M.T. Martínez, M.C. Sanjosé and A. Ballester

15 Diversity of Trees in the Mesoamerican Agroforestry System 455
Noel A. González-Valdivia, William Cetzal-Ix, Saikat Kumar Basu, Fernando Casanova-Lugo and Jesús F. Martínez-Puc

16 Native Trees and Shrubs for Ecosystems Services and the Redesign of Resilient Livestock Production Systems in the Mexican Neotropics 489
Francisco Solorio, Luis Ramírez, Saikat Basu, Liz Trenchard, Lucero Sarabia, Julia Wright, Carlos Aguilar, Baldomero Solorio, Armin Ayala and Juan Ku

Contributors

Norziha Abdullah Malaysian Palm Oil Board, Kuala Lumpur, Malaysia

Adel A. Abul-Soad Horticulture Research Institute, Agricultural Research Center, Giza, Egypt

Carlos Aguilar Campus de Ciencias Biológicas Y Agropecuarias, Universidad Autónoma de Yucatán, Mérida, Yucatán, Mexico

M.R. Ahuja New Paltz, NY, USA

Hülya Akdemir Faculty of Science, Department of Molecular Biology and Genetics, Gebze Technical University, Gebze, Kocaeli, Turkey

Shiferaw Alem Department of Forest Botany, Dendrology and Geo-Biocenology, Mendel University in Brno, Brno, Czech Republic

V. Arunachalam Section of Horticultural Science, ICAR-Central Coastal Agricultural Research Institute Ela, Old Goa, India

Armin Ayala Campus de Ciencias Biológicas Y Agropecuarias, Universidad Autónoma de Yucatán, Mérida, Yucatán, Mexico

A. Ballester Instituto de Investigaciones Agrobiológicas de Galicia, IIAG, CSIC, Santiago de Compostela, Spain

Saikat Basu Department of Biological Sciences, University of Lethbridge, Lethbridge, Canada

Saikat Kumar Basu Department of Biological Sciences, University of Lethbridge, Lethbridge, AB, Canada

Fernando Casanova-Lugo Instituto Tecnológico de la Zona Maya, Othón P. Blanco, Quintana Roo, Mexico

Carlos R. Castillo Martínez Laboratorio de Biotecnología Forestal del Centro de Investigación Disciplinaria en Conservación y Mejoramiento de Ecosistemas Forestales, INIFAP, Ciudad de México, Mexico

William Cetzal-Ix Instituto Tecnológico de Chiná, Colonia Centro Chiná, Campeche, Mexico

S. Changtragoon Forest and Plant Conservation Research Office, Department of National Parks, Wildlife and Plant Conservation, Chatuchak, Bangkok, Thailand

E. Corredoira Instituto de Investigaciones Agrobiológicas de Galicia, IIAG, CSIC, Santiago de Compostela, Spain

Carlos A. Cruz-Cruz Facultad de Ciencias Químicas, Universidad Veracruzana, Orizaba, VER, Mexico

Natalia Dolce Instituto de Botánica del Nordeste (IBONE), Universidad Nacional del Nordeste (UNNE)—Consejo Nacional de Investigaciones Científicas y Técnicas (CONICET), Facultad de Ciencias Agrarias, UNNE, Corrientes, Argentina

Kjersti Bakkebø Fjellstad Norwegian Genetic Resource Centre, Norwegian Institute of Bioeconomy Research, Ås, Norway

María Teresa González-Arnao Facultad de Ciencias Químicas, Universidad Veracruzana, Orizaba, VER, Mexico

María Elena González-Benito Departamento de Biotecnología-Biología Vegetal, Universidad Politécnica de Madrid, Ciudad Universitaria s/n, Madrid, Spain

Noel A. González-Valdivia Instituto Tecnológico de Chiná, Colonia Centro Chiná, Campeche, Mexico

Maizura Ithnin Malaysian Palm Oil Board, Kuala Lumpur, Malaysia

S. Mohan Jain Department of Agricultural Sciences, Helsinki University, Helsinki, Finland

R. Jalonen Bioversity International, Serdang, Selangor Darul Ehsan, Malaysia

Juan Pablo Jaramillo-Correa Instituto de Ecología, Universidad Nacional Autónoma de México, Ciudad Universitaria, Coyoacán, CDMX, Mexico

Mushtaque A. Jatoi Date Palm Research Institute, Shah Abdul Latif University, Khairpur, Sindh, Pakistan

A.S. Jesus Embrapa Coastal Tablelands, Aracaju, SE, Brazil

Juan Ku Campus de Ciencias Biológicas Y Agropecuarias, Universidad Autónoma de Yucatán, Mérida, Yucatán, Mexico

Ahmad Kushairi Malaysian Palm Oil Board, Kuala Lumpur, Malaysia

A.S. Ledo Embrapa Coastal Tablelands, Aracaju, SE, Brazil

A.J. Lowe Environment Institute and School of Biological Sciences, University of Adelaide, Adelaide, SA, Australia

Carlos Antonio López-Sánchez Instituto de Silvicultura e Industria de La Madera, Universidad Juárez del Estado de Durango, Durango, Dgo, Mexico

John K. Maingi Department of Geography, Miami University, Oxford, OH, USA

Samantha del Rocío Mariscal-Lucero Instituto de Silvicultura e Industria de La Madera, Universidad Juárez del Estado de Durango, Durango, Dgo, Mexico

M.T. Martínez Instituto de Investigaciones Agrobiológicas de Galicia, IIAG, CSIC, Santiago de Compostela, Spain

Jesús F. Martínez-Puc Instituto Tecnológico de Chiná, Colonia Centro Chiná, Campeche, Mexico

Kimberly E. Medley Department of Geography, Miami University, Oxford, OH, USA

Christine M. Mutiti Department of Biological and Environmental Sciences, Georgia College, Milledgeville, GA, USA

Muhammad Nakhooda Department of Biotechnology and Consumer Sciences, Faculty of Applied Sciences, Cape Peninsula University of Technology, Cape Town, South Africa; School of Life Sciences, College of Agriculture, Engineering and Science, University of KwaZulu-Natal, Durban, South Africa

Ahmet Onay Faculty of Science, Department of Biology, Dicle University, Diyarbakır, Turkey

M.K. Rajesh Division of Crop Improvement, ICAR-Central Plantation Crops Research Institute, Kasaragod, Kerala, India

Luis Ramírez Campus de Ciencias Biológicas Y Agropecuarias, Universidad Autónoma de Yucatán, Mérida, Yucatán, Mexico

M.C. Sanjosé Instituto de Investigaciones Agrobiológicas de Galicia, IIAG, CSIC, Santiago de Compostela, Spain

Lucero Sarabia Campus de Ciencias Biológicas Y Agropecuarias, Universidad Autónoma de Yucatán, Mérida, Yucatán, Mexico

Norhalida Mohamed Serdari Malaysian Palm Oil Board, Kuala Lumpur, Malaysia

A.V.C. Silva Embrapa Coastal Tablelands, Aracaju, SE, Brazil

Rajinder Singh Malaysian Palm Oil Board, Kuala Lumpur, Malaysia

Tore Skrøppa Department of Forest Genetics and Biodiversity, Norwegian Institute of Bioeconomy Research, Ås, Norway

A.N.R. Soares Embrapa Coastal Tablelands, Aracaju, SE, Brazil

Baldomero Solorio Campus de Ciencias Biológicas Y Agropecuarias, Universidad Autónoma de Yucatán, Mérida, Yucatán, Mexico

Francisco Solorio Campus de Ciencias Biológicas Y Agropecuarias, Universidad Autónoma de Yucatán, Mérida, Yucatán, Mexico

Cuauhtémoc Sáenz-Romero Instituto de Investigaciones Agropecuarias y Forestales, Universidad Michoacana de San Nicolás de Hidalgo, Tarímbaro Michoacán, Mexico

Amalendu Tikader Bangalore, Karnataka, India

Liz Trenchard Centre for Agroecology, Water and Resilience, Coventry University, Coventry, UK

J. Jesús Vargas-Hernández Programa Forestal, Colegio de Postgraduados, Puebla, Mexico

Kunjupillai Vijayan Central Silk Board, Bangalore, India

Christian Wehenkel Instituto de Silvicultura e Industria de La Madera, Universidad Juárez del Estado de Durango, Durango, Dgo, Mexico

Julia Wright Centre for Agroecology, Water and Resilience, Coventry University, Coventry, UK

Part I
Genetic Diversity and Conservation

Chapter 1
Coconut Genetic Diversity, Conservation and Utilization

V. Arunachalam and M.K. Rajesh

Abstract Coconut palm is an important economic plant species in fragile coastal and island ecosystems of the tropics. The palm is a multipurpose perennial crop whose diversity at spatial and trait levels has been exploited to improve yield, resistance to diseases and develop value-added products. Genetic erosion is rampant in coconut due to biotic, abiotic and anthropogenic threats and expansion in area under high yielding selections and hybrids. Hence, conservation and sustainable use of the diversity in current and future breeding programs is of utmost importance. In this chapter, we have focussed on the baseline information of known diversity at various geographical locations and at various traits. Morphological and molecular approaches, employed to assess the extent of genetic diversity, have been highlighted. An overview of state of conservation efforts and utilization of the existing diversity, in developing varieties/hybrids with one or more desirable traits, has also been presented in this review.

Keywords Coconut · Diversity · Genetics · Morphology · Markers

1.1 Introduction

Palms, members of the botanical family Arecaceae, form an important group of plants which serve mankind in numerous ways and are considered next only to grasses and legumes in importance and usefulness. They are harvested from wild

V. Arunachalam
Section of Horticultural Science, ICAR-Central Coastal Agricultural Research Institute Ela, Old Goa, Goa 403402, India
e-mail: v.arunachalam@icar.gov.in

M.K. Rajesh (✉)
Division of Crop Improvement, ICAR-Central Plantation Crops Research Institute, Kasaragod, Kerala 671124, India
e-mail: rajesh.mk@icar.gov.in

M.R. Ahuja and S.M. Jain (eds.), *Biodiversity and Conservation of Woody Plants*, Sustainable Development and Biodiversity 17, DOI 10.1007/978-3-319-66426-2_1

stands and forests, cultivated in plantations and home gardens or used in landscaping indoor and outdoor environments. Coconut (*Cocos nucifera* L.), oil palm (*Elaeis guineensis* Jacq.), date palm (*Phoenix dactylifera* L.), peach palm (*Bactris gasipaes* Kunth), babacu (*Attalea cohune* Mart), areca nut (*Areca catechu* L.) and palmyrah (*Borassus flabellifer* L.) are the major palm species cultivated on large scale.

Among these, coconut forms a major life-supporting species in tropical coastal and island ecosystems. The coconut palm is a perennial, multipurpose tree providing multifarious products for domestic and industrial purposes, both edible and non-edible. Grown both under plantation and homestead management system, coconut exerts a profound influence on the rural economy providing livelihood security to several millions of people across the world. Today, the role of coconuts as a health and nutraceutical food is increasingly being accepted.

Known since antiquity, coconut has spread across oceans, coconut populations have developed as a result of natural selection, and the history of coconut is interlinked with human migrations (Gunn et al. 2011). Coconut has no known wild relatives, and coconut accessions are considered to comprise of a solitary and a freely interbreeding gene pool. DNA-based investigations of around 1300 accessions, representing collections from around the globe, have revealed that coconut was brought under cultivation in two distinct regions, corresponding to the Indian Ocean and the Pacific Ocean basins (Gunn et al. 2011).

Predominantly a small and marginal holder's crop, coconut sector is currently facing numerous challenges which include fluctuating prices, lack of export/domestic market competitiveness, global warming as well as low levels of value-addition and product diversification. Coconut genetic diversity, which is the foundation for sustainable production, is facing severe threats due to genetic erosion brought about by natural and human factors. Therefore, conservation of these genetic resources holds vital significance.

Genetic diversity of coconut is exhibited in wide array of phenotypic traits mainly in stature of plant (Narayana and John 1949; Sugimura et al. 1997; Bourdeix et al. 2001; Arunachalam and Rajesh 2008) and composition of fruit (Harries 1978; Ashburner et al. 1997). Conservation of coconut genetic diversity is a major concern of national and international coconut research communities. Field gene banks and in situ conservation, complemented by cryopreservation, are the potential options to conserve the genetic diversity for future generations. International Coconut Genetic Resources Network (COGENT), under the Bioversity International, has been coordinating the collection, conservation and documentation of coconut germplasm, farmers' varieties and landraces worldwide (Batugal 2005).

Utilization of diversity in coconut breeding programs is limited to developing varieties resistant to lethal yellowing disease and/or exploiting hybrid vigour expressed by crossing dwarf and tall forms. *Makapuno*, the mutation resulting in jelly-like endosperm, has been gainfully employed in coconut research programs in the Philippines for obtaining value-added products. This chapter covers a gist of rich diversity existing in coconut and means to conserve them for their utilization in future coconut breeding programs.

1.2 Location-Specific Diversity

Diversity, spanning the global reach of coconut, is useful to enrich the gene pool. It helps to obtain alleles originating from the process of adaptation at locations differing in climatic, edaphic and ethnic environments and exposed to a range of pests and diseases. Although typically found in tropical shorelines, the coconut palm has adapted to hinterlands too. Some unexpected and interesting locations include the Dhofar coast of the Oman and Hainan island of China. Introduced coconut populations from Sri Lanka and Malaysia have survived in the dry regions of Oman in the Persian Gulf (Perera et al. 2011) and evolved under both natural and human selections. On the other extreme, coconut has also adapted to low temperatures of Hainan Island in China where Hainan Tall, the local cultivar, is cold and wind hardy (Mao and Lai 1993). Also, The Nilgiris of Tamil Nadu, South India, is a hill station at high altitude, but in normal tropical latitude and longitude where coconut plantations are located (Rahman et al. 2012). However, detailed diversity and adaptation studies of these populations are yet to be undertaken.

Bourdeix et al. (2005a) have reported geo-referencing of 710 distinct coconut accessions at the conservation sites with 579 collecting sites possessing unique latitude and longitude. They have also used geographical information system (GIS) tools to map the diversity at specific geo-locations. Bourdeix et al. (2005b) have suggested collecting from locations, which are not either represented in the gene banks or those suffering genetic erosion, to maintain the genetic diversity.

Being a coastal species and well-adapted to island ecology, hotspot of coconut diversity is found in Indian and Pacific Ocean islands. Size of the fruit and husk proportion in natural stands of coconut across the geographical locations (Harries 1978) have led to the identification of two major types: '*Niu Kafa*' wild types, which are angular and thick husked and presumed to have evolved naturally, and '*Niu Vai*' cultivated types, which are circular and thin husked. This seminal work gave invaluable insights on evolution of coconut by domestication, by selection for low husk content. Among the coconut populations of South Pacific region, a continuum of fruit size and husk content across West to East regions was reported by Ashburner et al. (1997). In the same study, clinal variation was observed to increase from small fruit and low husk content in the West to large fruit with high husk in East of the region. A set of case studies, highlighting geographical-level variations in phenotypic traits, is listed as Table 1.1.

1.3 Trait-Specific Diversity

There are two forms of coconut mainly based on stature: 'Talls' and 'Dwarfs' (Narayana and John 1949). Tall ('typica'), dwarf ('nana') and intermediate ('javanica') forms have been characterized in coconut germplasm from the Philippines using agro-morphological traits (Sugimura et al. 1997). Fruit component analyses, by dissecting the fruit and analysing the kernel, shell, husk contents, are an easy and

Table 1.1 Coconut diversity reports from different geographical locations

Country/region	Traits/populations	References
Vietnam	Landraces	Le et al. (1999)
Hainan (China)	Fruit colour	Xiaoping et al. (2004)
	Cold tolerance	Mao and Lai (1993)
The Philippines	–	Gruezo and Harries (1984)
Indonesia	Morphometric traits	Liyanage (1974)
Bangladesh	Morphometric traits	Islam et al. (2007)
Sri Lanka	Morphological traits	Liyanage (1958)
Lakshadweep Islands (India)	Fruit components	Jacob and Krishnamoorthy (1981), Samsudeen et al. (2006)
Goa-Konkan coast (India)	Morphometric traits	Ratambal et al. (2000)
Andaman and Nicobar Islands (India)	–	Abraham et al. (2008)
Odisha (East coast of India)	Morphometric traits	Kumaran et al. (2004)
Zanzibar (Tanzania)	Morphological traits	Krain et al. (1994)
Kenya	Morphological traits	Oyoo et al. (2015)
Comoro, Reunion, Seychelles	Morphometric traits	Kumaran et al. (2000)
Costa Rica	Fruit components	Vargas and Blanco (2000)
Baracoa (Cuba)	Morphological traits	Alonso et al. (2007)
Mexico	Fruit components	Villarreal et al. (1993, 2005)
Bahia (Brazil)	Fruit components	Romney and Dias (1979)
Keeling Island (Australia)	Fruit components	Leach et al. (2003)
Papua New Guinea	Fruit components	Ovasuru (1993)
South Pacific islands	Fruit components	Ashburner et al. (1997)

reliable means to classify the coconut into '*Niu Kafa*' or the wild type (large fruit, high husk, less kernel) and '*Niu Vai*' or the domesticated/cultivated type (medium fruit, less husk, more kernel) (Harries 1978). Foliar traits (Arunachalam et al. 2005) and fruit component traits (Ashburner et al. 1997) have also been used to describe coconut populations from the Pacific region. The trait-level variations in coconut were reviewed in our previous study (Arunachalam 1999).

Polyembryony has been reported in coconut (Davis et al. 1953; Arunachalam 2005) (Fig. 1.1), areca nut (Das 1966), palmyrah (Veerasamy 1982), date palm (Rao 1957) and needle palm (Clancy and Sullivan 1988). One of the weak partners of the twin has potential to develop into haploids in a few cases (Ninan and Raveendranath 1965).

Fig. 1.1 Twin coconut seedlings observed in West Coast Tall cultivar

Sprouting speed and degree of leaf splitting usually vary among the varieties; dwarfs sprout early and split quickly (Fernando et al. 2010). Harries (2012) considered germination rate as a key and significant characteristic determining coconut genetic diversity. Petiole colour is also a distinct trait useful in hybrid seedling identification (Saint and de Lamothe 1987; Arunachalam and Rajesh 2008). Table 1.2 list variations found at the seedling level in coconut palm, while the details of variations for stem, leaf and floral traits are given in Table 1.3. Quantitative trait loci (QTL) analysis for early germination trait, carried out by Herran et al. (2000), resulted in the identification of six loci; the identified QTLs were correlated with important characters viz., early flowering and yield.

Bulbil palms are characterized by production of bulbil shoots in leaf axils in place of normal inflorescence (Davis et al. 1981b; Jerard et al. 2014c) (Fig. 1.2) with the primary and secondary bulbil shoots possessing the ability to grow as individual palms.

Spicata palms possess exceptional floral traits with a large number of female flowers formed on the unbranched spadix and an obvious decrease in the number of male flowers (Fig. 1.3). In such palms, ratio of male to female flowers is 0.25: 1 in comparison to 1: 0.0036 in ordinary talls (Arunachalam et al. 2014b). The trait is controlled by a single gene in heterozygous state (Ss) (Ninan and Satyabalan 1963). Cytological studies of spicata palms have revealed several aberrations during meiosis (Ninan et al. 1960).

Importance of the diversity for fruit traits across the coconut germplasm for size, shape and colour is invaluable for the breeding programs (Table 1.4). The fruit size in coconut ranges from 31 g in Laccadive Mini Micro Tall of India (Jerard et al. 2014a) to nearly 2.3 kg in San Ramon Tall of the Philippines (Fernando 1998) (Figs. 1.4 and 1.5). The spectrum of variation of nearly 100-fold variation in fruit

Table 1.2 Range of variation for seedling traits in coconut

Trait	Range of variation	Population	References
Seedling/seed	One, two, three or more	Many	Arunachalam (2005), Davis et al. (1953)
Germination	Vivipary	Andaman Green Dwarf	Sankaran et al. (2012)
		Thai Tall	Dootson et al. (1989)
		Malayan Dwarfs	Shareefa et al. (2014)
Germination speed	Quick sprouting	Malayan Dwarf	Villarreal and Morín (1998)
	Slow and late sprouting	Tall varieties	Harries (1981)
Petiole colour	Green	Chowghat Green Dwarf, Malayan Green Dwarf	Many reports
	Brown	Sri Lankan Brown Dwarf	Perera et al. (1997)
	Orange/Red	Malayan Orange Dwarf, Chowghat Orange Dwarf	Many reports
	Yellow	Malayan Yellow Dwarf	Many reports
Albino		–	Louis (1991)
Splitting seedling leaves	Number of split leaves after 9 months of sowing	2 in dwarfs; 0.01 in talls and 0.3 in dwarf x tall	Fernando et al. (2010)
Seedling leaves	Rosette	Normal	Arunachalam et al. (2001)

size available within a single species offers scope to explore the adaptive mechanisms involved in evolution of fruit size. Some of the variations observed in fruit sizes in coconut populations from Minicoy, Lakshadweep islands, India, are given in Fig. 1.5. Baudouin et al. (2006) identified 52 putative QTLs for 11 fruit component traits, which included fruit composition and yield. Thirty-four of these QTLs were found to be grouped in six small clusters. QTLs underlying fruit component weight, endosperm humidity and fruit production were located at different positions in the genome.

Even though cultivars which are green fruited are common in coconut, a range of fruit colours from green, orange, yellow and to shades of red can be seen (Fig. 1.6).

Colour of the fruit could further be quantified by digital phenotyping techniques (Arunachalam et al. 2013) by latest image analysis software. The pink husk form (Kumaran et al. 2004) is an interesting material for investigating the inheritance and evolution of fruit colour in coconut.

The variability in coconut for fruit component traits (Table 1.5) has been the focus of investigation of many researchers to classify coconut populations. Husk content (Fig. 1.7) is important in the adaptation of coconut fruits to float in seawater

Table 1.3 Stem, foliar and floral traits in juvenile and adult palms

Trait	Novel variation	Population	References
Persistent leaves	Remains attached	Hutbay Island, Little Andamans (India)	Rao and Kumar (1998)
Leaf splitting	Plicata	Tiptur Tall (India)	Arunachalam et al. (2001)
Number of spathes	One, two, three or more	–	Michael (1963), Arunachalam et al. (2001)
Branching of inflorescence	No branching, primary and secondary branching	Tiptur Tall, East Coast Tall (India)	Arunachalam et al. (2001, 2014b)
Leafy shoots	Bulbils	–	Davis et al. (1981a), Jerard et al. (2014c)
Sex expression	Hermaphrodite	Nias Yellow Dwarf (Indonesia)	Davis et al. (1981b)
Sex expression	Monoecious, high degree of femaleness	Spicata (India)	Jacob (1940)
		Multiple spicata (India)	Sankaran et al. (2016)
Pollination habit	Self pollination	Dwarf varieties	Sangare et al. (1978)
Flowering/bearing	Precocious 3–4 years	Dwarf varieties	Manthiratna (1972)
Flowering in seedling	Midget-axillary Hapaxanthic terminal	–	Davis and Perianayagam (1985)
Inter- and intraspadix overlapping among male and female phases	Present	Dwarf varieties	Sangare et al. (1978)
	Less	Tall varieties	Ratnambal et al. (2003)

for several days before it reaches new island or coastal destinations (Harries 1978). In contrast, the water content is an important domesticated trait (Ashburner et al. 1997) preferred by humans, especially sea-farers, to carry coconuts during long voyages. Breaking of tough coconuts manually by hand is a famous gaming activity in festivals in many coconut-growing locations (Panda 1982). Farmers prefer to identify palms which provide coconut fruits with hard shell, and fruits from these palms are specifically used in games such as 'fighting nut' ('Ladahi'-in Hindi means fight).

Fig. 1.2 A coconut palm with numerous bulbils in its crown. [*Photo courtesy* Dr. S. Elain Apshara]

Fig. 1.3 Nuts (**a**) and inflorescence (**b**) of spicata palm. [*Photo courtesy* Dr. H.P. Maheswarappa]

Table 1.4 Fruit and tender coconut traits

Trait	New variation	Population	References
Shedding of fruits	Remain attached with plant	Takame	Gangolly et al. (1957)
Shape of fruit	Triangular	Lakshadweep Ordinary Tall (Lakshadweep Islands, India)	Jacob and Krishnamoorthy (1981)
Colour of fruit	Orange	Chowghat Orange Dwarf (India), Malayan Orange Dwarf (Malaysia), Kenthali Orange Dwarf (India)	Nair et al. (2016)
	Yellow	Malayan Yellow Dwarf (Malaysia), Nias Yellow Dwarf (Indonesia), Pemba Yellow Dwarf (Mauritius)	Nair et al. (2016)
	Green	Chowghat Green Dwarf (India), Malayan Green Dwarf (Malaysia), Pemba Green Dwarf (Mauritius)	Nair et al. (2016)
	Red	Cameroon Red Dwarf (Africa), Vanuatu Red Dwarf (Vanuatu)	Nair et al. (2016)
Husk texture	Soft and edible	Kaithathali Tall (Lakshadweep Islands, India)	Menon and Pandalai (1958)
Size of fruit	Smallest fruit	Laccadive Mini Micro (Lakshadweep Islands, India)	Jerard et al. (2014a)
	Largest fruit	San Ramon (The Philippines)	Fernando (1998)
Horns on fruit	Horns on pericarp	Horned Cocos (Andaman Islands, India)	Jerard et al. (2014b)
Pericarp/husk colour	Pink	Narangi (Odisha, India)	Kumaran et al. (2004)
Stylar end of fruit	Beaked	Wild coconut populations (Andaman and Nicobar Islands, India)	Balakrishnan and Nair (1979)
Husk attachment to nut	Easily removable	Kelapa Bawangi (Indonesia)	Davis et al. (1985)

1.4 Characterizing Diversity Using Molecular Markers

It been advocated that the best approach for the utilization and conservation of biodiversity is an integrated one involving the traditional approaches in combination with modern biotechnological applications, mainly to determine phylogenetic relationships, to recognize redundancies in a germplasm bank and to identify useful genes in the germplasm. Molecular markers constitute efficient and accurate tools for effective and rapid assessment of genetic diversity. The use of molecular markers is, however, expensive, and therefore, their employment for characterization of genetic diversity must be undertaken in a judicious manner.

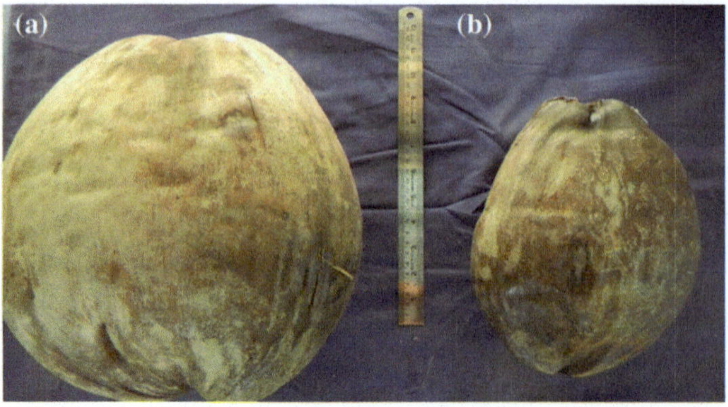

Fig. 1.4 Comparison of nut size of mature fruit of San Ramon Tall (**a**) with a normal nut (**b**). [*Photo courtesy* Dr. H.P. Maheswarappa]

Fig. 1.5 Variations for fruit size observed in coconut populations from Minicoy, Lakshadweep Islands, India. Mature and dehusked nuts of Laccadive Ordinary Tall (**a** and **b**), Laccadive Micro Tall (**b** and **e**) and Laccadive Mini Micro tall (**e** and **f**). [*Photo courtesy* Mr. M.I. Arif]

The advantages of use of molecular markers for characterization of genetic diversity in coconut were first mooted by Rohde (1993). Since then, a repertoire of DNA-based markers viz. inverse sequence tagged repeat (ISTR), restricted

Fig. 1.6 Diversity seen in fruit colour **a** Kenthali Orange Dwarf, **b** Malayan Yellow Dwarf, **c** Chowghat Green Dwarf, **d** Chowghat Orange Dwarf and **e** Malayan Green Dwarf. [*Photo courtesy* Dr. H.P. Maheswarappa]

fragment length polymorphism (RFLP), random amplified polymorphic DNA (RAPD), amplified fragment length polymorphism (AFLP), simple sequence repeats (SSR), inter simple sequence repeats (ISSR) and start codon targeted amplification (SCoT) have been utilized for genetic diversity assessment in coconut review by Niral et al. (2016).

Since their development (Rivera et al. 1999; Perera et al. 2000), the use of polymorphic microsatellites for assessing genetic diversity in coconut has been the most popular till date. A collaborative effort between CIRAD (Centre de Coopération Internationale en Recherche Agronomique pour le Développement, France) and COGENT (he International Coconut Genetic Resources Network) resulted in thc development of development of a set of 14 microsatellite markers ('CnCir' series) with sufficient discriminating power for undertaking genetic diversity studies of coconut (Baudouin and Lebrun 2002). Many studies using SSR markers have revealed the existence of a much higher genetic diversity in tall accessions compared to dwarfs (Rivera et al. 1999; Perera et al. 2000, 2003) which is a reflection of their breeding habits: talls being allogamous, while dwarfs generally autogamous. SSR profile of 25 genotypes of Tiptur Tall variety using the primer CnCir87 is provided in Fig. 1.8.

Island coconut populations have been reported to harbour high genetic diversity in coconut as revealed by SSR-based studies carried out in coconut accessions from Andaman and Nicobar islands, India (Rajesh et al. 2008b) and Lakshadweep Islands, India (Devakumar et al. 2010; Rajesh et al. 2014b). These results highlight the need for undertaking efforts for conservation of the rich genetic diversity in these islands and evaluation of these accessions for use in future breeding programs. In contrast to the above results, Perera et al. (2001) reported a high level of within population variation in coconut populations endemic to Sri Lanka which led to the conclusion of a common evolutionary history and a narrow genetic base for the native Sri Lankan coconut populations. A list of genetic diversity studies undertaken in different coconut populations utilizing microsatellite markers is provided in Table 1.6.

Table 1.5 Variability in fruit and component traits

Trait	Variation	Population	References
Husk	High proportion of husk	Niu Afa (1608 g/fruit)	Ashburner et al. (1997)
Water content	High water content in fruit	Niu Vai (998 g/fruit)	Ashburner et al. (1997)
Endosperm	Sweet and soft endosperm	*Mohachao Narel* (Maharashtra, India)	Samsudeen et al. (2013b)
	Soft and curdy endosperm	*Makapuno* (The Philippines)	Torres (1937)
		Thairu thengai (Andaman Islands, India)	Jerard et al. (2013)
		Dikiri Pol (Sri Lanka)	Peries (1995)
		Kopyor (Indonesia)	Santoso et al. (1996)
		Sap (Vietnam)	Le et al. (1999)
Endosperm	Pineapple fragrance	Dua (Vietnam)	Le et al. (1999)
Shell	Thick and hard	Ladhei (Odisha, India)	Panda (1982)
Embryo colour	Pink	Guelle Rose (Mauritius)	Kumaran et al. (2000)
Suitability as tender nut	High water content, high total soluble solids, low acidity, high potassium low sodium	Chowghat Orange Dwarf (Kerala, India)	Dhamodaran et al. (1993)
Size of the mature fruit	Large	Kappadam Tall (Kerala, India)	Narayana and John (1949)
Copra size and quality	Superior	Tall varieties	Rangaswami (1977)
Oil content	High (>72%)	Laccadive Micro Tall (Lakshadweep Islands, India)	Patil et al. (1993)

1.5 Conservation

Conservation of coconut genetic resources is important to maintain sustainability of the crop in the long run. The available coconut variability needs immediate conservation to protect them from threats of genetic erosion, expansion in area under high yielding selections and hybrids, climate change and sea level rise and emerging pests and diseases. Most germplasm collections in national coconut research programs suffer from lack of sufficient number of palms per accession

Fig. 1.7 Angular and thick husked fruits (**a** and **c**) and spherical and thin husked fruits (**b** and **d**) from Minicoy, Lakshadweep Islands, India. [*Photo courtesy* Mr. M.I. Arif]

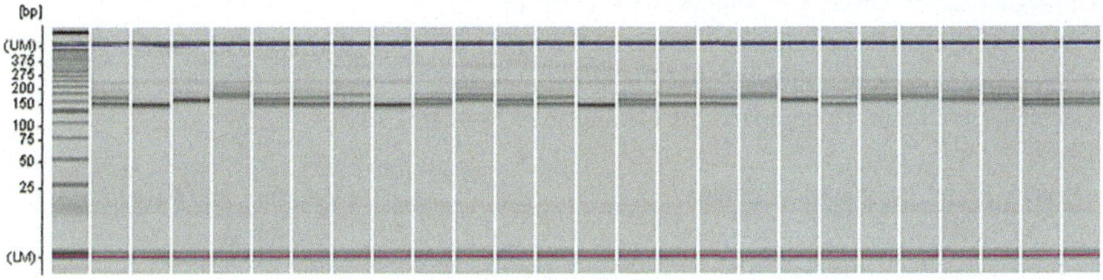

Fig. 1.8 SSR profile of 25 genotypes of Tiptur Tall variety generated in an automated electrophoresis unit using the primer CnCir87

leading to genetic bottleneck. To obtain a clear representation of coconut genetic diversity and for promotion of germplasm exchange, the CGRD (Coconut Genetic Resources Database) has been created. The database is an user-friendly computerized catalogue of coconut accessions comprising a large number of cultivars from all coconut-growing regions of the world (Hamelin et al. 2005).

A detailed strategy has been envisaged by Rao et al. (1998) for efficient conservation of coconut genetic resources. It has been emphasized that coconut accessions maintained in different gene banks have to be duplicated for both safety and promoting sustainable genetic resources conservation; one way of achieving this would be targeted exchanges between germplasm conservation centres (Bourdeix et al. 2005a). However, comprehensive exchange of coconut germplasm

Table 1.6 Genetic diversity studies using microsatellite markers

Population(s)	References
Coconut varieties/populations (Different geographical range)	Rivera et al. (1999)
Coconut varieties/populations (Different geographical range)	Teulat et al. (2000)
Tall coconut accessions (Sri Lanka)	Perera et al. (2001)
Coconut varieties/populations (Entire geographical range)	Perera et al. (2003)
Coconut accessions (Florida, USA)	Merrow et al. (2003)
Coconut landraces (Southern India)	Rajesh et al. (2008a)
Coconut populations (Andaman and Nicobar Islands, India)	Rajesh et al. (2008b)
Panama Tall, Malayan Yellow Dwarf, Jamaica Tall (Jamaica)	Baudouin et al. (2008)
Tall coconut populations (Brazil)	Ribeiro et al. (2010)
Coconut populations from Agatti and Kavaratti Islands (Lakshadweep, India)	Devakumar et al. (2010)
Coconut populations (Dominican Republic)	Martinez et al. (2009)
Tall coconut accessions of International Coconut Genebank for Africa and Indian Ocean (ICG-AIO), Côte d'Ivoire.	Yao et al. (2013)
Ecotypes (Kerala State, Southern India)	Thomas et al. (2013)
Tall ecotypes (Kerala State, Southern India)	Rajesh et al. (2014a), Manjula et al. (2014)
Yellow Dwarf forms (Sri Lanka)	Kamaral et al. (2014)
Coconut populations from Amini and Kadmat Islands (Lakshadweep, India)	Rajesh et al. (2014b)
Chowghat Green Dwarf and Malayan Green Dwarf	Thomas et al. (2015)
Coconut populations from coastal lowlands (Kenya)	Oyoo et al. (2016)
Tall coconut accessions of the International Coconut Genebank for Latin America and the Caribbean (ICG-LAC), Brazil	Loiola et al. (2016)
Dwarfs (Entire geographical range)	Perera et al. (2016)

among coconut-growing countries has been limited. For efficient and economic management of gene banks, use of morphological characterization complemented by molecular tools has been recommended for rationalization of large collections in gene banks so as to trim down the actual number of accessions in these centres (Bourdeix et al. 2005b). The ultimate rationale of conservation of germplasm is its utilization and, accordingly, any strategy for germplasm conservation should comprise of mechanisms which will guarantee germplasm access to all relevant stakeholders (Dulloo et al. 2005). Generally, two approaches of conservation are recognized in coconut viz. ex situ (field gene banks, seed banks and in vitro collections) and in situ (farm and home gardens), which are regarded complementary to each other (Rao et al. 1998).

1.5.1 Seed Banks

Even though seeds as a convenient form of storage, seed conservation is quite cumbersome in coconut given the allogamous nature (of tall accessions), large size of nuts and inherent recalcitrant nature (Dulloo et al. 2005).

1.5.2 Field Gene Banks

Field gene banks are preferred manner for conservation of coconut genetic diversity, even though they are expensive and time-consuming and require large land areas to meet sample size of each population, given the biological characteristics of coconut. Many efforts have been made to conserve the germplasm in the field gene banks in all major coconut-growing countries (Rao et al. 2005). These efforts have resulted in the establishment of five multisite International Coconut Genebanks (ICG) in five major coconut-growing regions viz. India, Indonesia, Côte d'Ivoire, Papua New Guinea and Brazil, under the International Coconut Genetic Resources Network (COGENT), representing South Asia, South East/East Asia, Africa/Indian Ocean, Pacific/Oceania and Latin America/Caribbean regions, respectively (Batugal and Jayashree 2005). The main objectives of establishing these ICG include conservation of identified regional/national/international coconut diversities, assessment of diversity and evaluation of the performance of coconut germplasm conserved in these gene banks and to provide germplasm to other interested countries, based on material transfer agreements (Batugal and Jayashree 2005).

The relative advantages of field gene banks include ease of characterization and requirement of simple infrastructure. However, some key issues which need to be addressed with respect to field gene banks include (i) deciding the least number of palms which would be required to preserve representative genetic diversity, (ii) standardization of field plot techniques for characterizing and evaluating the conserved germplasm, (iii) rationalization of collections, (iv) development of core collections and (v) working out the economics of maintenance of coconut collections (Dulloo et al. 2005). There is also an urgent need to updating and rationalizing the collection in the gene banks, some of which are quite old, and weed out germplasm duplicates (Dulloo et al. 2005). However, till date, exchange of coconut germplasm among the member countries has been restricted because of various issues (Bourdeix et al. 2005a).

1.5.3 In Vitro *Culture*

The traditional method of collecting seed nuts during germplasm expeditions from remote islands/regions had three major limitations: the bulkiness of the nuts (which

Fig. 1.9 Bunch of Horned Coconut

can restrict the number of accessions which can be collected from remote locations and increase the transport expenses), its recalcitrant nature and quarantine regulations. In vitro embryo culture has become an invaluable tool for collection, from remote locations, of coconut germplasm in the form of zygotic embryos and its subsequent retrieval in laboratory (Engelmann et al. 2005). This technique can overcome the limitations of collection of germplasm as seed nuts.

Embryo culture/rescue has become an invaluable tool to conserve rare mutants such as Laccadive Mini Micro Tall (the smallest known coconut; from Lakshadweep Islands, India), *Makapuno* (a soft, jelly-like endosperm type from the Philippines), *Thairu Thengai* (a soft, jelly-like endosperm type from Andaman Islands, India), Horned Coconut (with thick pericarp from Andaman Islands, India) (Fig. 1.9) and *Mohachao Narel* (a sweet and soft endosperm type from Western India) (Figs. 1.10 and 1.11). The nuts either fail to germinate under natural situations, due to limitations of food reserves in Laccadive Micro Tall and unfavourable endosperm in *Makapuno*, *Mohachao Narel* and *Thairu Thengai*, or take a long time to germinate, as in Horned Coconut because of the thick pericarp.

Coconut tissue culture has still not been standardized given its recalcitrant nature. Of the different explants tested, plumular tissues have responded the best to in vitro culture (Chan et al. 1998; Rajesh et al. 2005, 2014c).

Fig. 1.10 Nuts of normal and sweet kernelled coconut genotype (*Mohachao Narel*). [*Photo courtesy* Mr. P. Ajith Kumar]

Fig. 1.11 Retrieval of sweet kernelled coconut genotype (*Mohachao Narel*) through embryo rescue technique. [*Photo courtesy* Dr. Anitha Karun]

1.5.4 Cryopreservation

Cryopreservation of coconut zygotic embryos and pollen represents potential alternatives to long-term conservation of coconut germplasm, requiring only limited

Fig. 1.12 **a** Germination of pollen of West Coast Tall cultivar cryopreserved for 6 years **b** Nut set observed in Chowghat Orange Dwarf pollinated with pollen of West Coast Tall cryopreserved for 6 years. [*Photo courtesy* Dr. Anitha Karun]

space and maintenance costs. Assy-Bah and Engelmann (1992) were the first to successfully demonstrate use of liquid nitrogen (−196 °C) in cryopreservation of coconut zygotic embryos and subsequently, its application in long-term conservation of coconut germplasm. Sajini et al. (2011) have standardized cryopreservation of zygotic embryos by a simple vitrification protocol; the protocol has been successfully validated in diverse accessions. Long-term storage of coconut pollen under cryopreservation represents an important additional technique for genetic resources conservation, by allowing conservation of genes. Karun et al. (2014) have standardized coconut pollen cryopreservation, opening up avenues for establishment of pollen cryobanks as an adjunct technique of long-term germplasm conservation (Fig. 1.12a). Normal seed set could be observed from coconut pollen cryopreserved for up to 6 years (Fig. 1.12b). N'Nan et al. (2008) could achieve regeneration of leafy shoots from coconut plumules after cryopreservation by encapsulation–dehydration.

1.5.5 In Situ *and on Farm Conservation*

In situ or on farm conservation is the best method for conservation of coconut landraces as it presents an opportunity for the palm to evolve under natural/farmer-enforced situations. This method of conservation would be more effective if it involves active understanding and participation of local communities and ultimately be beneficial to these communities in the long run (Bourdeix et al. 2005a). The initial step in such an approach would be to collect sufficient data in situ (e.g. on fruit components and genetic erosion) and carry out molecular marker studies to assess the extent of genetic diversity/uniqueness of the selection population. Further analysis of the data obtained during the initial exploration phase could be carried out using GIS techniques. Based on the result obtained from these studies, novel/threatened/useful coconut genetic diversity can be identified for addition to the gene banks (Bourdeix et al. 2005a, b). Samsudeen et al. (2013c) have utilized the in situ approach to characterize two ecotypes from West coast of India viz. Bedakam and Kuttiyadi, to facilitate their on farm conservation.

Rajesh et al. (2008a) carried out microsatellite-based assessment of genetic diversity of 10 landraces from three coconut-growing communities of Southern India. Such studies could facilitate stakeholders to make use of useful germplasm for sustainable production, in addition to planning conservation strategies. Selected palms from these communities could then be utilized as source for production of seed nuts for planting and raising seedling nurseries managed by the local communities and could enhance income of the local communities.

As an alternative to conservation of coconut genetic diversity on farm or in home gardens, Bourdeix et al. (2011) mooted the concept of 'Polymotu' for conservation of coconut genetic diversity in remote islands, atolls, valleys and geographically isolated locations for breeding and seed nut production.

1.6 Germplasm Utilization

One would find only a few successful cases of utilization of coconut diversity despite the existence of large variations for various traits for coconut. The achievements include developing disease resistance, especially to Lethal Yellowing disease caused by phytoplasma, reduction in plant height, yield increase by exploitation of hybrid vigour and exploitation of '*Makapuno*' mutant as a value-added product. Available coconut varieties and methods of classical coconut improvement programs have been discussed in earlier reviews (Menon and Pandalai 1958; Ohler 1999).

1.6.1 Harnessing the Dwarf Trait and Hybrid Vigour

Selection of desirable coconut mother palms was the only method of crop improvement till hybridization was successful among two dwarf cultivars in Fiji (Marechal 1928). A breakthrough in coconut breeding came after the discovery of vigorous seedling hybrids of tall and dwarf palms (Patel 1938). Hybrids, involving dwarf and tall parents, were instrumental in obtaining palms which were early in bearing and high yielding. The knowledge provided by studies on inheritance of petiole colour by Bourdeix (1999), that yellow petiole/germ colour was governed by two recessive (*rrgg*) Mendelian factors, benefitted the hybrid coconut industry. It further facilitated the hybrid seedling identification by employing the yellow dwarf as female parent and green tall as male parent and removal of off-types in nursery by culling seedlings with yellow petiole. Molecular markers have been identified for differentiating tall and dwarf cultivars of coconut (Rajesh et al. 2013). Molecular markers have also been utilized for confirming the hybrids of crosses between dwarf and tall parents, which will facilitate supply of authentic hybrids to farmers (Perera 2010; Rajesh et al. 2012, 2014d; Preethi et al. 2016).

Perera et al. (2016), after carrying out detailed analysis of inheritance of height and the presence of bole in F2 population of a cross between dwarf and tall palms, concluded that a single co-dominant locus was responsible for the presence of the bole. A strong association was also detected between the presence of the bole and height, with the latter also under the control of a single co-dominant gene.

The inheritance of germ colour (Bourdeix 1999) involves only four petiole colours viz. brown, green, yellow and orange. It is not known how the pink colour of embryo in Guelle Rose (Kumaran et al. 2000) is inherited. Pink coloured embryos/husk could be favourably utilized as morphological marker once the genetic basis is understood. Differences in the trichome density in the leaf can be explored to identify hybrids at seedling stage. Dwarf varieties possess high trichome density, tall varieties low density and hybrids, intermediate density (Taulu et al. 1980). Dwarf varieties are also sensitive to exogenous application of copper (Schut 1975) at 1–5 g per plant. Dwarf varieties, infected with bud rot disease, exhibited phytotoxicity when sprayed with copper fungicide (Ochs et al. 1993). The differential responses of exogenous copper application in tall, hybrid and dwarf seedlings need to be verified. After conformation of the hypothesis, it is worth to test the response of dwarf seedlings to copper to discriminate the hybrid from dwarf by studying phototoxic symptoms.

Palms are large trees making the intercultural and harvest operations difficult due to their enormous heights. Dwarf varieties offer scope to farmers to carry out intercultural operations and harvest fruits easily. The success achieved in coconut by creating hybrids by crossing dwarf and tall parents has been favourably extended to other palm species. A dwarf *Areca* mutant (Naidu 1963) was observed in Hirehalli taluk of Tumkur District of Karnataka, India, which has been gainfully harnessed in areca nut breeding programs by crossing with high yielding *Mohitnagar* or *Sumangla* cultivars to evolve high yielding hybrids of short stature (Ananda 2002). In case of oil palm, short-statured palms were found in a small proportion in the interspecific hybrid progenies between the Old world African oil palms (*Elaeis guineensis* Jacq) crossed with new World American oil palms (*E. oleifera)* (Sunilkumar et al. 2015). Genetically dwarf cultivars and dwarf forms of wild species of *Phoenix* have also been employed to develop dwarf date palms (Jain 2012).

1.6.2 Sex Expression

Sex expression varies a great deal among the palms. Date palm and palmyrah are dioecious, where male and female palms are separate. Coconut, areca nut and oil palm are monoecious where the male and female flowers are present in the same individual. Hermaphrodite flowers (male and female organs in the same flower) have been reported, but are rare in coconut palms; the presence of such flowers was

Fig. 1.13 A multiple mutant palm with mutations at three traits: multispatheate, plicata and androgena

observed in low proportion among the Nias Yellow Dwarf palms of coconut which was observed by Davis et al. (1981b). The report is significant and suggests careful elimination of male organs from such flowers to avoid self pollination in hybridization programs. Male-dominated (androgena) and female-dominated (spicata) forms have been reported in coconut (Table 1.3). A multiple mutant (*TT-50*) with mutations at three traits (multispatheate, plicata and androgena) was reported in our earlier studies (Arunachalam 2011) (Fig. 1.13). Hybridization among the coconut forms differing sex expression (spicata x androgena) has been initiated (Arunachalam et al. 2014a) to understand their inheritance further. Suitable tall populations have also examined to exploit heterosis among the Tall x Tall hybrids (N'Cho et al. 1993) in coconut.

1.6.3 Resistance Breeding for Phytoplasmal Diseases

Variation in resistance/susceptibility in coconut germplasm has been exploited in developing resistance breeding approaches utilizing the resistant dwarf populations to lethal yellowing disease (Whitehead 1968; Harries 1974; Been 1981; Konan et al. 2007) and root (wilt) disease (Nair et al. 1996). Malayan Dwarf palms have provided a long-standing solution in lethal yellowing disease-prone areas of Latin America and Caribbean countries (Whitehead 1968; Been 1981; Harries 1974). Most coconut breeding programs around the world employ Malayan Dwarf, preferably yellow-fruited ones, as female parent and the local tall as male parent. Genetic diversity studies of Dery et al. (2008) revealed that coconut cultivars from the Pacific region, particularly dwarf cultivars, showed less susceptibility to Cape St. Paul Wilt Disease in comparison to cultivars from Indo-Atlantic region.

A comprehensive breeding program for evolving coconut varieties resistant/tolerant to root (wilt) disease is being implemented in India since 1988 (Nair et al. 2010). Three coconut varieties have been released for the root (wilt) disease prevalent area based on systematic evaluation trials carried out (Nair et al. 2016).

1.6.4 Resistance Breeding for Eriophyid Mite

Eriophyid mite (*Aceria guerreronis* Keifer) has emerged as serious pest of coconut. The perianth lobes/tepals offer a refugee for the mites to inhabit (Aratchige et al. 2007; review: Navia et al. 2013) but protect from predators and pesticidal sprays. Penetrometer was used to measure the toughness of fruit and perianth lobes of coconut. Resistance to eriophyid mite due to increased toughness of the perianth lobe (tepal) and fruit in some Benaulim palms was reported by our earlier studies (Arunachalam et al. 2013). da Silva et al. (2016) have demonstrated that artificially enlarging the perianth-rim-fruit space (40–120 μm) allowed earlier occurrence of larger predatory mites like *Neoseilus paspalivorus* beneath the perianth and subsequent lower count of coconut mites. Perianth lobe variation is observed in coconut such as Horned Coconut (Davis 1965; Balakrishnan and Nair 1979; Jerard et al. 2014b). Horns on coconut fruits display more distance of fruit rim to perianth lobe than normal varieties hence may allow the predatory mite to pass through. Horned Coconut, along with such small predatory mites, needs evaluation to evolve a suitable integrated pest management package. In an earlier study, Shalini et al. (2007) had reported identification of nine SSR and four RAPD markers with resistance to eriophyid mite using single marker analysis. Stepwise multiple regression analysis revealed that a combination of five markers could account for 100% of the association with eriophyid mite resistance.

1.6.5 Breeding for Drought Tolerance

Drought is a serious problem in coconut cultivation when the rainfall received is limited in quantity or its distribution is skewed over a few months. A review of ecophysiology of coconut indicated root production, less turgor loss in leaves, proline accumulation and abscisic acid production as drought-responsive traits in coconut (Gomes and Prado 2007). Well-distributed roots help the coconut palm to tolerate the impact of drought; based on this concept, coconut germplasm has been screened for distribution of root system (Cintra et al. 1993). The results of this study revealed that Polynesian Tall and Brazilian Tall develop deeper roots under water stress than other tall varieties studied.

Important physiological traits such as lipase and protease activity of the leaf lipid membranes (Repellin et al. 1997), content of stress-responsive proteins (Kumar et al. 2007), relative water content (Renju et al. 2015), leaf water potential (Rajagopal et al. 1988, 1990), epicuticular wax, chlorophyll fluorescence (Kasturibai et al. 2006) have been used in screening coconut diversity for drought tolerance. Riedel et al. (2009) have identified QTLs for cuticular wax composition in coconut. An in vitro assay was developed using zygotic embryos subjected to osmotic treatments in coconut to screen for drought tolerance (Karunaratne et al. 1991). Variety of single cross-hybrids have been evolved using drought-tolerant parents for imparting drought tolerance in hybrid progenies (Kasturibai et al. 2010).

1.6.6 Germplasm for Product Diversification

Coconut palm is known as the 'tree of heaven' with multitude of useful products for humanity. Utilization of coconut germplasm for developing several value-added products is a crucial strategy to boost coconut farmer's income. Mutants for endosperm traits play major role in boosting the profitability of coconut farmers. Kopyor coconuts, with soft endosperm, have been reported to fetch ten times the price of normal coconuts at Indonesia (Maskromo et al. 2016). The spontaneous mutant resulting in soft and curdy endosperm is available in different names at most coconut-producing countries [Sri Lanka: Peries (1995); Indonesia: Maskromo et al. (2016); India: Jerard et al. (2013)]. The knowledge of genetic basis of the Makapuno trait being controlled by a single recessive gene is known (Torres 1937). But, due to the requirement of embryo culture, this mutant type could not boost farmer's incomes only in the Philippines.

Fatty acid profiles of coconut germplasm, hybrids and parents have been generated (Kumar 2011; Laureles et al. 2002). Lauric acid rich (>50%) coconut hybrids viz. PCA 15-8 and PCA 15-9 have been identified to suit the needs of genotypes to suit the market for lauric acid (Laureles et al. 2002). Sheela et al. (2016) reported the cholesterol-lowering ability of lauric acid from coconut oil.

Tender coconut water is another product from coconut palm (DebMandal and Mandal 2011) with immense potential as health and sports drink. In India, Chowghat Green Dwarf has been identified as a potential dwarf cultivar suitable for tender nut water (Dhamodaran et al. 1993; Nadanasabapathy and Kumar 1999). Variety of single cross-hybrids, involving COD as one of the parent, have also been reported to be potentially suited for tender coconut water markets (Apshara et al. 2007). PB 121, an hybrid between Malayan Yellow Dwarf and West African Tall, has also been reported to be superior in tender nut water traits compared to its parents (Assa et al. 2013). Other dwarfs and semi-dwarf cultivars, suitable for tender nut water, include Galas Green Dwarf from the Philippines, Nam Wan and Nam Hom from Thailand, Brazilian Green Dwarf from Brazil and Malayan Dwarfs and King Coconut from Sri Lanka (Thampan and Gopalakrishnan 2010).

Unfermented inflorescence sap is gaining popularity as a healthy and nutritious drink, in addition to serving as a raw material for production of coconut sugar and jaggery. The glycemic index (GI) of coconut inflorescence sap sugar is low (around 35) which makes it an ideal sugar source for diabetic patients (Manohar et al. 2007). Tall varieties were reported to be better than dwarf varieties for inflorescence sap production (Samsudeen et al. 2013a). In Thailand, cultivars known for their high production of sap (viz. Tha-le Ba, Suricha, Sai Bua, Theung Bong, Kathi and Khi Kai) are exclusively maintained in coconut gardens for sugar production in the provinces of Samut Sakhon, Samut Songkhram and Samut Prakan. These palms are known to yield four litres of sweet toddy on an average (Thampan 2013).

In Thailand, a unique and rare coconut variety produces very thick nuts with three lobes; nuts from this variety, called Maphrao or 'fiddle sound box coconut' is used to make sound boxes, and these nuts are sold at exorbitant prices (Thampan and Gopalakrishnan 2010).

In oil palm, shell thickness is a trait controlled by single gene (Beirnaert and Vanderweyen 1941). High yielding thin-shelled tenera (*Sh/sh*) palms are heterozygous dominant for shell loci and evolved by hybridization of thick shelled dura (*Sh/Sh*) homozygous dominant with shell-less pisifera (*sh/sh*) (Singh et al. 2013). This trait has been favourably exploited in oil palm for breeding for enhanced oil yield and detection of non-tenera contamination (Ooi et al. 2016); such studies could be initiated in coconut for correlation of shell thickness with oil yield.

Coconut shell has potential as a source of activated carbon (Mozammel et al. 2002). It is not clear whether coconut forms varying for shell thickness (thick shelled 'Ladehi' types; Panda 1982) would exhibit heterosis and could be utilized to evolve high yielders or useful in fitting coconuts for market for activated carbon. Variety of cross-hybrids for possible improvements in spathe traits (Jayabose et al. 2010) has also been reported.

1.6.7 Phenotyping for Fungal Disease Resistance

Fungal diseases are serious threat to coconut cultivation. Screening of genetic diversity of coconut involves constraints of time and space due to delay in disease development and the requirement of large sick plots or container trials. Inoculation of the pathogen in the excised petiole of coconut could be a rapid way to phenotype disease resistance. Coconut varieties were successfully screened using a petiole inoculation technique to stem bleeding disease caused by *Thielaviopsis paradoxa* (Ramanujam et al. 1998).

1.7 Conclusion

Coconut palm is an important commodity for coastal and fragile island ecosystems. Undoubtedly, coconut genetic resources are composed of numerous traits of potential value to aid in the coconut crop improvement programs. Harnessing this genetic diversity is essential to evolve varieties or hybrids with high yield, suitability to product diversification and tolerance to pests and diseases. The chapter reviews the diversity of coconut available at geographical level and trait level and lists the proven cases of utilization and potential unexploited areas. Constraints in conservation of genetic resources are discussed along with potential solutions. The under-representation of coconut genetic resources in gene banks and the threats they face in their natural habitats warrants urgent concerted efforts at the global level, to guarantee their collection and appropriate conservation. The availability of genomic resources, though sequencing of coconut genome which is underway, could revolutionize analysis of genes governing key traits, carrying out robust population genetic studies and aid conservation and utilization initiatives for the benefit of human kind.

Acknowledgements We are indebted to the Indian Council of Agricultural Research (ICAR) for nurturing our knowledge on coconut palm.

References

Abraham Z, Senthilkumar R, John KJ, Sharma TVRS, Nair NV, Unnikrishnan M, Kumaran PM, George JK, Uma S, Latha M, Malik SS (2008) Collection of plant genetic resources from Andaman and Nicobar Islands. Genet Resour Crop Evol 55(8):1279–1289

Alonso M, Cueto JR, Santos Y, Romero W, Llauger R, Rohde W (2007) Variabilidad morfológica y molecular de una población de cocotero verde en la región de Baracoa. Cultivos Tropicales 28(3):69–76

Ananda KS (2002) Improved varieties and promising traditional cultivars of arecanut. ATIC Series Publication No. 7, CPCRI, Kasaragod, Kerala, India

Apshara SE, Arunachalam V, Jayabose C, Kumaran PM (2007) Evaluation of coconut hybrids for tender nut purpose. Indian J Hortic 64(3):320–323

Aratchige NS, Sabelis MW, Lesna I (2007) Plant structural changes due to herbivory: do changes in *Aceria*-infested coconut fruits allow predatory mites to move under the perianth? Exp Appl Acarol 43(2):97–107

Arunachalam V (2011) Coconut. Genomics of cultivated palms. Elsevier Insights Series, Elsevier, Amsterdam, pp 13–27

Arunachalam V (1999) Prospection and collection of coconut diversity. In: Ratnambal MJ, Kumaran PM, Muralidharan K, Niral V, Arunachalam V (eds) Improvement of plantation crops. CPCRI, Kasaragod, India, pp 12–19

Arunachalam V (2005) Phenotypic traits of genetic twins in coconut palm (*Cocos nucifera* L.) and their utility in seed production. Seed Tech 27(1):104–107

Arunachalam V, Rajesh MK (2008) Breeding of coconut palm. CAB Rev: Perspect Agric, Vet Sci, Nutr Nat Res 3(053):1–12

Arunachalam V, Jerard BA, Apshara SE, Jayabose C, Subaharan K, Ravikumar N, Palaniswami C (2013) Digital phenotyping of coconut and morphological traits associated with eriophyid mite infestation. J Plantn Crops 41(3):417–424

Arunachalam V, Jerard BA, Damodaran V, Ratnambal MJ, Kumaran PM (2005) Phenotypic diversity of foliar traits in coconut germplasm. Genet Resour Crop Evol 52(8):1031–1037

Arunachalam V, Jerard BA, Manju KP, Niral V (2014a) Alternate genetics and genomics of coconut. In: Thomas GV, Krishnakumar V, Jerard BA (eds) Improving productivity and profitability in coconut farming. Proceedings of international conference on coconut biodiversity for prosperity, Central Plantation Crops Research Institute, Kasaragod, Kerala, India, pp 97–101

Arunachalam V, Rajesh MK, Jerard BA, Jayabose C, Sairam CV (2014b) Characterization of a spicata mutant of coconut palm in India. J Plantn Crops 42(1):417–424

Arunachalam V, Jerard BA, Elangovan M, Ratnambal MJ, Dhanapal R, Rizal SK, Damodaran V (2001) Unexploited diversity in coconut palm (*Cocos nucifera* L.). Plant Genet Resour Newsl (Bioversity Int) 127:39–43

Ashburner GR, Thompson WK, Halloran GM, Foale MA (1997) Fruit component analysis of South Pacific coconut palm populations. Genet Resour Crop Evol 44(4):327–335

Assa RR, Prades A, Konan AG, Nemlin J, Konan JL (2013) Sensory evaluation and sugars contents of coconut (*Cocos nucifera* L.) water during nuts ripening. African J Food Sci 7:186–192

Assy-Bah B, Engelmann F (1992) Cryopreservation of immature embryos of coconut (*Cocos nucifera* L.). Cryo-Lett 13:67–74

Balakrishnan NP, Nair RB (1979) Wild populations of *Areca* and *Cocos* in Andaman and Nicobar Islands. Indian J For 2:350–363

Batugal P (2005) The International Coconut Genetic Resources Network (COGENT): Its history and achievements. In: Batugal P, Rao VR, Oliver J (eds) Coconut genetic resources. International Plant Genetic Resources Institute–Regional Office for Asia, the Pacific and Oceania (IPGRI-APO), Serdang, Selangor DE, Malaysia, pp 482–499

Batugal P, Jayashree K (2005) COGENT's multi-site International Coconut Genebank. In: Batugal, P, Rao VR, Oliver J (eds) Coconut genetic resources. International Plant Genetic Resources Institute–Regional Office for Asia, the Pacific and Oceania (IPGRI-APO), Serdang, Selangor DE, Malaysia, pp 106–114

Baudouin L, Lebrun P (2002) The development of microsatellite kit and dedicated software for use with coconuts. Burotrop Bull 17:16–20

Baudouin L, Lebrun P, Konan JL, Ritter E, Berger A, Billotte N (2006) QTL analysis of fruit components in the progeny of a Rennell Island Tall coconut (Cocos *nucifera* L.) individual. Theor Appl Genet 112:258–268

Baudouin L, Lebrun P, Berger A, Myrie W, Been BO, Dollet M (2008) The Panama Tall and the Maypan hybrid coconut in Jamaica: did genetic contamination cause a loss of resistance to lethal yellowing? Euphytica 161:353–360

Been BO (1981) Observations on field resistance to lethal yellowing in coconut varieties and hybrids in Jamaica. Ole´agineux 36:9–12

Beirnaert A, Vanderweyen R (1941) Contribution à l'étude génétique et biométrique des variétiés d'Elaeis guineensis Jacq. Publ Inst Nat Etude Agron Congo Belge Ser Sci 27:1–101

Bourdeix R (1999) Genetic determinism in dwarf coconut germ color. Ole´agineux 43:371–374

Bourdeix R, Baudouin L, Billotte N, Labouisse J, Noiret J (2001) Coconut. In: André C, Michel J, Serge H, Dominique N (eds) Tropical plant breeding. CIRAD, Montpellier, France, pp 106–127

Bourdeix R, Guarino L, Rao, VR and Baudouin L (2005a) Status, gaps and strategy in coconut germplasm. In: Batugal P, Rao VR, Oliver J (eds) Coconut genetic resources. International Plant Genetic Resources Institute–Regional Office for Asia, the Pacific and Oceania (IPGRI-APO), Serdang, Selangor DE, Malaysia, pp 44–64

Bourdeix R, Guarino L, Mathur PN, Baudouin L (2005b) Mapping of genetic diversity. In: Batugal P, Rao VR, Oliver J (eds) Coconut genetic resources. International Plant Genetic Resources Institute–Regional Office for Asia, the Pacific and Oceania (IPGRI-APO), Serdang, Selangor DE, Malaysia, pp 32–43

Bourdeix R, Johnson V, Baudouin L, Tuia VS, Kete T, Planes S, Lusty C, Weise S (2011) Polymotu: A new concept of island-based germplasm bank based on an old Polynesian practice. Ogasawara Res 37:33–51

Chan JL, Talavera C, Hornung R, Robert M, Oropeza C (1998) Regeneration of coconut (*Cocos nucifera* L.) from plumule explants through somatic embryogenesis. Plant Cell Rep 17:515–521

Cintra FLD, Passos EDM, da Silva Leal ML (1993) Evaluation of root system distribution in Tall coconut cultivars. Ole´agineux 48:453–461

Clancy KE, Sullivan MJ (1988) Some observations on seed germination, the seedling, and polyembryony in the needle palm *Rhapidophyllum hystrix*. Principes 32:18–25

da Silva FR, de Moraes GJ, Lesna I, Sato Y, Vasquez C, Hanna R, Sabelis MW, Janssen A (2016) Size of predatory mites and refuge entrance determine success of biological control of the coconut mite. Biocontrol. doi:10.1007/s10526-016-9751-2

Das NK (1966) Polyembryony in Areca. Sci Cult 32:469

Davis TA (1965) Addition to the theories on the morphology of horns in coconut fruits. Ceylon Cocon Quart 16:99–101

Davis TA, Perianayagam J (1985) Juvenile flowering in coconut. Coconuts Today 3:90–98

Davis TA, Anandan AP, Menon KPV (1953) A case of polyembryony in coconut. Indian Cocon J 6(2):67–71

Davis TA, Sudasrip H, Azis H (1981a) Bulbil-shoot production from clonally propagated coconuts. Principes 25:124–129

Davis TA, Sudasrip H, Aziz H (1981b) Significance of hermaphrodite flowers in coconut breeding. Proc Indian Natl Sci Acad Part B Biol Sci 4:527–535

Davis TA, Sudasrip H, Darwis SN (1985) An Overview of Research Activities. Coconut Research Institute, Manado, Indonesia, 165 p

DebMandal M, Mandal S (2011) Coconut (*Cocos nucifera* L.: Arecaceae) in health promotion and disease prevention. Asian Pacific J Trop Med 4:241–247

Dery SK, Philippe R, Baudouin L, Quaicoe RN, Nkansah-Poku J, Owusu-Nipah J, Arthur R, Dare D, Yankey N, Dollet M (2008) Genetic diversity among coconut varieties for susceptibility to Cape St. Paul Wilt Disease. Euphytica 164:1–11

Devakumar K, Niral V, Jerard BA, Jayabose C, Chandramohanan R, Jacob PM (2010) Microsatellite analysis of distinct coconut accessions from Agatti and Kavaratti Islands, Lakshadweep, India. Sci Hort 125:309–315

Dhamodaran S, Rathnambal MJ, Chempakam B, Pillai RV, Virakatamath BC (1993) Evaluation of tender nut water in coconut cultivars. In: Nair MK, Khan HH, Gopalasundaram P, Bhaskara Rao EVV (eds) Advances in coconut research and development. Oxford & IBH, New Delhi, pp 123–128

Dootson J, Thirakul A, Petchpiroon C, Rattanapruk M (1989) Early yields of a number of coconut varieties in Thailand. Ole´agineux 43:445–454

Dulloo ME, Rao VR, Engelmann F, Engels J (2005) Complementary conservation of coconuts. In: Batugal P, Rao VR, Oliver J (eds) Coconut genetic resources. International Plant Genetic Resources Institute–Regional Office for Asia, the Pacific and Oceania (IPGRI-APO), Serdang, Selangor DE, Malaysia, pp 75–90

Engelmann F, Malaurie B, N'Nan O, Borges M (2005) In: Batugal P, Rao VR, Oliver J (eds) Coconut genetic resources. International Plant Genetic Resources Institute–Regional Office for Asia, the Pacific and Oceania (IPGRI-APO), Serdang, Selangor DE, Malaysia, pp 142–148

Fernando WMU (1998) Genetic improvement of the coconut cultivar San Ramon. Cocos 13:30–39

Fernando WMU, Pieris TSG, Wickramaratne MRT (2010) Variation in seedling characters of three different coconut cultivars and their use in identification in the nursery. Cocos 9:23–29

Gangolly SR, Satyabalan K, Pandalai KM (1957) Varieties of coconut. Indian Cocon J 10:3–28

Gomes FP, Prado CH (2007) Ecophysiology of coconut palm under water stress. Braz J Plant Physiol 19(4):377–391

Gruezo WS, Harries HC (1984) Self-sown, wild-type coconuts in the Philippines. Biotropica 16 (2):40–147

Gunn BF, Baudouin L, Olsen KM (2011) Independent origins of cultivated coconut (*Cocos nucifera L.*) in the old world tropics. PLoS ONE 6(6):e21143

Hamelin C, Bourdeix R, Baudouin L (2005) The international coconut genetic resources database. In: Batugal P, Rao VR, Oliver J (eds) Coconut genetic resources. International Plant Genetic Resources Institute–Regional Office for Asia, the Pacific and Oceania (IPGRI-APO), Serdang, Selangor DE, Malaysia, pp 427–438

Harries HC (1974) Performance of F_1 hybrid coconuts in Jamaica. J Plantn Crops 2(2):15–20

Harries HC (1978) The evolution, dissemination and classification of *Cocos nucifera* L. Bot Rev 44(3):265–319

Harries HC (1981) Germination and taxonomy of the coconut. Ann Bot 48(6):873–883

Harries HC (2012) Germination rate is the significant characteristic determining coconut palm diversity. AoB Plants. doi:10.1093/aobpla/pls045

Herran A, Estioko L, Becker D, Rodriguez MJB, Rhode W, Ritter E (2000) Linkage mapping and QTL analysis in coconut (*Cocos nucifera* L.). Theor Appl Genet 101:292–300

Islam MN, Borromeo TH, Rzzaque MA (2007) Phenotypic diversity of coconut populations in Bangladesh. Philippine J Cocon Stud 30(1&2):13–22

Jacob KC (1940) A new variety of coconut palm *Cocos nucifera* L. var. spicata KC Jacob. J Bomb Nat Hist Soc 41:906–907

Jacob PM, Krishnamoorthy B (1981) Observations on the coconut genepool in Lakshadweep islands. In: Vishveshwara S (ed) Proceedings of PLACROSYM IV, 3–5 Dec 1981, CFTRI, Mysore, karnataka, India, pp 3–8

Jain SM (2012) Date palm biotechnology: current status and prospective-an overview. Emir J food Agric 24(5):386–399

Jayabose C, Ganesh S, Natarajan C, Arunachalam V, Arulraj S (2010) Variability and heterosis in spathe traits of coconut (*Cocos nucifera* L.). J Plantn Crops 38(2):157–160

Jerard BA, Damodaran V, Niral V, Samsudeen K, Rajesh MK, Sankaran M (2013) Conservation and utilization of soft endosperm coconut accession from Andaman Islands. J Plantn Crops 41 (1):14–21

Jerard BA, Niral V, Samsudeen K, Jacob PM, Devakumar K, Rajesh MK, Karun A, Thomas, GV (2014a) IND 331-Laccadive Mini Micro Tall (IC0598222; INGR13064), a coconut (*Cocos nucifera* L.) germplasm with distinct character of extremely small nuts with very low copra content, not found in any other coconut variety. Indian J Plant Genet Resour 27(1):77–77

Jerard BA, Niral V, Dhanapal R, Damodaran V, Arunachalam V, Rajesh MK, Devakumar K, Samsudeen K, Nair RV, Kumaran PM, Thomas GV (2014b) IND 221–Andaman Horned Cocos (IC0598221; INGR13063), a coconut (*Cocos nucifera* L.) germplasm with distinct character of horny nuts. Indian J Plant Genet Resour 27(1):76–77

Jerard BA, Rajesh MK, Elain SA, Sajini KK, Rahman S, Fayas TP, Karun A (2014c) Scope of novel and rare bulbiferous coconut palms (*Cocos nucifera* L.). Genet Resour Crop Evol 61 (1):1–6

Kamaral LCJ, Perera SACN, Perera KLNS, Dassanayaka PN (2014) Genetic diversity of the Sri Lanka yellow dwarf coconut form as revealed by microsatellite markers. Trop Agric Res 26 (1):131–139

Karun A, Sajini KK, Niral V, Amarnath CH, Remya P, Rajesh MK, Samsudeen K, Jerard BA, Engelmann F (2014) Coconut (*Cocos nucifera* L.) pollen cryopreservation. Cryo-Lett 35 (5):407–417

Karunaratne S, Santha S, Kovoor A (1991) An in vitro assay for drought-tolerant coconut germplasm. Euphytica 53(1):25–30

Kasturibai VK, Rajagopal V, Arunachalam V (2006) Assessment of diversity in coconut varieties for drought responsive physiological traits. J Plantn Crops 34(2):118–120

Kasturibai VK, Muralidharan K, Arunachalam V, Rajagopal V (2010) Heterosis for drought tolerant traits in coconut. J Plantn Crops 38(2):87–91

Konan KJN, Edmond KK, Louïs KJ, Lebrun P, Dery SK, Sangare A (2007) Microsatellite gene diversity in coconut (*Cocos nucifera* L.) accessions resistant to lethal yellowing disease. African J Biotech 6:341–347

Krain E, Issa JA, Kullaya A, Harries HC (1994) The coconut palm in East Africa 2. The Pemba dwarf in Zanzibar. Principes 38:138–141

Kumar SN (2011) Variability in coconut (*Cocos nucifera* L.) germplasm and hybrids for fatty acid profile of oil. J Agric Food Chem 59:13050–13058

Kumar SN, Bai KV, George J, Balakrishnan A, Thomas ST (2007) Stress responsive proteins in coconut seedlings subjected to water, high-light, flooding and high-temperature stresses. Indian J Hortic 64(4):373–380

Kumaran PM, Koshy PK, Arunachalam V, Niral V, Parthasasarathy VA (2000) Biometric clustering of coconut populations of Indian ocean islands. In: Muraleedharan N, Rajkumar R (eds) Recent advances in plantation crops. Allied Publishers, India, pp 73–81

Kumaran PM, Arunachalam V, Dash DK (2004) Collecting coconut diversity of Orissa, India. J Plantn Crops 32(Suppl.):39–45

Laureles LR, Rodriguez FM, Reaño CE, Santos GA, Laurena AC, Mendoza EMT (2002) Variability in fatty acid and triacylglycerol composition of the oil of coconut (*Cocos nucifera* L.) hybrids and their parentals. J Agric Food Chem 50:1581–1586

Le HT, Hancock JF, Trinh TT, Ho PH (1999) Germplasm resources in Vietnam: major horticultural and industrial crops. Hort Sci 34(2):175–180

Leach BJ, Foale MA, Ashburner GR (2003) Some characteristics of wild and managed coconut palm populations and their environment in the Cocos (Keeling) Islands, Indian Ocean. Genet Resour Crop Evol 50(6):627–638

Liyanage DV (1958) Varieties and phenotypes of coconut palms grown in Ceylon. Ceylon Cocon Quart 9:1–10

Liyanage DV(1974) Report on survey of coconut germplasm in Indonesia. Coconut Industry Development Project, UNDP/FAO Document. 26 p

Loiola CM, Azevedo AON, Diniz LEC, Aragão WM, Azevedo CDdO, Santos PHAD, Ramos HCC, Pereira MG, Ramos SRR (2016) Genetic relationships among tall coconut palm (*Cocos nucifera* L.) accessions of the International Coconut Genebank for Latin America and the Caribbean (ICG-LAC), evaluated using microsatellite markers (SSRs). PLoS ONE 11(3): e0151309. doi:10.1371/journal.pone.0151309

Louis HI (1991) Genetic variability in coconut palm. Madras Agric J 68:588–593

Manjula C, Samsudeen K, Rahman S, Rajesh MK (2014) Characterization of Kuttiyadi ecotype of coconut (*Cocos nucifera* L.) using morphological and microsatellite markers. J Plantn Crops 42:301–315

Manthiratna MAP (1972) The performance of dwarfs (*Cocos nucifera* L. variety nana) as a plantation crop in Ceylon. Ceylon Cocon Quart 23:92–99

Manohar EC, Kindipan NML, Sancha LV (2007) Coconut sap sugar production: from farm to market and from wealth to health. Philippine Entomologists 32:20

Mao Z, Lai Y (1993) The coconut germplasm of Hainan Island, China. Plant Genet Resour Newsl 91(92):53–57

Marechal H (1928) Observations and preliminary results on the coconut palm with a view to developing improved seed nuts for Fiji. Fiji Agric J 1(2):16–45

Martinez RT, Baudouin L, Berger A, Dollet M (2009) Characterization of the genetic diversity of the Tall coconut (*Cocos nucifera* L.) in the Dominican Republic using microsatellite (SSR) markers. Tree Genet Genom 6(1):73–81

Maskromo I, Larekeng SH, Novarianto H, Sudarsono S (2016) Xenia negatively affecting kopyor nut yield in Kalianda Tall kopyor and Pati Dwarf kopyor coconuts. Emir J Food Agric 28:644–652

Menon KPV, Pandalai KM (1958) The coconut. A Monograph. Indian Central Coconut Committee, Government Press, Madras, India

Merrow AW, Wisser RJ, Brown JS, Kuhn DN, Schnell RJ, Broschat TK (2003) Analysis of genetic diversity and population structure within Florida coconut (*Cocos nucifera* L.) using microsatellite DNA, with special emphasis on the Fiji Dwarf cultivar. Theor Appl Genet 106:715–726

Michael KJ (1963) A multi-spatheate coconut palm. Indian Cocon J 16(2):78–80

Mozammel HM, Masahiro O, Bhattacharya SC (2002) Activated charcoal from coconut shell using $ZnCl_2$ activation. Biomass Bioenergy 22:397–400

Nadanasabapathy S, Kumar R (1999) Physico-chemical constituents of tender coconut (*Cocos nucifera*) water. Indian J Agric Sci 69:750–751

Naidu CVG (1963) Seen a dwarf areca palm? Indian Farm 12:16–17

Nair MK, Koshy PK, Jacob PM, Nair RV, Bhaskara Rao EVV, Nampoothiri KUK, Iyer RD (1996) A root (wilt) disease resistant coconut hybrid and strategy for resistance breeding. Indian Cocon J 27:2–5

Nair RV, Jerard BA, Thomas RJ (2016) Coconut breeding in India. In: Jameel M Al-K, Jain SM, Johnson DV (eds) Advances in plant breeding strategies: agronomic, abiotic and biotic stress traits. Springer International Publishing, Switzerland, pp 257–279

Nair RV, Thomas RJ, Jacob PM (2010) Breeding for resistance to coconut root (wilt) disease. In: Thomas GV, Chandramohanan R, Jacob PM, Krishnakumar V (eds) Coconut Root (wilt) Management. CPCRI, Kasaragod, pp 58–71

Narayana GV, John CM (1949) Varieties and forms of the coconut. Madras Agric J 36:349–366

Navia D, Gondim MGC Jr, Aratchige NS, de Moraes GJ (2013) A review of the status of the coconut mite, *Aceria guerreronis* (Acari: Eriophyidae), a major tropical mite pest. Exp Appl Acarol 59(1–2):67–94

N'Cho YP, Sangaré A, Bourdeix R, Bonnot F Baudouin L (1993) Evaluation de quelques écotypes de cocotier par une approche biométrique. I. Etude des populations de Grands. Oléagineux 48 (3):121–132

Ninan CA, Pillai RV, Joseph J (1960) Cytogenetic studies of the genus Cocos 1. Chromosome number. *C. australis* Mart and *C. nucifera* L. vars. spicata and androgena. Indian Coconut J 13:129–134

Ninan CA, Raveendranath TG (1965) A naturally occurring haploid embryo in the coconut palm (*Cocos nucifera* L.). Caryologia 18(4):619–623

Ninan CA, Satyabalan K (1963) Cytogenetic studies in the genus *Cocos*. 2. Some observations on the 'spicata' character in coconuts. Indian Cocon J 16:109–114

Niral V, Rajesh MK, Parthasarathy VA (2016) Coconut. In: Parthasarathy VA, Nirmal Babu K, Aswath C, Senthilkumar R (eds) Breeding of horticultural crops, vol 1. Spices and plantation crops. Today and Tomorrow's Publishers and Printers, Daryaganj, New Delhi, India

N'Nan O, Hocher V, Verdeil JL, Konan JL, Ballo K, Mondeil F, Malaurie B (2008) Cryopreservation by encapsulation–dehydration of plumules of coconut (*Cocos nucifera* L.). Cryo- Lett 29(4):339–350

Ochs R, Bonneau X, Qusairi H (1993) Nutrition minerale en cuivre des cocotiers hybrides sous tourbe. Oléagineux 48:65–75

Ohler JG (1999) Modern coconut management: palm cultivation and products. Intermediate Technology Publication, Londres 458p

Ooi LC-L, Low E-TiL, Abdullah MO, Nookiah R, Ting NC, Nagappan J, Manaf MAA, Chan K-L, Halim MA, Azizi N, Omar W, Murad AJ, Lakey N, Ordway JM, Favello A, Budiman MA, van Brunt A, Beil M, Leininger MT, Jiang N, Smith SW, Brown CR, Kuek ACS, Bahrain S, Hoynes-O'Connor A, Nguyen AY, Chaudhari HG, Shah SA, Choo Y-M, Sambanthamurthi R, Singh R (2016) Non-tenera contamination and the economic impact of SHELL genetic testing in the Malaysian independent oil palm industry. Front Plant Sci doi:10.3389/fpls.2016.00771

Ovasuru T (1993) Preliminary analysis of coconut (*Cocos nucifera* L.) germplasm in Papua New Guinea. In: ACIAR PROCEEDINGS. Australian Centre for International Agricultural Research, pp 33–33

Oyoo ME, Muhammed N, Githiri SM, Ojwang PO, Muniu FK, Masha E, Owuoche JO (2015) In situ morphological characterization of coconut in the Coastal Lowlands of Kenya. Afr J Plant Sci 9(2):65–74

Oyoo ME, Muhammed N, Cyrus KN, Githiri SM (2016) Assessment of the genetic diversity of Kenyan coconut germplasm using simple sequence repeat (SSR) markers Afr. J Biotech 15 (40):2215–2223

Panda KC (1982) Coconut types of Orissa. Indian Cocon J 12(3):10–18

Patel JS (1938) Coconut breeding. Proc Assoc Biol 5:1–16

Patil JL, Haldanka PM, Jamadagni BM, Salv MJ (1993) Variability and correlation studies for nut characters in coconut. J Mahrashtra Agricul Univ 18(3):361–364

Perera L (2010) Hybrid testing and variety identification of coconut (*Cocos nucifera* L.) in Sri Lanka using microsatellite markers. CORD 26(1):39–46

Perera L, Baudouin L, Bourdeix R, Fadhil AB, Hountondji FC, Al-Shanfari A (2011) Coconut palms on the edge of the desert: genetic diversity of *Cocos nucifera* L. in Oman. CORD 27:9–19

Perera L, Baudouin L, Mackay I (2016) SSR markers indicate a common origin of self-pollinating dwarf coconut in South-East Asia under domestication. Sci Hort 211:255–262

Perera L, Peries RRA, Padroasiri MHL (1997) The brown dwarf: a recent addition to the varieties and forms of the coconut palm. Cocos 12:82–84

Perera L, Russell JR, Provan J, Powell W (2000) Use of microsatellite DNA markers to investigate the level of genetic diversity and population genetic structure of coconut (*Cocos nucifera* L.). Genome 43:15–21

Perera L, Russell JR, Provan J, Powell W (2001) Levels and distribution of genetic diversity of coconut (*Cocos nucifera* L., var. Typica form typica) from Sri Lanka assessed by microsatellite markers. Euphytica 122:381–389

Perera L, Russell JR, Provan J, Powell W (2003) Studying genetic relationships among coconut varieties/populations using microsatellite markers. Euphytica 132:121–128

Peries RRA (1995) Dikiri-Pol: Fallacies, facts and the future. CORD XI (2): 50–58

Preethi P, Rajesh MK, Rahul CU, Jerard BA, Samsudeen K, Thomas RJ, Karun A (2016) Identification and utilization of informative EST-SSR markers for genetic purity testing of coconut hybrids. J Plantn Crops 44(2):77–84

Rahman PM, Varma RV, Sileshi GW (2012) Abundance and diversity of soil invertebrates in annual crops, agroforestry and forest ecosystems in the Nilgiri biosphere reserve of Western Ghats, India. Agrofor Syst 85(1):165–177

Rajagopal V, Kasturibai KV, Voleti SR (1990) Screening of coconut genotypes for drought tolerance. Ole´agineux 45(5):215–223

Rajagopal V, Shivishankar S, Kasturibai KV, Voleti SR (1988) Leaf water potential as an index of drought tolerance in coconut *Cocos nucifera*. Plant Physiol Bioch 15:80–86

Rajesh MK, Arunachalam V, Nagarajan P, Lebrun P, Samsudeen K, Thamban C (2008a) Genetic survey of ten Indian coconut landraces by simple sequence repeats (SSRs). Sci Hort 118:282–297

Rajesh MK, Nagarajan P, Jerard BA, Arunachalam V, Dhanapal R (2008b) Microsatellite variability of coconut accessions (*Cocos nucifera* L.) from Andaman and Nicobar Islands. Curr Sci 94:1627–1631

Rajesh MK, Radha E, Sajini KK, Karun A, Parthasarathy VA (2005) Plant regeneration through organogenesis and somatic embryogenesis from plumular explants of coconut (*Cocos nucifera* L.). J Plantn Crops 33(1):9–17

Rajesh MK, Jerard BA, Preethi P, Thomas RJ, Fayas TP, Rachana KE, Karun A (2013) Development of a RAPD-derived SCAR marker associated with tall-type palm trait in coconut. Sci Hort 150(4):312–316

Rajesh MK, Samusdeen K, Rejusha P, Manjula C, Rahman S, Karun A (2014a) Characterization of Annur and Bedakam ecotypes of coconut from Kerala state, India, using microsatellite markers. Int J Biodiv. doi:10.1155/2014/260895

Rajesh MK, Samusdeen K, Jerard BA, Rejusha P, Karun Anitha (2014b) Genetic and phylogenetic relationships of coconut populations from Amini and Kadmat Islands, Lakshadweep (India). Emir J Food Agric 26(10):898–906

Rajesh MK, Radha E, Sajini KK, Karun A (2014c) Polyamine induced somatic embryogenesis and plantlet regeneration *in vitro* from plumular explants of dwarf cultivars of coconut (*Cocos nucifera* L.). Ind J Agric Sci 84(4):527–530

Rajesh MK, Jerard BA, Preethi P, Thomas RJ, Karun A (2014d) Application of RAPD markers in hybrid verification in coconut. Crop Breed Appl Biot 14:36–41

Rajesh MK, Thomas RJ, Rijith J, Shareefa M, Jacob PM (2012) Genetic purity assessment of D x T hybrids in coconut with SSR markers. Ind J Genet Plant Breed 72(4):472–474

Ramanujam B, Nambiar KKN, Ratnambal MJ (1998) Screening of coconut cultivars/hybrids against *Thielaviopsis paradoxa* (de Seynes) Hohnel, using petiole inoculation technique. In: Mathew NM, Kuruvilla Jacob C, Licy J, Joseph T, Meenattoor JR, Thomas KK (eds) Developments in plantation crops research. Proceedings of the 12th symposium on plantation crops, PLACROSYM XII, Kottayam, India, Nov 27–29 1996, pp 284–286

Rangaswami G (1977) Palm tree crops in India. Outlook Agric 9:167–173

Rao JS (1957) Polyembryony in *Phoenix dactylifera* L. Sci Cult 22:686–687

Rao PSN, Kumar VS (1998) Some botanical curiosities. Curr Sci 75:91–92

Rao VR, Batugal P, Oliver J (2005) Coconut genetic resources. IPGRI—Regional Office for Asia, the Pacific and Oceania, Serdang, Malaysia, 779 p

Rao VR, Riley KW, Engels JMM, Engelmann F, Diekmann M (1998) Towards a coconut conservation strategy. In: Rao VR, Batugal P (eds) Proceedings of the COGENT regional coconut genebank planning workshop, Feb 26–28 1996, Pekanbaru, Riau, Indonesia. IPGRI-APO, Serdang, Selangor, Malaysia, pp 4–20

Ratambal MJ, Krishnan M, Devadas K (2000) Coconut cultivars of Goa. Indian Cocon J 31(2):1–3

Ratnambal MJ, Arunachalam V, Krishnan M (2003) Floral biology of some coconut accessions. J Plantn Crops 31(1):14–22

Renju NA, Stephen R, Manju RV (2015) Evaluation of selectively fertilized hybrids of coconut (*Cocos nucifera* L.) for water use efficiency under moderate water stress. J Plant Sci Res 31 (2):197

Repellin A, Thi AP, Tashakorie A, Sahsah Y, Daniel C, Zuily-Fodil Y (1997) Leaf membrane lipids and drought tolerance in young coconut palms (*Cocos nucifera* L.). Eur J Agron 6(1): 25–33

Ribeiro FE, Baudouin L, Lebrun P, Chaves LJ, Brondani C, Zucchi MI, Vencovsky R (2010) Population structures of Brazilian tall coconut (*Cocos nucifera* L.) by microsatellite markers. Genet Mol Biol 33(4):696–702

Riedel M, Riederer M, Becker D, Herran A, Kullaya A, Arana-López G, Pena-Rodríquez L, Billotte N, Sniady V, Rohde W, Ritter E (2009) Cuticular wax composition in *Cocos nucifera* L.: Physicochemical analysis of wax components and mapping of their QTLs onto the coconut molecular linkage map. Tree Genet Genom 5:53–69

Rivera R, Edwards KJ, Barker JHA, Arnold GM, Ayad G, Hodgkin T, Karp A (1999) Isolation and characterization of polymorphic microsatellites in *Cocos nucifera* L. Genome 42:668–675

Rohde W (1993) Improvement of coconut by biotechnology: Application of molecular techniques to breeding and crop protection. In: "La Recherche Européene au Service du Cocotier - Actes du Séminaire - Septembre 8–10 1993, Montpellier". CIRAD (Collection: Colloques du CIRAD), Montpellier, France, pp 41–52

Romney DH, Dias BC (1979) Coconut varieties in Bahia state. In: Session V of the FAO Technical Working Party on Coconut Production, Protection and Processing, Manila (Philippines), 3 Dec 1979

Saint LJP, de Lamothe NM (1987) Les hybrides de cocotiers Nains: performance et intérêt. Oléagineux 42(10):353–362

Sajini KK, Karun A, Amarnath CH, Engelmann F (2011) Cryopreservation of coconut (*Cocos nucifera* L.) zygotic embryos by vitrification. Cryo-Lett 32(4):317–328

Samsudeen K, Jacob PM, Niral V, Kumaran PM, Salooja R, Moosa H (2006) Exploration and collection of coconut germplasm in Kadmat and Amini Islands of Lakshadweep, India. Genet Resour Crop Evol 53(8):1721–1728

Samsudeen K, Niral V, Jerard BA, Kumar Mukesh, Sugatha P, Hebbar KB (2013a) Influence of variety and season in the inflorescence sap production in *Cocos nucifera* L. J Plantn Crops 41:57–61

Samsudeen K, Rajesh MK, Nagwaker DD, Reshmi R, Kumar PA, Devadas K, Karun A (2013b) Diversity in *Mohachao Narel*, a sweet endosperm coconut (*Cocos nucifera* L.) population from Maharashtra, India. Natl Acad Sci Lett 36(3):319–330

Samsudeen K, Thamban C, Niral V, Jerard BA, Rajesh MK, Manjula C, Devadas K, Karun A (2013c) In situ approach for rapid characterization to aid on farm conservation of coconut germplasm- A case study of two ecotypes from West coast of India. J Plantn Crops 41(3):357–363

Sangare A, Rognon F, de Nuce de Lamothe M (1978) Male and female phases in the inflorescence of coconut: Influence of mode of reproduction. Ole'agineux 33:609–615

Sankaran M, Damodaran V, Jerard BA, Abirami K (2016) Multiple Spicata Coconut (MSC): a rare type of coconut in Andaman Islands. Transcriptomics 3:123. doi:10.4172/2329-8936.1000123

Sankaran M, Damodaran V, Singh DR, Jerard BA (2012) Vivipary in *Cocos nucifera* L. var. Andaman Green Dwarf. Curr Sci 103(10):1139–1140

Santoso U, Kubo K, Ota T, Tadokoro T, Maekawa A (1996) Nutrient composition of kopyor coconuts (*Cocos nucifera* L.). Food Chem 57:299–304

Schut B (1975) Possible phytotoxic effect of copper on dwarf coconut palms. Surinaamse Landbouw 23(1):26–30

Shareefa M, Thomas RJ, Nampoothiri CK, Karun A (2014) Studies on vivipary in dwarf coconut cultivars. Indian J Hortic 71(4):464–468

Shalini KV, Manjunatha S, Lebrun P, Berger A, Baudouin L, Pirany N, Ranganath RM, Prasad DT (2007) Identification of molecular markers associated with mite resistance in coconut (*Cocos nucifera* L.). Genome 50:35–42

Sheela DL, Nazeem PA, Narayanankutty A, Manalil JJ, Raghavamenon AC (2016) In silico and wet lab studies reveal the cholesterol lowering efficacy of lauric acid, a medium chain fat of coconut oil. Plant Foods Hum Nutr. doi:10.1007/s11130-016-0577-y

Singh R, Low ET, Ooi LC, Ong-Abdullah M, Ting NC, Nagappan J, Nookiah R, Amiruddin MD, Rosli R, Manaf MA, Chan KL, Halim MA, Azizi N, Lakey N, Smith SW, Budiman MA, Hogan M, Bacher B, Van Brunt A, Wang C, Ordway JM, Sambanthamurthi R, Martienssen RA (2013) The oil palm *SHELL* gene controls oil yield and encodes a homologue of SEEDSTICK. Nature 500:340–344

Sugimura Y, Itano M, Salud CD, Otsuji K, Yamaguchi H (1997) Biometric analysis on diversity of coconut palm: cultivar classification by botanical and agronomical traits. Euphytica 98(1–2):29–35

Sunilkumar K, Mathur RK, Sparjanbabu DS, Pillai RSN (2015) Evaluation of interspecific oil palm hybrids for dwarfness. J Plantn Crops 43(1):29–34

Taulu DB, Rompas TM, Sudasrip N (1980) Coconut trichomes: their significance in classification and insect-resistance [Study conducted in Indonesia]. Philipp J Coconut Stud 5(2):39–46

Teulat B, Aldam C, Trehin R, Lebrun P, Barker JHA, Arnold GM, Karp A, Baudouin L, Rognon F (2000) An analysis of genetic diversity in coconut (*Cocos nucifera*) populations from across the geographic range using sequence tagged microsatellites (SSRs) and AFLPs. Theor Appl Genet 100:764–771

Thampan PK (2013) Production of Neera and coconut sugar deserves encouragement. Indian Cocon J 56(2):11–14

Thampan PK, Gopalakrishnan R (2010) Suitable coconut varieties for multipurpose uses. In: Ananda Bose (ed) In a Nutshell- Essays on coconut. Coconut Development Board, Publication No. 151, pp 107–111

Thomas RJ, Rajesh MK, Kalavathi S, Krishnakumar V, George DJ, Jose M, Nair RV (2013) Analysis of genetic diversity in coconut and its conservation in root (wilt) disease affected areas of Kerala: a community participatory approach. Ind J Genet Plant Breed 73(3):295–301

Thomas RJ, Rajesh MK, Jacob PM, Jose M, Nair RV (2015) Studies on genetic uniformity of Chowghat Green Dwarf and Malayan Green Dwarf varieties of coconut using molecular and morphometric methods. J Plantn Crops 43(2):89–96

Torres J (1937) Some notes on Makapuno coconut and its inheritance. Philipp J Agric 8:27–37

Vargas A, Blanco FA (2000) Fruit characterization of *Cocos nucifera* L. (Arecaceae) cultivars from the Pacific coast of Costa Rica and the Philippines. Genet Resour Crop Evol 47(5):483–487

Veerasamy S (1982) Polyembryony and twin seedlings in *Borassus flabellifer* L. (Palmae). Bot J Linn Soc 85:147–152

Villarreal DZ, Barrera MF, Hernández NT, Marín PCG (2005) Morphological variation of fruit in Mexican populations of *Cocos nucifera* L. (Arecaceae) under *in situ* and *ex situ* conditions. Genet Resour Crop Evol 52(4):421–434

Villarreal DZ, Morín JA (1998) Germination patterns in coconut populations (*Cocos nucifera* L.) in Mexico. Genet Resour Crop Evol 45(5):465–473

Villarreal DZ, Roque FH, Harries HC (1993) Coconut varieties in Mexico. Econ Bot 47(1):65–78

Whitehead RA (1968) Selecting and breeding coconut palms (*Cocos nucifera* L.) resistant to lethal yellowing disease. A review of recent work in Jamaica. Euphytica 17(1):81–101

Xiaoping Y, Jiabao W, Yunzhi Y (2004) Two new types of coconut (*Cocos nucifera* L.) in Hainan Island. Chinese J Trop Crops 4:003

Yao SDM, Konan KJL, Konan KJN, Issali AE, Sie RS, Zoro BIA (2013) Assessment of the genetic diversity conservation in three tall coconut (*Cocos nucifera* L.) accessions regenerated by controlled pollination, using microsatellite markers. African J Biotech 12(20):2808–2815

Chapter 2
Genetic Diversity and Conservation of Mexican Forest Trees

Christian Wehenkel, Samantha del Rocío Mariscal-Lucero, Juan Pablo Jaramillo-Correa, Carlos Antonio López-Sánchez, J. Jesús Vargas-Hernández and Cuauhtémoc Sáenz-Romero

Abstract Over the last 200 years, humans have impacted the genetic diversity of forest trees. Because of widespread deforestation and over-exploitation, about 9000 tree species are listed worldwide as threatened with extinction, including more than half of the ~600 known conifer taxa. A comprehensive review of the floristic-taxonomic literature compiled a list of 4331 recorded tree species in Mexico. The highest diversity of pine and oak worldwide is located in the Mexican temperate forests. Because species and genetic diversity are often positively associated, a very high trans-specific genetic diversity in Mexican tree species is thus expected. Contrasting with its high species and genus richness, studies of genetic diversity in Mexican forest trees are rather scarce, and often biased to particular families, like the Pinaceae. Moreover, even within those particular families the available surveys have a penchant for specific genus. The markers used in most of

C. Wehenkel (✉) · S.d.R. Mariscal-Lucero · C.A. López-Sánchez
Instituto de Silvicultura e Industria de La Madera, Universidad Juárez del Estado de Durango, Aptdo. Postal 741 Zona Centro, Durango, Dgo C.P. 34000, Mexico
e-mail: wehenkel@ujed.mx

S.d.R. Mariscal-Lucero
e-mail: sam_mar9@hotmail.com

C.A. López-Sánchez
e-mail: calopez@ujed.mx

J.P. Jaramillo-Correa
Instituto de Ecología, Universidad Nacional Autónoma de México, Ciudad Universitaria, Circuito Exterior S/N. Apartado Postal 70-275, Coyoacán, CDMX 04510, Mexico
e-mail: jaramillo@ecologia.unam.mx

J.J. Vargas-Hernández
Programa Forestal, Colegio de Postgraduados, Km. 36.5 Carretera México-Texcoco Montecillo Edo. de México, Puebla C.P. 56230, Mexico
e-mail: vargashj@colpos.mx

C. Sáenz-Romero
Instituto de Investigaciones Agropecuarias y Forestales, Universidad Michoacana de San Nicolás de Hidalgo, Km 9.5 Carr. Morelia-Zinapécuaro, Tarímbaro Michoacán 58880, Mexico
e-mail: csaenzromero@gmail.com

M.R. Ahuja and S.M. Jain (eds.), *Biodiversity and Conservation of Woody Plants*, Sustainable Development and Biodiversity 17, DOI 10.1007/978-3-319-66426-2_2

these studies include the traditional and "universal" isozymes and chloroplast microsatellites and, to a lesser extent, the anonymous SSRs, AFLPs, and RAPDs. More studies on more varied taxa and using more advanced technologies and markers seem thus necessary. Because of the poor comparability of the genetic diversity estimates among the studied Mexican tree species, it is extremely difficult to discern general trends across species or regions. We thus recommend that genetic diversity should be measured across species with an identical type of genetic marker, by surveying similar numbers of loci, individuals and populations, and using identical indices of genetic diversity, relevant to conservation of trees.

Keywords Tree species diversity · Trans-specific genetic diversity
Species distribution · in situ · ex situ conservation · Genetic erosion
Neotropical species

2.1 Introduction: The Importance of Genetic Diversity

Biodiversity or "biological diversity" can be defined as the variety and variability among living organisms and the ecological complexes in which they occur. It can be subdivided in four hierarchical categories of biological organization: regional landscape, community-ecosystem, population-species, and genetic (Noss 1990). In this context, genetic diversity is the foundation of all diversity levels (Frankham et al. 2010), as it is directly related to the adaptability of biological systems (Gregorius 2001). For instance, in the short term it is vital for increasing population fitness by reducing inbreeding depression in the short term, and, in the longer term, to develop new local adaptations in response to environmental changes (Reed and Frankham 2003; Kremer et al. 2012).

Genetic diversity also affects ecological processes such as primary productivity, population recovery from disturbances, interspecific competition, community structure, and fluxes of energy and nutrients. Thus, genetic diversity is an important factor for the populations, communities, and ecosystems health, whose effects on long-term evolution capability are comparable in magnitude to those of species diversity. Genetic diversity should have the most ecological impact on the four non-exclusive conditions: (1) when one or a few main habitat-providing species dominate a community or ecosystem (like in forests or coral reef ecosystems), (2) when the abundance or distribution of a keystone species is controlled by genetic diversity in other species (like in host parasite interactions), (3) when keystone species show measurable genetic diversity within populations for important traits, and/or (4) when species evolve in highly variable environments or in those subjected to rapid anthropogenic change (Hughes et al. 2008).

The level of genetic diversity in natural populations is influenced by evolutionary forces such as mutation, genetic drift, natural selection, and gene flow/migration (Futuyma 2013), along with other factors including the type of mating system (Schoen and Brown 1991), population size (Frankham 1996), inbreeding, spatial distribution, social organization, and behavior traits (Amos et al. 1998). For instance in plants, Hamrick et al. (1992) reported in a meta-analysis of 322 woody taxa that genetic diversity within species was higher within populations than among populations, and that genetic diversity was higher in outcrossers and in species with large geographic ranges or that were wind pollinated or dispersed than in inbreeders or in taxa with restricted distributions or that were pollinated/dispersed by animal vectors. They also found that differences in life history and ecological traits among species explained a significant proportion of their genetic diversity (for the genetic parameters measured), the specific evolutionary history of each species has a major role in determining the level and distribution of genetic diversity (i.e., the genetic structure).

Due to the importance of genetic diversity for the successful conservation of long-lived species, like most forest trees (Kahilainen et al. 2014), and for the long-term maintenance of all other forms of diversity (Riggs 1990), Brown et al. (1997) proposed seven indicators to monitor natural genetic diversity in carefully selected indicative taxa: (1) the number of sub-specific taxa; (2) population size, number, and physical location; (3) the environmental amplitude of each population; (4) genetic diversity at marker loci within individuals and populations; (5) quantitative genetic variation; (6) interpopulation genetic structure; and (7) mating system.

2.2 Genetic Diversity and Conservation of Tree Species

Over the last 200 years, humans have impacted the genetic diversity of forest trees by converting their original habitats for agricultural and urban uses, changing the demographic structure of forests (for instance, by forest management practices), fragmenting wildlands, exploiting species (e.g., timber exploitation), introducing exotic competitors and pests, degrading the environment with soil and atmospheric contaminants, and/or domesticating favored species (Ledig 1992).

Because of widespread deforestation and over-exploitation, about 9000 tree species are listed worldwide as threatened with extinction, including more than half of the ~600 known conifer taxa (Newton et al. 1999). Thus, efforts to conserve and manage tree species have been increasing around the world since the late 80s (Farnsworth and Sahotra 2007). Such conservation activities have encouraged studies of within-species genetic variation and distribution to enrich current conservation strategies (Newton et al. 1999). Indeed, such surveys can help designing evolutionarily significant units (ESU) or management units (MU) through the identification of unique and worth-conserving populations and/or hotspots of genetic diversity, while phylogeographic or genomic studies aiming to pinpoint adaptive variation should contribute to the development of new strategies and objectives for conservation (Bowen 1999; Newton et al. 1999). Analyses with molecular markers in tree species have, for example, demonstrated a far higher genetic differentiation between populations than initially shown by isozymes (Newton et al. 1999), while the distribution of cytoplasmic (both chloroplast and mitochondrial) lineages has allowed the identification of refugial areas and potential postglacial migration paths (e.g., Soltis et al. 2007; Jaramillo-Correa et al. 2009). Refugia are of particular importance for conservation because they often shelter higher genetic diversity than newly colonized areas (Newton et al. 1999). However, despite its importance for the long-term viability and the evolutionary potential of tree species and functioning of ecosystems, the genetic diversity of populations is seldom given explicit consideration in conservation programs (Mace and Purvis 2008; Walpole et al. 2009; GEO BON 2011; Kahilainen et al. 2014).

Because genetic diversity is found at various levels of organization, from the ecosystem, through species, and their provenances and family groups, to individual genotypes and alleles at the molecular level, it is essential that all levels are considered in conservation activities (Namkoong 1990). Moreover, the functioning of the ecosystem and methods for the conservation of genetic variation of individual species are dependent on the organization and structure of genetic diversity at each level (Riggs 1990). In this sense, and whenever it is applicable, in situ methods (conservation of genetic resources in natural populations) should be preferred to the ex situ strategies, such as conservation in botanical gardens, test plantations, seed orchards, clone banks, or tissue cultures. Indeed, in situ programs provide the opportunity for not only preserving individuals or species but also ongoing evolutionary processes, like gene flow or the fixation of new variants through local adaptation (Frankel 1981).

Some of the major problems of ex situ activities include the lack of long-term and continuous funding, which can affect the success of the conservation programs, or cause adverse changes in the preserved materials, which can compromise the species survival or regeneration. Some ex situ options, such as botanical gardens and arboreta, which are very common, only harbor a few specimens per taxa, which lead to conserve a low percentage of the genetic diversity of each species. Provenance trials, which compare offspring performance from a range of populations of the same species, maintain a broader genetic base than arboreta, but are generally restricted to commercial or potentially commercial taxa, and have a limited lifetime that rarely exceeds 50–100 years. Moreover, these tests are usually not self-perpetuating and occur in alien environments for most of the populations included. Seed orchards are generally established for commercial seed production of important timber species, and often use grafted clones, which can be easily replaced at a short-time period by the next generation of breeding (approx. 20 years). Thus, they usually bear a subset of the total genetic variation of a species, with an important bias for genotypes with desired commercial traits (wood quality, fast growth, pest resistance, etc.). Clone banks, on the other hand, have a broader genetic base and a longer lifespan than seed orchards, but they are commonly established with grafted clones, which also diminish their capacity for preserving an adequate percentage of genetic diversity. Seed banks are an obvious way to preserve genetic diversity, but not all seeds from all species can be stored under current storage conditions (i.e. low light and temperature) and germination rates tend to decline with time. Finally, tissue cultures might be of limited use given that mutations can accumulate at faster rates in disorganized tissues and because not all genotypes are prone to form somatic embryos or regenerate whole plants from cultured tissues (Ledig 1988).

Despite its limitations, ex situ conservation becomes especially important due to climatic change. The progressive decoupling between natural forest tree populations and the site where the climate for which they have evolved and adapted will occur, possess an immense challenge (Tchevakova et al. 2005; Sáenz-Romero et al. 2010). It is necessary to conduct an ex situ conservation program that includes the realignment of the natural populations to their suitable climate, by planting individuals in advance to the sites where their corresponding suitable climatic habitat will occur in the future. This has been named assisted migration (Ledig et al. 2010; Ledig 2012; Rehfeldt et al. 2014).

Conservation of genetic diversity in tree species is also essential for the sustainable and productive management of the forest ecosystems in which they occur. Indeed, it can be seen as one of the services of natural forests under multiple-use management. Thus, the reconcilability of economic and genetic conservation objectives may be easier to realize if appropriate levels of sustainable timber production are included as management objectives (http://www.fao.org/docrep/006/t0743e/T0743E02.htm).

2.3 Mexican Forests and Their Genetic and Species Richness of Vascular Plants and Tree Species

The total Mexican land area is ~196.7 million ha and about 72% consists of woodland (http://www.fao.org/docrep/meeting/x4702e.htm). SEMARNAP-UACH (1999) reported that the Mexican forests (excluding bushlands and grasslands) cover up to 55.3 million ha, including 0.721 million ha of mangroves and 0.163 million ha of gallery (riverine) forest. According to Palacio-Prieto et al. (2000), the main vegetation types in Mexico are temperate forests (which include the montane cloud, pine, and oak forests) that cover about 32 million hectares (ha), the tropical dry forests (encompassing both thorn and tropical deciduous forests) with 16.9 million ha, the tropical humid forests (including tropical rain, sub-evergreen tropical, and semi-deciduous tropical forests) that spread over 14 million ha, and the xeric shrublands with some 55.4 million ha.

Mexico is considered one of the countries with the largest diversity of plants worldwide (WCMC 1992; Mittermeier and Goettsch 1992). Although a complete inventory of its flora is still lacking, estimates range between 18,000 and 30,000 species, with more than 50% of them being restricted to its national territory (Rzedowski 1991a, b; Villaseñor 2003). An updated account of the generic richness of Mexican vascular plants found 2804 native genera, distributed in 304 families (Villaseñor 2004). Subsequently, Villaseñor and Ibarra-Manríquez in Cué-Bär et al. (2006) reported that at least 3639 species of Mexican native flora are angiosperm trees, which are grouped in 728 genera and 128 families. Families notable for their species richness include Fagaceae (270 species), Mimosaceae (268 species), and Fabaceae (205 species). However, some of them might be even richer. For instance, Sousa et al. (2001, 2003) found 623 tree species of the Fabaceae (Leguminosae) in Mexico. These authors also indicated that endemism at the generic level was marginal (3%), contrasting with at the species level (42.1%). However, a comprehensive review by Villaseñor (2013) of the floristic-taxonomic literature of Mexico compiled a list of 4331 recorded tree species. The highest diversity of pine and oak worldwide with 55 and 161 species, respectively, is located in the Mexican temperate forests (Galicia et al. 2015). This floristic richness is mainly caused by a wide range of climatic conditions, the numerous isolated mountains and mountain ranges, and large arid and semiarid expanses (Rzedowski 1994).

Figures 2.1 and 2.2 show that the tree species and genus richness in Mexico are not equally distributed and are principally functions of climate variables such as minimum temperature of coldest month, precipitation of driest month, and precipitation seasonality. The highest tree species and genus richness occur in the tropical rain forests of the states of Quintana Roo, Veracruz, and Chiapas, and the lowest in the Chihuahuan desert. Because species and genetic diversity are often positively associated (Bergmann et al. 2013; Kahilainen et al. 2014), a distribution of transspecific genetic diversity (Gregorius 2003; Wehenkel et al. 2006) similar to the Mexican tree species diversity shown in Fig. 2.1 is expected.

2.4 Studies About Genetic Diversity in Mexican Forest Trees

Studies of genetic diversity in Mexican forest tree are rare in proportion to the large species diversity in this country. To date (by 2015), the genetic variation has been studied in about only five tree families, ten genera and 52 species (mostly endemic). This represents only 1.2% of all Mexican tree species. Of these species, 58% belong to the genus *Pinus*, 15% to *Quercus*, 10% to *Abies,* and 6% to *Picea*. Thus, most of the information (80%) concerning the genetic diversity of Mexican trees comes from a single family of conifers (Pinaceae) that includes only 2% of all Mexican tree taxa. Moreover, from the studied tree species, approximately 38% are listed as endangered or vulnerable in the Red List of the International Union for Conservation of Nature and Natural Resources (IUCN 2013; Farjon et al. 2017), which indicates that, with the exception of a few species of *Pinus* and *Quercus*, there is virtually no genetic information available to design management plans for most tree taxa native to Mexico. Indeed, genetic diversity has been mostly assessed either in endemic species with narrow or discontinuous distribution, species facing fragmentation, habitat loss or other human-related threats, species with unique habitat or controversial taxonomic status, and only in a few commercially valuable,

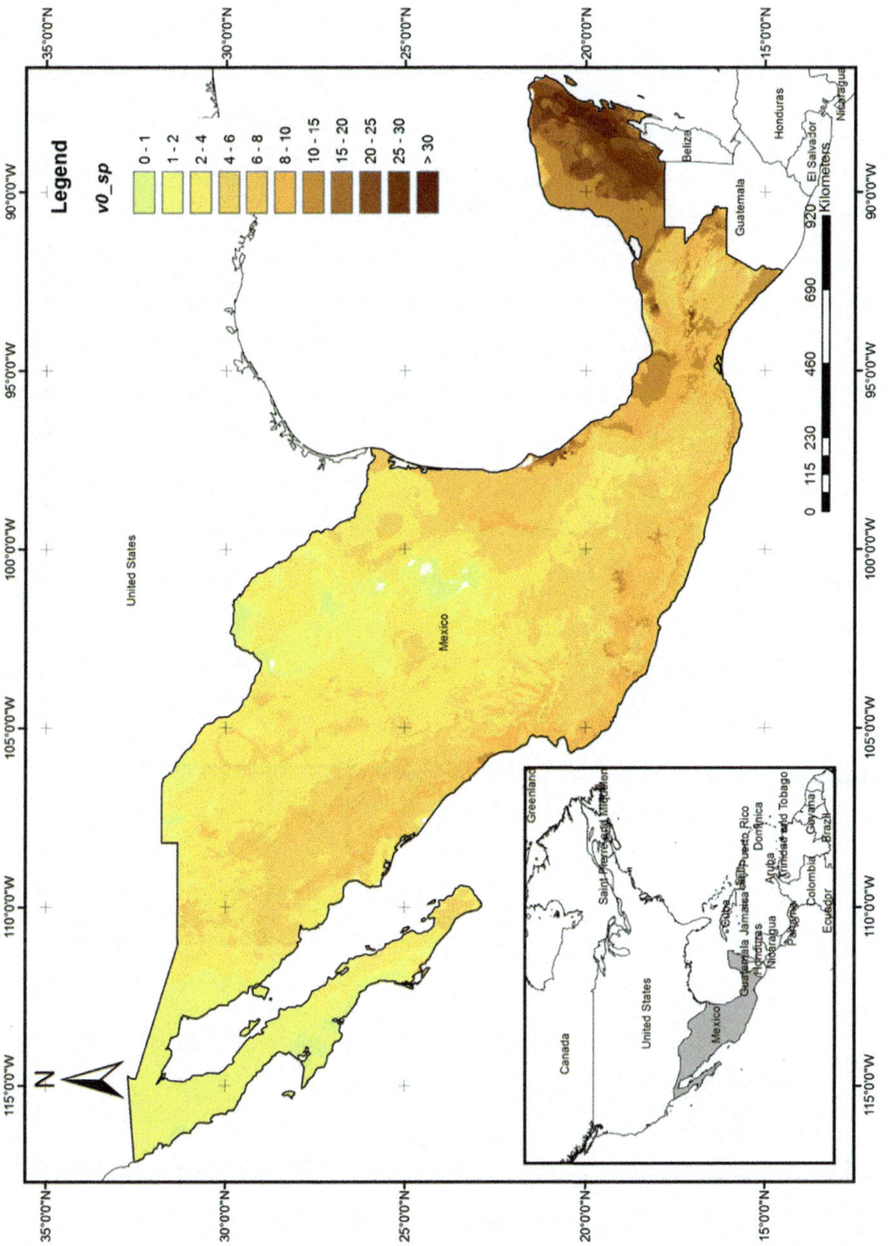

Fig. 2.1 Map of the tree species richness (v_{0_sp}) (in 1600 m^2) in Mexico predicted by minimum temperature of coldest month, precipitation of driest month, and precipitation seasonality ($R^2 = 0.59$, standard error = 5.68). *Yellow* colors indicate *lower* richness, *brown* colors *highest* richness

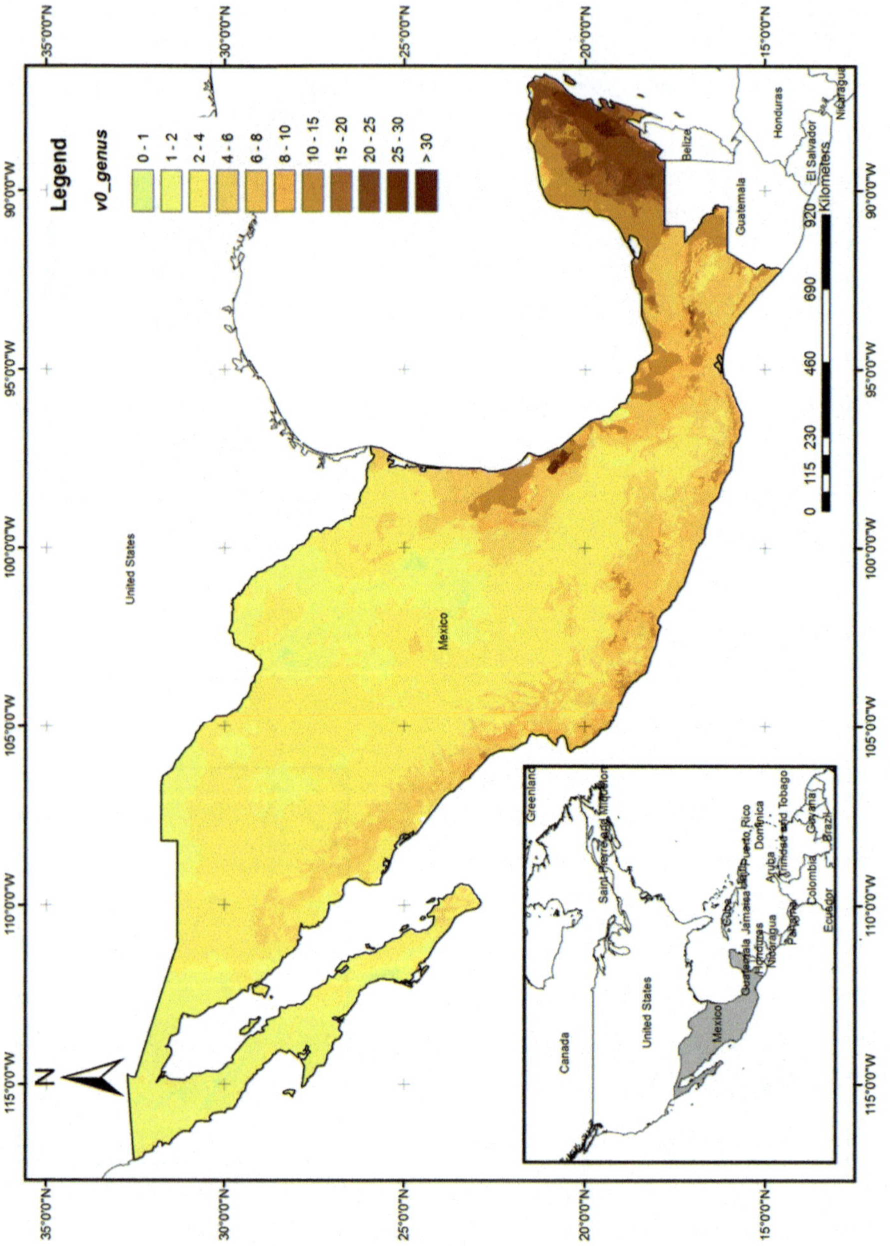

Fig. 2.2 Map of the tree genus richness (v_{0_genus}) (in 1600 m^2) in Mexico predicted by minimum temperature of coldest month, precipitation of driest month, and precipitation seasonality (R^2 = 0.64, standard error = 4.99). *Yellow* colors indicate *lower* richness, *brown* colors *highest* richness

wide-ranging timber species, such as those belonging to the genus *Pinus* and *Quercus* (Farjon and Styles 1997; Galicia et al. 2015).

Different molecular marker and isozymes have been traditionally used (Tables 2.1, 2.2, 2.3, 2.4, 2.5, 2.6, 2.7, 2.8, and 2.9), including simple sequence repeats or microsatellites (SSR), chloroplast microsatellites (cpSSR), random amplified polymorphic DNA (RAPD), amplified fragment length polymorphisms (AFLP), polymerase chain reaction-restriction fragment length polymorphisms (PCR-RFLP), and sequence variation in both mitochondrial DNA (mtDNA), and chloroplast DNA (cpDNA) regions. Different genetic diversity parameters have been usually calculated, such as the expected heterozygosity (H_e), the percentage of polymorphic loci (% P), the Shannon diversity index (S), the total haplotype diversity (H), the nucleotide diversity (π), the observed degrees of Gregorius' total differentiation (δ_T), and mean genetic diversity ($v_{mean,2}$) among many others (see references in the tables for details). However, given that different genetic markers, laboratories, numbers of loci, individuals, populations, and indices of genetic diversity have been reported, an accurate comparison of the genetic diversity between the studied Mexican tree species is difficult, although most of the general trends are summarized below for particular tree families (Tables 2.1, 2.2, 2.3, 2.4, 2.5, 2.6, 2.7, 2.8, and 2.9).

2.4.1 Family Pinaceae

Genetic diversity studies in the family Pinaceae have centered on the genus *Pinus*, although some individual reports in *Abies*, *Picea,* and *Pseudotsuga* have also been performed (Tables 2.1, 2.2, 2.3, 2.4, 2.5, 2.6, and 2.7).

2.4.1.1 Genus *Pinus*

Pinus is particularly relevant because more than half of the species of this genus occur naturally in Mexico, most of which are endemic, thus making of Mexico one of its main centers of pine diversity (Piñero et al. 2008). Only 14% of pine species studied at genetic level so far is listed as endangered in the Red List of the IUCN. Tables 2.1, 2.2, 2.3, and 2.4 show summary statistics of genetic diversity estimated with molecular markers for different Mexican species of *Pinus*. The majority of the studies were performed using isoenzymes, while the least used type molecular markers to determine the genetic variation in this genus was the RAPD; only two species include this type of marker: *Pinus chiapensis* and *Pinus culminicola* (Newton et al. 2002; Favela-Lara 2010).

On average the expected heterozygosity (H_e) for *Pinus* populations equals 0.21 using isozyme and 0.60 with microsatellite (SSR). While the average genetic diversity assessed by AFLP markers is d_g = 1.40. The estimated diversity for the Mexican species of *Pinus* is comparable to that of their European counterparts

Table 2.1 Genetic diversity obtained in Mexican species of *Pinus* using isoenzymes

Genus *Pinus* Isozymes									
Species	MM	P	I	L	Genetic diversity			Conservation status (IUCN 2013)	References
					He	% P	S		
Pinus cembroides (lagunae)	Iso	4	140	14	0.386	81.6	—	LC	Molina-Freaner et al. (2001)
Pinus coulteri	Iso	1	35	13	0.193	48.5	—	NT	Ledig (2000)
Pinus greggii	Iso	19	366	15	0.123	31.9	—	VU	Parraguirre-Lezama et al. (2002)
Pinus hartwegii	Iso	4	107	12	0.120	—	—	LC	Viveros-Viveros et al. (2010)
Pinus johannis	Iso	4	130	16	0.245	88.0	—	LC	García-Gómez et al. (2014)
Pinus maximartinezii	Iso	1	100	26	0.122	30.3	—	EN	Ledig et al. (1999)
Pinus muricata	Iso	3	66	33	0.346	87.7	—	VU	Molina-Freaner et al. 2001; Sáenz-Romero and Tapia-Olivares (2003)
Pinus oocarpa	Iso	5	49	12	0.102	—	—	LC	Sáenz-Romero and Tapia-Olivares (2003)
Pinus pinceana	Iso	8	132	27	0.174	—	—	LC	Ledig et al. (2001)
Pinus pinceana	Iso	5	114	13	0.374	96.9	—	LC	Molina-Freaner et al. (2001)
Pinus pseudostrobus	Iso	8	80	14	0.100	78.6	0.22	LC	Viveros-Viveros et al. (2014)
Pinus rzedowskii	Iso	9	295	14	0.219	46.8	—	VU	Delgado et al. (1999)

MM molecular marker, *Iso* isozymes, *P* number of populations, *I* number of individuals included per populations, *L* number of loci, H_e expected heterozygosity, *% P* percent polymorphic loci, *S* Shannon diversity index, *IUCN* Union for Conservation of Nature and Natural Resources, *LC* least concern, *EN* endangered, *VU* vulnerable, *NT* near threatened

Table 2.2 Genetic diversity obtained in Mexican species of *Pinus* using cpSSR

Genus *Pinus* cpSSR									
Species	MM	PP	P	I	L	Genetic diversity		Conservation status (IUCN 2013)	References
						H	H_{eSEQ}		
Pinus pseudostrobus	cpSSR	6	3	48	6	0.972	—	LC	Delgado et al. (2007)
Pinus ayacahuite	cpSSR	12	7	75	12	—	0.510	LC	Moreno-Letelier and Piñero (2009)
Pinus leiophylla var. *chihuahuana*	cpSSR	6	9	101	6	0.511	—	LC	Rodríguez-Banderas et al. (2009)
Pinus leiophylla var. *leiophylla*	cpSSR	6	15	222	6	0.869	—	LC	Rodríguez-Banderas et al. (2009)
Pinus montezumae	cpSSR	6	4	93	6	0.858	—	LC	Delgado et al. (2007)
Pinus nelsonii	cpSSR	11	9	232	11	0.648	—	EN	Cuenca et al. (2003)
Pinus strobiformis	cpSSR	12	17	312	12	—	0.661	LC	Moreno-Letelier and Piñero (2009)

cpSSR chloroplast microsatellite, *MM* molecular marker, *PP* number of primer-pair combinations, *P* number of populations, *I* number of individuals included per populations, *L* number of loci, H_e expected heterozygosity, H_{eSEQ} genetic diversity of data corrected for molecularly accessible size homoplasy, *IUCN* Union for Conservation of Nature and Natural Resources, *LC* least concern, *EN* endangered

Table 2.3 Genetic diversity obtained in Mexican species of *Pinus* using SSR

Genus *Pinus* SSR								
Species	MM	PP	P	I	L	H_e	Conservation status (IUCN 2013)	References
Pinus ayacahuite	SSR	5	—	—	5	0.654	LC	Villalobos-Arámbula et al. (2014)
Pinus ayacahuite var. veitchii	SSR	4	—	—	4	0.810	LC	Villalobos-Arámbula et al. (2014)
Pinus chiapensis	SSR	5	—	—	5	0.659	EN	Villalobos-Arámbula et al. (2014)
Pinus montezumae	SSR	5	1	27	5	0.274	LC	Delgado et al. (2013)
Pinus oocarpa	SSR	11	27	266	11	0.630	LC	Dvorak et al. (2009)
Pinus patula	SSR	3	6	108	3	0.802	LC	Alfonso-Corrado et al. (2014a)
Pinus patula	SSR	11	6	60	11	0.559	LC	Dvorak et al. (2009)
Pinus pseudostrobus	SSR	5	1	37	5	0.277	LC	Delgado et al. (2013)
Pinus radiata	SSR	19	2	62	19	0.685	EN	Karhu et al. (2006)
Pinus strobiformis	SSR	5	—	—	4	0.654	LC	Villalobos-Arámbula et al. (2014)

SSR simple sequence repeat microsatellite, *MM* molecular marker, *PP* number of primer-pair combinations, *P* number of populations, *I* number of individuals included per populations, *L* number of loci, H_e expected heterozygosity, *IUCN* Union for Conservation of Nature and Natural Resources, *LC* least concern, *EN* endangered

Table 2.4 Genetic diversity obtained in Mexican species of *Pinus* using AFLP and RAPD

Genus *Pinus*
AFLP & RAPD

Species	MM	PP	P	I	L	Genetic diversity				Conservation status (IUCN 2013)	References
						% P	S	*dg*	$v_{mean,2}$		
Pinus arizonica	AFLP	1	8	280	319	—	—	1.36	—	LC	Wehenkel et al. (2015)
Pinus cembroides	AFLP	1	3	105	319	—	—	1.50	—	LC	Wehenkel et al. (2015)
Pinus cooperi	AFLP	1	5	35	319	—	—	1.37	—	VU	Wehenkel et al. (2015)
Pinus discolor	AFLP	1	1	175	319	—	—	1.54	—	NA	Wehenkel et al. (2015)
Pinus durangensis	AFLP	1	5	315	319	—	—	1.36	—	NT	Wehenkel et al. (2015)
Pinus engelmannii	AFLP	1	9	105	319	—	—	1.37	—	LC	Wehenkel et al. (2015)
Pinus herrerae	AFLP	1	3	315	319	—	—	1.37	—	LC	Wehenkel et al. (2015)
Pinus leiophylla	AFLP	1	9	175	319	—	—	1.38	—	LC	Wehenkel et al. (2015)
Pinus lumholtzii	AFLP	1	5	70	319	—	—	1.36	—	NT	Wehenkel et al. (2015)
Pinus oocarpa	AFLP	1	2	37	319	—	—	1.47	—	LC	Wehenkel et al. (2015)
Pinus strobiformis	AFLP	1	12	129	243	—	—	—	1.542	LC	Simental-Rodríguez et al. (2014)
Pinus teocote	AFLP	1	9	37	319	—	—	1.36	—	LC	Wehenkel et al. (2015)
Pinus chiapensis	RAPD	8	11	138	49	—	0.546	—	—	EN	Newton et al. (2002)
Pinus culminicola	RAPD	5	2	60	72	57.3	0.569	—	—	EN	Lara (2010)

RAPD random amplified polymorphic DNA, *AFLP* amplified fragment length polymorphism, *Iso* isozymes, *MM* molecular marker, *P* number of populations, *I* number of individuals included per populations, *L* number of loci, *% P* percent polymorphic loci, *S* Shannon diversity index, d_g genetic diversity $v_{mean,2}$ mean genetic diversity, *IUCN* Union for Conservation of Nature and Natural Resources, *LC* least concern, *EN* endangered, *VU* vulnerable, *NT* near threatened

Table 2.5 Genetic diversity obtained in Mexican species of *Abies*

Genus *Abies*											
Species	MM	PP	P	I	L	Genetic diversity				CV (IUCN 2013)	References
						H_e	% P	H	H_s		
Abies flinckii	cpSSR	7	6	105	7	—	—	0.802	—	VU	Jaramillo-Correa et al. (2008)
Abies guatemalensis	cpSSR	7	10	160	7	—	—	0.934	—	EN	Jaramillo-Correa et al. (2008)
Abies hickelii	cpSSR	7	7	112	7	—	—	0.937	—	EN	Jaramillo-Correa et al. (2008)
Abies religiosa	cpSSR	7	11	159	7	—	—	0.908	—	LC	Jaramillo-Correa et al. (2008)
Abies religiosa	cpDNA	2	17	128	2	—	—	0.395	0.187	LC	Heredia-Bobadilla et al. (2013)
Abies religiosa	Iso	—	11	30	16	0.108	31.8	—	—	LC	Aguirre-Planter et al. (2000)
Abies flinckii	Iso	—	6	29	16	0.113	30.2	—	—	VU	Aguirre-Planter et al. (2000)
Abies guatemalensis	Iso	—	10	30	16	0.069	20.0	—	—	EN	Aguirre-Planter et al. (2000)
Abies hickeli	Iso	—	6	33	16	0.100	28.2	—	—	EN	Aguirre-Planter et al. (2000)
Abies religiosa	mtDNA	2	17	128	2	—	—	0.319	0.157	LC	Heredia-Bobadilla et al. (2013)
Abies flinckii	mtDNA regions sequence	20	6	105	20	—	—	0	—	VU	Jaramillo-Correa et al. (2008)
Abies guatemalensis	mtDNA regions sequence	20	10	160	20	—	—	0.036	—	EN	Jaramillo-Correa et al. (2008)
Abies hickelii	mtDNA regions sequence	20	7	112	20	—	—	0.047	—	EN	Jaramillo-Correa et al.(2008)
Abies religiosa	mtDNA regions sequence	20	11	159	20	—	—	0	—	LC	Jaramillo-Correa et al. (2008)
Abies religiosa	SSR	2	17	128	2	—	—	0.297	0.158	LC	Heredia-Bobadilla et al. (2013)

cpSSR chloroplast microsatellite, *cpDNA* chloroplast DNA, *mtDNA* mitochondrial DNA, *Iso* isozymes, *SSR* simple sequence repeat microsatellite, *MM* molecular marker, *PP* number of primer-pair combinations, *P* number of populations, *I* number of individuals included per populations, *L* number of loci, H_e expected heterozygosity, *H* total haplotypic diversity, H_s average within-population haplotype diversity, *% P* percent polymorphic loci, *CV* conservation status, *IUCN* Union for Conservation of Nature and Natural Resources, *LC* least concern, *EN* endangered, *VU* vulnerable, *NT* near threatened

Table 2.6 Genetic diversity obtained in Mexican species of *Picea*

Genus *Picea*												
Species	MM	PP	P	I	L	Genetic diversity					Conservation status (IUCN 2013)	References
						H_e	% P	H	δT	$v_{mean,2}$		
Picea chihuahuana	mt DNA	16	16	156	16	—	—	0	—	—	EN	Jaramillo-Correa et al. (2006)
Picea chihuahuana	cpSSR	6	16	156	6	—	—	0.41	—	—	EN	Jaramillo-Correa et al. (2006)
Picea chihuahuana	AFLP	1	14	669	243	—	—	—	—	1.5	EN	Simental-Rodríguez et al. (2014)
Picea chihuahuana	AFLP	1	5	254	319	—	—	—	0.3	—	EN	Wehenkel and Sáenz-Romero (2012)
Picea chihuahuana	Iso	—	10	164	24	0.09	27	—	—	—	EN	Ledig et al. (1997)
Picea martínezii	Iso	—	2	54	22	0.11	32	—	—	—	EN	Ledig et al. (2000b)
Picea mexicana	Iso	—	3	82	18	0.12	35	—	—	—	EN	Ledig et al. (2002)

cpSSR chloroplast microsatellite, *mtDNA* mitochondrial DNA, *Iso* isozymes, *AFLP* amplified fragment length polymorphism, *MM* molecular marker, *PP* number of primer-pair combinations, *P* number of populations, *I* number of individuals included per populations, *L* number of loci, H_e expected heterozygosity, *H* total haplotypic diversity, *% P* percent polymorphic loci, δT Gregorius' total differentiation, $v_{mean,2}$ mean genetic diversity, *IUCN* Union for Conservation of Nature and Natural Resources, *EN* endangered

Table 2.7 Genetic diversity obtained in Mexican species of *Pseudotsuga menziesii*

Pseudotsuga menziesii

Species	MM	PP	P	I	L	Genetic diversity					Conservation status (IUCN 2013)	References
						H_e	% P	H	π	$v_{mean,2}$		
Pseudotsuga menziesii	cpDNA segments sequence	2	11	129	2	—	—	0.79	0.0008	—	LC	Gugger et al. (2011)
Pseudotsuga menziesii	cpDNA segments sequence	1	9	107	1	—	—	0.75	0.0008	—	LC	Wei et al. (2011)
Pseudotsuga menziesii	cpSSR	3	11	129	3	—	—	0.91	—	—	LC	Gugger et al. (2011)
Pseudotsuga menziesii	mtDNA segments sequence	2	7	55	2	—	—	0.59	0.0005	—	LC	Gugger et al. (2011)
Pseudotsuga menziesii	AFLP	1	6	63	1	—	—	—	—	1.6	LC	Simental-Rodríguez et al. (2014)
Pseudotsuga menziesii	Iso	—	11	170	18	0.08	28	—	—	—	LC	Cruz-Nicolas et al. (2011)

cpSSR chloroplast microsatellite, *cpDNA* chloroplast DNA, *mtDNA* mitochondrial DNA, *AFLP* amplified fragment length polymorphism, I*so* isozymes, *MM* molecular marker, *PP* number of primer-pair combinations, *P* number of populations, *I* number of individuals included per populations, *L* number of loci, H_e expected heterozygosity, *% P* percent polymorphic loci, *H* total haplotypic diversity, *π* nucleotide diversity, $v_{mean,2}$ mean genetic diversity, *IUCN* Union for Conservation of Nature and Natural Resources, *LC* least Concern

Table 2.8 Genetic diversity obtained in the family Fagaceae

Family Fagaceae											
Species	MM	PP	P	I	L	Genetic diversity					References
						H_e	% P	S	H	H_s	
Fagus grandifolia subsp. *mexicana*	Iso	—	7	164	10	0.20	63	—	—	—	Montiel-Oscura et al. (2013)
Fagus grandifolia subsp. *mexicana*	RAPD	18	4	96	62	—	39	0.20	0.135	—	Rowden et al. (2004)
Quercus laurina	PCR-RFLP	6	6	50	—	—	—	—	—	0.411	González-Rodríguez et al. (2004)
Quercus eduardii	RAPD	2	4	120	58	0.33	95	0.50	—	—	Alfonso-Corrado et al. (2004)
Quercus potosina	RAPD	2	3	90	44	0.35	97	0.53	—	—	Alfonso-Corrado et al. (2004)
Quercus crassipes	RAPD	18	4	80	121	0.37	—	—	—	—	Tovar-Sánchez et al. (2015)
Quercus rugosa	RAPD	18	4	80	121	0.28	—	—	—	—	Tovar-Sánchez et al. (2015)
Quercus rugosa	SSR	3	4	80	3	0.48	—	—	—	—	Tovar-Sánchez et al. (2015)
Quercus crassipes	SSR	3	4	80	3	0.42	—	—	—	—	Tovar-Sánchez et al. (2015)
Quercus sideroxyla	SSR	4	9	150	4	0.85	—	—	—	—	Alfonso-Corrado et al. (2014b)
Quercus sideroxyla	SSR	7	4	44	7	0.82	—	—	—	—	Peñaloza-Ramírez et al. (2010)
Quercus scytophylla	SSR	7	4	39	7	0.80	—	—	—	—	Peñaloza-Ramírez et al. (2010)
Quercus Hypoleucoides	SSR	7	2	21	7	0.75	—	—	—	—	Peñaloza-Ramírez et al. (2010)

Iso isozymes, *RAPD* random amplified polymorphic DNA, *SSR* simple sequence repeat microsatellite, *PCR-RFLP* polymerase chain reaction-restriction fragment length polymorphism, *MM* molecular marker, *PP* number of primer-pair combinations, *P* number of populations, *I* number of individuals included per populations, *L* number of loci, H_e expected heterozygosity, *% P* percent polymorphic loci, *S* Shannon diversity index, *H* total haplotypic diversity, H_s average within-population haplotype diversity

Table 2.9 Genetic diversity obtained in the family Meliaceae, family Salicaceae and family Cupressaceae

Species	MM	PP	P	I	L	Genetic diversity					Conservation status (IUCN 2013)	References
						H_e	*h*	S	π	$v_{mean,2}$		
Family Meliaceae												
Cedrela odorata	RAPD	8	3	34	—	—	—	0.07	—	—	VU	Navarro et al. (2005)
Swietenia macrophylla	RAPD	10	3	31	102	—	—	0.30	—	—	VU	Gillies et al. (1999)
Swietenia macrophylla	SSR	8	4	94	8	0.78	—	—	—	—	VU	Alcalá et al. (2014)
Swietenia macrophylla	SSR	7	1	40	7	0.68	—	—	—	—	VU	Novick et al. (2003)
Family Salicaceae												
Populus tremuloides	AFLP	1	7	76	243	—	—	—	—	1.4	NA	Simental-Rodríguez et al. (2014)
Populus tremuloides	SSR	8	1	13	8	0.610	—	—	—	—	NA	Callahan et al. (2013)
Family Cupressaceae												
Juniperus blancoi	Nuclear loci	6	8	74	8	0.345	—	—	0.0014	—	NT	Moreno-Letelier et al. (2014)
Callitropsis guadalupensis	Plastid DNA	11	1	54	11	—	0.6	—	0.0007	—	EN	Escobar et al. (2011)
Callitropsis forbesii	Plastid DNA	11	6	100	11	—	0.3	—	0.0002	—	EN	Escobar et al. (2011)

RAPD random amplified polymorphic DNA, *AFLP* amplified fragment length polymorphism, *SSR* simple sequence repeat, microsatellite, *Iso* isozymes, *MM* = molecular marker, *PP* number of primer-pair combinations, *P* number of populations, *I* number of individuals included per populations, *L* number of loci, H_e expected heterozygosity, *h* haplotypic diversity. *S* Shannon diversity index, π nucleotide diversity, $v_{mean,2}$ mean genetic diversity, *IUCN* Union for Conservation of Nature and Natural Resources, *EN* endangered, *VU* vulnerable, *NT* near threatened, *NA* none assigned

(H_e = 0.211, Müller-Starck et al. 1992), and it is slightly higher than that found by Hamrick et al. (1992) in studies published from 1968 to 1990 that reported estimates of allozyme variation for gymnosperms and angiosperms (H_e = 0.136 in *Pinus*). Galicia et al. (2015) suggested that most Mexican pine species have high genetic diversity and relatively low genetic differentiation among populations. For example, some species with a high genetic diversity are *Pinus oocarpa, P.* leiophylla, *P. johannis, P. pinceana, P. cembroides, and P. muricata* (Dvorak et al. 2009; Rodríguez-Banderas et al. 2009; Molina-Freaner et al. 2001; García-Gómez et al. 2014). Such high levels of genetic diversity have allowed the selection of high-yielding genotypes for reforestation programs and the establishment of commercial plantations in Mexico and other countries (Molina-Freaner et al. 2001). Although, it must also be noted that many exceptions with low genetic diversity figures also exist, such as some endemics and taxa with fragmented distributions, like *P. culminicola* (Favela-Lara 2010), *P. maximartinezii* (Ledig et al. 1999), *P. chiapensis* (Newton et al. 2002), and *P.greggii* (Parraguirre-Lezama et al. 2002).

Wehenkel et al. (2015) found in a study of seed stands of 11 *Pinus* in the Sierra Madre Occidental, that pine species and populations within species that were exposed to prolonged cold periods and low Mg proportion of the cation-exchange capacity in the soil in their distribution range possess lower AFLP diversity. However, Dvorak et al. (2009) reported that *Pinus oocarpa* shows no significant changes in genetic diversity in populations across its geographic range of 3000 km in Mesoamerica.

2.4.1.2 Genus *Abies*

Studies on the genus *Abies* mostly have focused in the southern taxa, namely *Abies flinckii, A. guatemalensis, A. hickelii,* and *A. religiosa* (Table 2.5). These four species are in the Red List of the IUCN. Although widely distributed, *A. guatemalensis* is considered an endangered species, given that it is distributed in various but isolated populations (Aguirre-Planter et al. 2000). Both *A. flinckii* and *A. hickelii* have restricted distributions and are also considered as endangered (Bello and López-Mata 2001), while *A. religiosa* has a mostly continuous and wide natural range, but its recurrent exploitation for timber has put it at risk (Wallace et al. 2015). In general, the average expected heterozygosity (H_e) in this genus in Mexico, as estimated with isoenzymes, equals 0.10 (Aguirre-Planter et al. 2000), which is far lower than the average for gymnosperms determined by Hamrick et al. (1992; H_e = 0.151) or than that found in the European species *Abies alba* (H_e = 0.41, Müller-Starck et al. 1992). On the other hand, the average total haplotypic diversity (*H*) measured with chloroplast microsatellites was 0.80, which is not different from the values found for other temperate firs in Europe or North America (Jaramillo-Correa et al. 2008).

Unlike other conifers, the genus *Abies* in Mexico present low genetic diversity within, but high genetic differentiation among populations; studies of these species have demonstrated that *Abies* populations have likely passed through genetic bottlenecks that decreased their genetic diversity and steered to interpopulation differentiation (Aguirre-Planter et al. 2000). Furthermore, Jaramillo-Correa et al. (2008) suggested that the four species of *Abies* in Mesoamerica share a recent common ancestor. Given these patterns of genetic structure in the *Abies* populations it has been suggested to implement conservation strategies for these Mexican conifers, proposing in situ conservation of populations along the Transverse Volcanic Belt in central México where most of population differentiation occurs and where they are more threatened by human activities (Jaramillo-Correa et al. 2008).

2.4.1.3 Genus *Picea*

Studies in the genus *Picea* have covered all three Mexican species: *Picea chihuahuana, P. martínezii,* and *P. mexicana.* All three taxa are listed as endangered in the Red List of the IUCN (Table 2.6). The average expected heterozygosity (H_e) in this genus estimated by using isoenzymes equals 0.11. This level of genetic diversity is similar to that found in the rare congeners *Picea breweriana*, native of North America (H_e = 0.129) and *P. asperata*, native of Southwest China (Western china H_e = 0.096) (Luo et al. 2005; Ledig and Johnson 2005). On the other hand, when compared with the values found in the European spruce *P. abies* (H_e = 0.371) and in North American black spruce (*P. mariana,* H_e = 0.300) these diversity values are much lower (Lundkvist 1979; Isabel et al. 1995).

Picea chihuahuana has been the most genetically studied of all Mexican spruces (Ledig et al. 1997; Jaramillo-Correa et al. 2006; Wehenkel and Sáenz-Romero 2012; Simental-Rodríguez et al. 2014). This species is distributed in the Sierra Madre Occidental in the states of Chihuahua and Durango, forming only 40 populations (Wehenkel and Sáenz-Romero 2012). Its genetic diversity is within the range reported for conifers, although Ledig et al. (1997) found a high fixation index and a 45% of empty seeds, which indicates that inbreeding may be a serious problem for Chihuahua spruce. In addition, Jaramillo-Correa et al. (2006) found only eight chloroplast DNA haplotypes and two mitochondrial DNA haplotypes, which is less than the amount for more boreal spruces. The distribution of mitotypes showed two areas, differentiating the northern populations from the central and southern ones, which suggests that these two areas could represent different ancestral populations. The study also detected recent bottlenecks in some stands and suggested the conservation of southern populations that exhibit high levels of genetic diversity. In a more recent study, Wehenkel and Sáenz-Romero (2012) found significant genetic erosion only in a very small population of 120 individuals and concluded that the loss of genetic diversity per se does not explain the relict status of *P. chihuahuana.* Unexpectedly, Dominguez-Guerrero et al. (2017) found that higher mean temperature in the warmest month are associated to larger mean AFLP diversity.

P. mexicana has three small populations isolated and fragmented on sky islands from the Sierra Madre Oriental and the Sierra Madre Occidental (Sierra el Coahuilón, Cerro Mohinora and Sierra la Marta) and are threatened by global warming, fire, and grazing (Ledig et al. 2002). The fire affected Mexican spruce in 1975 and most individuals of Sierra la Marta disappeared (Rushforth 1986; Gordon 1991). Ledig et al. (2002) concluded that the populations of *P. mexicana* are genetically viable and their main threat is the environment, while Jaramillo-Correa et al. (2015a) estimated the bottleneck that affects the species to be rather ancient, which suggests that this species has developed some kind of tolerance to stochastic processes. However, Flores-Lopez et al. (2005) found that the reproductive state in the three populations is critical and suggested ex situ conservation, in addition to protecting natural stands along with the establishment of strict measures of management, and protection against grazing and forest fires.

Picea martínezii (Martínez spruce) is a conifer with only four known relict populations of less than 800 trees each (Mendoza-Maya et al. 2015). Ledig et al. (2000b) studied two of those populations, finding levels of genetic diversity comparable to those of *P. chihuahuana* and *P. mexicana*; they further detected that the mating system of Martínez spruce was characterized by a high frequency of selfing, which indicates that this species could be an example of facultative selfing and survival in interglacial refugia. Therefore, for species like *Picea martínezii,* ex situ conservation, including establishment of seed banks and botanical gardens seems required (Ledig et al. 2000b).

Based on the genetic diversity and structure of this three Mexican species of *Picea*, Mendoza-Maya et al. (2015) suggested several in situ conservation activities, including protection, and enlargement of genetic diversity through the expansion of specific populations, by planting individuals originated from seed collected in different locations. The aim of such activities would be getting a genetically viable minimum population size of 1035–3836 individuals. For ex situ conservation, it was recommended the establishment of populations at sites outside the natural range, where the suitable climatic habitat for each species is predicted to occur. For example, such conditions might be encountered in the near future at higher altitudes (which, paradoxically, sometimes occur at southern latitudes in México) at the volcano Cofre de Perote, Veracuz (*P. mexicana*), in the municipality of Guanaceví, Durango (*P. chihuahuana*), and in the region of Tlatlauquitepec, Puebla (*P. martinezii*).

2.4.1.4 Genus *Pseudotsuga*

Douglas-fir (*Pseudotsuga menziesii*, Mirb (Franco)) is an ecologically and economically important coniferous tree species in North America, with a broad natural distribution in western Canada and USA that spreads south into Mexico (Hermann and Lavender 1999). In Mexico, Douglas-fir is distributed along three different mountain ranges, throughout the northwest, northeast, and into the central and southern regions of the country, having an extremely fragmented and discontinuous distribution (Cruz-Nicolas et al. 2011). Even though in the past a controversy was raised about the number of *Pseudotsuga* species growing in this country (Martinez 1963; Little 1979), recent studies using morphological, biochemical, and molecular markers have shown that all Mexican Douglas-fir populations are indeed

Pseudotsuga menziesii (Debreczy and Racz 1995; Reyes-Hernández et al 2006; Wei et al. 2011; Adams et al. 2012, 2013).

The studies of genetic diversity for the Mexican populations of this species showed that its average total haplotypic diversity (H) equals 0.76 (Table 2.7). This genetic diversity is comparable to that of the Canadian (H = 0.815) and American stands (H = 0.745) (Wei et al. 2011). However, a study using isozymes showed a much lower genetic diversity (H_e = 0.08), particularly for the small and isolated populations from the mountains of central Mexico (Cruz-Nicolas et al. 2011). Several studies have shown that *P. menziesii* may be affected by climate change in the future, so it is important to conserve this and other temperate species in Mexico (Gugger et al. 2011). A recent study evaluated the extinction risk of Douglas-fir populations in central Mexico, considering anthropogenic, genetic, and ecological criteria in order to prioritize in situ and ex situ conservation activities, including protection of some natural populations and germplasm collection for assisted gene flow and migration to reduce inbreeding, and increase genetic diversity and population size (López-Upton et al. 2015).

2.4.2 *Family* Fagaceae

Table 2.8 shows genetic diversity estimates for some species within family Fagaceae, the majority of which focused on the genus *Quercus*. Oaks are one of the most important forest resources of Mexico and represent an important source of firewood and wood for furniture (Galicia et al. 2015). SSR and RAPD molecular markers have been used for the study of this genus. In general, the average heterozygosity (H_e) found for *Quercus* equals 0.33 and 0.69, as estimated with RAPDs and SSRs, respectively. The highest heterozygosity (H_e) was found in *Quercus sideroxyla* (0.86), while the lowest was found in *Quercus rugosa* (H_e = 0.28) (Table 2.8). The genetic diversity of the genus *Quercus* in Mexico is comparable with that of *Quercus virginiana* in USA (H_e = 0.712) (Cavender-Bares et al. 2010), and it is higher than that of *Quercus petraea* in Europe (H_e = 0.275) (Müller-Starck et al. 1992). Some oaks may be susceptible to climate change and forestry exploitation; for example, *Q. sideroxyla* has been exploited for over 500 years to obtain firewood, charcoal for local uses and used for wood production in Mexico's forestry industry, which is thought to have largely depleted its original populations. Therefore, it is important to preserve its genetic diversity, by developing strategies for working with local people in conservation programs and devising conservation policies (Alfonso-Corrado et al. 2014b).

Using RAPD and SSR markers, Tovar-Sánchez et al. (2015) found a positive and significant relationship between genetic diversity and canopy arthropod diversity in *Quercus crassipes* and *Quercus rugosa*. The results of this study represent a potent instrument to predict the effects of the genetic diversity of host-plant species on species diversity, which allows establishing a novel conservation status to foundation species based on their genetic variation.

Another taxa studied within the family Fagaceae is *Fagus grandifolia* (Mexican beech). This species only grows in 11 small and isolated populations in Mexico. The timber harvesting, overgrazing, firewood collection, and seeds for consumption are possible causes that have influenced the reduction in size of its populations (Montiel-Oscura et al. 2013). This species is protected by the Mexican standard NOM-059-SEMARNAT-2010, within the risk category of species in danger of extinction (Semerant 2010), but is not listed in the Red List. The genetic diversity of this species was analyzed with isoenzymes ($H_e = 0.20$) and RAPD markers ($H = 0.14$) (Table 2.8). Rowden et al. (2004) and Montiel-Oscura et al. (2013), respectively, found a moderate to high genetic diversity in *Fagus grandifolia*. The average value of expected heterozygosity (H_e) for this species is comparable with those found in other species of the genus *Fagus* (0.186) (Kitamura and Kawano 2001) and is larger than the average value estimated for angiosperms (0.143) (Hamrick et al. 1992). This variability should be considered in the development of urgent conservation strategies to prevent the extinction of this taxon.

2.4.3 Other Species

The genetic diversity of tropical species has been evaluated only in *Cedrela odorata* and *Swietenia macrophylla* (family Meliaceae), which represent important commercial wood sources in Mexico. These two species are in the Red List of UICN and protected by the Mexican standard NOM-059-SEMARNAT-2010, as a consequence of uncontrolled logging and wood trade. The studies of variation in these species were based on RAPD and SSR markers. Alcalá et al. (2014) found an H_e of 0.78 in four populations of *S. macrophylla* in Mexico, while Novick et al. (2003) found an H_e of 0.68 (Table 2.9). This genetic diversity is comparable to that found in populations of these species in South America ($H_e = 0.65$) (Novick et al. 2003). Conservation suggestions included the preservation of the populations of *S. macrophylla* located in Veracruz or within the Yucatan peninsula, because of their potential adaptation to drier and warmer conditions that could alter the future distribution of Mexican tropical forest (Alcalá et al. 2014).

Finally, other important forest species, *Populus tremuloides* and *Juniperus blancoi*, have been studied to determine their levels of genetic diversity (Table 2.9). *Populus tremuloides* (Quaking aspen) is distributed in North America (Canada, USA, and Mexico) and is a commercially important source of wood fiber. The genetic diversity of Quaking aspen in Mexico ($H_e = 0.61$) is lower than that found in USA and Canada populations ($H_e = 0.79$) (Callahan et al. 2013). Quaking aspen populations from southwestern Mexico have a high risk of mortality with climate change; therefore, it is important to preserve their genetic diversity. *Juniperus blancoi* is an endemic juniper of Mexico with only nine populations located so far. The expected heterozygosity of *J. blancoi* ($H_e = 0.345$) was lower than that observed in junipers from the Tibetan Plateau ($H_e = 0.641$; Li et al. 2012).

2.5 Conclusions

Contrasting with its high species and genus richness, studies of genetic diversity in Mexican forest trees are rather scarce, and often biased to particular families, like the Pinaceae or the Fagaceae. Moreover, even within those particular families the available surveys have a penchant for specific genus, namely *Pinus* and *Quercus*, respectively. The markers used in most of these studies include the traditional and "universal" isozymes and chloroplast microsatellites and, to a lower extent, the anonymous SSRs, AFLPs, and RAPDs. Although some exceptions have been arising in recent years (like Moreno-Letelier et al. 2014, 2015), there is a rampant lack of studies using sequence variation on nuclear genes or more advanced next-generation technologies, which somehow hampers the estimation of genetic variation linked to adaptive processes, such as performed in Europe or northern North America for more than a decade (reviewed in Jaramillo-Correa et al. 2015b). More studies on more varied taxa and using more advanced technologies (e.g., next generation sequencing) and markers seem thus necessary. For instance, because of the poor comparability of the genetic diversity estimates among the studied Mexican tree species, it is extremely difficult to discern general trends across species or regions. Not to mention across taxa with different life-history traits, (such as made in earlier works with allozymes; Hamrick et al. 1992). We thus recommend that genetic diversity should be measured across species with an identical type of genetic marker, and by surveying similar numbers of loci, individuals and populations and using identical indices of genetic diversity.

In spite of such a lack of genetic studies, some trends could be found. For instance, most populations of Mexican forest trees appear to have reduced levels of genetic diversity when compared to their boreal or neotropical counterparts. Given that Mexico represents the southern limit of the distribution of the boreal taxa and the northern bound of the neotropical species, modern populations of forest trees in Mexico are either the rear limit or the advancing colonization front of many genera, which in both cases translates in increased magnitudes of genetic drift over other evolutionary forces. Moreover, most Mexican forest tree species are distributed in fragmented ranges, either because they inhabit fragmented landscapes, like mountain ranges (i.e., most conifers or oaks), or because they have limited dispersal capabilities, which result in scattered stretches of individuals occurring in rather continuous habitats, like most tropical species. This translates in strong population differentiation, which complicates many of the currently established conservation programs.

Based on the available studies it could be recommended that most conservation efforts on Mexican forest trees concentrate in reducing genetic erosion and ensuring the long-term viability of populations and species. Such efforts should focus, among other things, on protecting or encouraging natural regeneration in surrounding areas of current populations, promoting programs of assisted migration, or the establishment of new populations in areas suitable for specific species under future climate change scenarios. In addition, it would be prudent to create real management programs of Mexico's genetic resources that make good use of current technologies such as genomic selection of preadapted individuals at the earlier stages of development. A national plan headed by the government in such programs for the conservation of forest trees would be more than necessary.

Acknowledgements This study was supported by joint funding from the Mexican Council of Science and Technology (CONACyT) and the Ministry of Education (SEP) (SEP-CONACyT: Project CB-2010-01 158054 to CW and CB-2010-153305-B to JPJC; CONACyT: postdoctoral fellowship in Instituto de Silvicultura e Industria de la Madera, Durango, Mexico to SDRML). This paper is an undertaking of the Forest Genetic Resources Working Group/North American Forest Commission/Food and Agricultural Organization of the United Nations.

References

Adams RP, Vargas-Hernández JJ, González-Elizondo MS, Hunter G, Fairhall TA, Thornburg D, Callahan F (2012) Geographic variation in leaf essential oils of Douglas fir (*Pseudotsuga menziesii*). Phytologia 94(2):199–218

Adams RP, Vargas-Hernández JJ, González-Elizondo MS, Hunter G, Fairhall TA, Thornburg D, Callahan F (2013) Taxonomy of douglas fir (*Pseudotsuga menziesii*) infraspecific taxa: vars. menziesii, glauca and oaxacana: nrDNA, and cpDNA sequences and leaf essential oils. Phytologia 95(1):1–9

Aguirre-Planter E, Furnier GR, Eguiarte LE (2000) Low levels of genetic variation within and high levels of genetic differentiation among populations of species of *Abies* from southern Mexico and Guatemala. Am J Bot 87(3):362–371

Alcalá RE, Salazar H, Guitérrez-Granados G, Snook LK (2014) Genetic structure and genetic diversity of *Swietenia macrophylla* (Meliaceae): implications for sustainable forest management in Mexico. J Trop For Sci 142–152

Alfonso-Corrado C, Campos-Contreras J, Sánchez-García G, Monsalvo-Reyes A, Clark-Tapia R (2014a) Manejo forestal y diversidad genética de *Pinus patula* Schiede ex Schltdl & Cham, en Sierra Juárez. Oaxaca. Madera y Bosques 20(2):11–22

Alfonso-Corrado C, Clark-Tapia R, Monsalvo-Reyes A, Rosas-Osorio C, González-Adame G, Naranjo-Luna F, Campos JE (2014b) Ecological-genetic studies and conservation of endemic *Quercus sideroxyla* (Trel.) in Central Mexico. Nat Resour 5:442–453

Alfonso-Corrado C, Esteban-Jiménez R, Clark-Tapia R, Piñero D, Campos JE, Mendoza A (2004) Clonal and genetic structure of two Mexican oaks: *Quercus eduardii* and *Quercus potosina* (Fagaceae). Evol Ecol 18(5–6):585–599

Amos W, Harwood J (1998) Factors affecting levels of genetic diversity in natural populations. Philos T Roy Soc B 353:177–186

Bello CHÁ, López-Mata L (2001) Distribución y análisis estructural de *Abies hickelii* (Flous & Gaussen) en México. Interciencia 26(6):244–251

Bergmann F, Gregorius H-R, Kownatzki D, Wehenkel C (2013) Different diversity measures assess species-genetic diversity relationships differently: A marker-based case study in forest tree communities. Silvae Genet 62(1–2):25–38

Bowen BW (1999) Preserving genes, species, or ecosystems? Healing the fractured foundations of conservation policy. Mol Ecol 8:S5–S10

Brown A, Young A, Burdon J, Christides L, Clarke G, Coates D, Sherwin W (1997) Genetic indicators for state of the environment reporting. State of the Environment Technical Paper Series (Environmental Indicators), Department of Environment, Sport and Territories, Canberra ACT, Australia

Callahan CM, Rowe CA, Ryel RJ, Shaw JD, Madritch MD, Mock KE (2013) Continental-scale assessment of genetic diversity and population structure in quaking aspen (Populus tremuloides). J Biogeog 40(9):1780–1791

Cavender-Bares J, Gonzalez-Rodriguez A, Pahlich A, Koehler K, Deacon N (2010) Phylogeography and climatic niche evolution in live oaks (*Quercus* series Virentes) from the tropics to the temperate zone. J Biogeogr 38:962–981

Cruz-Nicolás J, Vargas-Hernández JJ, Ramírez-Vallejo P, López-Upton J (2011) Genetic diversity and differentiation of *Pseudotsuga menziesii* (Mirb.) Franco populations in Mexico. Rev Fitotec Mex 34(4):233–240

Cué-Bär EM, Villaseñor JL, Arredondo-Amezcua L, Cornejo-Tenorio G, Ibarra-Manríquez G (2006) La flora arbórea de Michoacán, México. Bol Soc Bot México 78:47–81

Cuenca A, Escalante AE, Piñero D (2003) Long-distance colonization, isolation by distance, and historical demography in a relictual Mexican pinyon pine (Pinus nelsonii Shaw) as revealed by paternally inherited genetic markers (cpSSRs). Mol Ecol 12(8):2087–2097

Debreczy Z, Racz I (1995) New species and varieties of conifers from Mexico. Phytologia 78:217–243

Delgado Valerio P, Núñez Medrano J, Rocha Granados M, Muñoz Flores HJ (2013) Variación genética de dos áreas semilleras de pino establecidas en el estado de Michoacán. Rev Mex Cien For 4(18):104–115

Delgado P, Piñero D, Chaos A, Pérez-Nasser N, Alvarez-Buylla ER (1999) High population differentiation and genetic variation in the endangered Mexican pine *Pinus rzedowskii* (Pinaceae). Am J Bot 86(5):669–676

Delgado P, Salas-Lizana R, Vázquez-Lobo A, Wegier A, Anzidei M, Alvarez-Buylla ER, Piñero D (2007) Introgressive hybridization in *Pinus montezumae* Lamb and *Pinus pseudostrobus* Lindl. (Pinaceae): morphological and molecular (cpSSR) evidence. Int J Plant Sci 168(6):861–875

Dominguez-Guerrero IK, del Rocío Mariscal-Lucero S, Hernández-Díaz JC, Heinze B, Prieto-Ruiz JÁ, Wehenkel C. (2017) Discrimination of Picea chihuahuana Martinez populations on the basis of climatic, edaphic, dendrometric, genetic and population traits. PeerJ 5:e3452

Dvorak WS, Potter KM, Hipkins VD, Hodge GR (2009) Genetic diversity and gene exchange in *Pinus oocarpa*, a Mesoamerican pine with resistance to the pitch canker fungus (*Fusarium circinatum*). Int J Plant Sci 170(5):609–626

Escobar PR, Gernandt DS, Piñero D, Garcillán PP (2011) Plastid DNA diversity is higher in the island endemic Guadalupe cypress than in the continental Tecate cypress. PLoS ONE 6(1): e16133

Farjon A, Styles BT (1997) Flora Neotropica. New York Botanical Garden, Pinus (Pinaceae)

Farjon A, Gardner M, Thomas P (2017) Conifer Database (version Jan 2014). In: Roskov Y, Abucay L, Orrell T, Nicolson D, Bailly N, Kirk PM, Bourgoin T, DeWalt RE, Decock W, De Wever A, Nieukerken E van, Zarucchi J, Penev L. Species 2000 and ITIS Catalogue of Life, 26th July 2017. Digital resource at http://www.catalogueoflife.org/col. Species 2000: Naturalis, Leiden, the Netherlands. ISSN 2405-8858.

Farnsworth E, Sahotra S (2007) Conservation and management of rare plant species. In: Cutler J Cleveland Environmental Information Coalition, National Council for Science and the Environment (ed) Encyclopedia of Earth, Washington, DC

Flores-Lopéz C, López-Upton J, Vargas-Hernández JJ (2005) Indicadores reproductivos en poblaciones naturales de *Picea mexicana* Martínez. Agrociencia 39:117–126

Frankel OH (1981) Evolution in jeopardy; the role of nature reserves. In: Atchley WR, Woodruff D (eds) Evolution and Speciation. Cambridge University Press, Cambridge

Frankham R (1996) Relationship of genetic variation to population size in wildlife. Cons Biol 10 (6):1500–1508

Frankham R (2010) Challenges and opportunities of genetic approaches to biological conservation. Biol Conserv 143(9):1919–1927

Futuyma, DJ (2013) Evolution. Sinauer Associated Inc. New York, p 656

Galicia L, Potvin C, Messier C (2015) Maintaining the high diversity of pine and oak species in Mexican temperate forests: a new management approach combining functional zoning and ecosystem adaptability. Can J For Res 45(10):1358–1368

García-Gómez V, Ramírez-Herrera C, Flores-López C, López-Upton J (2014) Diversidad y estructura genética de *Pinus johannis*. Agrociencia 48(8):863–873

GEO BON (2011) Adequacy of Biodiversity Observation Systems to Support the CBD 2020 Targets. http://www.earthobservations.org/documents/cop/bi_geobon/2011_cbd_adequacy_report.pdf

Gillies ACM, Navarro C, Lowe AJ, Newton AC, Hernandez M, Wilson J, Cornelius JP (1999) Genetic diversity in Mesoamerican populations of mahogany (*Swietenia macrophylla*), assessed using RAPDs. Heredity 83(6):722–732

González-Rodríguez A, Bain JF, Golden JL, Oyama K (2004) Chloroplast DNA variation in the *Quercus affinis–Q. laurina* complex in Mexico: geographical structure and associations with nuclear and morphological variation. Mol Ecol 13(11):3467–3476

Gordon AG (1991) Genetics, genecology and breeding in the genus Picea L. (spruce). In: Magnussen S, Lavereau J, Boyle TJ (eds) Proceeding of the Twenty-third Meeting and Activity Report of the Canadian Tree Improvement Association. Ontario, Ottawa, pp 86–88

Gregorius HR (2001) The notion of stability in open dynamical systems from an ecological perspective. Int J Gen Syst 30(3):347–378

Gregorius HR, Bergmann F, Wehenkel C (2003) Analysis of biodiversity across levels of biological organization: a problem of defining traits. Perspect Plant Ecol Evol Syst 5:209–218

Gugger PF, González-Rodríguez A, Rodríguez-Correa H, Sugita S, Cavender-Bares J (2011) Southward Pleistocene migration of Douglas-fir into Mexico: phylogeography, ecological niche modeling, and conservation of 'rear edge' populations. New Phytol 189(4):1185–1199

Hamrick JL, Godt MJW, Sherman-Broyles SL (1992) Factors influencing levels of genetic diversity in woody plant species. New Forest 6:95–124

Heredia-Bobadilla RL, Gutiérrez-González G, Franco-Maass S, Arzate-Fernández AM (2012) Genetic variability of sacred fir (*Abies religiosa*) in the Nevado de Toluca National Park. Int J Biodivers 4(3):130–136

Hermann RK, Lavender DP (1999) Douglas-Fir planted forests. New. Forest 17:53–70

Hughes AR, Inouye BD, Johnson MT, Underwood N, Vellend M (2008) Ecological consequences of genetic diversity. Ecol Lett 11(6):609–623

Isabel N, Beaulieu J, Bousquet J (1995) Complete congruence between gene diversity estimates derived from genotypic data at enzyme and random amplified polymorphic DNA loci in black spruce. P Natl Acad Sci 92(14):6369–6373

IUCN Red List of Threatened Species. (2013). Available at http://www.iucnredlist.org/details/32479/0 (accessed on November 2015).

Jaramillo-Correa JP, Aguirre-Planter E, Khasa DP, Eguiarte LE, Piñero D, Furnier GR, Bousquet J (2008) Ancestry and divergence of subtropical montane forest isolates: molecular biogeography of the genus *Abies* (Pinaceae) in southern México and Guatemala. Mol Ecol 17 (10):2476–2490

Jaramillo-Correa JP, Beaulieu J, Ledig FT, Bousquet J (2006) Decoupled mitochondrial and chloroplast DNA population structure reveals Holocene collapse and population isolation in a threatened Mexican-endemic conifer. Mol Ecol 15(10):2787–2800

Jaramillo-Correa JP, Beaulieu J, Khasa DP, Bousquet J (2009) Inferring the past from the present phylogeographic structure of North American forest trees: seeing the forest for the genes. Can J For Res 39(2): 286–307.

Jaramillo-Correa JP, Gérardi S, Beaulieu J, Ledig FT, Bousquet J (2015a) Inferring and outlining past population declines with linked microsatellites: a case study in two spruce species. Tree Genet Genomes 11(1):1–12

Jaramillo-Correa JP, Prunier J, Vázquez-Lobo A, Keller SR, Moreno-Letelier A (2015b) Molecular signatures of adaptation and selection in forest trees. In: Plomion C, Adam-Blondon AF. Genomes of photosynthetic organisms and fungi—Land Plants: Trees. Advances in Botanical Research 74: 265–306. Elsevier

Kahilainen A, Puurtinen M, Kotiaho JS (2014) Conservation implications of species–genetic diversity correlations. Glob Ecol Conserv 2:315–323

Kitamura K, Kawano S (2001) Regional differentiation in genetic components for the American beech, *Fagus grandifolia* Ehrh. in relation to geological history and mode of reproduction. J Plant Res 114:353–368

Karhu A, Vogl C, Moran GF, Bell JC, Savolainen O (2006) Analysis of microsatellite variation in *Pinus radiata* reveals effects of genetic drift but no recent bottlenecks. J Evolutionary Biol 19 (1):167–175

Kremer A, Ronce O, Robledo-Arnuncio JJ, Guillaume F, Bohrer G, Nathan R, Bridle JR, Gomulkiewicz R, Klein EK, Ritland K, Kuparinen A (2012) Long-distance gene flow and adaptation of forest trees to rapid climate change. Ecol Lett 15(4):378–392

Lara SF (2010) Population variation in the endemic *Pinus culminicola* detected by RAPD. Polibotánica 30:55–67

Ledig FT (1998) Genetic variation in Pinus. In: Richardson DM (ed) Ecology and Biogeographyof *Pinus*. Cambridge University Press, Cambridge, pp 251–280

Ledig FT (1992) Human impacts on genetic diversity in forest ecosystems. Oikos 63:87–108

Ledig FT, Conkle MT, Bermejo-Velazquez B, Eguiluz-Piedra T, Hodgskiss PD, Johnson DR, Dvorak WS (1999) Evidence for an extreme bottleneck in a rare Mexican pinyon: genetic diversity, disequilibrium, and the mating system in *Pinus maximartinezii*. Evolution 91–99

Ledig FT, Jacob-Cervantes V, Hodgskiss PD, Eguiluz-Piedra T (1997) Recent evolution and divergence among populations of a rare Mexican endemic, Chihuahua spruce, following Holocene climatic warming. Evolution 1815–1827

Ledig FT (1998) Genetic variation in *Pinus*. In: Richardson DM (ed) Ecology and Biogeography of Pinus. Cambridge University Press, Cambridge, pp 251–280

Ledig FT (2000) Founder effects and the genetic structure of Coulter pine. J Hered 91(4):307–315

Ledig FT, Bermejo-Velázquez B, Hodgskiss PD, Johnson DR, Flores-López C, Jacob-Cervantes V (2000) The mating system and genic diversity in Martinez spruce, an extremely rare endemic of Mexico's Sierra Madre Oriental: an example of facultative selfing and survival in interglacial refugia. Can J Forest Res 30(7):1156–1164

Ledig FT, Capó-Arteaga MA, Hodgskiss PD, Sbay H, Flores-López C, Conkle MT, Bermejo-Velázquez B (2001) Genetic diversity and the mating system of a rare Mexican piñon, *Pinus pinceana*, and a comparison with *Pinus maximartinezii* (Pinaceae). Am J Bot 88 (11):1977–1987

Ledig FT, Hodgskiss PD, Jacob-Cervantes V (2002) Genetic diversity, mating system, and conservation of a Mexican subalpine relict Picea mexicana Martinez. Conserv Genet 3(2):113–122

Ledig FT, Hodgskiss PD, Johnson DR (2005) Genic diversity, genetic structure, and mating system of Brewer spruce (Pinaceae), a relict of the Arcto-Tertiary forest. Am J Bot 92 (12):1975–1986

Ledig FT, Rehfeldt GE, Sáenz-Romero C, Flores-López C (2010) Projections of suitable habitat for rare species under global warming scenarios. Am J Bot 97(6):970–987

Ledig FT (2012) Climate Change and Conservation. Acta Silv. Lign. Hung. 8:57–74

Li Z, Zou J, Mao K, Lin K, Li H, Liu J, Källman T, Lascoux M (2012) Population genetic evidence for complex evolutionary histories of four high altitude juniper species in the Qinghai-Tibetan Plateau. Evolution 66(3):831–845

Little E L Jr (1979) Checklist of United States Trees (native and naturalized). USDA Forest Service. Agricultural Handbook No541, p 375

López-Upton J, Valdez-Lazalde JR, Ventura-Ríos A, Vargas-Hernández JJ, Guerra-de-la-Cruz V (2015) Extinction risk of *Pseudotsuga menziesii* populations in the central region of Mexico: an AHP analysis. Forests 6:1598–1612

Lundkvist K (1979) Allozyme frequency distributions in four Swedish populations of Norway spruce (*Picea abies* K.). Hereditas 90(1):127–143

Luo JX, Wang YH, Korpelainen H, Li CY (2005) Allozyme variation in natural populations of *Picea asperata*. Silva Fennica 39(2):167–176

MaceGM Purvis A (2008) Evolutionary biology and practical conservation: bridging a widening gap. Mol Ecol 17:9–19

Martínez M (1963) Las Pináceas Mexicanas, 3rd edn. UNAM, México, DF, pp 27–74

Mendoza-Maya E, Espino-Espino J, Quiñones-Pérez CZ, Flores-López C, Wehenkel C, Vargas-Hernández JJ, Sáenz-Romero C (2015) Propuesta de conservación de tres especies mexicanas de picea en peligro de extinción. Rev Fitotec Mex 38(3):235–247

Mittermeier RA, Goettsch MC (1992) La importancia de la diversidad biológica de México. In: Sarukhán J, Dirzo R (eds) México ante los retos de la biodiversidad. Comisión Nacional para el Conocimiento y Uso de la Biodiversidad, México, DF, pp 63–73

Molina-Freaner F, Delgado P, Piñero D, Perez-Nasser N, Alvarez-Buylla E (2001) Do rare pines need different conservation strategies? Evidence from three Mexican species. Can J Bot 79 (2):131–138

Montiel-Oscura D, Ramírez-Herrera C, Ángeles-Pérez G, López-Upton J, Antonio-López P (2013) Variación aloenzimática y tamaño de poblaciones de Haya mexicana (*Fagus grandifolia subsp. mexicana*) en la Sierra Madre Oriental. Rev Fitotec Mex 36(4):413–420

Moreno-Letelier A, Piñero D (2009) Phylogeographic structure of *Pinus strobiformis* Engelm. across the Chihuahuan Desert filter-barrier. J Biogeogr 36(1):121–131

Moreno-Letelier A, Mastretta-Yanes A, Barraclough TG (2014) Late Miocene lineage divergence and ecological differentiation of rare endemic *Juniperus blancoi*: clues for the diversification of North American conifers. New Phytol 203(1):335–347

Moreno-Letelier A, Barraclough TG (2015) Mosaic genetic differentiation along environmental and geographic gradients indicate divergent selection in a white pine species complex. Evol Ecol 29:733–748

Müller-Starck G, Baradat P, Bergmann F (1992) Genetic variation within European tree species. New Forest 6(1–4):23–47

Namkoong G (1990) Biological diversity: its conservation and use for sustainable agricultural, forestry and fisheries development. FAO Working Paper of the Inter-Departmental Sub-Group on Biological Diversity. FAO, Rome (Italy)

Navarro C, Cavers S, Pappinen A, Tigerstedt P, Lowe AJ, Merila J (2005) Contrasting quantitative traits and neutral genetic markers for genetic resource assessment of Mesoamerican *Cedrela odorata*. Silvae Genet 54(6):281–292

Newton AC, Allnutt TR, Gillies ACM, Lowe AJ, Ennos RA (1999) Molecular phylogeography, intraspecific variation and the conservation of tree species. Tree 14(4):140–145

Newton AC, Allnutt TR, Dvorak WS, Del Castillo RF, Ennos RA (2002) Patterns of genetic variation in *Pinus chiapensis*, a threatened Mexican pine, detected by RAPD and mitochondrial DNA RFLP markers. Heredity 89(3):191–198

Noss RF (1990) Indicators for monitoring biodiversity: a hierarchical approach. Conservation biology, 355–364

Novick RR, Lemes MR, Navarro C, Caccone A, Bermingham E (2003) Genetic structure of Mesoamerican populations of Big-leaf mahogany (Swietenia macrophylla) inferred from microsatellite analysis. Mol Ecol 12(11):2885–2893

Palacio-Prieto JL, Bocco G, Velázquez A, Mas JF, Takaki-T F, Victoria A, Luna-González L, Gómez-Rodríguez G, López-García J, Palma-M M, Trejo-Vázquez I, Peralta-H A, Prado-Molina J, Rodríguez-Aguilar A, Mayorga-Saucedo R, González-M F (2000) La condición actual de los recursos forestales en México: resultados del Inventario Forestal Nacional 2000. Investigaciones Geográficas, Boletín del Instituto de Geografía, UNAM 43:183–203

Parraguirre-Lezama C, Vargas-Hernández JJ, Ramírez-Vallejo P, Aspiroz-Rivero HS, Jasso-Mata J (2002) Estructura de la diversidad genética en poblaciones naturales de *Pinus greggii* Engelm. Rev Fitotec Mex 25:279

Peñaloza-Ramírez JM, González-Rodríguez A, Mendoza-Cuenca L, Caron H, Kremer A, Oyama K (2010) Interspecific gene flow in a multispecies oak hybrid zone in the Sierra Tarahumara of Mexico. Ann Bot 105(3):389–399

Piñero D, Caballero-Mellado J, Cabrera-Toledo D (2008) La diversidad genética como instrumento para la conservación y el aprovechamiento de la biodiversidad: estudios en especies mexicanas, en Capital natural de México, vol. I: Conocimiento actual de la biodiversidad. CONABIO, México, p 437–494

Reed DH, Frankham R (2003) Correlation between fitness and genetic diversity. Conserv Biol 17 (1):230–237

Rehfeldt GE, Jaquish BC, Sáenz-Romero C, Joyce DG, Leites LP, StClair JB, López-Upton J (2014) Comparative genetic responses to climate for the varieties of *Pinus ponderosa* and *Pseudotsuga menziesii*: Reforestation. Forest Ecol Manag 324:147–157

Reyes-Hernández VJ, Vargas-Hernández JJ, López-Upton J, Vaquera-Huerta H (2006) Similitud fenotípica de poblaciones mexicanas de *Pseudotsuga* Carr. Agrociencia 40:545–556

Riggs LA (1990) Conserving genetic resources on-site in forest ecosystems. Forest Ecol Manag 35:45–68

Rodríguez-Banderas A, Vargas-Mendoza CF, Buonamici A, Vendramin GG (2009) Genetic diversity and phylogeographic analysis of Pinus leiophylla: a post-glacial range expansion. J Biogeogr 36(9):1807–1820

Rowden A, Robertson A, Allnutt T, Heredia S, Williams-Linera G, Newton AC (2004) Conservation Genetics of Mexican Beech. Fagus grandifolia var. mexicana. Conserv Genet 5 (4):475–484

Rushforth K (1986) México's spruces—rare members of an important genus. Curtis's Botanical Magazine 3(3):119–124

Rzedowski J (1991a) Diversidad y orígenes de flora fanerogámica de México. Acta Bot Mex 14:3–21

Rzedowski J (1991b) El endemismo en la flora fanerogámica mexicana: una apreciación analítica preliminar. Acta Bot Mex 15:47–64

Rzedowski J (1994) Vegetación de México. Editorial LIMUSA, México

Sáenz-Romero C, Tapia-Olivares BL (2003) *Pinus oocarpa* isoenzymatic variation along an altitudinal gradient in Michoacán. México. Silvae Genet 52(5–6):237–240

Sáenz-Romero C, Rehfeldt GE, Crookston NL, Pierre D, St-Amant R, Beaulieu J, Richardson B (2010) Spline models of contemporary, 2030, 2060 and 2090 climates for Mexico and their use in understanding climate-plant impacts on vegetation. Clim Change 102:595–623

Schoen DJ, Brown AH (1991) Intraspecific variation in population gene diversity and effective population size correlates with the mating system in plants. P Natl Acad Sci 88(10):4494–4497

Secretaria del Medio Ambiente, Recursos Naturales y Pesca (SEMARNAT) (2010) Norma Oficial Mexicana NOM-059-SEMARNAT-2010. Protección ambiental-especies nativas de flora y fauna silvestres-Categorías de riesgo y especificaciones para su inclusión, exclusión o cambio-lista de especies en riesgo

SEMARNAP-UACH (1999) Atlas Forestal de México. Universidad Autónoma Chapingo. México, DF, Secretaría de Medio Ambiente Recursos Naturales y Pesca (SEMARNAP)

Simental-Rodríguez SL, Quiñones-Pérez CZ, Moya D, Hernandez-Tecles E, Lopez-Sanchez CA, Wehenkel C (2014) The Relationship between Species Diversity and Genetic Structure in the Rare *Picea chihuahuana* Tree Species Community. Mexico. Plos One 9(11):e111623

Sousa-S M, Ricker M, Hernández H (2003) An index for the tree species of the family Leguminosae in Mexico. Harvard Papers in Botany 7:198–381

Sousa-S M, Ricker M, Hernández H (2001) Tree species of the family Leguminosae in Mexico. Harvard Pap Bot 6:339–365

Tchebakova NM, Rehfeldt GE, Parfenova EI (2005) Impacts of climate change on the distribution of *Larix* spp. and Ledeb. and *Pinus sylvestris* and their climatypes in Siberia. Mitig Adapt Strat GL 11:861–882

Tovar-Sánchez E, Valencia-Cuevas L, Mussali-Galante P, Ramírez-Rodríguez R, Castillo-Mendoza E (2015) Effect of host-plant genetic diversity on oak canopy arthropod community structure in central Mexico. Rev Chil Hist Nat 88(1):12

Villalobos-Arámbula AR, de la Rosa JP, Arias A, Rajora OP (2014) Cross-species transferability of eastern white pine (*Pinus strobus*) nuclear microsatellite markers to five Mexican white pines. Genet Mol Res 13(3):7571–7576

Villaseñor JL (2001) Catálogo de autores de plantas vasculares de México. Instituto de Biología, Universidad Nacional Autónoma de México, México, DF, p 40

Villaseñor JL (2003) Diversidad y distribución de las Magnoliophyta de México. Interciencia 28 (3):160–167

Villaseñor JL (2004) Los géneros de plantas vasculares de la flora de México. B Soc Bot Mex 75:105–135

Villaseñor JL (2013) Informe final* del Proyecto JE012 La flora arbórea de México, www.conabio.gob.mx

Viveros-Viveros H, Tapia-Olivares BL, Sáenz-Romero C (2014) Variación isoenzimática de *Pinus pseudostrobus* Lindl. a lo largo de un gradiente altitudinal en Michoacán, México. Agrociencia 48(7):713–723

Viveros-Viveros H, Tapia-Olivares BL, Sáenz-Romero C, Vargas-Hernández JJ, López-Upton J, Santacruz-Varela A, Ramírez-Valverde G (2010) Variación isoenzimática de *Pinus hartwegii* Lindl. en un gradiente altitudinal en Michoacán, México. Agrociencia 44(6):723–733

Wallace J, Aquilé N, Archambault C, Carpentier S, Francoeur X, Greffard M-H, Laforest I, Galicia L, Messier C (2015) Present forest management structures and policies in temperate forests of Mexico: Challenges and prospects for unique tree species assemblages. Forest Chron 91:306–317

Walpole M, Almond REA, Besançon C (2009) Tracking progress toward the 2010 biodiversity target and beyond. Science 325:1503–1504

WCMC (World Conservation Monitoring Centre) (1992) Global biodiversity. Status of the Earth´s living resources. Chapman & Hall, London. p 594

Wehenkel C, Bergmann F, Gregorius HR (2006) Is there a trade-off between species diversity and genetic diversity in forest tree communities? Plant Ecol 185(1):151–161

Wehenkel C, Sáenz-Romero C (2012) Estimating genetic erosion using the example of *Picea chihuahuana* Martínez. Tree Genet Genomes 8(5):1085–1094

Wehenkel C, Simental-Rodríguez LS, Silva-Flores R, Hernández-Díaz C, López-Sánchez CA, Antúnez P (2015) Discrimination of 59 seed stands of various Mexican pine species based on 43 dendrometric, climatic, edaphic and genetic traits. Forstarchiv 86(6):194–201

Wei XX, BeaulieuJ Khasa DP, Vargas-Hernández JJ, López-Upton J, Jaquish B, Bousquet J (2011) Range-wide chloroplast and mitochondrial DNA imprints reveal multiple lineages and complex biogeographic history for Douglas-fir. Tree Genet Genomes 7(5):1025–1040

Chapter 3
Climate Change, Genetic Diversity, and Conservation of Paleoendemic Redwoods

M.R. Ahuja

Abstract Global climate has always been changing in the past and will continue to change in the future. However, according to current predictions, the climate has been changing more rapidly and has impacted species distributions, requiring strategies to conserve genetic resources in forest trees. Conservation of genetic resources of four endemic redwoods, *Sequoia sempervirens*, *Sequoiadendron giganteum*, *Metasequoia glyptostroboides*, and *Fitzroya cupressoides*, from the family Cupressaceae are discussed in this paper. All four genera are monospecific, share a number of common phenotypic traits, including red wood, and are threatened in their natural habitats. Although fossil history of the redwoods can be traced back to more than 100 million years ago in the Cretaceous Period, these redwoods were widespread during the Tertiary period (7–65 million years ago) in the northern and southern hemisphere. Following the geological upheavals and climate changes, the redwoods have become living fossils or paleoendemics, and are now restricted in their native narrow ranges in USA, China, and South America. Therefore, it is necessary to conserve the genetic resources in these paleoendemic redwoods and, at the same time, maintain an appropriate level of genetic diversity in the redwood species and populations for their future survival. In situ and ex situ strategies for the conservation of genetic resources of redwoods are discussed in this paper. Although these redwoods are protected in the national parks, reserves, and in privately owned forests in their habitats, it would be desirable to conserve them in new ex situ reserves, and by other ex situ strategies involving biotechnological approaches to preserve seed, tissues, and DNA in gene banks for future exploitations in the face of climate change.

Keywords Climate change · Redwoods · *Sequoia* · *Sequoiadendron* *Metasequoia* · *Fitzroya* · Paleoendemics · Genetic diversity · Genetic resources Conservation

---

M.R. Ahuja (✉)
60 Shivertown Road, New Paltz, NY 12561, USA
e-mail: mrahuja@hotmail.com

M.R. Ahuja and S.M. Jain (eds.), *Biodiversity and Conservation of Woody Plants*, Sustainable Development and Biodiversity 17, DOI 10.1007/978-3-319-66426-2_3

3.1 Introduction

Forest decline and loss of genetic diversity are a global concern. Causes for the loss of forests and genetic diversity are manifold, both natural and anthropogenic. Natural disasters leading to forest destruction include fires, hurricanes, floods, drought, and pests among others. However, for the main part, human activity involving logging and conversion of forest lands for agriculture, industry and housing has led to forest tree decline. In addition, air pollution and spewing of greenhouse gases in the atmosphere have contributed to global warming and climate change, also leading to forest degradation, fragmentation, and loss of genetic diversity at an alarming rate. Global climate change has occurred in the past and will continue in the future. What is new is that according to current estimates by meteorologists global warming has been occurring at an accelerated rate, that has impacted the redistribution and decline and loss of genetic diversity of plant species, including forest trees. Although meteorologists are uncertain about the actual amount of global warming, the Assessment Report 4 (AR 4) by Intergovernmental Panel on Climate Change (IPCC) predicted an increase in mean annual global temperature, as a result of human activity, would range between 2.4 and 6.4 °C (IPCC 2007), accompanied by significant changes in the rainfall cycles (Allan and Soden 2008; O'Gorman and Schneider 2009; Sugiyama et al. 2010; Trenberth 2011) by the end of the current century. However, the recent AR 5 suggests that anthropogenic emissions of greenhouse gases are the highest in history (IPCC 2014), resulting in rapid climate change, have had widespread impacts on humans and natural systems. The increase in global mean surface temperature by the end of the twenty-first century (2081–2100) is predicted to rise between 2.6 and 4.8 °C under very high greenhouse emission scenario (IPCC 2014).

Rapid climate change, most likely, a result of carbon dioxide (CO_2)-induced global warming, may impact plant/tree species in a number of different ways by causing: (1) range shifts, migration, and geographical distribution (Parmesan and Yohe 2003; Root et al. 2005; Iverson et al. 2005, 2008; McKenney et al. 2007; Kelly and Goulden 2007; Thuiller et al. 2008; Pautasso 2013; Iverson and McKenzie 2013); (2) shifts in phenology (e.g., flowering time) (Franks et al. 2007; Craufurd and Wheeler 2009; Anderson et al. 2012; Iler et al. 2013; Gaira et al. 2014; Wang et al. 2015); (3) threat to biodiversity (Botkin et al. 2007; Ibáñez et al. 2006; Bellard et al. 2012; Javeline et al. 2013); (4) ecosystem disturbance (Dale et al. 2001; Littell et al. 2010; Staudt et al. 2013; Weed et al. 2013; Bellard et al. 2014; Wylie et al. 2014); (5) forest growth and mortality (Battles et al. 2008; Quirk et al. 2013); (6) changes in genetic and evolutionary response (Davis et al. 2005; Hoffmann and Willi 2008; Hoffmann and Sgro 2010; Rehfeldt et al. 2014); and (7) extinction risks (Thomas et al. 2004; Parmesan 2006; Foden et al. 2008; Pearson et al. 2014; Urban 2015).

Conservation of germplasm and maintenance of genetic diversity are important considerations for the survival of forest trees, in particular, endemics, in the face of global forest decline and climate change. In this paper, we review the range of genetic diversity and strategies for conservation of genetic resources in the paleoendemic redwoods in the face of global climate change.

3.2 Paleoendemic Redwoods

The redwoods discussed in this paper, namely coast redwood (*Sequoia sempervirens* (D. Don) Endl.), giant sequoia or Sierra redwood (*Sequoiadendron giganteum* (Lindl.) Buchholz), and dawn redwood (*Metasequoia glyptostroboides* Hu & Cheng) (Fig. 3.1), and South American redwood, commonly known as alerce (*Fitzroya cupressoides* (Mol.) Johnst.), are endemics in USA, China, and South America, respectively (Olson et al. 1990; Weatherspoon 1990; Chu and Cooper 1999; Allnutt et al. 1999). All four genera belong to the family Cupressaceae, are monospecific, and share the same basic chromosome number of $x = 11$ (Gadek et al. 2000; Ahuja 2009). Although polyploidy is rare in conifers (Khoshoo 1959), two of the four genera, *Sequoia* and *Fitzroy*, are polyploids (Ahuja 2005). *Sequoia* is a hexaploid ($2n = 6x = 66$) and is the only hexaploid conifer (Stebbins 1948; Schlarbaum and Tsuchiya 1984a; Saylor and Simons 1970; Ahuja and Neale 2002; Ahuja 2005), while *Fitzroya* is a tetraploid ($2n = 4x = 44$) (Hair 1968). The other two genera, namely *Sequoiadendron* and *Metasequoia*, are diploids ($2n = 22$) (Schlarbaum and Tsuchiya 1984b; Ahuja 2005, 2009). All four redwood genera are long-lived and have been around for millions of years. Although fossil history of redwoods can be traced back to more than 100 million years ago in the Cretaceous

Fig. 3.1 Branches, cones, and seeds of *Metasequoia glyptostroboides* (*left*), *Sequoia sempervirens* (*center*), and *Sequoiadendron giganteum* (*far right*). The material was collected at the arboretum of the Institute of Forest Genetics, USDA, Placerville, CA

Table 3.1 Current distribution, ages, ploidy level and genome size of redwoods

Characters	*Sequoia*	*Sequoiadendron*	*Metasequoia*	*Fitzroya*
Endemic distribution	~720 × 8–56 km range along the fogbelt of northern California and southern Oregon, USA	~420 × 24 km range in the Sierra Nevada, California, USA	~25 km × 1.5 km range in Hupeh province and Szechuan Valley in Central China	~360 × 200 km range in coastal Cordillera in Chile, and western and eastern slopes of Andes in Chile/Argentina
Elevation range (m)	0–915	1400–2000	800–1500	100–1200
Maximum height (m)	110	98.5	40	60
Oldest known tree (years)	2200	3300	–	3620
Ploidy level	Hexaploid	Diploid	Diploid	Tetraploid
Chromosome number	$2n = 6x = 66$	$2n = 22$	$2n = 22$	$2n = 44$
Genome size (MB)	31,500	10,800	9700	–

Sources: *Sequoia* (Olson et al. 1990; *Sequoiadendron* (Weatherspoon 1990); *Metasequoia* (LePage et al. 2005); *Fitzroya* (Premoli et al. 2000; Lara and Villalba 1993); Ploidy levels (Ahuja 2005; Ahuja and Neale 2005)

Period, these redwoods were widespread during the Tertiary period (7–65 million years ago) in northern and southern hemisphere. Following the geological upheavals and climate changes, the redwoods have become living fossils or paleoendemics and are now restricted in their native narrow ranges in USA, China, and South America (Table 3.1). According to IUCN categories of threatened species, *Sequoia* and *Sequoiadendron* have been classified as vulnerable, while *Fitzroya* is listed as endangered and *Metasequoia* as a critically endangered forest tree species (IUCN 2013).

The native range of *Sequoia* extends from southwestern corner of Oregon to Santa Lucia Mountains of southern end of Monterey County in California (Table 3.1). The redwood forests are confined to a narrow coastal fog belt that is 720 km long and 8–56 km wide. The spatial pattern of *Sequoia* forests is mostly continuous, but in portions of their range, they are fragmented (Olson et al. 1990). *Sequoia* has undergone both expansions and contractions in its range in the past 10,000 years, and currently, it appears that *Sequoia* is retreating from its southern range and expanding northward (Sawyer et al. 2000). There has been 33% reduction in fog since in the *Sequoia* forest region in the past century, and climate change is impacting the growth of new *Sequoia* trees (Johnstone and Dawson 2010). Furthermore, because of reduction in fog and global changes in the temperature in the future, near-term shifts are expected in the habitat suitable for coast redwood (Fernandez et al. 2015).

The natural range of *Sequoiadendron* (Table 3.1) consists of 75 groves scattered over 420-km-long and 24-km-wide belt extending from Tulane to Placer counties on the western slopes, within an altitude range of 1400–2000 m, in the Sierra Nevada region in central California, USA (Hartesveldt et al. 1975; Weatherspoon 1990). The third redwood, *Metasequoia*, is now confined to its native range of $\sim$25 km $\times$ 1.5 km, within an altitude range of 700–1350 m, in western Hubei, eastern Chogqing, and northern Hunan provinces in south-central China (LePage et al. 2005; Leng et al. 2007). The fourth redwood, *Fitzroya*, is also an endemic to the temperate rain forests, within an altitude range of 100–1200 m, in southern South America; it grows in discontinuous populations within an area of $\sim$360 $\times$ 200 km (Table 3.1) in the coastal Cordilleras and central depression in Chile, and on the western slopes of Andes in Chile and Argentina (Allnutt et al. 1999; Premoli et al. 2000). All four redwoods are vulnerable to climate change in their endemic ranges, and it is imperative that conservation of their genetic resources should be a priority for their survival.

3.3 Genetic Diversity

Genetic diversity is the keystone for the survival and evolution of species. Genetic variation within a species is important for its ability to adapt to a changing environment. Species having larger levels of genetic diversity have a better chance of adaptation, survival, and deployment over a wide range of environmental conditions (Ledig 1988, 2012). Appropriate levels of genetic variation should be maintained in the populations of a species for conservation planning (Ledig 1986, 1988). The conservation of genetic resources should be based on the genetic architecture and phenology, and how genetic and phenotypic variation is organized and distributed within and among populations of a species. Population genetic investigations have shown that variation in qualitative and quantitative traits is essential because of its beneficial effects for adaptive genetic diversity. It is important to monitor genetic variation at both phenotypic and biochemical/molecular levels. In this context, allozyme and DNA-based genetic markers have provided relatively more precise estimates of genetic variation in the populations of a species.

The paleoendemic redwoods exhibit different levels of genetic diversity in their populations. Based on polymorphic loci and number of alleles per locus, allozyme studies have revealed that *Sequoia*, a hexaploid, seems to have higher levels of genetic diversity as compared to other redwoods (Table 3.2). The percentage of polymorphic loci and alleles per locus in redwoods measured by allozyme studies are: *Sequoia* (92%, and 3.1; Rogers 1997, 1999, 2000), *Sequoiadendron* (50% and 1.5; Fins and Libby 1982), *Metasequoia* (20% and 1.7; Kuser et al. 1997), and the tetraploid alerce (23% and 1.5; Premoli et al. 2000). All endemic redwoods showed relatively greater levels of genetic diversity within populations as compared to between populations (Rogers 2000; Fins and Libby 1994; Libby et al. 1996; Allnutt et al. 1999; Premoli et al. 2000; Li et al. 2005). With the exception of the hexaploid *Sequoia*, other three endemic redwoods, *Metasequoia*, *Sequoiadendron*, and *Fitzroya*, exhibited considerably lower genetic diversity compared to other gymnosperms (Table 3.2). In spite of the fact that *Sequoia* is also an endemic, its genetic diversity levels compare at par or even higher than other widespread gymnosperms (Libby et al. 1996; Sawyer et al. 2000).

Molecular genetic markers (microsatellites) have been employed to determine the levels of genetic diversity in redwoods. These studies have revealed that levels of genetic diversity are relatively higher in *Sequoia* populations north of San Francisco Bay, and rather lower in the fragmented southern populations (Brinegar

Table 3.2 Levels of genetic diversity based on allozymes within genera of redwoods, as compared to other gymnosperms (modified from Ahuja 2009)

Genus	No. of alleles/locus	Polymorphic loci (%)	Expected heterozygosity	Reference
Sequoia	3.1	92	–	Rogers (2000)
Sequoiadendron	1.5	50	–	Fins and Libby (1982)
Metasequoia	1.7	20	0.091	Kuser et al. (1997)
Fitzroya	1.5	23	0.074	Premoli et al. (2000)
Other gymnosperms	1.83	71	0.169	Hamrick et al. (1992)

2012). By using highly polymorphic microsatellite loci, up to six alleles per individual were scored in a large number of individuals in *Sequoia*, suggesting that coast redwood could either be an autohexaploid (AAAAAAA), segmental allohexaploid ($A_1A_1A_1A_1A_2A_2$), or autoallohexaploid (AAAABB). These polyploidy patterns would be expected when the parental genomes involved in the ancestry of hexaploid *Sequoia* were at least partially homologous (Brinegar et al. 2007; Douhovnikoff and Dodd 2011). These possibilities on the origin of *Sequoia* had been proposed in earlier publications (Stebbins 1948; Ahuja and Neale 2002; Ahuja 2005). Earlier cytological and morphological studies (Li 1987) and recent molecular phylogenetic study of Cupressaceae have supported the idea that *Sequoia* possibly originated by recombination/hybridization between giant sequoia and dawn redwood (Yang et al 2012). However, the enigmatic nature of polyploidy still remains unresolved, as *Sequoia* may have originated by genomic duplication of an ancestral diploid species of coast redwood (Ahuja and Neale 2002), or by hybridization between living or extinct related species, including *Sequoiadendron* and *Metasequoia* (Ahuja 2005, 2009; Yang et al. 2012). Molecular studies involving genome sequencing in coast redwood and other redwoods, including dawn redwood and giant sequoia, may be able to resolve or shed some light on the ancestry of hexaploid *Sequoia*.

Molecular genetic studies involving random amplified polymorphic DNAs (RAPDs) and microsatellites markers have been employed to find out the levels of genetic diversity in *Metasequia*. The study with RAPDs has shown that the number of alleles per locus was ranged from 1.57 to 1.65, and there was relatively higher level of polymorphic loci (81.3–87.5%) and heterozygosity levels (0.298–0.350) in *Metasequoia* (Chen et al. 2003). Microsatellite genetic markers revealed that observed and expected levels of heterozygosity ranged from 0.1842 to 0.8421 and 0.2389–0.6639, respectively, and the number of alleles per locus varied from 2 to 6 in *Metasequoia* (Cui et al. 2010). More recent studies (Jin et al. 2015) employing expressed sequence tag–simple sequence repeat (EST-SSR) genetic markers showed that observed and expected heterozygosity levels ranged from 0 to 1.0 and 0.117 to 0.813, respectively, and number of polymorphic alleles ranged from 2 to 8 in *Metasequoia*. Compared to DNA-previous study (Kuser et al. 1997) based markers allozymes genetic markers have revealed low levels of polymorphic loci (20%) and heterozygosity (0.083–0.091) in *Mertasequoia*. Apparently, molecular marker tends to show a higher level of genetic diversity compared to studies with allozymes.

3.4 Conservation of Genetic Resources

Genetic diversity is essential for the survival and preservation of a species in different sets of environments. Greater the levels of genetic diversity in a species, better are the chances for its survival and deployment over a wide range of environments (Ledig 1986, 1988). It is essential to maintain an appropriate level of genetic diversity in the species and populations for conservation planning, especially under a rapidly changing climate (Aitkin et al. 2008; St. Claire and Howe 2011; Ledig 2012). It is now well known that species ranges are shifting in the face of climate change (Iverson et al. 2008; Iverson and McKenzie 2013). Therefore, it is important to have an understanding of the genetic architecture and how genetic variation is organized within and among populations of a species. In the initial stages, it would be worthwhile to focus on those populations of a species that have higher levels of genetic diversity. Those species with small and fragmented populations would be faced with the following three unpredictable outcomes: (1) persistence through adaptations to somewhat new conditions in the current locations; (2) persistence through natural or assisted migrations to a new environment; and (3) an unavoidable extinction (Aitkin et al. 2008).

Allozyme genetic markers have revealed that hexaploid *Sequoia* seems to have a greater level of genetic diversity (Libby et al. 1996; Rogers 1997, 2000), as compared to other three diploid redwoods, namely *Metasequoia*, *Sequoiadendron*, and a tetraploid *Fitzroya* (Kuser et al. 1997; Premoli et al. 2000; Chen et al. 2003; Ahuja 2009). Redwoods are all endemic, restricted, and threatened species. Therefore, knowledge on the levels of genetic diversity is essential for their deployment in environmental locations outside their endemic ranges.

Conservation of genetic resources can be accomplished by in situ and ex situ methods, which take into account the maintenance of genetic diversity (Fig. 3.1) and also promote ecosystem stability. The strategy for conservation of genetic sources has to be holistic that is based on the entire gene pool of populations as well as species. One of the major objectives of conservation genetics is to ensure that functionally adaptive alleles will be available in the future for breeding programs and evolution of the species (Ledig 1988, 2012). In order to cope with the changing climate, it would be necessary to maximize genetic diversity by selecting heterozygous individuals in a population that exhibit hybrid vigor (Ledig and Kitzmiller 1992; Geburek and Konrad 2007). Since it is uncertain that global climate is likely to stabilize in the near future, it would be desirable to maintain adaptive genetic diversity in a species for its survival in a changing climate.

3.5 In Situ Conservation of Germplasm

In situ conservation of a forest tree includes preservation and maintain ace of stands and populations via regeneration by generative and vegetative methods within the area of natural occurrence (Fig. 3.2). In other words, it involves propagation and maintenance of the tree species within its habitat. This type of conservation has been in practice in the national parks, reserves, and research institutions for the redwoods in USA, China, Chile, and Argentina, and is primarily based on ecological approaches that ensures not only the conservation of the redwoods but also the associated animal and plant species in the ecosystem (Noss et al. 2000; Evarts and Popper 2001; Li 1999; Parker and Donoso 1993; Premoli et al. 2003).

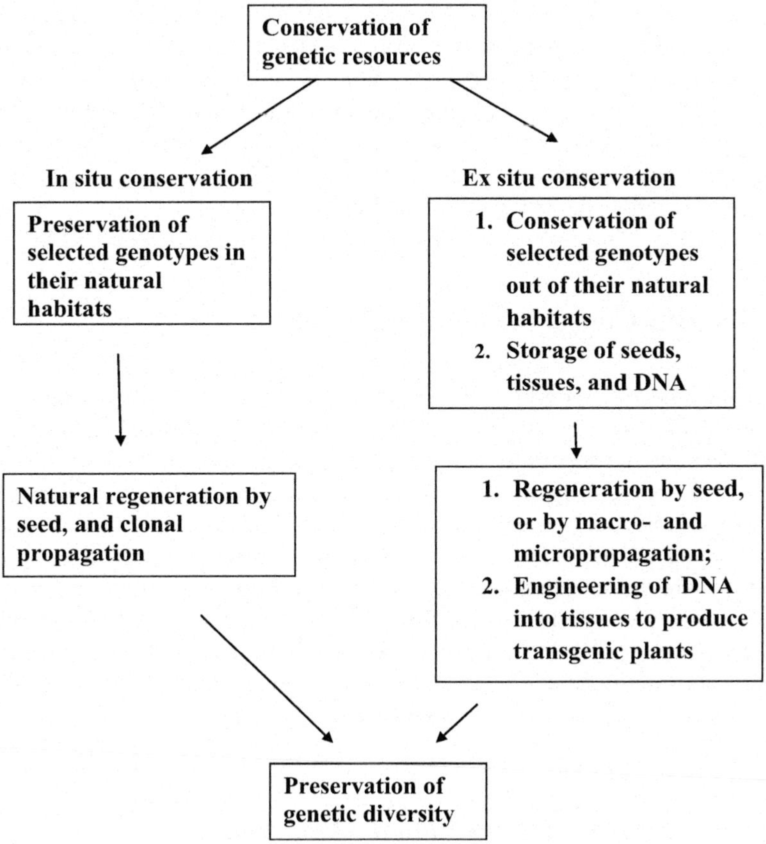

Fig. 3.2 Genetic conservation involves preservation of genetic resources by in situ and ex situ strategies that aim at conservation of genetic diversity in the forest ecosystem

For in situ preservation, it is necessary to regenerate populations from controlled crosses within and conserved diverse population to enhance hybrid vigor that would be advantageous for an overall conservation of genetic diversity in the redwoods in the face of global climate change. In this context, a preservation of genetic resources should include not only old growths and secondary growth populations, but also maintaining adequate genetic diversity to ensure adaptation to the new and diverse environments. For the in situ conservation redwoods, it is important to recognize the species composition, its abundant pattern, genetic architecture, ecological dynamics, genetic diversity, and ecosystem (Libby et al. 1996; Noss et al. 2000). Each redwood species has a unique endemic environmental niche in its native country, and understanding the soil, nutritional, and companion forest and other vegetation can provide direction of maintenance and conservation of the individual redwood in the present and future environments in the face of global climate change. For contingency planning for the future of redwoods in the rapid climate change scenarios, it would be necessary to explore new locations outside the endemic ranges of redwoods where genetically diverse populations can be safely planted. Of the four redwoods, *Sequoia*, endowed with higher levels of genetic diversity, would have a better chance of survival in new habitats. For this reason, it would be prudent to explore new locations for genetically diverse populations of other three redwoods, *Sequoiadendron*, *Metasequoia*, and *Fitzroya*, and also *Sequoia*, before the anticipated rapid climate change.

3.6 Ex Situ Conservation of Germplasm

The ex situ methods conserve the genetic resources out of the natural range of distribution of a species (Fig. 3.2). These include: (1) seed orchards and clonal plantations under nursery and field conditions, (2) preservation of tissues in culture, and (3) cold storage of germplasm. It should be mentioned here that implementing ex situ measures could be expensive, time consuming and may have some drawbacks. Among others, ex situ planning requires subjecting plants to a new environment where the selection pressure may be aliens to the plants, compared to the habitat where the original populations evolved (Ledig 1986). Some of these problems can be resolved by regularly checking variation and genetic fidelity of the populations using molecular genetic techniques. We examine the status of these ex situ preservation methods in redwoods in the following sections.

3.6.1 Seed Orchards and Clonal Plantations

In forest tree species, seed orchards and clonal orchards under nursery and field conditions are a routine for the ex situ conservation of germplasm (Melchior et al. 1986; Ledig 1986; Millar 1993; Hattemer 1995; Pritchard et al. 2014; FAO 2014a). Redwoods seed and clonal plantations have also been tested for their ex situ performance. *Sequoia* has been successfully grown out of its fogbelt in California several different locations in the USA for more than 100 years out its endemic range at the elevation of 1690 m above sea level at Placerville in the foothills of Sierra Nevada; and other locations in Seattle, Washington, Hawkinsville, Georgia, USA. It is now known that *Sequoia* can be grown and survive in many parts of the world, including New Zealand, Chile, Spain, England, South Africa, and Tasmania as well as several other countries and Victoria, British Columbia, Canada (Kuser 1981). It is now known that *Sequoia* can be grown and can survive in many parts of the world. These include Western Europe Turkey, Crimea, New Zealand, Chile, South Africa, and Tasmania (Kuser et al. 1995). An international provenance trial, that involved 180 clones from 90 locations throughout the natural range of *Sequoia*, was carried out in three plantation sites in USA, two in France, and plots in Spain, England, and New Zealand (Kuser et al. 1995). Early results from this study suggest that the continued survival of *Sequoia* outside their native ranges depends on the climate/soil conditions, somewhat comparable to original endemics sites, when planted in the ex situ locations. Therefore, differences in the origin of the materials are an important consideration for ex situ plantations of *Sequoia*.

Ex situ plantations of *Sequoiadendron* have been field-tested since the 1960s in Blodgett Forest Research Station, which is operated by the University of California, Berkeley. Early results suggested that height and diameter growth patterns of 18-year-old *Sequoiadendron* trees equaled or exceeded other conifers of that native site (Heald 1986). Another study with 22-year-old *Sequoiadendron* plantations Blodgett Forest Research Station also showed good growth and restoration potential of the species outside its endemic range (York et al. 2013). *Sequoiadendron* has been planted for more than a century ago outside its range in 591 locations in 25 European countries in Europe (Hartesfeldt 1969; Knigge 1994). In several countries (France, Hungary, Greece, Belgium, Netherlands, Denmark, Norway, and Germany), solitary trees in Arboretums/Parks or stands of *Sequoiadendron* have survived, while in other countries (Yugoslavia and Romania) *Sequoiadendron* has not done well (Libby 1981). Although healthy *Sequoiadendron* trees/stands still exist in some countries of Europe, the species is sensitive to frost and disease damage. In one study, frost tolerance of *Sequoiadendron* was tested in the 2-year-old seedling from 22 provinces (from the entire endemic range of *Sequoiadendron* from California) at subzero ranging temperatures ranging from −5 to −14 °C in the greenhouse studies in Germany. Two studies tested frost tolerance in *Sequoiadendron* at the seedling stage (Guinon et al. 1982). Significant differences were observed in the degree of frost tolerance, as measured by the freezing test. In a second study, growth performance of 14-year-old *Sequoiadendron* trees from four provinces tested at different northern and southern locations in Germany (Melchior and Hermann 1987). Significant differences were also observed in the survival and growth performance of 14-year-old *Sequoiadendron* grown under field conditions. In the northern location at Grosshansdorf, Germany, *Sequoiadendron* was badly damaged by frost and infection by *Armillaria mellea*. The authors recommended the use of frost-tolerant genotypes of *Sequoiadendron* for suitable locations in Germany (Melchior and Hermann 1987). On the other hand, *Sequoiadendron* trees performed relatively batter in southwest Germany.

Metasequoia has been widely introduced in different parts of the world outside its endemic location in China. Solitary and multiple trees of *Metasequoia* have also been in China and different countries of the world in Arboretums/Parks or stands. In China, *Metasequoia* trees were planted and have survived, outside its endemic range, in the parks of Tongii University and East China Normal University in Shanghai (Chen et al. 2003). *Metasequoia* trees have survived heavy snows in the botanical gardens of Hamilton and Montreal, Canada, Alaska, USA, Oslo, Norway, and high summer temperatures of Adelaide, Australia (Satoh 1999; Williams 2005), and perhaps in other regions of the world with similar climates. In excess of 50,000 *Metasequoia* trees have been cultivated in the cool climate of Denmark (Hendricks and Søndergaard 1998). *Metasequoia* was introduced in the USA more than 60 years ago, and with its moderate reservoir of genetic diversity, it has shown good growth in plantations in the eastern and western USA (Kuser 1999). Ex situ plantation of 40-year-old *Metasequoia* showed that that tree growth improved (based on tree rings chronology) in the highest levels of air pollution area near the Steelwork factory, and reduced growth was recorded when air pollution decreased in Krakow, Poland (Wilczynski et al. 2014). It would appear that *Metasequoia* seems to be relatively more resilient to environmental extremes, and possibly climate change, than the other redwoods.

Although ex situ stands and plantations of the other tree redwoods, *Sequoia*, *Sequoiadendron*, and *Metasequoia*, have been established outside their endemic ranges in several countries, it is not clear if any ex situ stands or plantations of *Fitzroya* have been established outside Chile. Single or multiple trees have been planted in the botanical gardens/parks in some countries, and only small experimental stands of *Fitzroya* have been mainly restricted to some locations in Chile and Argentina (Claudio Donoso, private communications). In Chile 17% of the *Fitzroya* forests (47,440 ha) are protected within the National Reserves and Parks, and 83% (264,993 ha) are in the private domain. On the other hand, in Argentina more than 80% *Fitzroya* forests are in protected areas (IUCN 2013; Premoli et al. 2013). *Fitzroya* is one of the endangered tree species and must be conserved in ex situ plantations and by other ex situ approaches for the conservation of its genetic resources. Genetic diversity in *Fitzroya* is much lower than other conifers (Premoli et al. 2000), but still has a moderate level of diversity, similar to *Metasequoia* and *Sequoiadendron* (Kuser et al. 1997; Fins and Libby 1982), and should be able to adapt to different environments. Therefore, it is recommended that, like other three redwoods, ex situ plantations of *Fitzroya* are also established in suitable locations outside its endemic range, to new locations in Chile, Argentina, and other parts of the world for its conservation and future survival.

3.6.2 Preservation of Tissues in Vitro

Bud meristems and shoots have been routinely used for micropropagation of trees (Ahuja 1987, 1993). In vitro explants are normally maintained in culture at 4–25 °C over short or long periods of time (Aitkin-Christie and Singh 1987: Ahuja 1994, 1999). Clones derived from tissue culture can be used as a resource for the conservation of germplasm. Interest in micropropagation of *Sequoia* goes back more than 75 years, when Earnst Ball (1950) was able to induce buds on callus cultures. In later years, Boulay (1978) employed juvenile basal sprouts from *Sequoia* for clonal the propagation. Micropropagation has been accomplished by using meristens, nodal stem explants, and needles juvenile explants, as well as tissues from mature trees up to 90-year-old *Sequoia* (Boulay 1978; Ball et al. 1978; Arnaud et al. 1993; Bon et al. 1994; Sul and Korban 2005; Korban and Sul 2007). Somatic embryogenesis and organogenesis were also achieved by culturing needles of *Sequoia* in vitro (Liu et al. 2006). In general, stump shoots from the base of mature trees are more responsive than shoots from the crown of the same mature tree to in vitro differentiation and organogenesis.

In a *Sequoia* plantation near Cologne, Germany, most the redwood trees were exposed to the severe winter of 1985–1986, with temperature as low as −20 °C, and were badly damaged. However, several redwoods survived, and we received

Fig. 3.3 Micropropagated *Sequoia sempervirens* clones derived by tissue culture of mesristems from frost-tolerant trees grown in the glasshouse and nursery at the Institute of Forest Genetics, Grosshansdorf, Germany

branches from three of the surviving old trees. One of the three trees was 8 years old, while other two trees were 25 years old. Meristems from the three trees were cultured in a modified woody plant medium for clonal propagation at the Institute of Forest Genetics, Grosshansdorf, Germany. Several thousand clones were produced by tissue culture from three trees (Fig. 3.3), and several hundred clones were tested under glasshouse, nursery (Fig. 3.4) and field conditions (starting 1987–1987) for their over-wintering capacity (Ahuja 1996). Field trials of clones from three frost-tolerant trees were tested for many years in Grosshansdorf and Trenthorst, in northern Germany. Trees in Grosshansdorf arboretum (Fig. 3.5) seem to be frost-tolerant and have survived more than 25 winters in Germany. However, only those *Sequoia* clones survived in the field trial in Trenthorst that were sheltered by the tree canopy during early growth. Therefore, sheltering of *Sequoia* seedlings/clones that are derived from frost-tolerant trees may be necessary during several years of early growth for their survival in climates with harsh winters (Ahuja 1996).

Tissue culture studies in *Sequoiadendron* have been carried to clonally propagated tissues derived from juvenile and mature trees. Plants have been regenerated by meristems cultured in vitro from juvenile 2-year-old plants of *Sequoiadendron* (Monteuuis 1987). On the other hand, tissues from mature trees (90–100 years old) were less responsive to regeneration in vitro. Mature branches from mature trees

Fig. 3.4 Micropropagated clones from three frost-tolerant trees of *Sequoia sempervirens* grown in the nursery at the Institute of Forest Genetics, Grosshansdorf, Germany

Fig. 3.5 Tissue culture-derived frost-tolerant *Sequoia sempervirens* trees (~25 years old) growing in the arboretum of the Institute of Forest Genetics, Grosshansdorf, Germany (Photograph by Dr. Georg von Wühlisch)

were grafted on the young *Sequoiadendron* seedlings and grown to induce a degree of rejuvenation. Meristems from such grafted scions were cultured in vitro to obtain regenerated plants (Monteuuis 1987; Monteuuis and Bon 1989). While plants could be readily regenerated from meristem tissues of 2-year-old plants in culture, in vitro regeneration from meristems from mature trees (90–100 years old) of

Sequoiadendron grafted on young seedlings were somewhat recalcitrant, but plant regeneration was possible after several passages in tissue culture (Bon and Monteuuis 1991; Monteuuis 1991). Compared to *Sequoia* where juvenile explants are readily available from the basal sprouts from mature trees, *Sequoiadendron* does not produce juvenile basal sprouts, and therefore rejuvenation of mature trees has to be accomplished before clonal propagation in tissue culture (Monteuuis et al. 2008). Tissue culture offers opportunities for germplasm preservation from selected genotypes of redwoods in tissue culture gene banks.

3.6.3 Cold Storage of Germplasm

Cold storage of seeds and tissues offers prospects for short-term and long-term preservation of forest tree germplasm. Sequoia seeds containing 6–10% moisture content have been stored in airtight sealed bottles at 5 °C for 3 years retained 14% viability, but rapidly lost viability to zero percentage after 16 years (Schubert 1952). Seeds of *Sequoiadendron* with 18% germination capacity stored at 5 °C lost their viability to 8% after 14 years of storage at that temperature (Schubert 1952). Dry seeds of *Metasequoia* have been successfully stored at 2–4 °C in airtight bottles for a period of time (Johnson 1974).

When Sequoia seeds were stored at −2 to −4 °C, they retained germination viability for 1 year, but rapidly lost viability after removal from subzero temperatures (Metcalf 1924). Boe (1974) reported that Sequoia seeds stored at −16 °C for 7 years retained 12–15% viability. Storage of meristems, embryos, dormant buds, cells, pollen, seeds, and DNA at ultra-low temperatures of liquid nitrogen (−196 ° C) (liquid nitrogen at −196 °C) offers potential for cryopreservation of genetic resources (Fig. 3.1) in plants, including forest trees (Adams 1997; Ahuja 1986, 1989, 1994, 2011; Bonner 1990; Rice et al. 2006; Engelmann 2004; Cruz-Cruz et al. 2013; Gonzalez-Arnao et al. 2014). In this direction, gene banks (FAO 2014b) present new opportunities for storage of germplasm (seeds, tissues, pollen, and DNA) of the redwoods for future exploitation in time and space. Shoot tips excised from tissue cultures of *Sequoia* were successfully cryopreserved and subsequently after retrieval from liquid nitrogen were able to regenerated shoots in vitro (Halmagyi and Deliu 2011; Ozudogru et al 2011).

3.7 Survival of Redwoods

The paleoendemic redwoods are threatened species and their conservation is essential. Although these redwoods are protected in the national parks, reserves, and in privately owned redwood forests in their habitats, we have to consider the conservation in view of global warming and climate change. We can use prudent means to conserve them in situ, and ex situ locations that have comparable environmental conditions to their endemic climate regions based on recent GIS (Geographic Information System) search for model conservation sites (Noss et al. 2000), and by other ex situ strategies for conservation of their germplasm. In this direction, ex situ conservation planning should consider relocation in protected areas that are managed to promote biodiversity and ecosystem stability (Hannah et al. 2007; Thuiller et al. 2008). Therefore, it is necessary to conserve the genetic resources in these paleoendemic redwoods and, at the same time, maintain an appropriate level of genetic diversity in the redwood species and populations for their future survival. In situ and ex situ strategies for the conservation of genetic resources of redwoods are discussed in this paper. Although these redwoods are protected in the national parks, reserves, and in privately owned forests in their habitats, it would be desirable to conserve them in new ex situ reserves, and by other ex situ strategies involving biotechnological approaches, to preserve seed, tissues, and DNA in gene banks for future exploitations.

In addition to in situ and ex situ conservation of genetic resources in redwoods, we have to consider the response of trees to environmental changes caused by global warming (Peters 1990; Hamrick 2004). A number of recent studies have suggested that species ranges are shifting and will continue to do so in the next century under the rapid climate change scenario (Aitkin et al. 2008; St. Clair and Howe 2011; Iverson et al. 2008; Iverson and McKenzie 2013). If the climate changes slowly and the southern areas in the northern hemisphere start becoming hot and dry, the tree species from such climatic zones may have to move north to colonize areas that are suitable for their survival. The reverse would be the case in species in the southern hemisphere. However, if the earth warms 2 °C in 100 years, the tree species may have to move ~3.6 km every year. Tree species whose seeds are dispersed by birds may be able to spread at that rate (US Environmental Protection Agency 2000). However, species whose seeds are dispersed by wind (unless strong winds) may not able to spread more than a few hundred meters. The redwoods, though have winged seeds, their dispersal is usually very close to their plantations. However, *Sequoia*, which reproduces both by seed and vegetative regeneration from stump sprouts, may have a better chance of survival than other redwoods. What will be the fate of redwoods in the future with global warming looming on our planet? Will these paleoendemic redwoods become extinct under a rapid climate change scenario? That scenario seems highly unlikely because of human intervention to locate the redwoods in suitable climate zones. However, the spatially fragmented populations of redwoods still remain vulnerable to climate change in native ranges in the foreseeable future.

What is the other alternative for the survival of redwoods? Although anthropogenic barriers, including extensive logging, land use for agriculture and industry, would be impediments for species migration in response to climate change, self-migration and assisted migration would be a viable alternative for the survival of these paleoendemic redwoods. Assisted poleward migration would be required for these ancient redwoods, *Sequoia*, *Sequoiadendron*, and *Metasequoia* toward the

Arctic Region, and *Fitzroya* migration toward the Antarctic Region for their survival (Ahuja 2011). Fossil history suggests that during the earlier geologic Periods (Cretaceous and Tertiary) ancestors of three redwoods (*Sequoia*, *Sequoiadendron*, and *Metasequoia*), part of the so-called Arcto-Tertiary Geoflora (Chaney 1950; Ornduff 1998), were widespread more than 100 million years ago in near the earth's Polar Region. In response to global climate change, this Arcto-Tertiary Geoflora has gradually moved southward. The composition of this flora changed during further geological upheavals, some becoming extinct, while the three paleoendemic redwoods surviving in the present endemic locations. The same fate may have occurred with *Fitzroya* in the southern hemisphere. The Cenozoic climate changes may have possibly shaped the evolutionary history of members of Cupressaceae (Pitterman et al. 2012; Quirk et al. 2013), including the current distributions of redwoods in their current restricted ranges. A highly speculative "back-to-the-future" scenario for the return of the redwoods to their previously widespread occurrence has been considered by Thomas and LePage (2011). According to these authors, there may be a remote chance that previous climatic conditions, that prevailed when redwoods were largely widespread as part of the Arctic Geoflora, could occur in the future. And getting back there in the Polar Regions that are so devoid of forests would be the resurgence of these paleoendemic redwoods (Thomas and LePage 2011). This is an interesting speculation, but at present, we have to act to preserve the redwoods.

3.8 Conclusions and Prospects

What would be the fate of the paleoendemic redwoods in a rapidly changing global climate? Although extinction of these endemic redwoods seems unlikely in the foreseeable future, because of human intervention involving assisted migration, they are still vulnerable to climate change in their restricted endemic regions in the world. In the face of rapid and sometimes unpredictable climate change, it might be useful to deploy a broad mixture of seeds from widely divergent populations from different environments (Ledig and Kitzmiller 1992) as a resource for seed orchards and subsequently planting materials for future climate change scenarios. In an uncertain climate change, one of best options would be to use flexible approaches to promote adaptive genetic biodiversity and conservation of genetic resources in the endemic redwoods. Finally, it is imperative that genetic conservation of redwoods must proceed unabatedly, with more emphasis on ex situ strategies to 'Save the Redwoods'. Global approaches by in situ and ex situ measures, including biotechnological approaches, should be fully integrated for the conservation of the genetic resources of these threatened redwoods in the face of rapid climate change. Genomic research is lacking in the redwoods and should be pursued for a better understanding of their genome structure, function and gene conservation in these paleoendemic redwoods.

References

Adams RP (1997) Conservation of DNA: DNA banking. In: Callow JA, Ford-Lloyd BV, H. J. Newbury HJ (eds) Biotechnology and plant genetic resources, conservation and Use. CABI Publishing, London, pp 163–174

Ahuja MR (1986) Storage of forest tree germplasm in liquid nitrogen (-196°C). Silvae Genet 35:249–251

Ahuja MR (1987) In vitro propagation of poplars and aspens. In: Bongs JM, Durzan DJ (eds) Cell and tissue culture in forestry, vol 3. Martinus Nijhoff Publishers, Dordrecht, pp 207–223

Ahuja MR (1989) Storage of forest tree germplasm at sub-zero temperatures. In: Dhawan V (ed) Application of biotechnology in forestry and horticulture. Plenum Press, New York, pp 215–228

Ahuja MR (1993) Micropropagation of woody plants. Kluwer Academic Publishers, Dordrecht

Ahuja MR (1994) Reflections on germplasm preservation of trees. In: Pardos JA, Ahuja MR, Rosello RE (eds) Biotechnology of trees. Investgacion Agraria Sistemas y Recursos Forestales, Madrid, pp 227–233

Ahuja MR (1996) Micropropagation and field testing of frost-tolerant *Sequoia sempervirens* genotypes. In: LeBlanc J (ed) Proceeding of the conference on coast redwood forest ecology and management. Humboldt State University, Arcata, pp 153–155

Ahuja MR (1999) Biotechnology in forest tree gene banks. In: Edwards DGW, Naithani SC (eds) Seed and nursery technology of forest trees. New Age International (P) Limited Publishers, New Delhi, pp 23–36

Ahuja MR (2005) Polyploidy in gymnosperms: revisited. Silvae Genet 54:59–69

Ahuja MR (2009) Genetic constitution and diversity in four narrow endemic redwoods from the family Cupressaceae. Euphytica 165:5–19

Ahuja MR (2011) Strategies for conservation of germplasm in endemic redwoods in the face on climate change: a review. Plant Genet Resour 9:411–422

Ahuja MR, Neale DB (2002) Origins of polyploidy in coast redwood (*Sequoia sempervirens* (D. Don) Endl.) and relationship of coast redwood to other genera of Taxodiaceae. Silvae Genet 51:93–100

Ahuja MR, Neale DB (2005) Evolution of genome size in conifers. Silvae Genet 54:126–137

Aitkin-Christie J, Singh AP (1987) Cold storage of tissue cultures. In: Bonga JM, Durzan DJ (eds) Cell and tissue culture in forestry, vol 1. Martinus Nijhoff Publishers, Dordrecht, pp 285–304

Aitkin SN, Yeaman S, Holiday JA, Wang T, Curtis-McLane S (2008) Adaptation, migration or extirpation: climate change outcomes of tree populations. Evol Appl 1:95–111

Allan RP, Soden BJ (2008) Atmospheric warming and the amplification of precipitation extremes. Science 321:1481–1494

Allnutt TR, Newton AC, Lara A, Premoli A, Armesto JJ, Vergara R, Gardner M (1999) Genetic variation in *Fitzroya cupressoides* (alerce), a threatened South American conifer. Mol Ecol 8:975–987

Anderson JT, Inouye DW, McKinney AM, Colauti RI, Mitchell-Olds (2012) Pheotypic plasticity and adaptive evolution contribute to advancing flowering phenology in response to climate change. Proc R Soc B 279:3843–3852

Arnaud Y, Franclet A, Tranvan H, Jacques M (1993) Micropropagation and rejuvenation of *Sequoia sempervirens* (Lamb) Endl.): a review. Annales des Sciences Forestieres 50:273–295

Ball EA (1950) Differentiation in a callus culture of *Sequoia sempervirens*. Growth 14:295–325

Ball EA, Morris DM, Reydelius JA (1978) Cloning of *Sequoia sempervirens* from mature trees through tissue culture. Round Table Conference. In Vitro Multiplication of Woody Species, Gamboloux, Belgium, pp. 181–226

Battles JJ, Robards T, Das A et al (2008) Climate change impacts on forest growth and tree mortality: a data-driven modeling study in the mixed-conifer forest of Sierra Nevada, California. Clim Change 87(Supplement 1):S193–S213

Bellard C, Bertelsmeier C, Leadley P, Thuiller W, Courchamp F (2012) Impacts of climate change on the future of biodiversity. Ecol Lett 15:365–377

Bellard C, Leclerc C, Leroy B et al (2014) Vulnerability of biodiversity hotspots to global change. Global Ecol Biogeogr. doi:10.1111/geb.12228

Boe KN (1974) *Sequoia sempervirens* (D.Don) Endl. - Redwood. In: Schopmeyer CS (ed) Seeds of woody plants in the United States. Forest Service, USDA, Washington DC, pp 764–766

Bon MC, Monteuuis O (1991) Rejuvenation of a 100-year old *Sequoiadendron giganteum* through in vitro meristem culture. I. Organogenesis and morphological arguments. Physiol Plant 81:111–115

Bon MC, Riccari F, Monteuuis O (1994) Influence of phase change within a 90-year old *Sequoia sempervirens* on its in vitro organogenic capacity and protein patterns. Trees 8:283–287

Bonner FT (1990) Storage of seeds: potential and limitations for germplasm conservation. For Ecol Manage 35:35–43

Botkin DB, Saxe H, Arauju MB et al (2007) Forcasting the effects of global warming on biodiversity. Bioscience 57:227–236

Boulay M (1978) Multiplication rapide du *Sequoia sempervirens* en culture in vitro. Annales AFOCEL, pp 37–66

Brinegar C (2012) Rangewide genetic variation in coast redwood populations at a chloroplast microsatellite locus. In: Standiford RB, Weller TJ, Piirto DD, Stuart JD (eds) Proceedings of the coast redwood forests in a changing California: a symposium for scientists and managers. USDA, Pacific Southwest Research Station, Gen Tech Rep PSW-GTR-238, Albany, CA, pp 241–249

Brinegar C, Bruno D, Kirkbride R, Glavas S, Udransky U (2007) Applications of redwood genotyping by using microsatellite markers. In: Standiford RB, Giusti GA, Valachovic Y, et al (eds) Proceedings of the redwood region forest science symposium: what does the future hold? USDA, Forest Service, Gen Tech Rep PSW-GTR-194, Albany, CA, pp 47–56

Chaney RW (1950) A revision of fossil *Sequoia* and *Taxodium* in western North Amerixca based on the recent discovery of *Metasequoia*. Trans Am Philos Soc N Ser 40:171–263

Chen XY, Li YY, Wu TY, Zhang X, Lu HP (2003) Size-class differences in genetic structure of *Metasequoia glyptostroboides* Hu et Cheng (Taxodiaceae) plantations in Shanghai. Silvae Genet 52:107–109

Chu K, Cooper SC (1999) An ecological reconnaissance in the native home of *Metasequoia glyptostroboides*. Arnoldia 59:40–46

Craufurd PQ, Wheeler TR (2009) Climate change and flowering time in annual crops. J. Exptl Bot 60:2529–2539

Cruz-Cruz CA, Gonzalez-Arnao MT, Engelmann F (2013) Biotechnology and conservation of plant biodiversity. Resources 2:73–95

Cui MY, Yu S, Liu M, Li YY (2010) Isolation and characterization of polymorphic microsatellite markers in *Metasequoia glyptostoboides* (Taxodiaceae). Conserv Genet Resour 2:19–21

Dale VH, Joyce LA, McNulty S et al (2001) Climate change and forest disturbances. Bioscience 51:723–734

Davis MB, Shaw RG, Etterson JR (2005) Evolutionary responses to climate change. Ecology 86.1704

Douhovnikoff V, Dodd RS (2011) Linkage divergence in coast redwoos (*Sequoia sempervirens*), detected by a new set of nuclear microsatellite loci. Am Midl Nat 165:22–37

Engelmann F (2004) Plant cryopreservation: progress and prospects. In Vitro Cell Dev Biol Plants 40:427–433

Evarts J, Popper M (2001) Conservation and management of redwood forests. In: Evarts J, Popper M (eds) Coast redwood a natural and cultural history. Cachuma Press, Los Olivos, CA, pp 165–205

FAO (2014a) The state of world's forest genetic resources. Rome

FAO (2014b) Genebank standards for plant genetic resources for food and agriculture. Rev. ed. Rome

Fernandez M, Hamilton HH, Kueppers LM (2015) Back to the future: using historical climate variation to project near-term shifts in habitat suitable for coast redwood. Global Change Biol. doi:10.111/geb13027

Fins L, Libby WJ (1982) Population variation in Sequoiadendron: seed and seedling studies, vegetative propagation and isozyme variation. Silvae Genet 31:102–110

Fins L, Libby WJ (1994) Genetics of giant sequoia. USDA Forest Serv Gen Tech Rep PSW 151:65–68

Foden W, Mace G, Vié JC et al (2008) Species susceptibility to climate change impacts. In: Vié J-C, Hilton-Taylor C, Stuart SN (eds) The 2008 review of the IUCN red list of threatened species. IUCN Gland, Switzerland, pp 1–11

Franks SJ, Sim S, Weis AE (2007) Rapid evolution of flowering time by an annual plant in response to a climate fluctuation. Proc Natl Acad Sci USA 104:1278–1282

Gadek PA, Alpers DL, Heslewood MM, Quinn CJ (2000) Relationship within Cupressaceae sensu latu: a combined morphological and molecular approach. Am J Bot 87:1044–1057

Gaira KS, Rawal RS, Rawat B, Bhatt ID (2014) Impact of climate chanhe on the flowering of *Rhododendron arboretum* in central Himalaya, India. Curr Sci 106:1735–1738

Geburek T, Konrad H (2007) Why the conservation of forest genetic resources has not worked. Conserv Biol 22:267–274

Gonzalez-Arnao MT, Mratinez-Montero ME, Cruz-Cruz CA, Engelmann F (2014) Advances in cryogenic techniques for the long-term preservation of plant biodiversity. In: Ahuja MR, Ramawat KG (eds) Biotechnology and biodiversity. Springer, Berlin, pp 129–170

Guinon M, Larson JB, Spethmann W (1982) Frost resistance and early growth of Sequoiadendron giganteum seedlings of different origin. Silvae Genet 31:173–178

Hair JB (1968) The chromosomes of the Cupressaceae. I. Tetraclineae and Actinostrobeae (Callitroideae). NZ J Bot 6:277–284

Halmagyi A, Deliu C (2011) Conservation of redwood (*Sequoia sempervirens* (D.Don) Endl.) shoot apices by encapsulation-dehydration. Contributti Botanice 46:117–125

Hamrick JL (2004) Response of forest trees to global environmental changes. For Ecol Manage 197:323–335

Hamrick JL, Godt MJW, Sherman-Boyles SL (1992) Factors influencing levels of genetic diversity in woody plant species. New Forest 6:95–124

Hannah L, Midgley G, Andelman S et al (2007) Protected areas needs in a changing climate. Front Ecol Environ 5:131–138

Hartesfeldt RJ (1969) *Sequoia* in Europe with a review of their discovery and their resultant importation into Europe. Final Contract Report to the National Park Service. Contract # 14-10-0434, 22 p

Hartesveldt RJ, Harry HT, Schellhammer HS, Stecker RR (1975) The giant sequoia of the Sierra Nevada. U.S. Department of Interior, National Park Service, Washington, D.C., p 180

Hattemer HH (1995) Concepts and requirements in the conservation of forest genetic resources. Forest Genet 2:125–134

Heald RC (1986) Management of giant sequoia at Blodgett Forest Research Station. USDA Gen Tech Rep PSW-95, Berkeley, CA, pp 37–39

Hendricks DR, Søndergaard P (1998) *Metasequoia glyptostroboides* 50 years out of China. Observations from the United States and Denmark. Dansk Dendrologisk Årsskrift 16:6–24

Hoffmann AA, Willi Y (2008) Detecting genetic response to environmental change. Nat Rev Genet 9:421–432

Hoffmann AA, Sgro CM (2011) Climate change and evolutionary adaptation. Nature 470:479–485

Ibáñez I, Clark JS, Dietze MC et al (2006) Predicting biodiversity change: outside the climate envelop, beyond the species area curve. Ecology 87:1896–1906

Iler AM, Høye TK, Inouye DW, Schmidt NM (2013) Nonlinear flowering responses to climate: are species approaching their limits of phonological change? Phil Trans R Soc B 368:20120489

IPCC (2007) Climate change 2007. The physical science basis. Summary for policymakers. www.ipcc.ch

IPCC (2014) Climate change 2014: Impacts, adaptation, and vulnerability. Summary for policymakers. www.ipcc.ch
IUCN (2013) The IUCN list of threatened species. http://www.iucnredlist.org/details/30926/0
Iverson LR, Prasad AM, Schwartz MW (2005) Predicting potential changes in suitable habitat and distribution by 2100 for tree species in the eastern United States. J Agric Meteorol 61:29–37
Iverson LR, Prasad AM, Matthews SN, Peters M (2008) Estimating potential habitat for 134 Eastern US tree species under six climate scenarios. For Ecol Manage 254:390–406
Iverson LR, McKenzie D (2013) Tree-species range shifts in the changing climate: detecting, modeling, assisting. Landscape Ecol 28:879–889
Javeline D, Hellmann JJ, Cornejo RC, Shufeldt G (2013) Expert opinion on climate change and threats to biodiversity 63:666–673
Jin Y, Bi Q, Guan W, Mao JF (2015) Development of 23 novel polymorphic EST-SSR markers for the endangered relict conifer *Metasequoia glyptostroboides*. Appl Plant Sci 3(9):150038
Johnson LC (1974) *Metasequoia glyptostroboides* Hu and Cheng - Dawn Redwood. In: Schopmeyer CS (ed) Seeds of woody plants in the United States. Forest Service, USDA, Washington DC, pp 540–542
Johnstone JA, Dawson TE (2010) Climate context and ecological implications of summer fog decline in the cost redwood region. Proc Natl Acad Sci USA 107:4533–4538
Kelly AE, Goulden ML (2007) Rapid shifts in plant distribution with recent climate change. Proc Natl Acad Sci USA 105:11823–11826
Khoshoo TN (1959) Polyploidy in gymnosperms. Evolution 13:24–39
Knigge W (1994) Giant sequoia (*Sequoiadendron giganteum* (Lindl.) Buchholz) in Europe. USDA Gen Tech Rep PSW 151:28–48
Korban SS, Sul IW (2007) Micropropagation of coast redwood (*Sequoia sempervirens*) In: Jain SM, Häggman (eds) Protocols for micropropagation of woody trees and fruits. Springer, Berlin, pp 23–32
Kuser JE (1981) Redwoods around the world. Am Forest 87:30–32
Kuser JE (1999) *Metasequoia glyptostroboides*: Fifty years of growth in North America. Arnoldia 59:76–79
Kuser J E, Bailly A, Franclet A, Libby WJ et al (1995) Early results of a range-wide provenance test of *Sequoia sempervirens*. Forest Genetics Resources, FAO, Rome, No. 23:21–25
Kuser JE, Sheely DL, Hendricks DR (1997) Genetic variation in two *ex situ* collections of the rare *Metasequoia glyptostroboides* (Cupressaceae). Silvae Genet 46:258–264
Lara A, Villalba R (1993) A 3,820-year temperature record from alerce tree rings in southern South America. Science 260:1104–1106
Ledig FT (1986) Conservation strategies for forest gene resources. For Ecol Manage 14:77–90
Ledig FT (1988) The conservation of diversity in forest trees. Bioscience 38:471–479
Ledig FT (2012) Climate change and conservation. Acta Slv Hung 8:57–74
Ledig FT, Kitzmiller JH (1992) Genetic strategies for reforestation in the face of global climate change. For Ecol Manage 50:153–169
LePage BA, Yand H, Matsumoto M (2005) The evolution and biogeographic history of *Metasequoia*. In: Le Page BA, Williams CJ, Yang H (eds) The geobiology and ecology of *Metasdequoia*. Springer, Berlin, pp 3–114
Leng Q, Fan SH, Wang L, Yang H et al (2007) Database of native *Metasequoia glyptostroboides* trees in China based on new consensus surveys and expeditions. Bull Peabody Mus Nat Hist 48:185–233
Li J (1999) *Metasequoia*: an overview of its phylogeny, reproductive biology, and ecotypic variation. Arnoldia 59:54–59
Li L (1987) The origin of Sequoia sempervirens (Taxodiaceae) based on karyotype. Acta Botanica Yunnancia 99:192–197
Li YY, Chen XY, Zhang X, Wu TT, Lu HP, Cai YW (2005) Genetic differences between wild and artificial populations of *Metasequoia glyptostroboides*: implications of species recovery. Conserv Biol 19:224–231

Libby WJ (1981) Some observations on *Sequoiadendron* and *Calocedrus* in Europe. Calif Forest Forest Prod N0(49):1–12
Libby WJ, Anekonda TS, Kuser JE (1996) The genetic architecture of coast redwood. In: Leblanc J (ed) Proceeding of the conference on coast redwood forest ecology and management. Humboldt State University, Arcata, pp 147–149
Littell JS, Oneil EE, McKenzie D et al (2010) Forest ecosystems, disturbance, and climate change in Washington State, USA. Clim Change 102:129–158
Liu C, Xia X, Yin W, Huang L, Zhou J (2006) Shoot regeneration and somatic embryogenesis from needles of redwood (*Sequoia sempervirens* (D. Don) Endl.). Plant Cell Rep 25:621–628
McKenney DW, Pedlar JH, Lawrence K, Campbell K, Hutchison MF (2007) Potential impacts of climate change on the distribution of North American trees. Bioscience 57:939–948
Melchior GH, Hermann S (1987) Differences in growth performance of four provenances of giant sequoia (*Sequoiadendron giganteum* (Lindl) Buchh.). Silvae Genet 38:65–68
Melchior GH, Muhs HJ, Stephan BR (1986) Tactics for forest tree resources in the Federal Republic of Germany. For Ecol Manage 17:73–81
Metcalf W (1924) Artificial reproduction of redwood (*Sequoia sempervirens*). J Forest 22:873–893
Millar CI (1993) Conservation of germplasm of forest trees. In: Ahuja MR, Libby WJ (eds) Clonal forestry II. Conservation and application. Springer, Berlin, pp 42–65
Moneuuis O (1987) *In vitro* meristem culture of juvenile and mature *Sequoiadendron giganteum*. Tree Physiol 3:265–272
Monteuuis O (1991) Rejuvenation of a 100-year-old *Sequoiadendron giganteum* through in vitro meristem culture. I. Organogenic and morphological arguments. Physiol Plant 81:111–115
Monteuuis O, Bon MC (1989) Rejuvenation of a 100 yr old giant sequoia (*Sequoiadendron giganteum* Buchholz) through *in vitro* meristem culture. Ann Sci For 46 (Supplement): 183s–186s
Monteuuis O, Doulbeau S, Verdeil JL (2008) DNA methylation in different origin clonal offspring from a mature *Sequoiadendron giganteum* genotype. Trees 22:779–784
Noss RF, Strittholt JR, Heilman GE, Frost PA, Sorensen M (2000) Conservation planning in the redwoods region. In: Noss RF (ed) The redwood forest. History, ecology, and conservation of redwoods. Save-the-Redwoods League. Island Press, Washington DC, pp 201–228
O'Gorman PA, Schneider T (2009) The extreme basis for increase in precipitation in simulations of the 21st-century climate change. Proc Natl Acad Sci USA 106:14773–14777
Olson DF, Roy DF, Walters GA (1990) *Sequoia sempervirens* (D. Don) Endl. Redwood. In: Burns RM, Honkala BH (eds) Silvics of North America. Vol 1. Conifers. Agriculture Handbook 654. US Department of Agriculture, Forest Service, Washington DC, pp 541–551
Ornduff R (1998) The *Sequoia sempervirens* (coast redwood) forest of the pacific coast, USA. In: Lederman AD (ed) Coastally restricted forests.Oxford University Press, New York, pp 221–236
Ozudogru EA, Kirdok E, Kaya E, Capuana M, Beneli C, Engelmann F (2011) Cryopreservation of redwood (*Sequoia sempervirens* (D.Don.) *in vitro* buds using vitrification-based techniques. CryoLetters 32:99–110
Parker T, Donoso C (1993) Natural regeneration of *Fitzroya cupressoides* in Chile and Argentina. For Ecol Manage 59:63–85
Parmesan C (2006) Ecological and evolutionary responses to recent climate change. Annu Rev Ecol Evol Syst 37:637–669
Parmesan C, Yohe G (2003) A globally coherent fingerprint of climate change impacts across natural systems. Nature 421:37–42
Pautasso M (2013) Forest ecosystems and global change: the case study of insurbia. Annali di Botanica 3:1–29
Pearson RG, Stanton JC, Shoemaker KT et al (2014) Life history and spatial traits predict extinction risk due to climate change. Nat Clim Change 4:217–221
Peters RL (1990) Effects of global warming on forests. For Ecol Manage 35:13–33

Pitterman J, Stuart SA, Dawson TE, Moreau A (2012) Cenozoic climate change shaped the evolutionary ecophysiology of the Cupressaceae conifers. Proc Natl Acd Sci USA 109:9647–9652

Premoli AC, Kitzberger T, Veblen TT (2000) Conservation genetics of the endangered conifer *Fitzroya cupressoides* in Chile and Argentina. Conserv Genet 1:57–66

Premoli AC, Vergara R, Souto CP, Lara A, Newton AC (2003) Lowland valleys shelter the ancient conifer *Fitzroya cupressoides* in the Central Depression of southern Chile. J Roy Soc NZ 33:623–631

Premoli A, Quiroga P, Souto C, Gardner M (2013) *Fitzroya cupressoides*. The IUCN Red List of Threatened Species 2013: e. T30926A279874

Pritchard HW, Moat JF, Ferraz JBS et al (2014) Innovative approaches to the preservation of forest trees. For Ecol Manage 333:88098

Quirk J, McDowell NG, Leake JR, Hudson PJ, Beerling DJ (2013) Increased susceptibility to drought-induced mortality in *Sequoia sempervirens* (Cupressaceae) trees under Cenozoic atmospheric carbon dioxide starvation. Am J Bot 100:582–591

Rehfeldt GE, Jaquish BC, Sáenz-Romero C et al (2014) Comparative genetic response to climate in the varieties of *Pinus ponderosa* and *Pseudotsuga menziesii*: restoration. For Ecol Managm 324:147–157

Rice N, Cordeiro G, Shepherd M et al (2006) DNA banks and their role in facilitating the application of genomics to plant germplasm. Plant Genet Resour 4:64–70

Rogers DL (1997) Inheritance of allozymes from seed tissues of the hexaploid gymnosperm, *Sequoia sempervirens* (D.Don) Endl. (Coast redwood). Heredity 78:166–175

Rogers DL (1999) Allozyme polymorphisms discriminate among coast redwood (*Sequoia sempervirens*) siblings. J Hered 90:429–433

Rogers DL (2000) Genotypic diversity and clones size in old-growth populations of coast redwood (*Sequoia sempervirens*). Can J Bot 78:1408–1419

Root TL, MacMynowski DP, Mastrandrea MD, Schneider SH (2005) Human modified temperature induce species changes: joint attribution. Proc Natl Acad Sci USA 102:7465–7469

Satoh K (1999) *Metasequoia* travels the globe. Arnoldia 59:72–75

Sawyer JO, Gray J, West J, Thorburgh DA, Noss RF, Engbeck JH, Marcot BG, Raymond R (2000) History of redwoods and redwood forests. In: Noss RF (ed) The redwood forest. History, ecology, and conservation of redwoods. Save-the-Redwoods League. Island Press, Washington DC, pp 81–118

Saylor LC, Simons HA (1970) Karyology of *Sequoia sempervirens*: karyotype and accessory chromosomes. Cytologia 35:294–303

Schlarbaum SE, Tsuchiya T (1984a) Chromosome study of coast redwood, Sequoia sempervirens (D.Don) Endl.). Silvae Genet 33:56–62

Schlarbaum SE, Tsuchiya T (1984b) Cytotaxonomy and phylogeny in certain species of *Taxodiaceae*. Plant Syst Evol 147:29–54

Schubert G H (1952) Germination of various coniferous seeds after cold storage. USDA Forest Service Research Note PSW-83, 7 p

Staudt A, Leidner AK, Howard J et al (2013) The added complications of climate change: understanding and managing biodiversity and ecosystems. Front Ecol Environ 11:494–501

St. Clair JB, Howe GT (2011) Strategies for conserving forest genetic resources in the face of climate change. Turk J Bot 35:403–409

Stebbins GL (1948) The chromosomes and relationship of *Metasequoia* and *Sequoia*. Science 108:95–98

Sugiyama M, Shiogama H, Emori S (2010) Precipitation extreme changes exceeding moisture content increases in MIROC and IPCC climate models. Proc Natl Acad Sci USA 107:571–575

Sul IW, Korban SS (2005) Direct shoot organogenesis from needles of three genotypes of *Sequoia semperviren*. Plant Cell Tissue Org 80:353–358

Thomas CD, Cameron A, Green RE et al (2004) Extinction risk from climate change. Nature 427:145–148

Thomas P, LePage BA (2011) The end of an era—the conservation status of redwoods and other members of the former Taxodiaceae in the 21st century. Jpn J Hist Bot 19:89–100

Thuiller W, Albert C, Araújo MB et al (2008) Predicting global change impacts on plant species distributions: future challenges. Perspect Plant Ecol Evol Syst 9:137–152

Trenberth KE (2011) Changes in precipitation with climate change. Climate Res 47:123–138

Urban MC (2015) Accelerating extinction risk from climate change. Science 348:571–573

US Environmantal Protection Agency (2000) Global warming—impacts: forests. Document, Washington

Wang B, Liu DL, Asseng S, Macadam I, Yu Q (2015) Impact of climate change on wheat flowering time in eastern Australia. Agric Forest Meteorol 209:11–21

Weatherspoon CP (1990) *Sequoiadendron giganteum* (Lindl.) Buchholz, Giant Sequoia.. In: Burns RM, Honkala BH (eds) Silvics of North America. Vol 1. Conifers. Agriculture Handbook 654. US Department of Agriculture, Forest Service, Washington DC, pp 552–562

Weed AS, Ayres MP, Hicke JA (2013) Consequences of climate change for biotic disturbances in North American forests. Ecol Monogr 83:441–470

Wilczynski S, Mutter E, Wertz B (2014) The application of the tree ring chronology in assessing ecological requirements of *Metasequoia glyptostroboides* growing in southern Poland. Geochronometria 41:129–135

Williams CJ (2005) Ecological characteristics of *Metasequoia glyptostroboides*. In: LePage BA, Williams CJ, Yang H (eds) The geobiology and ecology of *Metasequoia*. Springer, Berlin, pp 285–304

Wylie B, Rigge M, Brisco B et al (2014) Effects of disturbance and climate change on ecosystem performance in Yukon river basin boreal forest. Remote Sens 6:9145–9169

Yang ZY, Ran JH, Wang XQ (2012) Three genome-based phylogeny of Cupressaceae s.i.: further evidence for the evolution of gymnosperms and southern hemisphere biogeography. Mol Phylogenet Evol 64:452–470

York RA, O'Hara KL, Battles JJ (2013) Density effects on giant sequoia (*Sequoiadendron giganteum*) growth thru 22 years: implications for restoration and plantation management. West J Appl For 28:30–36

Chapter 4
Mulberry (*Morus* Spp.) Genetic Diversity, Conservation and Management

Amalendu Tikader and Kunjupillai Vijayan

Abstract Mulberry is a perennial tree being cultivated for sericulture purposes besides being used for multipurposes such as fruit, timber, etc. So far 68 species have been recognized; only a few species such as *M. alba, M. bombycis, M. indica, M. latifolia* and *M. multicaulis* are cultivated for foliage and *M. nigra* for fruit. The remaining species along with several landraces of the cultivated species are considered wild, which were largely neglected. As the arable land area is reducing due to urbanization and human interference, it is necessary to utilize the available land optimally for conservation of genetic resources by adopting established practices of conservation and management. The climate change is another major threat to biological systems and it needs proper adoption of multidisciplinary approaches to mitigate it. In India, four species of mulberry have been reported and described briefly in this chapter including cultivated and wild collections. Two species are endemic to higher altitude of Himalayan belt and are used for sericulture as well as other uses such as timber, fruit, etc. Mulberry is conserved in a traditional way, but in recent times, modern methods are also adopted to maintain the variation in cultivated and wild species in different countries including India. The methods are applied depending on the materials and its uses.

Keywords *Morus* spp · Diversity · Characterization · Evaluation
Management · Conservation

A. Tikader (✉)
#203 Aranya Enclave, 11th Main, 3rd Cross, Kakappa Layout Begur Main Road, Hongasandra, Bangalore 560 068, Karnataka, India
e-mail: atikader_csgrc@yahoo.co.in

K. Vijayan
Central Silk Board, BTM Layout, Madiwala, Bangalore 560068, India
e-mail: kvijayan01@yahoo.com

M.R. Ahuja and S.M. Jain (eds.), *Biodiversity and Conservation of Woody Plants*, Sustainable Development and Biodiversity 17, DOI 10.1007/978-3-319-66426-2_4

4.1 Introduction: The Importance of Mulberry in Sericulture

Mulberry (*Morus* sp.) is a perennial deciduous tree, and reported to have originated in the lower Himalayan belt of Indo–China region, the primary centre of plant origin (Vavilov 1951) and migrated to various places. The genus *Morus* is now available in "China–Japan" region, including China, Korea and Japan, in the warm and moist climatic zones between 50°N latitude and 10°E longitude. Linnaeus (1953) introduced plant classification based on the presence or absence of flowers, nature of flowers, number of sex organs, etc. The genus has been classified into many species and sub-species based on morphological and floral characters (Linneaus 1753; Koidzumi 1917; Hotta 1954; Katsumata 1972a, b). Kadambi (1949) reported important types of mulberry grown in India for rearing of silkworm, namely *Morus alba* var. *multicaulis* as bush crop and *Morus alba* var. *atropurpurea* as trees. Further, several mulberry varieties were introduced into India from Europe, China and probably from Japan and Philippines as well. A good number of wild mulberries were collected through survey and exploration from different Indian states (Dandin et al. 1993, 1995; Parkinson 1923; Ravindran et al. 1997, 1999; Sanjappa 1989) and presently maintained in germplasm bank, Central Sericultural Germplasm Resources Centre, Hosur, India.

Biodiversity means the variability among living organisms available in ecosystems. The ecological complexes include diversity within species between species and of ecosystems. So, the biodiversity represents all living forms. India is one of the mega-biodiversity centres in the world and has two of the world's 18 "biodiversity hotspots" located in the Western Ghats and in the Eastern Himalayas. The country is estimated to have 49,219 (12.5%) plant species and 81,251 (6.6%) animal species representing the total world diversity. The centre of diversity is the geographic region having maximum available resources. The primary centre of diversity represents the original forms of living materials, and secondary centre developed by spreading the materials from primary centre. The plant diversity is an important component and has a crucial role to strengthen the food and nutritional security for ever increasing population. Over 50% of the world's plant species and 42% of all terrestrial vertebrate species are endemic to the 34 biodiversity hotspots. Among the 34 hotspots of the world, two have been identified in India—the Eastern Himalayas and the Western Ghats (Kanjilal et al. 1940). They are particularly rich in floral wealth and endemism, not only in flowering plants but also reptiles, amphibians, butterflies and mammals.

The leaf yield of mulberry variety has gone up from 30 MT/ha/yr in 1960 to 70 MT/ha/year in 2016. Researchers continue to evaluate germplasm for new crop development and for value addition. Benefits are also derived by conserving mulberry biodiversity of indigenous plant species. In addition to commonly occurring species, threatened, rare or endangered species are made available for both studying and small habitat restoration programmes. Three tools that improve the use of plant genetic resources are: better knowledge of the characteristics of germplasm accessions; making this knowledge more accessible to germplasm users; and germplasm enhancement programmes which integrate new genes and trait into genetic backgrounds more frequently used by plant breeders.

Sericulture in India is practised in diverse climatic conditions that require evolution of large number of region-specific mulberry varieties. To develop these varieties, germplasm from diverse climatic conditions such as temperate, sub-tropical and tropical has been maintained and utilized.

4.2 Classification of Existing Mulberry Germplasm

The contribution of gene banks to crop improvement is direct as it provides the plant breeders with regular supply of new breeding materials. When searching for a desirable trait, breeders usually turn first to their working genetic material collections due to their superiority of agronomic traits and fulfil farmer's needs (Tikader et al. 2014). The genetic resource centres undertake the responsibility of the active collections, base collections and duplicate base collections.

4.3 Characterization of Mulberry Genetic Resources

The mulberry germplasm is maintained through survey and exploration, collection and conserved in nurseries. For better utilization, the collected resources need characterization, evaluation and documentation in a systematic manner. In fact, characterization is a part of primary evaluation, and visual recording of highly heritable phenotypic characters. The characterization includes study of morphology, anatomy, reproductive, biochemical following standard methods (Tikader et al. 2014). For characterization purpose, small tree plantation, 1.2–1.3 m tall, maintaining at a spacing of 2.4–3.0 m is used. The data generated in this way provide the basic information of a plant and its initial performance for future crop improvement programme (Dandin and Jolly 1986). The underground and above-ground data for propagation parameters are also recorded for understanding the rooting and survival behaviour of different mulberry germplasm (Chloupek and Rod 1992). The disease data are recorded for diseases that are prevalent in the field. The descriptors are listed for recording mulberry germplasm characterization data (Table 4.1, Fig. 4.1).

The descriptor selected for evaluation and characterization are mentioned in catalogues (Dandin et al. 1987a; Dandin and Kumar 1989; Machii et al. 1997, 2001; Thangavelu et al. 1997, 2000; Tikader et al. 2006). The characters were recorded on morphology (Tikader et al. 2000b, 2002), reproductive biology (Machii et al. 1999b; Tikader et al. 2000a), anatomy, and growth traits, rooting percentage and breeding aspects. The data from each accession were recorded from three plants after 90 days of pruning. The following parameters have been used for generating the data.

Table 4.1 List of descriptors for characterization of mulberry germplasm

Morphology	Reproductive	Propagation parameters
Branching nature	Sex	Survival %
Curve/straight of branch	Male inflorescence length (cm)	Leaves/sapling
Colour of young shoot	Female inflorescence length (cm)	Leaf length (cm)
Colour of mature shoot	Bisexual inflorescence length.(cm)	Leaf width (cm)
Phyllotaxy	Male inflorescence width (cm)	Fresh leaf weight/sapling (g)
Stipule nature	Female inflorescence width (cm)	Dry leaf weight/sapling (g)
Stipule duration	Bisexual inflorescence width (cm)	Shoot length (cm)
Lenticels density/cm^2	Flowers/catkin (male)	Fresh shoot weight/sapling (g)
Leaf lobation type	Flowers/catkin (female)	Dry shoots weight/sapling (g)
Leaf lobation number	Flowers/catkin (bisexual)	Shoot diameter (cm)
Leaf nature	Male inflorescence ped. length (cm)	Total fresh biomass/sapling (g)
Leaf colour	Female inflorescence ped. length (cm)	Total dry biomass/sapling (g)
Leaf surface	Bisexual inflores. ped. length (cm)	Roots/sapling
Leaf texture	Stamen length (mm)	Fresh root weight/sapling (g)
Leaf apex	Anther length (mm)	Dry root weight/sapling (g)
Leaf margin	Pollen diameter (μm)	Longest root length/sapling (cm)
Leaf base	Pollen viability (%)	Root volume/sapling (ml)
Leaf shape	Style length (mm)	
Leaf length (cm)	Stigma length (mm)	
Leaf width (cm)	Stigma type	
Petiole length (cm)	Stigma nature	Diseases
Petiole width (cm)	Fruit length (cm)	
	Fruit width (cm)	Powdery mildew
Anatomy	Fruit weight (g)	Leaf spot
	Fruit colour	Bacterial blight
Stomata size (sq.μm)	Fruit taste	
Stomata size (sq.μm)		
Idioblast length (μm)		
Idioblast width (μm)	Growth Traits	Biochemical parameters
Idioblast frequency (no./sq.mm)		
Palisade thickness (μm)	Branches/plant	Chlorophyll a (mg/g fresh wt.)
Spongy thickness (μm)	Longest shoot length (cm)	Chlorophyll b (mg/g fresh wt.)
Palisade-Spongy ratio (P: S)	Total shoot length (cm)	Total chlorophyll (mg/g fr. wt.)
Upper cuticle thickness (μm)	Internodal distance (cm)	Chlorophyll a/b
Lower cuticle thickness (μm)	Hundred leaf weight (g)	Protein soluble (fresh wt%)
Upper epidermis thickness (μm)	Leaf area (cm^2)	Protein soluble (dry wt%)
Lower epidermis thickness (μm)	Lamina weight (g)	Carbohydrate soluble (fr. wt%)
Leaf thickness (μm)	Laminar index (%)	Carbohydrate soluble (dry wt%)
Chloroplast/stomata	Leaf petiole ratio by length	

(continued)

Table 4.1 (continued)

Morphology	Reproductive	Propagation parameters
	Leaf petiole ratio by weight	
	Leaf yield/plant (kg)	
	Leaf shoot ratio (L: S)	
	Moisture content (%)	
	Moisture after 6 h (%)	
	Moisture retention (%)	

Fig. 4.1 Mulberry being maintained in the germplasm

4.4 Morphological Variations in Mulberry

After pruning, the morphological data both qualitative (non-parametric) and quantitative (parametric) are recorded visually (ordinal scale), and measurements are made (in metric scale) from three plants per accession (Minamizawa 1963). Altogether 22 morphological characters comprising 12 parametric and 10 non-parametric are recorded. The leaf showed great variability within and between species (Fig. 4.2) The descriptor list is presented in Table 4.2.

4.4.1 Anatomy

Three months after pruning, fully expanded leaves (5–7th position) in descending order are collected in between 9–10 am. A small rectangular piece was taken out

Fig. 4.2 Variability in the leaf of mulberry (*Morus* spp.)

Table 4.2 List of descriptors of mulberry morphological characters

Descriptor	Growth range and time of data recording	Classification	Method
Growth nature	Data recorded at full growth stage after 90 days of pruning of plants	1.1 Erect 1.2 Spreading 1.3 Drooping	Observe visually the angle between the foliage branch and main trunk
Branching nature	Data recorded at full growth stage after 90 days of pruning of plants	2.1 Straight 2.2 Slightly curve 2.3 Curve	Observe visually the straight or curviness of the branch and record
Young shoot colour	Data recorded after 30 days of pruning of 1/3rd of young shoots from top	3.1 Green 3.2 Greenish purple 3.3 Purple	Visually observe the young shoot colour and record
Mature shoot colour	Data recorded after 90 days of pruning in the lower portion of the longest shoot	4.1 Brown 4.2 Grey 4.3 Greyish green 4.4 Greyish brown 4.5 Purplish brown 4.6 Greenish brown	Visually observe the mature shoot colour and record
Phyllotaxy	Data recorded after 70 days of pruning in the middle portion of the longest shoot	5.1Distichous (1/2) 5.2 Tristichous (1/3) 5.3 Pentastichious (2/5)	First and third leaf in the same direction on the stem First and fourth leaf in same direction on the stem First and sixth leaf in same direction on the stem
Lenticels density /sq.cm	Data recorded after 90 days of pruning in the middle portion of the longest shoot	6.1 Sparse (<5) 6.2 Medium (>5–10) 6.3 Dense (>10)	Put one sq.cm cut film role around the stem surface and count the number of lenticels in that area
Lenticels shape	Data recorded after 90 days of pruning in the middle portion of the longest shoot	7.1 Round 7.2 Elliptical 7.3 Oval	Observe mature lenticels from the middle portion of the longest branch and record
Bud shape	Data recorded after 90 days of pruning of the longest shoot	8.1 Round 8.2 Acute triangle 8.3 Long triangle 8.4 Spindle shape	Buds are circular shape Bud tip is acute Bud tip is elongated Bud base is tapering
Leaf apex	Data recorded after 60 days of pruning in the middle portion of longest shoot. Observe the leaf bounded by upper 25% of the leaf margin	9.1 Acute 9.2 Acuminate 9.3 Caudate 9.4 Obtuse	The tip portion of leaf forms an angleless than 90° Leaf tip slightly extended Leaf extended and forms a tail-like structure Straight to convex margin forming an angle more than 90°
Leaf margin	Data recorded after 60 days of pruning in the middle portion of longest shoot.	10.1 Crenate 10.2 Serrate 10.3 Dentate 10.4 Repand	Crenations are smoothly rounded without a pointed apex Serrations are pointed forming oblique angle

(continued)

Table 4.2 (continued)

Descriptor	Growth range and time of data recording	Classification	Method
			Dentations are pointed perpendicular to the axis Margin forming a smooth line or arc without projections
Leaf base	Data recorded after 60 days of pruning in the middle portion of longest shoot. Observe the leaf bounded by upper 25% of the leaf margin	11.1 Truncate 11.2 Cordate 11.3 Lobate	Terminate abruptly and the margin perpendicular to mid vein Leaf base-like heart shaped Leaf base with two large pointed lobes
Leaf surface	Data recorded after 60 days of pruning in the middle portion of longest shoot	12.1 Smooth 12.2 Slightly rough 12.3 Rough 12.4 Hairy (Pubescent)	Rub the upper portion of the leaf and feel the leaf surface to record the classification as per feeling
Leaf texture	Data recorded after 60 days of pruning in the middle portion of longest shoot	13.1 Membranous 13.2 Charatacious 13.3 Coriaceous	The leaves are thin and transparent like thin membrane The leaves are opaque and like a writing paper The leaves are thick, leathery and stiff
Stipule nature	Data recorded after 60 days of pruning in the middle portion of longest shoot	14.1 Bud scale 14.2 Free-lateral 14.3 Foliaceous	Leaf stipules are very short, slender and pointed Stipules are freely hangs from both sides of node Stipules are broad and look like miniature leaves
Leaf texture	Data recorded after 60 days of pruning in the middle portion of longest shoot	13.1 Membranous 13.2 Charatacious 13.3 Coriaceous	The leaves are thin and transparent like thin membrane The leaves are opaque and like a writing paper The leaves are thick, leathery and stiff
Stipule nature	Data recorded after 60 days of pruning in the middle portion of longest shoot	14.1 Bud scale 14.2 Free-lateral 14.3 Foliaceous	Leaf stipules are very short, slender and pointed Stipules are freely hangs from both sides of node Stipules are broad and look like miniature leaves
Stipule duration	Data recorded after 60 days of pruning in the middle portion of longest shoot	15.1 Cauducous 15.2 Persistent	The stipules fall off with maturity of leaves. The stipules stay for longer period and attached with the leaf

(continued)

Table 4.2 (continued)

Descriptor	Growth range and time of data recording	Classification	Method
Shape of leaf scar	Data recorded after 90 days of pruning in the middle portion of the longest shoot after leaf harvest	16.1 Circular 16.2 Elliptical 16.3 Triangular	Observe leaf scar after harvest of leaves from mature lower portion of the stem and record
Leaf nature	Data recorded after 60 days of pruning in the middle portion of longest shoot	17.1 Homophyllous Lobed Unlobed 17.2 Heterophyllous	All the leaves are similar in nature in a plant; when all lobed or unlobed The leaves are mixed type either lobed and unlobed
Leaf lobation number	Data recorded after 60 days of pruning in the middle portion of longest shoot	18.1 0-lobed 18.2 1-2 lobed 18.3 3-4 lobed 18.4 Multilobed	The lobe number is counted on individual and classified
Leaf colour	Data recorded after 60 days of pruning in the middle portion of longest shoot	20.1 Light green 20.2 Green 20.3 Dark green	Natural colour of leaves is recorded visually using the standard colour chart and classified
Leaf length (cm)	Data recorded after 70 days of pruning from 7th-9th position of leaf in longest shoot excluding petiole	22.1 Narrow (<10) 22.2 Medium (10–15) 22.3 Broad (>15)	The lamina width is recorded from the widest point from both sides of leaf margin
Leaf shape (cm)	The length and breadth of the leaves are recorded from 7th-9th position of fully expanded leaf in longest shoot	23.1 Lanceolate 23.2 Narrow ovate 23.3 Ovate 23.4 Wide ovate 23.5 Cordate	Leaf length/width ratio = 1:3 Leaf length/width ratio = 2:1 Leaf length/width ratio = 1.5:1 Leaf length/width ratio = 1.2:1 Leaf length/width ratio = <1:1
Petiole length (cm)	Data recorded after 70 days of pruning from 7th–9th position of leaf in longest shoot	24.1 Small (<2) 24.2 Medium (2-4) 24.3 Large (>4)	The petiole portion is detached from the base of leaf and measure the length
Petiole groove	Data recorded after 70 days of pruning from 7th–9th position of leaf in longest shoot	26.1 Present 26.2 Absent	Observe the lower portion of the petiole and record its presence or absence

from the middle portion of the leaf blade—avoiding vein and veinlets (Metcalf and Chalk 1979), preserved in FAA solution (formalin 5 ml; glacial acetic acid 5 ml; 70% ethanol 90 ml). Stomatal studies are conducted by applying a thin layer of Wimbley's quick fix on the abaxial (lower) and adaxial (upper) leaf surface dried and peeled out the thin layer with the impression of stomata. This layer is later fixed on a slide, covered with cover slip and observed under Leica Leitz, DMRB Wetzlar microscope. The size and number of the stomata and number of idioblasts are counted per unit area, and the frequency per mm^2 is calculated. For observing the cuticular, epidermal, palisade, spongy mesophyll, leaf thickness and idioblast length, width, thin hand-cut sections are made and observed under microscope after staining with 1% safranin and mounted on 50% glycerin. Chloroplast number per stomata is counted from 3rd–5th leaf of a shoot in descending order from the fully open leaf at the top. Ventral surface peelings of the freshly collected leaf samples were used after staining and mounted in potassium iodide–iodine (2% I + KI) solution following the method of Chaudhuri and Barrow (1975). In each case, nine microscopic fields are observed for data recording. A total of 11 parameters are recorded in leaf anatomical study. The data are analyzed for general statistics, analysis of variance, correlation, genetic study and divergence analysis, and the result observed in different mulberry species are tabulated and presented (Table 4.3).

4.4.2 Reproductive Biology

Sex expression is studied in normal flowering season (spring) as well as after pruning of plants from January through March–May. Sex expression is recorded as male, female (unisexual or dioecious group), bisexual (male and female sex organs in same flowers, i.e. complete flower), monoecious (male and female flower on separate catkin or on same catkin in one plant) (Table 4.4). The stamen length and anther length are observed under microscope and measured in millimetre. In each accession, nine observations are made and 27 observations for three replications. To estimate pollen viability, male catkin from field grown plants is dehisced over a glass slide stained in a drop of 2% acetocarmine and covered with a cover slip. After 5 min, the percentage of fully stained grain is recorded (Hussain and Williams 1997). Other flower characters are recorded from catkin itself. The length of style and stigma is observed under microscope and measured in millimetre. Sorosis characters, i.e. fruit length, width, weight, taste and colour are recorded after ripening of fruits (Jolly et al. 1987; Tikader et al. 1995). Fruits showed vide variability among the species and within species (Fig. 4.3).

Table 4.3 List of anatomical descriptors observed for data recording

Descriptor	unit	Classification	Method	Descriptor
Stomata size (sq.μm)	1.1 Small (<200) 1.2 Medium (200-400) 1.3 Large (>400)	Length and breadth of stomata is recorded using ocular micrometre under oil immersion and converted data into μm.	Stomata size (sq.μm)	1.4 Small (<200) 1.5 Medium (200–400) 1.6 Large (>400)
Stomata frequency (no/sq.mm)	2.1 Low (<400) 2.2 Medium (400–800) 2.3 High (>800)	The observation is recorded under low magnification (10 × 45). Total number stomata per unit area is counted and converted into one sq.mm.	Stomata size (sq.μm)	2.1 Low (<400) 2.2 Medium (400–800) 2.3 High (>800)
Idioblast frequency (no/sq.mm)	3.1 Small (<15) 3.2 Medium (15–30) 3.3 High (>30)	The observation is recorded under low magnification (10 × 10). Total number idioblast per unit area is counted and converted into one sq.mm.	Stomata size (sq.μm)	3.1 Small (<15) 3.2 Medium (15–30) 3.3 High (>30)
Idioblast projection length (μm)	4.1 Less (<5) 4.2 Medium (5–15) 4.3 Large (>15)	The observation is recorded under ocular magnification (10 × 45). Total length of idioblast projection protrudes from upper epidermis is recorded and measured.	Stomata size (sq.μm)	4.1 Less (<5) 4.2 Medium (5–15) 4.3 Large (>15)
Palisade thickness (μm)	5.1 Thin (<50) 5.2 Medium (50–75) 5.3 Thick (>75)	The observation is recorded under ocular magnification (10 × 45) and the data converted into μm.	Stomata size (sq.μm)	5.1 Thin (<50) 5.2 Medium (50–75) 5.3 Thick (>75)
Spongy thickness (μm)	6.1 Thin (<50) 6.2 Medium (50–75) 6.3 Thick (>75)	The observation is recorded under ocular magnification (10 × 45) and the data converted into μm.	Stomata size (sq.μm)	6.1 Thin (<50) 6.2 Medium (50–75) 6.3 Thick (>75)
Upper cuticular thickness (μm)	7.1 Thin (<4) 7.2 Medium (4–8) 7.3 Thick (>8)	The observation is recorded under ocular magnification (10 × 45) and the data converted into μm.	Upper cuticular thickness (μm)	7.1 Thin (<4) 7.2 Medium (4–8) 7.3 Thick (>8)
Lower cuticular thickness (μm)	8.1 Thin (<4) 8.2 Medium (>4–8) 8.3 Thick (>8)	The observation is recorded under ocular magnification (10 × 45) and the data converted into μm.	Lower cuticular thickness (μm)	8.1 Thin (<4) 8.2 Medium (>4–8) 8.3 Thick (>8)
Upper epidermal thickness (μm)	9.1 Thin (<20) 9.2 Medium (20–40) 9.3 Thick (>40)	The observation is recorded under ocular magnification (10 × 45) and the data converted into μm.	Upper epidermal thickness (μm)	9.1 Thin (<20) 9.2 Medium (20–40) 9.3 Thick (>40)
Lower epidermal	10.1 Thin (<5) 10.2 Medium (5–10)	The observation is recorded under ocular magnification	Lower epidermal	10.1 Thin (<5) 10.2 Medium (5–10)

(continued)

Table 4.3 (continued)

Descriptor	unit	Classification	Method	Descriptor
thickness (μm)	10.3 Thick (>10)	(10 × 45) and the data converted into μm.	thickness (μm)	10.3 Thick (>10)
Total leaf thickness (μm)	11.1 Thin (<150) 11.2 Medium (150–175) 11.3 Thick (>175)	Add the each layer of thickness from upper cuticular to lower cuticular to obtain the total leaf thickness	Total leaf thickness (μm)	
Chloroplast/ stomata	12.1 Less (<12) 12.2 Medium (12–16) 12.3 High (>16)	To count chloroplast/guard cells, the lower leaf surface is peeled and observed under microscope with iodine solution and counted the number in both guard cells.	Chloroplast / stomata	
Cystolith length (μm)	13.1 Small (<50) 13.2 Medium (50–75) 13.3 Large (>75)	The idioblast used for measuring projection length can be used for measurement of cystolith length in ocular magnification (10x45) and convert into μm.	Cystolith length (μm)	13.1 Small (<50) 13.2 Medium (50–75) 13.3 Large (>75)
Idioblast type	15.1 Blunt 15.2 Bleak	Idioblast type is recorded the type of idioblast projection from upper epidermal cells of leaf surface and grouped.	Idioblast type	15.1 Blunt 15.2 Bleak

4.4.3 Growth Parameters:

Growth attributes such as height, total shoot length, number of primary branches of pruning are recorded following standard procedures (Machii et al. 1997). For moisture content, 25 leaves from 5–9th position in descending order on a stem are collected between 9–10 am in polythene bags, and fresh weight is recorded. Moisture loss is recorded at room temperature after 6 h of storage to know the moisture retention up to 6 h storage. Leaves are further dried at 80 °C for 48 h following standard procedures, and weight of oven-dried samples is recorded to calculate the moisture percentage (Dandin and Kumar 1989; Machii et al. 1997; Tikader et al. 2014; Vijayan et al. 1997 a, b). Moisture content (MC) and moisture retention capacity (MRC) are calculated using the standard formula. The descriptor list is presented in Table 4.5.

Table 4.4 List of descriptors of mulberry reproductive characters

Descriptor	Growth range and time of data recording	Classification	Method
Sex	Data recorded during favourable season before pruning of plants (February-April) and unfavourable season October-November)	1.1 Dioecious (male) 1.2 Dioecious (female) 1.3 Monoecious 1.3.1 Male and bisexual 1.3.2 Female and bisexual 1.3.3 Male and female 1.3.4 Male, female and bisexual	The plant bears only male catkin The plant bears only female catkin The plant bears male, female and bisexual catkin in same plant The plant bears predominantly male catkin with few bisexual catkin The plant bears predominantly female catkin with bisexual catkin The plant bears male and female catkin with bisexual catkin The plant bears male, female and bisexual catkin
Inflorescence per meter length of branch	Data recorded from longest branch of the plant in main flowering season	2.1 Low (<20) 2.2 Medium (20-40) 2.3 Profuse (>40)	The inflorescences is counted in one meter branch and grouped as per classification
Inflorescence length (cm)	Data recorded from longest branch of the plant in main flowering season and selecting mature inflorescence	3.1 Small (<2) 3.2 Medium (2–4) 3.3 Large (>4)	The full bloomed inflorescence of male (before anthesis), female (receptive stage) and bisexual catkins are recorded as observed and grouped as per classification
Inflorescence width (cm)	Data recorded from longest branch of the plant in main flowering season and selecting mature inflorescence	4.1 Small 4.2 Medium 4.3 Large	The full bloomed inflorescence of male (before anthesis), female (receptive stage) and bisexual catkins are recorded as observed and grouped as per classification
Inflorescence peduncle length and width (cm)	Data recorded from longest branch of the plant in main flowering season and selecting mature inflorescence	5.1 Small 5.2 Medium 5.3 Large	The full bloomed inflorescence of male (before anthesis), female (receptive stage) and bisexual catkins are recorded as observed and grouped as per classification
Flowers per Inflorescence	Data recorded from longest branch of plant in main flowering season and selecting mature inflorescence	6.1 Low (<25) 6.2 Medium (25–50) 6.3 Profuse (>50)	The inflorescence used for measurement can be used for counting flowers per catkin. The male, female and bisexual flowers can be counted separately to group
Stamen length (mm)	Data recorded from longest branch of the plant in main flowering season and selecting mature inflorescence	7.1 Small (<4) 7.2 Medium (4–6) 7.3 Large (>6)	Fully bloomed male catkin collected before anthesis and observed under microscope. Stamen length is recorded under ocular micrometre and converted to mm.
Anther length (mm)	Data recorded from longest branch of the plant in main flowering	8.1 Small (<0.5) 8.2 Medium (0.5–1.0)	Fully bloomed male catkin collected before anthesis and observed under

(continued)

Table 4.4 (continued)

Descriptor	Growth range and time of data recording	Classification	Method
	season and selecting mature inflorescence	8.3 Large (>1)	microscope. Anther length is recorded lengthwise
Pollen diameter (μm)	Data recorded from longest branch of the plant in main flowering season and selecting mature inflorescence	9.1 Small (<10) 9.2 Medium (15–30) 9.3 Large (>30)	Fully stained pollen grains are measured in ocular micrometre (10 x 45). In case higher ploidy, macro and micro pollen is recorded separately
Pollen viability (%)	Data recorded from longest branch of the plant in main flowering season and selecting mature inflorescence	10.1 Low (<50) 10.2 Medium (50–75) 10.3 High (>75)	Fully stained pollen grains are observed for pollen viability and calculated % of viability as per formula
Germ pores per pollen grain	Data recorded from longest branch of the plant in main flowering season and selecting mature inflorescence	11.1 Monozoroporate 11.2 Dizonoporate 11.3 Trizonoporate 11.4 Tetrazonoporate 11.5 Pentazonoporate	Pollen grains having single pore to five pore single grain. Data are recorded and classified in different groups.
Style length (mm)	Data recorded from longest branch of the plant in main flowering season and selecting mature inflorescence	12.1 Short (<0.5) 12.2 Medium (0.5–0.75) 12.3 Long (>0.75)	Receptive stigma is observed before ovary is fertilized. Data recorded from tip of ovary to base of stigma under ocular micrometre and grouped
Stigma length (mm)	Data recorded from longest branch of the plant in main flowering season and selecting mature inflorescence	13.1 Short (<2) 13.2 Medium (2–4) 13.3 Long (>4)	Receptive stigma is observed. Both the bi-fid stigma is observed separately and added to get the total length. Data ocular micrometre and grouped
Stigma type	Data recorded from longest branch of the plant in main flowering season and selecting mature inflorescence	14.1 Erect 14.2 Spreading 14.3 Divericate 14.4 Twisted	Receptive stigma is erect and straight Receptive stigma is spreading Receptive stigma is bent and curvet Receptive stigma is twisted with each other
Stigma nature	Data recorded from longest branch of the plant in main flowering season and selecting mature inflorescence	15.1 Pubescent 15.2 Papillate	Stigma surface is covered with long hairs Stigma surface is covered with no hairs or minute hairs

(continued)

Table 4.4 (continued)

Descriptor	Growth range and time of data recording	Classification	Method
Fruit length (cm)	Data recorded from longest branch of the plant in main flowering season and selecting mature inflorescence	16.1 Small (<2) 16.2 Medium (2–4) 16.3 Long (>4 – <8) 16.4 Very long (>8)	Data are recorded of mature ripened fruit. Fruit length is recorded including peduncle
Fruit breadth (cm)	Data recorded from longest branch of the plant in main flowering season and selecting mature inflorescence	17.1 Small (<1) 17.2 Medium (1–1.5) 17.3 Long (>1.5)	Data are recorded of mature ripened fruit. Fruit breadth is recorded from the same fruit used for length
Fruit weight (g)	Data recorded from longest branch of the plant in main flowering season and selecting mature inflorescence	18.1 Low (<2) 18.2 Medium (2–4) 18.3 High (>4)	Data are recorded of mature ripened fruit. Fruit weight is recorded from the same fruit used for length and breadth
Fruit colour	Data recorded from longest branch of the plant in main flowering season and selecting mature inflorescence	19.1 Black 19.2 White 19.3 Purple 19.4 Greenish white	Data are recorded of mature ripened fruit and grouped on visual observation. Fruit colour is recorded from same fruit used for other measurements
Fruit taste	Data recorded from longest branch of the plant in main flowering season and selecting mature inflorescence	20.1 Sour 20.2 Less sweet 20.3 Sweet	Data are recorded of mature ripened fruit and grouped on visual observation. Fruit taste is recorded from same fruit used for other measurements
Seed colour	Data recorded from longest branch of the plant in main flowering season and selecting mature inflorescence	21.1 Light yellow 21.2 Light brown 21.3 Yellowish brown 21.4 Dark brown 21.5 Blackish brown	Data are recorded of mature ripened fruit after seed is harvested under stereomicroscope and grouped on visual observation.
Seed weight (mg/100 seed)	Data recorded from longest branch of the plant in main flowering season and selecting mature inflorescence	22.1 Low (<100) 22.2 Medium (100–200) 22.3 High (>200)	Data are recorded of mature ripened fruit after seed is harvested and dried under stereomicroscope and grouped as per weight.
Seed set (%)	Data recorded from longest branch of the plant in main flowering season and selecting mature inflorescence	23.1 Low (<50) 23.2 Medium (50–75) 23.3 High (>75)	Data are recorded of mature ripened fruit after seed is harvested. It is the ration between total healthy seeds per fruit to total number of achene's per fruit

Fig. 4.3 Variation in the fruits in different mulberry species

4.4.4 Biochemical Parameters

The biochemical analysis of mulberry leaf samples is done for chlorophyll, protein and carbohydrate, etc. for quality assessment as growth and development of silkworm is greatly influenced by the quality of mulberry leaf. The chlorophyll molecule contains porphyrin (tetrapyrrole) with a chelated magnesium atom at the centre and long chain hydrocarbon (phytyl) attached through a carboxylic group. The chlorophyll is extracted with dimethyl sulfoxide (DMSO) as per the method of Hiscox and Israelstam. The absorbance at 663 and 645 nm is recorded in spectrophotometer. The quantity of chrlorophyll a, b and total chlorophyll are estimated. In higher plants, proteins mainly form the enzymes (functional), membrane proteins (structural) and stored proteins (nutritional). The quantity and quality of proteins determine the quality of leaves. The soluble protein (fresh and dry) is estimated by following Folin–Ciocalteau reagent method. Carbohydrates are the important component, which exists as free sugar and polysaccharides. The basic unit of carbohydrates is the monosaccharides. The carbohydrate content can be measured by hydrolysis into simple sugar by acid hydrolysis, and the resultant monosaccharides are estimated. The estimation is made following Anthrone method and absorbance read at 620 nm. The descriptor list is presented in Table 4.6.

Table 4.5 List of descriptor of mulberry growth and yield characters

Descriptor	Classification	Method
Number of branches	1.1 Low (<20) 1.2 Medium (20–40) 1.3 High (>40)	Data are recorded after 70th day of pruning by counting total number of branch per plant
Length of longest shoot (cm)	2.1 Low (<75) 2.2 Medium (75–150) 2.3 High (>150)	Data are recorded after 70th day of pruning by measuring the longest branch of the plant
Total shoot length (cm)	3.1 Low (1500) 3.2 Medium (1500–3000) 3.3 High (>3000)	Data are recorded after 70th day of pruning by measuring the length of all shoots of individual plant
Internodal distance (cm)	4.1 Less (<3) 4.2 Medium (3–6) 4.3 High (>6)	Data are recorded after 70th day of pruning by counting the total number of nodes of longest shoot divided by total shoot length
Hundred leaves weight (g)	5.1 Low (<300) 5.2 Medium (300–600) 5.3 High (>600)	Data are recorded after 70th day of pruning by weighing equal number of base, middle and top leaves of a branch excluding tender leaves of longest shoot
Lamina weight (g)	6.1 Low (>4) 6.2 Medium (4–10) 6.3 High (>10)	Data are recorded after 70th day of pruning by weighing equal number of base, middle and top leaves of a branch excluding tender leaves and grouped
Petiole weight (g)	7.1 Low (>0.05) 7.2 Medium (0.05–1.5) 7.3 High (>1.5)	Data are recorded after 70th day of pruning by weighing equal number of base, middle and top leaves of a branch excluding tender leaves and grouped
Laminar index (%)	8.1 Min (75.0) 8.2 Max (90.0)	It is calculated using the formula = Lamina weight/Leaf weight x 100
Lamina petiole ratio by length	9.1 Low (<2.5) 9.2 Medium (2.5–5.0) 9.3 High (>5)	Data are recorded after 70th day of pruning by measuring length of fully-grown leaves from 7–9th position on longest shoot. It is the ratio of lamina length/petiole length
Lamina petiole ratio by weight	10.1 Low (<5) 10.2 Medium (5–10) 10.3 High (>10)	Data are recorded after 70th day of pruning by measuring length of fully-grown leaves from 7–9th position on longest shoot. It is the ratio of lamina weight/ petiole weight
Leaf area (sq. cm)	11.1 Small (<200) 11.2 Medium (200–400) 11.3 High (>400)	Data are recorded after 70th day of pruning by measuring equal number of base, middle and top leaves of a branch excluding tender leaves of longest shoot. The data can be recorded with leaf area metre
Leaf yield/ plant (Kg)	12.1 Low (<2) 12.2 Medium (2–4) 12.3 High (>4)	Data are recorded after 75–90th day of pruning by harvesting all leaves of a plant and weighed in kilogram
Leaf shoot ratio by weight	13.1 Low (<1) 13.2 Medium (1–1.5) 13.3 High (>1.5)	Data are recorded after 75–90th day of pruning by harvesting all leaves of a plant and weighed in kilogram and also recorded shoot weight per plant. It is the ratio of leaf weight/shoot weight
Moisture content (%)	14.1 Low (<60) 14.2 Medium (60–70) 14.3 High (>70)	Data are recorded after 70th day of pruning. Fresh leaves of tender, medium and mature are collected and weight is recorded. The leaves are dried in hot air oven

(continued)

Table 4.5 (continued)

Descriptor	Classification	Method
		at 60 °C for 72 h and dry weight recorded. Moisture % is calculated using the formula
Moisture after 6 h (%)	15.1 Low (<60) 15.2 Medium (60–70) 15.3 High (>70)	Data are recorded after 70th day of pruning. Fresh leaves of tender, medium and mature are collected and weight is recorded after 6 h in room temperature. Moisture % after 6 h is calculated using the formula
Moisture retention capacity (%)	16.1 Low (<60) 16.2 Medium (60–70) 16.3 High (>70)	Data are recorded after 70th day of pruning. It is the capacity of leaves to retain moisture after complete dry and calculated using the formula

Table 4.6 List of descriptor of mulberry biochemical characters

Descriptor	Classification	Method
Chlorophyll-a (mg/gm)	1.1 Low (<1.5) 1.2 Medium (1.5–3) 1.3 High (>3)	Data are recorded after 60 days of pruning by recording fresh weight of 7–9th position of leaf from longest shoot/plant
Chlorophyll-b (mg/gm)	2.1 Low (<0.5) 2.2 Medium (0.5–1.0) 2.3 High (>1)	Data are recorded after 60 days of pruning by recording fresh weight of 7–9th position of leaf from longest shoot/plant
Total chlorophyll (mg/gm)	3.1 Low (<2) 3.2 Medium (2–4) 3.3 High (>4)	Data are recorded after 60 days of pruning by recording fresh weight of 7–9th position of leaf from longest shoot/plant
Total carbohydrate (%)	4.1 Low (<10) 4.2 Medium (10–15) 4.3 High (>15)	Data are recorded after 60 days of pruning by recording from 7–9th position of leaf on dry weight basis from longest shoot/plant
Total protein (%)	5.1 Low (<10) 5.2 Medium (10–20) 5.3 High (>20)	Data are recorded after 60 days of pruning by recording from 7–9th position of leaf on dry weight basis from longest shoot/plant
Fresh leaf weight (g)	6.1 Low (>3) 6.2 Medium (3–6) 6.3 High (>6)	Data are recorded after 60 days of pruning by weighing equal number of base, middle and top leaves of a branch excluding tender leaves and grouped

4.4.5 Disease Parameters

Occurrence of disease in mulberry plant is regularly observed in the field gene bank. Leaf spot, powdery mildew and rust are the most common diseases of mulberry in India. Incidence of foliar fungal disease in mulberry varies from place to place depending upon the climatic conditions of the area, pathogen load and susceptibility of the genotype. These diseases cause considerable damage to mulberry leaf quality and yield, thus affecting the silkworm rearing and cocoon production. The mulberry accessions were screened to ascertain the infection status/ susceptibility to these diseases under field conditions in field gene bank.

The infected leaves were observed per plant covering all branches, covering tender, medium and mature leaves (Tikader et al. 2014).

4.4.6 Fungal Diseases in Mulberry

There are many diseases, which are caused by fungi in mulberry. The intensity and severity vary based on climatic factors, geographical distribution of mulberry and other related factors. The most common foliar fungal diseases are leaf spot (*Cerocospora moricola* Cooke), powdery mildew (*Phyllactina corylea* (Pers.) Karst) and leaf rust (*Cerotelium fici* (Cast) Arth).

4.4.7 Bacterial Diseases in Mulberry

Bacterial diseases are found in rainy season in most of the mulberry growing areas. The bacteria can survive in high altitude where temperature ranges from 10–40 °C where optimum is 20–35 °C. The bacterial blight is a common disease in mulberry. Besides this, some minor diseases are root rot, twig blight, leaf scorch and shoot rot. The infection status or the disease index was graded for per cent of leaf area infected on 0–5 scales, and the percent disease index (PDI) was calculated. Data on disease tolerance/susceptibility of mulberry germplasm accessions were studied and presented in PDI rank as resistant (0.1–5.0%), moderately resistant (5.1–20.0%), susceptible (20.1–50.0%) and highly susceptible (51.0–100%) for different accessions. PDI = total no. of infected leaves × grade value given/total no. of leaves × maximum grade value × 100.

4.5 Evaluation of Mulberry Genetic Resources

The germplasm evaluation involves different activities and opportunities of rejuvenation of the major species at germplasm centre. The plant breeder utilizes the germplasm in developing new hybrids and breeding lines (Kulkarni et al. 2002; Pan 2003; Ono 1983). The germplasm centre is allowed to concentrate on preservation technology, genetic diversity analysis, development of core collection, distribution, database development, etc. Germplasm is evaluated for a number of desirable agronomic traits such as tolerance to drought, saline and alkaline stresses, early sprouting or winter hardiness, resistance to pests and diseases (Tikader and Kamble 2007).

4.5.1 Maintenance of Mulberry Germplasm

Mulberry germplasm is conserved and maintained in ex situ or in situ. Ex situ conservation is the maintenance of the collected materials in an identified area or plot, which includes collection from different sources, diverse geographical areas and exotics (Machii et al. 1997; Tikader et al. 2014). The germplasm is maintained for long-term utilization and should be under spaced plantation to provide scope for proper growth and reduced competition between plants. The germplasm can be maintained as bush with spacing 1.5 × 1.5 m and can be used for working germplasm. CSGRC, Hosur, maintains the mulberry germplasm in high bush/tree form with spacing 2.4 × 2.4 m for characterization and introduction of new collections time to time and also bush form in 0.9 × 0.9 m for preliminary evaluation. In temperate region, the plantation is maintained as tree with spacing 3.0 × 3.0 m. Normally, 5–10 plants in each accession are maintained in the gene bank. In Japan, the mulberry plantation is maintained following different spacing, i.e. 2.4 × 0.7 m in Fukushima, 2 × 0.7 m in Tokyo, 2 × 0.6 m in Shinjo. For bush type plantation, a stump height of 0.6–0.9 m and for tree type 1.20 m stump heights is recommended. In general, the germplasm is pruned once in a year, after establishment, during monsoon. Manures and fertilizers are applied as per the recommended dose for the particular area where the germplasm has been established.

4.5.2 Factors for Core Collections

There are five factors that determine core collections they are as follows: (i) Size of core collections: The size of the core depends on the total collections available in gene bank. It is advocated that a minimum of 10% of the total collection should be considered as core collection, which retains more than 70% of the alleles. (ii) Grouping of accessions into categories of genetic similarity or commonality. The grouping begins with taxonomy, i.e. species, sub-species, races, etc. then geography, climatic or agro-ecological regions, etc. Clustering could be done within the broad geographical group to sort accessions into clusters using hierarchical clustering methods (Ward 1963). (iii) Core collection should be the representative of entire collection, its diversity, which is decided, based on various parametric and non-parametric statistical methods. (iv) The core accessions may be multiplied, conserved, evaluated and kept ready for dispatch to the researchers on short notice. (v) The core collection may be grouped as global core collection, regional core collection and demand-based core collection.

4.5.3 Experimental Design to Manage the Germplasm

The role of experimental design is to collect the data in a systematic way and analyze the data with suitable statistical package for logical interpretation. With the progress in agricultural research technique, statistics have become an active tool of planning the experiment. Randomization of treatments has been suggested by Fisher, which enables each treatment to set such a manner to take care of soil heterogeneity. Randomization of treatments or genotypes is followed with random number available in random number table. Modern statistical methods, planning of experiments, observation, presentation and interpretation of the results are used in present day-to-day field experiment. Reliability of an experiment supports the design is logical, correct, consistent and meet the objective. While formulating or setting up an experiment, some precautions are to be taken for management of errors in the experiment which may be accidental, random, systematic and blunder. The other factors like environment, soil fertility, preparation of experimental site number of treatments, replication and replicates, size and shape of the plot, orientation of plot, guard rows also influence the results of the experiment. The record data without brassiness, application of inputs and other unforeseen factors influencing yield and yield attributes should be minimized while handling the experiment. The statistical methods will minimize the error, and with large number of observation random error can be minimized. In general, for mulberry breeding and evaluation of mulberry, germplasm follows two design of experiment, i.e. (i) randomized block design (RBD) and (ii) augmented randomized complete block design (ARCBD). Other designs are also followed based on the necessity of experiment as and when required.

4.5.4 Bioassay

The selected mulberry varieties are subjected to quality test as well as bioassay to determine the suitability of the accession/varieties. The practice is strictly followed before recommending the mulberry varieties for commercial exploitation.

4.6 Utilization of Mulberry Germplasm

Basically, mulberry is a perennial tree, which is maintained as bush, high bush and even dwarf trees depending on the cultural practices and method of silkworm rearing. The mulberry trees are grown from tropical to temperate regions in many countries of the world predominantly in Eastern, Southern and South Eastern Asia, Southern Europe, Southern North America, North Western South America and some parts of Africa.

Mulberry varieties grow in various environments from sea level to altitudes of 4000 m, and domestication of mulberry must have been started several thousands of years ago for silkworm raring (FAO 1990). Mulberry also grows from humid tropics to semi-arid lands like in the near East with 250 mm of annual rainfall and South West of the USA. The conservation of mulberry genetic resources has become very much essential to meet the desired objectives of long-term management and utilization.

The wide distribution of mulberry indicates its adaptability and genetic plasticity to various environments. Mulberry is supposed to be a native of Indo–Chinese border area and is distributed in the northern and southern lower slopes of Sub-Himalayan region up to the elevation of 3300 m (Dhar 2002). Most of the Indian varieties belong to *M. indica* and *M. laevigata, M. serrata* (Singh et al. 2006).

4.6.1 Global Position of Mulberry Utilization

Mulberry is widely distributed and used for different purposes all over the world (Datta 2000; Tzenov 2002). Though the primary uses are to feed silkworms, but it is being used for multipurpose. Due to decline of sericulture in Japan, Korea, USSR the scientists are diverting their attention for using mulberry for fodder, fruit, medicinal use, etc. and other purpose (Sanchez 2000). In Asian countries, mulberry is being utilized in different ways.

It is also available in different African countries and is being maintained and utilized for various purposes. The following countries are engaged in sericulture and other related activities (Sanchez 2000). The African countries are involved in sericulture, fruit, and agronomy and to some extent animal nutrition. In American countries, it is being maintained and utilized for multipurpose including forage, agronomy, breeding/selection and animal nutrition (Sanchez 2000). Eleven countries are involved in utilization of mulberry as forage. In European countries, it is utilized for multipurpose (Sanchez 2000).

4.7 Conservation of Mulberry Germplasm

The mulberry genetic materials are to be conserved to prevent loss of genetic resources due to destruction of habitat, human over exploitation to meet the population growth and greed, pollution and further exploitation by the stakeholders for future generation.

4.7.1 Conservation Practices

The maintenance of biodiversity depends on the nature of the material to be conserved its ecological status, life cycle, mode of reproduction and population size. The conservation strategies can be either in situ conservation of genetic resources within their ecosystem or natural habitat or ex situ conservation of components of genetic materials outside their natural habitat. The conservation of mulberry is a holistic concept that encompasses a wide spectrum of activities ranging from establishment of protected areas to building of DNA libraries. The basis of all conservation actions is sufficient knowledge of the diversity of the plant concerned and the ecosystem in which they occur. There are many conservation methods but some of the most important strategies are (i) in situ, (ii) ex situ, (iii) field gene bank, (iv) on-farm participatory, (v) in vitro and (v) cryopreservation.

4.7.2 In Situ *Conservation*

In situ conservation is defined as the conservation of plants within their ecosystem and natural habitat. It means conservation of biodiversity in nature by setting aside the natural reserve, where species are allowed to grow in their natural habitat by maintenance of ecological continuum. However, there is always a threat of a species becoming extinct or declining due to genetic drift and inbreeding, demographic and environmental variations, habitat loss, competition from invasive alien species, disease or over exploitation, human disturbances, etc. In such circumstances, the only way to conserve directly among and within species for posterity is to maintain them as ex situ collection. In situ conservation of plant genetic resources covers biosphere reserve, national parks and gene sanctuary.

4.7.2.1 Habitat Conservation

Many species survive and best perpetuate only in their own niche or microclimate available in the wild habitat itself. The advantages of such approach are that it does not require detailed knowledge of the flora available in that habitat. The Namdapha Biosphere Reserve of India in Arunachal Pradesh for *Coptis teeta*, Demabeyang Valley in North Sikkim for *Panax* pseudo ginseng and Nanda Devi Biosphere for Rhododendron, pine including *Morus serrata,* which is available in that area.

4.7.2.2 Biosphere Reserves

The biosphere reserve establishes a balanced relationship between human and biosphere. The biosphere is the site of excellence for foster economic and human development, which is socio-culturally and ecologically sustainable with a logistic support for demonstration projects, environmental education and training cum research. The national committee on environmental planning and co-ordination (NCEPC) and man and biosphere (UNESCO) are involved in designated areas as biosphere reserves. Out of the 14 biosphere reserves in India, Nandadevi (Uttarkhand), Namdapha (Arunachal Pradesh), Kaziranga (Assam), Manas (Assam), Nokrek (Meghalaya), North Andaman and Great Nicobar (A&N islands) for in situ conservation of mulberry. *M.rubra* is found in Hamilton's Royal Botanical Gardens, Ball's Falls Conservation Area, Niagara Glen, Rondeau Provincial Park, Point Pelee National Park, Fish Point Provincial Nature Reserve, Pelee Island, Middle Island and East Sister Island.

4.7.2.3 Gene Sanctuaries/National Parks

The significant approaches for in situ conservation are declaring the highly rich communities and habitats under protected areas. The protected area can be easily maintained with the help of communities residing in that area. India has one of the largest networks of protected areas including national park (81), wild life sanctuaries (441) and wetland and mangroves (31) providing in situ conservation of many plant species including *Morus* species. Five of the protected areas have been designated as world's heritage sites under UNESCO's world heritage programme, namely Kaziranga National Park, Keoladeo Ghana National Park, Manas Wildlife Sanctuary, Nanda Devi National Park and Sunderban National Park.

4.7.2.4 Sacred Grooves

From time immemorial plant is worshiped like Gods or their blessings in India. Many tribal communities live in complete harmony with nature; their feeling to cut a plant might cause evil effects on their family. These communities have developed their own system of conservation forests or habitats by naming sacred grooves. These are untouched virgin forests with a taboo that even taking a dead wood or fallen fruits may cause harm to the person. There exist more than 500 sacred grooves in the tribal inhabited area of North Eastern region; Maharastra, Western Ghats, Nilgiri, Orissa and Uttarakhand, Khasi and Jaintia hills and Baster area in central India are rich in such diversity. *Morus serrata* is being worshiped in Uttarakhand; Himachal Pradesh as sacred grooves and don't uses its wood, leaves and fruits, etc. The oldest mulberry tree of about 1200 years old is being worshiped in Joshimath of Uttarakhand state (Rau 1967).

4.7.3 Ex-Situ *Conservation*

Ex situ conservation is the preservation of material outside their natural habitat. The ex situ conservation refers to the man-made gene bank conservation that includes ex situ seed conservation in seed gene bank; ex situ plant conservation in field gene bank; in vitro conservation of explants or organs band and cryo-bank, DNA library and DNA bank. Conservation of plants in botanical gardens, on-farm conservation by farmers with traditional agricultural systems, arboreta and field gene banks all comes under the umbrella of ex situ conservation.

4.7.3.1 Ex Situ Field Gene Bank Management in India

The management of collections is important, as it has to be preserved for posterity. The utility of germplasm is only realized if the germplasm stored in has been clearly defined, evaluated, characterized scientifically and documented (Tikader et al. 2014). The development of an effective data management, therefore also forms gene bank manager's responsibility. Central Sericultural Germplasm Resources Centre (CSGRC), Hosur (India), is situated around 12.45°N and 77.51°E, altitude of 942 m with dry tropical climate. The average rainfall ranges from 700–1000 mm per annum. CSGRC, Hosur, maintains 1254 accessions (Ind-984; Exo-270) in the field gene bank as ex situ conservation. All the accessions are maintained as dwarf tree with spacing of 2.4 × 2.4 m. Pruning is followed once in a year for conservation and twice for recording data.

4.7.4 On-Farm Participatory Conservation

The other form of conservation is the on-farm conservation linked with farmers' participatory breeding (FPB), which gives special emphasis on sustaining and utilizing on-farm biodiversity by the farmers. In India, rich *Morus* diversity exists under managed habitats, i.e. in the backyards, kitchen gardens, farmhouses, horticultural gardens, agricultural lands and roadside plantations. These are the first-hand selections of the farmers and tribals for varied utilizations; hence, conservation of potentially interesting alleles and development of diversity is promoted. In mulberry, the wild species like *M. laevigata* and *M. serrata* and others do not get attention in the formal sector for cultivation for sericulture purposes. However, these wild species have been used for other non-sericultural purposes such as horticulture and agro forestry. Farmers/aboriginals largely use fruits and timbers of these species as a livelihood. Thus, the biodiversity of these species are conserved through the on-farm participation of aboriginals and farmers.

4.7.5 Botanical Garden/Herbaria

Botanical gardens are being maintained from thousands of years when herbal doctors, healers, sages maintained gardens of their own for medicinal plants. Even today, one can notice many ashrams in and around Haridwar, Rishikesh (Uttarakhand) and Himalayan region where medicinal plants have been maintained since ages including mulberry. The major botanical gardens in India are Tropical Botanical Garden and Research Institutes (TBGRI), Thiruvananthapuram (Kerala); Medicinal and Aromatic Plant Garden and Herbarium, Pune; Lal Bagh Botanical Garden (LBG), Bangalore; Royal Botanical Garden (RBG), Kolkata; Llyod Botanical Gardens (BG), Darjeeling (West Bengal). The Government of India and Provincial Government together run and maintain 33 Botanical gardens, which maintain the diversity in the form of plants or plant population. Biodiversity has also been preserved in the form of herbarium specimens. Botanical Survey of India (BSI) has the largest holding of 1,500,000 specimens. There are many more herbaria in India maintained in different research institutes where *Morus* species is also stored. The National Germplasm on mulberry also maintained more than 1000 samples of different *Morus* species.

4.7.6 Arboreta

Traditionally, arboreta have been regarded as assemblages of tree for scientific purposes generally with some sort of economic imperative in the not too distant back ground. The arboreta and botanical gardens had more utilitarian purpose—to find species that would benefit the new colonies and to establish which trees would provide wood, fruit, foliage for future needs. Community can get use of such arboretum in different ways. The arboreta help in maintaining genetic conservation of valuable rare and endangered trees. Arboreta can reflect changing values in society where they are becoming something for all to enjoy rather than the domain of select few and serve the purpose of society.

4.7.7 In Vitro *Conservation*

In vitro conservation indicates maintenance under aseptic condition using culture of plants. In vitro technology offers unique strategy for clonal multiplication and conservation, particularly for vegetatively propagated crop plant species. Preservation of germplasm diversity using in vitro technology can be accomplished in the vegetatively propagated crop plants using meristem/shoot tip cultures, axillary buds and nodal segments. Slow growth in vitro can be accomplished either by allowing cultures to survive on minimal media or by reduction in sucrose, thus restricting the availability of nutrients to cultures. Use of osmotic agent such as sorbitol or mannitol also helps in achieving the reduction of growth. In vitro culture techniques have been extensively developed and applied to more than 1000 species including many tropical species (Engelmann 2000; Vijayan et al. 2011). Many national and international organizations have used in vitro techniques as a complementary conservation method to field gene bank for conservation of germplasm of various vegetatively propagated and recalcitrant seed species. Over 37,600 accession of germplasm of various crops have been covered worldwide (FAO 1994).

4.7.7.1 Use of Tissue Culture Technology for Conservation

The rich diversity in plant species would lay a strong foundation for development of new cultivars presently and more so in future allowing alteration and reconstruction of existing ones to suit the changing human needs and environment. It is satisfying that germplasm value in the third world is also very impressive, and Indian gene centre is uniquely rich in this respect (Chandel and Bhat, 1992). Thus, a worldwide interest in germplasm collection conservation and utilization using conventional breeding approaches as well as new emerging techniques of tissue culture and biotechnology has assumed importance. Efforts are being made currently at global level to preserve biodiversity. Today, efforts are being directed worldwide to diversify methods, and new strategies are being integrated to achieve genetic conservation. Tissue culture technology and cryopresrvation (freeze preservation) employing liquid nitrogen at ultra-low temperature (−146–196 °C) is being used (Chaudhury et al. 2008).

4.7.8 DNA Library

Conserving the extracted DNA is a very recent approach. The genome of the accession is the main source of the genes required in various crop improvements. The present day DNA banking offers tremendous opportunities of practical and academic value. In fact, DNA as a gene bank resource has emerged out of revolution in genomic information. The use DNA libraries, collection of segments of DNA containing several copies of the part of genome, which includes cDNA, cosmid, PAC (plasmid derived artificial chromosome), etc. can offer important resources for several applications, i.e. characterization of source materials and understanding genetic and evolutionary relationship between taxa, functional analysis of genes comparative genomic and plant breeding. DNA bank is a particular type of genetic resources bank that preserves and distributes the DNA samples and associated information (Vijayan et al. 2006).

4.7.8.1 DNA Storage

DNA is a highly stable molecule, and the degradation kinetic models suggest that fully hydrated DNA can be kept at room temperature, and it takes about 10,000 years to depolymerise into small fragments. An alternative approach is to store cells and tissues rather than purified DNA. Further, stored cells and tissues have added advantages of providing a continuous supply of DNA and enabling biochemical and molecular studies of living cells. Depending upon the available facilities, the reference sample may be in the form of live plant in field repository, a propagule conserved in a gene bank, which can be recovered into a full plant or a herbarium specimen. DNA banks around the world are maintained at Australia, Brazil, USA, Great Britain and South Africa. The DNA bank at Royal Botanical Gardens (RBG), Kew (UK), maintained nearly 23,000 samples of plant genomic DNA stored at -800 °C. The bank has a large collection of orchid. DNA samples and samples of rare and endangered species are maintained.

4.7.9 Cryopreservation of Germplasm

Cryopreservation of plant materials has proven to be a potentially ideal method for long-term preservation, because it requires minimal space, labour, medium and maintenance. The techniques for cryopreservation that are currently in use varied greatly and include the older classical techniques based on freeze-induced dehydration of cells as well as newer techniques based on vitrification (Engelmann 2000). Classical cryopreservation techniques involve slow cooling down at a controlled rate (usually 0.1–4 °C/min) down to about −40 °C, followed by rapid immersion of samples in liquid nitrogen. They are generally operationally complex, as they require the use of sophisticated and expensive programmable freezers. In the new vitrification-based procedures, cell dehydration is performed prior to freezing by physical or osmotic dehydration of explants. This is followed by ultra-rapid freezing which results in vitrification of intracellular solutes, i.e. formation of an amorphous glassy structure without occurrence of ice crystals, which are detrimental to cellular structural integrity. These techniques are less complex and do not require a programmable freezer, hence are suited for use in any laboratory with basic facilities for tissue culture. Cryopreservation involves storage of plant material at ultra-low temperatures in liquid nitrogen (−196 °C). At this temperature, cell division and metabolic activities remain suspended and the material can be stored without changes for long periods. Thus, cryopreservation method ensures genetic stability of the mulberry germplasm and requires limited space, protects material from contamination, involves very little maintenance and is considered a cost-effective method for conservation of mulberry germplasm. In fact, cryopreservation is the only available method for long-term conservation of vegetative propagated plant like mulberry. In mulberry, the most appropriate material for cryopreservation was found to be winter buds, though embryonic axes, pollen, synthetic seeds have also been used (Niino 1995; Niino and Sakai 1992; Niino et al. 1993) Keeping this in view, many laboratories across the world has established cryopreservation laboratories. For instances, the cryopreservation facilities at Central Sericultural Germplasm Resources Center (CSGRC), Hosur, is actively involved in preservation of its 908 mulberry germplasm accessions (Rao et al. 2009). Success has been achieved in the cryopreservation of several accessions belonging to *M. indica*, *M. alba*, *M. latfolia*, *M. cathayana*, *M. laevigata*, *M. nigra*, *M. australis*,*M. bombycis*, *M. sinensis*, *M multicaulis* and *M. rotundiloba*. Likewise, with the epoch making research on cryopreservation, Japan has undertaken cryopreservation of mulberry in a large scale. About 450 germplasm accessions within several species have been cryopreserved in liquid nitrogen tanks in mulberry gene bank now. Shoot tips of pre-frozen winter buds of *M. bombycis* Koidz. were able to withstand long storage in liquid nitrogen. The general procedure for cryopreservation of shoot tips is that the shoot segments were first pre-frozen at −3 °C for 10 days, −5 °C for three days, −10 °C for one day and −20 °C for one day before immersion in liquid nitrogen. Buds were cultured on MS medium after thawing in air at 0 to 20 °C. Survival rate was 55 to 90%. Prior to pre-freezing at −20 °C partial dehydration of the bud up to 38.5% has found improving the recovery rates. The survival rates of the winter buds stored in liquid nitrogen up to 3–5 years did not change significantly. Encapsulation of winter-hardened shoot tips of many mulberry species with calcium alginate coating was also tested successfully. In addition, the scientists conducted experiments on cryopreservation of intact vegetative buds of mulberry *(M. bombycis)* attached to shoot segments by pre-freezing and storing in liquid nitrogen. The buds were later thawed, and the meristems were excised for culture on Murashige and Skoog's medium supplemented with 1 mg I^{-1} BA to regenerate plants. Either pre-freezing at −10 °C or −20°C along with rapid thawing at 37 °C or pre-freezing at −20 °C or −30 °C along with slow thawing at 0 °C was a suitable condition for high percentage of survival and shoot regeneration Rao et al. (2009). The conservation may be of different types based on the need of crops, i.e. (i) long term storage = −18 °C without Relative humidity maintenance; (ii) medium term storage = 4 °C and 40–45% RH and (iii) short term storage = 25 °C and 40–45% RH.

The conservation of plant genetic resources and their wild relatives in their natural habitats is of utmost importance and deserves top priority in any conservation programme (Tikader and Mukherjee 1999). Plants have been ancient times for food, in performing rituals by way of culture and on religious ground. A large number of trees are treated as sacred and protected in various ways. The forest is preserved in the name of forest Gods and removal of any plant material (even a piece of dead wood, dry leaf, etc.) is considered as taboo even today. Thus, it is pertinent to advocate the importance of endemic taxa not only due to their unique distribution pattern, but more importantly, the manifestation they carry along, be it taxonomic, ecological, economic, evolutionary, cytological or genetic. An endemic taxon can possess multiple range of attribute such as small population size, high economic value, specific habitat and low genetic diversity. Dominance of rare/endemic elements at high altitude especially in trans/north west and west

Himalayan belt suggests high conservation value of the zone (Dhar 2002). Uttarakhand state is rich in "green gold" and this treasure is to be protected at all costs. Garhwal Himalayas have been identified as hotspots for endemic species and genetic diversity in India and as such more attention should be paid to prevent loss of habitat of endemic plants.

4.8 Conclusion

Plant genetic resources (PGR) are highly valuable for the present and future utilization for maintaining sustainable food production for the benefit of human growing population. They have been used worldwide for genetic improvement of crops and contributed to major increases in crop productivity and resistance to pests, diseases and adverse climatic growing conditions.

The mulberry germplasm resources are used for (i) building blocks, gene pool for genetic improvement, enhancement; (ii) genes for adaptation, endurance for biotic, abiotic stresses/environments; (iii) contribute to develop high yielding varieties; (iv) contribute to tackle pest and disease management; (v) reduce dependency on external inputs and (vi) conserved future use and posterity. To meet the demand, large number of mulberry accessions is required to maintain with planned approach and should be utilized in crop improvement programmes. The efficient utilization depends on proper characterization, evaluation, documentation and conservation of the mulberry genetic resources.

References

Brandis D (1906) Indian trees. Constable and Co., London, pp 612–613

Chandel KPS, Bhat SR (1992) Conservation of biodiversity and prospect of crop improvement in developing countries. pp 253–272. In: Srivastava HC (ed) Biotechnological applications for food security in developing countries. Oxford and IBP Pub. Co. New Delhi, India

Chaudhury R, Rao AA, Malik SK, Kumar Susheel (2008) Successful cryopreservation of mulberry genetic resources. ICAR News, New Delhi

Chaudhuri HK, Barrow JR (1975) Identification of cotton haploid by stomata chloroplast technique. Crop Sci 15:760–763

Chloupek O, Rod RJ (1992) The root system as a selection criterion. Plant Breed Abstr 62 (12):1337–1341

Chen S, Liu HC, Chu YN, Lu HX, Zhong J, Wu KJ, Lou K, Yang W (2007) Determination and analysis of 1-DNJ content in leaf of12 mulberry varieties. Acta Sericoloca Sinica 33:637–641

Chinnaswamy KP, Doreswamy C, Devaiah MC, Venkatesha M (1993) Pharmaceutical value of mulberry. Indian Silk 31(11):47–48

Chinnaswamy KP, Hariprasad KB (1998) Utilization of mulberry crop residue as fuel wood. Indian Silk 37(1):17–18

Dandin SB, Jolly MS (1986) Mulberry descriptor Sericologia 26:465–475

Dandin SB, Kumar R, Ravindran S, Jolly MS (1987) Crossability studies in mulberry. Ind. J. Sericulture 29(1):1–4

Dandin SB, Kumar R (1989) Evaluation of mulberry genotypes for different growth and yield parameters. In: Sengupta K, Dandin SB (eds) Genetic Resources of mulberry and utilization. Central Sericultural Research and Training Institute, Central Silk Board, Mysore, pp 143–151
Dandin SB, Basavaiah Kumar R, Venkateshaiah HV (1993) Phyto-geographical studies in *Morus* L. I: Geographical distribution and natural variation of *Morus serrata* Roxb. Indian J Plant Genet Resour 7:223–226
Dandin SB, Basavaiah Kumar R, Mallikarjunappa RS (1995) Phyto-geographical studies in *Morus*L. II: Geographical distribution and natural variation of *Morus laevigata* Wall. Ex Brandis. Indian J Plant Genet Resour 8:129–131
Datta R.K. (2000) Mulberry Cultivation and Utilization in India. FAO Electronic conference on mulberry for animal production (*Morus* L). http://www.fao.org/ OCREP/005/X9895E/ 9895e04.htm#Top of page)
Dhar Upendra (2002) Conservation implications of plant endemism in high-altitude Himalaya. Curr Sci 82(2):141–148
Engelmann F (2000) Importance of cryopreservation for the conservation of plant genetic resources. In: Engelmann F, Takagi H (eds) Cryopreservation of tropical plant germplasm. Current research progress and application, IPGRI, Rome, Italy, pp 8–20
FAO (1990) Sericulture training manual. FAO Agricultural Services Bulletin, 80 Rome, p 117
FAO (1994) Survey of existing data on ex-situ collection of plant genetic resources for food and agriculture. Food and Agricultural Organizations of United Nations, Rome, Italy
Fotadar RK, Bindroo BB, Bisaria A, Khan MA (2003) Utilization of mulberry twigs for basket making. Indian Silk 42(8):9–10
Hooker JD (1885) Flora of British India, L. Reeve and Co. Ltd., The East Book House, Ashford, Kent, UK, pp 491–493.
Hotta T (1954) Taxonomical study on the cultivated Mulberry in Japan. Faculty of Textile Fibres, Kyoto University, Kyoto, Japan, p 94
Hussain SW, Williams (1997) Evidence of functional unreduced gametes in *Trifolium repens* L. Euphytica 97:21–24
Jolly MS, Dandin SB, Ravindran S, Kumar R (1987) Sexual polymorphism in genus *Morus* L. Proc Ind Acad Sci (Plant Sci) 96(4):315–320
Kadambi K (1949) *Morus alba* L. (The common mulberry). Indian Forester 75(11):459–466
Katsumata F (1972a) Mulberry species in West Jawa and their peculiarities. J Sericulture Sci 41 (3):213–222
Katsumata F (1972b) Relationship between length of style and the shape of idioblast in mulberry leaves with special reference to the classification of mulberry. J Sericulture Sci 41(5): 387–395
Kanjilal UN, Kanjilal PC, Das A, De RN, Bor N L (1940) Flora of Assam, vol 4, Government Press, Shillong, pp 272–274
Kim HR, Patel KR, Thorpe TA (1985) Regeneration of mulberry plantlets through tissue culture. Bot Gazzette 146:335–340
Kimura T, Nakagawa K, Saito Y, Yamagishi K, Suzuki M, Yamaki K, Shinmoto H, Miyawaza T (2004) Determination of 1-deoxynojirimycin in mulberry leaves using hydrophilic interaction chromatography with evaporative light scattering detection. J Agric Food Chem 52:1415–1418
Kulkarni VN, Khadi BM, Sangam VS (2002) Pre-breeding efforts for low gossypol seed and high gossypol plant in *G. herbaceum* L. cotton utilizing *G. australis* Mueller. Curr Sci 82:434–439
Koidzumi G (1917) Taxonomy and phytogeography of the genus *Morus*. Bull Imperial Sericulture Stat 11:1–50
Linneaus C (1753) Species Plantarum, Stockholm, p 986
Machii H, Koyama A, Yamanouchi H, Katagiri K (1997) Manual for characterization and evaluation of mulberry genetic resources. Miscellaneous Publ Natl Inst Sericulture Entomoll Sci 22:105–124
Machii H, Koyama A, Yamanouchi H (1999a) Fruit traits of genetic mulberry resources. J Sericulture Sci 68(2):45–155

Machii H, Koyama A, Yamanauchi H (1999b) A list of genetic mulberry resources maintained at National Institute of Sericultural and Entomological Science. Miscellaneous Publ Natl Inst Sericulture Entomoll Sci 26:1–77

Machii H, Koyama A, Yamanouchi H, Matsumoto K, Kobayashi S, Katagiri K (2001) Morphological and agronomical traits of mulberry. Miscellaneous Publ Natl Inst Sericulture Entomoll Sci 29:1–307

Metcalfe CR, Chalk L (1979) Anatomy of dicotyledons. Systematic anatomy of leaf and stem, with a brief history of the subject, vol I, 2nd edn. Clarendon Press, Oxford, p 276

Minamizawa K (1963) Experimental studies on the sex differentiation in mulberry. Bull Fac Agric 7:4–47

Niino T, Sakai A (1992) Cryopreservation of alginate - coated in-vitro grown shoot tips of apple, pear and mulberry. Plant Sci 87:199–206

Niino T, Koyaman A, Shirata K, Ohuchi S, Suguli M, Sakai A (1993) Long-term storage of mulberry winter buds by cryopreservation. J Sericulture Sci 62:431–434

Niino T (1995) Cryopreservation of germplasm of mulberry (*Morus* sps.). In: Bajaj YPS (ed) Biotechnology in agriculture and forestry, vol 32. Springer, Berlin, pp 102–106

Ono M (1983) Cultivation and breeding of mulberry in China. Japan J Breed 33(3):337–340

Pan YL (2003) Popularization of good mulberry varieties and sericulture development. Acta Sericologic Sinica 1:1–6

Parkinson CE (1923) A Forest Flora of Andaman Islands. Pp. 155. Bishen Singh and Mahendra Pal Singh, Dehradun, p 155

Rao AA, Chaudhury R, Malik SK, Kumar S, Radhakrishnan R, Qadri SMH (2009) Mulberry biodiversity conservation through cryopreservation. In Vitro Cell Dev Biol. doi:10.1007/s11627-008-9185-3

Rao AA, Chauhan SS, Radhakrishnan R, Tikader A, Borpuzari MM, Kamble CK (2011) Distribution, variation and conservation of mulberry (*Morus* spp.) genetic resources in the arid zone of Rajasthan, India. Bio Biodiver Bioavailab, 5(1):52–62

Rau MA (1967) The sacred mulberry tree of Joshimath, UP. Indian Forester 93(8):333–335

Ravindran S, Rao AA, Naik VG, Tikader A, Mukherjee P, Thangavelu K (1997) Distribution and variation in mulberry germplasm. Indian J Plant Genet Resour 10(2):233–242

Ravindran S, Tikader A, Naik VG, Rao AA, Mukherjee P (1999) Distribution of mulberry species in India and its utilization. Indian J Plant Genet Resour 2:163–168

Sanjappa M (1989) Geographical distribution and exploration of the genus *Morus* L. (Moraceae). In: Sengupta K, Dandin SB (eds) Genetic resources of mulberry and utilization, Jwalamukhi Job Press, Bangalore, India, pp 4–7

Sanchez MD (2000) World distribution and utilization of mulberry, potential for animal feeding. FAO Electronic Conference on mulberry for animal production, World Anim Rev 93:1–11, FAO, Rome

Singhvi NR, Sinha AK (1991) Mulberry: The Kalpavriksha of the twenty first century. Indian Silk 30(2):24–26

Singh KK, Pandey AB, Tiwari DK (2006) Potential of *Morus serrata*a wild mulberry variety as a alternate feed for late age silkworm and additional crop in Gohar Block of Mandi districts of Himachal Pradesh. In Proceedings of regional seminar on Prospects and Problem of sericulture as an economic Enterprise in North West India, November, 11–12, 2006, Dehradun, Uttarakhand, India, pp 10–14

Tewary PK, Rao GS (1990) Mulberry by-products and their utilization. Indian Silk 29(3):43–44

Thangavelu K, Mukherjee P, Tikader A, Ravindran S, Goel AK, Rao AA, Naik VG, Sekar S (1997) Catalogue on mulberry (*Morus* spp.) germplasm. vol 1. Published by Silkworm and Mulberry Germplasm Station, Hosur, Tamil Nadu, India

Thangavelu K, Tikader A, Ramesh S R, Rao AA, Naik VG, Sekar S, Deole AL (2000) Catalogue on mulberry (*Morus* spp.) germplasm. vol 2. Published by Central Sericultural Germplasm Resources Centre Tamil Nadu, India 2:1–225

Tikader A, Vijayan K, Raghunath MK, Chakroborti SP, Roy BN, Pavankumar T (1995) Studies on sexual variation in mulberry (*Morus* spp.). Euphytica 84:115–120

Tikader A, Rao AA, Mukherjee P (1999a) Ex-situ conservation of the oldest mulberry tree. Indian Silk 38:17–18

Tikader A, Rao AA, Mukherjee P (1999b) Stomatal variation and its relation with ploidy levels in mulberry (*Morus*spp.). Indian J Sericulture 38:160–162

Tikader A, Rao AA, Mukherjee P (2000a) Pollen morphology studies in mulberry (*Morus* spp.). Indian. J Sericulture 39(2):160–162

Tikader A, Rao AA, Mukherjee P (2000b) Phenotypic characterization of mulberry (*Morus* spp.) germplasm. In: Proceedings of 4th International Silk Conference, CISC–4, May 2000, China, pp 114–125

Tikader A, Rao AA (2001) Morpho-anatomical and pollen studies in mulberry germplasm. Sericologia 41:69–76

Tikader A, Rao AA (2002) Phenotypic variation in mulberry (*Morus* spp.) germplasm. Sericologia 42:221–233

Tikader A, Chandrasekhar M, Borpuzari MM, Saraswat RP, Rao AA, Sekar S (2006) Catalogue on mulberry (*Morus* spp.) germplasm. vol 3 and 4. Published by Central Sericultural Germplasm Resources Centre Tamil Nadu, India 2: 1–172

Tikader A, Kamble CK (2007) Evaluation of mulberry germplasm (*Morus* spp.) for leaf yield and quality through bioassay. Int J Ind Entomol 14(2):87–92

Tikader A, Saratchandra B, Vijayan K, Singh RN (2014) Mulberry germplasm management and utilization. A.P.H. Publishing Corporation, 4435-36/7 Ansari Road. Driyaganj, New Delhi 110002, pp 1–494

Tzenov PI (2002) Conservation status of mulberry germplasm resources in Bulgaria. Paper contributed to Expert Consultation on Promotion of Global Exchange of Sericulture Germplasm Satellite Session of XIXth ISC Congress, September 21st–25th Bangkok, Thailand. http://www.fao.org/DOCREP/005/AD107E/e01.htm

Vavilov N I (1951) The origin, immunity and breeding of cultivated plants. Translated from the Russia by Chester KS. Cherenica Botanica vol 1/6

Vijayan K, Raghunath MK, Das KK, Tikader A, Chakroborti SP, Roy BN, Qadri SMH (1997a) Studies on leaf moisture of mulberry germplasm varieties. Indian J Sericulture 36:55–157

Vijayan K, Tikader A, Das KK, Chakroborti SP, Roy BN (1997b) Correlation studies in mulberry (*Morus* spp.). Indian J Genet Plant Breed 57(4):455–460

Vijayan K, Tikader A, Kar PK, Srivastava PP, Awasthi AK, Thangavelu K, Saratchandra B (2006) Assessment of genetic relationships between wild and cultivated mulberry (*Morus*) species using PCR based markers. Genet Resour Crop Evol 53:873–882

Vijayan K, Tikader A, da Silva AJT (2011) Application of tissue culture techniques for propagation and crop improvement in mulberry (*Morus* spp.). Tree Forest Science and. Biotechnology 5(spl):1–13

Ward JH (1963) Hierarchical grouping to optimize an objective function. J Am Statististics Assoc 58:236–244

Chapter 5
Conservation of Forest Genetic Resources in Norway in a Climate Change Perspective

Tore Skrøppa and Kjersti Bakkebø Fjellstad

Abstract Forests and wooded land cover 39% of the land area of Norway, with two conifer species, *Picea abies* and *Pinus sylvestris,* dominating the forest area. Twenty-five of 35 native forest tree species have their northern limit in this country. The genetic resources of 18 species are considered to be vulnerable or threatened either at a local or national level. Genetic information is available for 13 of the native species, with *Picea abies* being the species that has been most thoroughly characterised. The National Programme for Forest Genetic Resources is administered by the Norwegian Genetic Resource Centre. This programme covers four major areas: generating knowledge and monitoring processes influencing genetic resources; in situ and ex situ conservation activities; sustainable use and development of forest genetic resources; and networking, coordination and dissemination of knowledge. In situ conservation of genetic resources of forest tree species is carried out in nature reserves. Twenty-three gene conservation units, covering ten species, have been established in such reserves. Ex situ conservation of forest genetic resources is achieved through collections in arboreta and botanical gardens and in the long-term field plantations of research and breeding programmes. In addition, seed samples of selected forest tree species are stored at Svalbard Global Seed Vault. Forests in Norway are regenerated both by natural and artificial means. A revised tree breeding strategy, with emphasis on *Picea abies,* has been developed to improve climatic adaptation, growth and quality, without decreasing the genetic diversity in future forests or the potential for adaptation to future climatic conditions.

Keywords Genetic diversity · In situ conservation · Ex situ conservation Sustainability · Tree breeding

---

T. Skrøppa (✉)
Department of Forest Genetics and Biodiversity, Norwegian Institute of Bioeconomy Research, 1431, Ås, Norway
e-mail: skt@nibio.no

K.B. Fjellstad
Norwegian Genetic Resource Centre, Norwegian Institute of Bioeconomy Research, 1431, Ås, Norway

M.R. Ahuja and S.M. Jain (eds.), *Biodiversity and Conservation of Woody Plants*, Sustainable Development and Biodiversity 17, DOI 10.1007/978-3-319-66426-2_5

5.1 Introduction

Norway is Europe's northernmost country, ranging over some 1750 km between 58°N and 71°N. The country's total area is 323,787 km^2 (excluding the islands of Svalbard and Jan Mayen). The population is 5.2 million, with a population density of 17 people per km^2.

The total area covered by forests and wooded land is 13.4 million hectares, or 39% of Norway. Of this, 8.6 million hectares are productive forest land, that is to say forest areas that can produce more than one cubic metre per hectare per year. Built-up areas make up less than one percentage of the country's area, and cultivated land covers just 3%. The rest of the country comprises mountains, heathland, moors, wetlands and lakes.

Norway has substantial north–south and east–west climate gradients. Inland areas in northern and eastern Norway have a typical continental climate, with warm summers and cold winters. The entire coastline is characterised by a maritime climate, with relatively cool summers and mild winters.

Annual precipitation also varies. The zone with the highest annual rainfall lies about 30–40 km inland from the coast. The driest areas are the inland regions of Finnmark (in the far north), as well as parts of the valleys of eastern Norway. The length of the growing season, defined as the number of days with a mean temperature of more than 5 °C, varies between 200 days in south-western Norway and 100 days along the coast of eastern Finnmark. In the alpine regions, the growing season is even shorter.

In addition to climatic conditions, soil types and topography have had a great impact on the extent of the forests in Norway, their species composition and tree growth. The vast majority of the forest is boreal coniferous forest with principal species Norway spruce (*Picea abies*) and Scots pine (*Pinus sylvestris*), and with downy birch (*Betula pubescens*) and silver birch (*B. pendula*) as the dominating deciduous tree species. Boreal deciduous forests are an important component of the forests at high altitudes and in the northern areas. Hardwood forests, which constitute just one percentage of the forest area, occur in the southern part of the country and in particular along the coast, while the coniferous forests dominate inland.

The current forest composition is greatly influenced by different human–forest interactions. For several centuries, until the mid-twentieth century, there was widespread deforestation in Norway, and much of the present forest is the result of human-induced regeneration and various silvicultural treatments. The species composition and structure of the present forests in all ecological zones is thus significantly different from primeval forests (Fig. 5.1).

In 2011, the total growing stock in Norwegian forests was more than 900 million cubic metres, with a yearly increment of 24 million cubic metres. The total annual harvest is considerably lower than the yearly increment and in recent years has been between 8 and 11 million cubic metres. The growing stock and its increment in 2011 was more than twice the level documented by the first National Forest

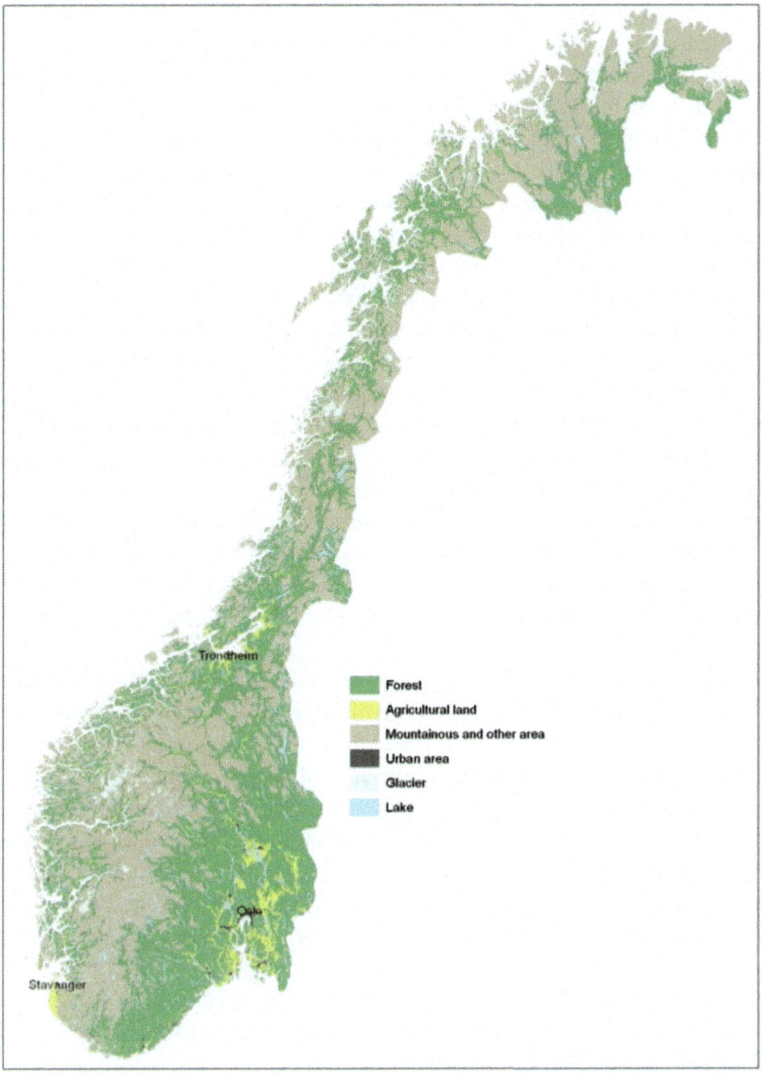

Fig. 5.1 Map showing the forest area and other types of land in Norway. *Source* Norwegian Institute of Bioeconomy Research

Inventory in 1932. The amounts of dead wood, old forest and deciduous trees, which are important for biological diversity, have increased considerably during the same period.

The forests are of great importance for Norwegian society. They provide a whole range of ecosystem services that contribute to social welfare and quality of life, as well as economic development. The forest resources are of great historical importance and have played a major role in developing trade and industry.

The Norwegian Ministry of Agriculture and Food is the national authority responsible for forest policy, which is based on a wide range of measures, including legislation, taxation, financial support schemes, research and advisory bodies. The main objective of the Forestry Act of 2005 is to promote sustainable forest management, stimulating local and economic development, while at the same time securing biological diversity, landscape values, outdoor recreation and the cultural values associated with the forest. The Forestry Act applies to all categories of forest ownership. The Act includes provision for a particular type of forest protection ("vernskog"), designed to prevent damage to climatically vulnerable forests or to forests that have a protective role such as reducing erosion, avalanche or flooding risk. Approximately 20% of the Norwegian forest area has been designated under this category of protection, the majority being forest bordering mountain areas.

Norwegian forest policy, as well as voluntary environmental standards associated with forest certification, emphasises environmental considerations, such as maintaining and developing biological diversity, and the social and cultural functions of forests. Biodiversity-rich habitats are registered and mapped in forest management plans, according to a standardised and well-documented system. However, the share of virgin forests is small in Norway. Today, there is considerable public interest as to whether Norwegian forestry is environmentally sustainable and takes sufficient consideration of biological diversity and threatened habitats.

Growing forests capture CO_2, and active management of forest resources may play an increasing part in reducing CO_2 emissions. The Norwegian forests annually capture and store more than 50% of the national CO_2 emissions. The governmental forest policy has a view to increase the use of forest resources to mitigate CO_2 emissions by means of sustainable, active forestry. This has resulted in increased funding to establish forest plantations at a denser spacing and to encourage forest tree breeding for the production of genetically superior reproductive material.

Forest genetic resources are considered important both as a component of the biological diversity that should be conserved for future generations and as a basis for the supply of forest reproductive material for the regeneration of forest after harvest. In the present Norwegian forest policy, these genetic resources may contribute to the success of the high-priority-targeted programmes: forests and climate, increased use of wood and bioenergy.

5.2 Forest Genetic Resources

5.2.1 Species Composition and Distribution

Norwegian forests can be broadly classified into three major types: coniferous evergreen boreal forest, broadleaved forest and mixed forest. The forests and other wooded land cover 13.4 million hectares. The forests are to a large extent formed by two conifers, *Picea abies* (Norway spruce) and *Pinus sylvestris* (Scots pine), and the two birch species *Betula pendula* and *B. pubescens*. The two conifers are economically the most important species and are the only native species actively managed for wood production through commercial forestry; together they cover 71% of the forest area and 97% of the annual harvest.

Pinus sylvestris is mainly naturally regenerated, while *Picea abies* is planted on approximately 50% of the harvested forest area. The number of spruce seedlings planted annually is at present 29 million, which is less than 40% of the number planted 40 years ago. The planted spruce seedlings are primarily of native origin. Exotic conifer species have been tested in experiments since the beginning of the twentieth century, but except for the planting of *Picea sitchensis* and the hybrid *P. sitchensis* x *P. glauca* along the coast in central and northern Norway, no exotic species are used to any large extent in commercial forestry. Very few broadleaved trees are planted for wood production. However, several of these species are used for landscaping, along roads and in parks and gardens.

Species composition and distribution of forest trees in Norway are largely determined by the invasion of tree species after the Ice Age, subsequent climatic changes and human activities. It was earlier assumed that the first tree species to establish after the ice retreated more than 10,000 years ago were birch (*Betula pubescens*), poplar (*Populus tremula*) and Scots pine (*Pinus sylvestris*). These species spread fast and to altitudes 200–300 m higher than the present timber line. Recently, however, analyses of ancient DNA from lake sediments from north-western and central Norway suggest that Norway spruce was also present along the Atlantic coast during the Last Glacial Maximum (17,700 year ago), and in the Trøndelag region, central Norway, 10,300 years ago (Parducci et al. 2012). During the warm and dry period that later followed, species demanding high temperatures, such as lime (*Tilia cordata*), ash (*Fraxinus excelsior*) and oak (*Quercus robur, Q. petraea*), spread and formed forests in the southern and south-western part of the country. Small remnants of these forests still exist. These and other deciduous tree species (e.g. *Fagus sylvatica, Ulmus glabra, Acer platanoides, Prunus avium)* that occur as scattered trees in mixed stands with other species have their main distribution in warmer climates at more southern latitudes and occur in Norway today at the northernmost border of their natural range.

It was not until approximately 2500 years ago, during a cooler and more humid period that the conifer *Picea abi*es started to form forests in the Norwegian landscape. The *Picea abies* spruce populations originated on the Russian plains and most likely also in the Baltic area. During a period of 7000 years, the species spread through Finland and northern Sweden, and also from the Baltic area through southern Sweden to southern Norway. The colonisation of the south-eastern lowland area started 3000 years ago, but the migration up the valleys to the species' present altitudinal boundary was not completed until the period 1000–1500 AD. The coastal spruce forest in central Norway established rather late (approx. 1300 AD). The present natural occurrence of *Picea abies* is in south-eastern Norway from sea level up to 1000 m, and at decreasing altitudes from central to northern Norway, north to latitude 67°N. Outside this area, the species has been planted during the last century both in Western Norway and north of its natural boundary in northern Norway. In both regions, it has become an important timber species.

5.2.2 The State of Forest Genetic Resources

An evaluation has been made of the native forest trees species in Norway with a description of life history traits such as geographic range (limited or widespread), type of occurrence (scattered or in stands), pollination vector and seed dispersal (Myking and Skrøppa 2001). Based on this information and available genetic knowledge, the genetic resources of each species were characterised as *vital*, *uncertain*, *vulnerable* or *threatened*. This information, modified based on new knowledge on the *Sorbus* species (Grundt and Salvesen 2011) and new developments for some species (*Fraxinus excelsior*), is presented in Table 5.1. Special for Norway is that 28 of the 35 native species have their northern limit in this country. Of the *Sorbus* species, eight are endemic in Norway. Eleven of the species characterised as vulnerable or threatened are included in the Norwegian Red List for Species (Henriksen and Hilmo 2015).

Twelve widely distributed species with generally effective dispersal of pollen and seeds are considered *vital* (e.g. *Betula* spp., *Alnus incana*, *Pinus sylvestris*, *Picea abies*) and as such have no particular conservation requirements. Four species are considered *uncertain* (*Quercus* spp., *Acer platanoides*, *Fagus sylvatica*) because of limited ranges, scattered occurrences and possibly less effective dispersal of seeds and/or pollen than the former group. *Sorbus intermedia* is also classified as uncertain due to uncertainty regarding its native origin. Fifteen species are considered vulnerable (nine *Sorbus* spp., *Malus sylvestris*, *Prunus avium*, *Tilia cordata*, *Taxus baccata*, *Ilex aquifolium*, *Fraxinus excelsior*) owing to marginal occurrences, reliance on insect-pollination (all except for *T. baccata* and *I. aquifolium*), limited sexual reproduction (*T. cordata*), diseases (*F. excelsior)* and endemism (some *Sorbus* spp.). *S. neglécta* and *S. lancifólia* are considered threatened due to very marginal occurrences. *Ulmus glabra* is classified as *threatened* because of Dutch Elm disease that may reduce genetic variability at the population level. Human influence has minor impact on the above classification.

5.2.3 Genetic Knowledge

The first provenance and species trials with both native and introduced tree species, in particular conifers, were planted in Norway approximately 100 years ago. Since then, short- and long-term field trials with both native and introduced species have provided knowledge about genetic differences between species, and of the within-species genetic variability among provenances, populations within

Table 5.1 Native forest tree species in Norway and their characteristics

Species scientific name	Geographic range	Occurrence	Pollination vector	Seed dispersal	Northern limit in Norway?	Genetic resource category
Picea abies	Widespread	Stand	Wind	Wind		Vital
Pinus sylvestris	Widespread	Stand	Wind	Wind	Yes	Vital[b]
Juniperus communis	Widespread	Scattered	Wind	Birds	Yes	Vital
Taxus baccata	Limited	Scattered	Wind	Birds	Yes	Vulnerable
Salix caprea	Widespread	Scattered	Insect	Wind	Yes	Vital
Populus tremula	Widespread	Stand/scattered	Wind	Wind		Vital
Betula pendula	Widespread	Stand/scattered	Wind	Wind		Vital
Betula pubescens	Widespread	Stand/scattered	Wind	Wind		Vital
Alnus incana	Widespread	Stand/scattered	Wind	Water/wind		Vital
Alnus glutinosa	Medium	Stand/scattered	Wind	Water/wind		Vital
Corylus avellana	Medium	Stand/scattered	Wind	Mammals	Yes	Vital
Prunus padus	Widespread	Scattered	Insect	Birds	Yes	Vital
Fagus sylvatica	Marginal	Stand/scattered	Wind	Birds	Yes	Uncertain
Quercus robur	Limited	Stand/scattered	Wind	Mammals/birds	Yes	Uncertain
Quercus petraea	Limited	Stand/scattered	Wind	Mammals/birds	Yes	Uncertain
Acer platanoides	Limited	Scattered	Insect	Wind	Yes	Uncertain
Fraxinus excelsior	Limited	Stand/scattered	Wind	Wind	Yes	Vulnerable
Ilex aquifolium	Limited	Scattered	Wind	Birds	Yes	Vulnerable
Malus sylvestris	Limited	Scattered	Insect	Mammals/birds	Yes	Vulnerable
Prunus avium	Marginal	Scattered	Insect	Birds	Yes	Vulnerable
Tilia cordata	Limited	Stand/scattered	Insect	Wind	Yes	Vulnerable
Ulmus glabra	Medium	Stand/scattered	Wind	Wind	Yes	Threatened
Sorbus aucuparia	Widespread	Scattered	Insect	Birds		Vital

(continued)

Table 5.1 (continued)

Species scientific name	Geographic range	Occurrence	Pollination vector	Seed dispersal	Northern limit in Norway?	Genetic resource category
Sorbus hybrida	Limited	Scattered	Insect	Birds	Yes	Vulnerable
Sorbus meinichii[a]	Marginal	Scattered	Insect	Birds	Yes	Vulnerable
Sorbus subsimilis[a]	Marginal	Scattered	Insect	Birds	Yes	Vulnerable
Sorbus subpinnata[a]	Marginal	Scattered	Insect	Birds	Yes	Vulnerable
Sorbus subarranensis[a]	Marginal	Scattered	Insect	Birds	Yes	Vulnerable
Sorbus neglécta[a]	Marginal	Scattered	Insect	Birds	Yes	Threatened
Sorbus lancifólia[a]	Marginal	Scattered	Insect	Birds	Yes	Threatened
Sorbus sognensis[a]	Marginal	Scattered	Insect	Birds	Yes	Vulnerable
Sorbus norvegica[a]	Marginal	Scattered	Insect	Birds	Yes	Vulnerable
Sorbus rupicola	Limited	Scattered	Insect	Birds	Yes	Vulnerable
Sorbus intermedia	Marginal	Scattered	Insect	Birds	Yes	Uncertain
Sorbus aria	Marginal	Scattered	Insect	Birds	Yes	Vulnerable

[a]Species that are considered to be endemic in Norway

[b]*Pinus sylvestris* ssp. *lapponica* is rated as near threatened in The Norwegian Red List for Species

provenances and within populations. Studies have focused in particular on traits that characterise adaptation to climatic conditions. More recently, molecular genetic studies have been initiated, at present in eight native species. Forest tree species, for which genetic knowledge is available, are listed in Table 5.2. *Picea abies* is the only species that has a sufficient genetic characterisation at the provenance, family and individual level.

Generally, there is a lack of knowledge of the importance and implications of factors that may influence the genetic diversity of forest tree species. **Fragmentation** of the landscape reduces gene flow among individuals and populations which may lead to a smaller effective population size and increased degree of inbreeding. Yet gene flow among populations has been characterised for very few tree species with a fragmented distribution in Norway. For many tree species, **regeneration** is hindered by browsing from increasing population sizes of wild animals such as moose and red deer, e.g. browsing on *Taxus baccata*. **Change in land use** and **logging** influence growth conditions, in particular for hardwood broadleaved species, and may change the competitive environment. **Pests** and

Table 5.2 Native forest tree species for which genetic variability has been evaluated at different genetic levels

Species	Genetic level of material evaluated	Morphology traits	Adaptive and production characters assessed	Molecular characterisation
Picea abies	Populations, families, clones	X	X	X
Pinus sylvestris	Populations, families	X	X	
Betula pendula	Populations, families	X	X	X
Alnus glutinosa	Populations, families	X	X	
Acer platanoides	Populations, families	X	X	X
Fraxinus excelsior	Populations, individuals			X
Sorbus aucuparia	Populations, families	X	X	
Ulmus glabra	Populations, families	X	X	X
Malus sylvestris	Individuals	X		X
Taxus baccata	Populations			X
Fagus sylvatica	Populations			X
Quercus petraea *Q. robur*	Populations	X	X	

diseases, which may be more common due to an increasingly warmer climate at northern latitudes, may lead to loss of populations and thereby reduced diversity for some species. The implications of **climate change** on the forest genetic resources are not well understood, in particular as the prediction of the future climatic conditions is uncertain. More information should be generated about the influence of these factors, and their interactive effects, on forest genetic diversity. In general, prospects for future conditions are good, but pests and diseases (e.g. ash decline), perhaps associated with climate change, and browsing pressure appear to be the main causes for concern today.

5.2.3.1 Genetic Diversity in Picea Abies

Recent molecular studies confirm that the vast northern range of *Picea abies* was colonised from a single Russian refugium and that the expansion westward took place along two main migration routes (Tollefsrud et al. 2008). Populations in southern Norway show relatively high levels of diversity compared to the northern ones (Tollefsrud et al. 2009). In the north, limited seed and pollen production may have increased inbreeding and decreased diversity, reflecting the marginality of the species in the north.

Genetic information characterising adaptation of *Picea abies* to climatic conditions is available from provenance, progeny and clonal trials. Measurements have, in particular, been made of annual growth rhythm traits: the timing and duration of the annual growth period, frost hardiness development in the autumn and dehardening in the spring, and the occurrence of climatic damage under field conditions. All studies demonstrate a clinal variation in growth rhythm traits in natural populations from the south to the north and from low to high altitudes. The southern and lowland populations have the longest duration of growth season, and as a consequence, the highest growth potential. They also develop latest autumn frost hardiness. The only well-known characterisations of the adaptive process of spruce populations are the responses to temperature and photoperiod. Within natural populations, a large genetic diversity is present, also for traits that show clinal variation at the provenance level and in populations at the geographic margin of the species.

In addition, a number of studies have shown that trees of *Picea abies* can adjust their performance in adaptive traits by a rapid epigenetic mechanism, through a kind of long-term memory of temperature sum and photoperiod during reproduction (see Johnsen et al. 2009 and references therein). Traits that are affected include the timing of dehardening and bud burst in the spring, leader shoot growth cessation in the summer and cold acclimation in the autumn. This memory may help the conifer to cope with the anticipated rapid change in climatic conditions. It will have importance for practical forest tree breeding and for the deployment of seedlings produced in seed orchards. The molecular mechanism behind this striking epigenetic memory mechanism is not yet unravelled, but transcriptional changes are clearly involved (Yakovlev et al. 2014). The research on epigenetic effects on adaptive traits in *Picea abies* has high priority and is in particular being followed up by molecular methods.

5.2.3.2 Genetic Diversity in Other Species

Pinus sylvestris colonised early after the ice retreated in Scandinavia and has had a much longer period for adaptation than the other main conifer, *Picea abies*. Provenance trials have shown that populations of *Pinus sylvestris* show a clinal variation in adaptive traits depending on latitude, altitude and distance from the coast, with a corresponding variation in growth traits.

Short-term tests of families from several populations of broadleaved tree species (Table 5.2) have demonstrated large genetic variability within populations for phenology, growth and morphology. For most species, similar variation is also present among populations, and the large-scale variability patterns can in most cases be related to climatic parameters of the populations characterised by latitude, altitude and whether they are from coastal or inland areas. In *Taxus baccata*, isozyme analyses demonstrate less genetic variability in the northernmost populations, indicating a higher level of inbreeding in the most marginal populations. A study of molecular markers (microsatellites) in *Malus sylvestris* show that the trees of this species are genetically differentiated from those further south in Europe and indicate no decrease in genetic diversity in the northern range of the species. The population genetic history of *Fraxinus excelsior* has been characterised both by maternally inherited cpDNA and nuclear microsatellites. The analyses show that the ash populations in the northern area are genetically different from those in the more southern areas of Europe. Within Norway, there are genetic differences between eastern and western populations and the genetic diversity decreases to the north.

5.3 National Programme for Forest Genetic Resources

Forest genetic resources, and their conservation and use, are explicitly mentioned in several national policy documents related to forestry and climate challenges in the agricultural sector, both in general terms and in recommendations related to the production and use of forest reproductive material and in the implementation of important measures to mitigate climate change. Genetic resources are also specifically treated in the Nature Diversity Act adopted by the Parliament in 2009. This Act regulates the conservation, access and use of genetic resources, and also the import and release of alien organisms in Norwegian nature.

Thus, the importance of forest genetic resources was recognised, and in 2001, an advisory committee was appointed. It approved the first National Programme for the period 2003–2006. In 2006, the Norwegian Genetic Resource Centre was established by the Ministry of Agriculture and Food to coordinate and promote activities related to the conservation and sustainable use of national genetic resources in farm animals, crop plants and forest trees. Together with advisory committees, the Centre develops and conducts national programmes for conservation and sustainable use of genetic resources in these three sectors. During the first two four-year periods from 2003, high priority was given to organisation and establishment of a national network, documentation of knowledge and to initiate research about national forest genetic resources.

The current action plan for forest genetic resources, for the period 2016–2019, focuses on four major areas. These are basically the same as the priority areas in the FAO Global Plan of Action for the Conservation, Sustainable Use and Development of Forest Genetic Resources (FAO 2014):

- Generate knowledge and monitor processes influencing forest genetic resources.
- In situ and ex situ conservation activities.
- Sustainable use and development of forest genetic resources.
- Networking, coordination and dissemination of knowledge about forest genetic resources and raising public awareness.

5.3.1 Major Challenges in the National Programme

For most of the national tree species, genetic knowledge is scarce and the generation of better genetic information should have high priority. This is particularly important in view of the changing climatic conditions, both for the development of adapted reproductive material for the present commercial species, but also for broadleaved species that may become more important under future conditions, and for the management of forest tree genetic resources in general. Monitoring of the development of changes in genetic diversity is also a priority, in order to be able to actuate specific conservation activities when needed. This is especially important for rarer forest tree species and for their conservation in situ, which requires sustainable management in nature reserves, or specific ex situ conservation actions. A large number of actors collaborate in the National Programme, including nature and forest management and administration, research institutions, tree breeders and the users of genetic resources. Agreements on common objectives and how to achieve these objectives are needed for the success of the Programme. Governmental funding is available for projects and activities that contribute to the follow-up of the action plan.

5.4 The State of In situ Genetic Conservation

In situ conservation of forest tree species comprises the conservation of viable populations in their natural environment, be that a production forest or a protected area. The term is often applied to naturally regenerating wild populations, but can also apply to managed production and multiple-use forests. The aim of in situ conservation is often to conserve the functioning of an ecosystem and evolutionary processes rather than just a given species. Under certain conditions, areas protected by the Nature Diversity Act provide significant potential for in situ conservation of forest genetic resources. Norway has chosen a strategy of establishing in situ gene conservation units in protected areas for some target species. This forms part of our national contribution to the common European project EUFGIS (www.eufgis.org), which created an online information system for forest genetic resource inventories in Europe, focusing on improving documentation and management of dynamic conservation units of forest trees.

The Norwegian Nature Diversity Act provides mainly for three different types of protected areas in forest, which differ in size, objectives (i.e. what is protected) and management regulations. The three types are national parks, nature reserves and protected landscapes. Altogether 3.3% of the productive forest area is protected in national parks and nature reserves, covering approximately 285,000 hectares. In cooperation with the Norwegian Environment Agency and the environmental units of the County Governor, the Genetic Resource Centre has selected candidate nature reserves for being considered as in situ gene conservation areas for specific species and to be part of the EUFGIS network. Assessments have been made in the field of the suitability of the selected reserves. This allows the possibility to combine in situ conservation of genetic resources with other protection objectives in already protected areas. Nature reserves were considered to be the most relevant protection category for in situ conservation of forest genetic resources because such areas are quite well documented as regards species content, and the conservation regime is relatively strict, development in such areas is to some extent monitored and some management activities can be allowed. An online searchable database of the protected areas in forests and a listing of both main and associated tree species growing there have been established by the Norwegian Genetic Resource Centre. It is based on, and linked to, the database of all protected areas managed by the Norwegian Environment Agency. The main aim of the searchable database is to show how general protection of forest in protected areas also contributes to the protection of forest genetic resources, in addition to more specifically planned gene conservation units.

Certain requirements must be fulfilled in order for a nature reserve to qualify as a gene conservation unit. The requirements relate to population size, number of reproducing trees, sex ratio and whether trees are growing in stands or scattered. Minimum requirements for a given species will depend heavily on a number of factors including its reproductive biology and growth, ecology and kind of genetic threats it is currently facing or will most likely face in the near future. Conservation of the genetic resources of the specific species must be in accordance with the original objectives for establishing the nature reserve. This is particularly important for future management of the genetic resources in the area.

During the last 5 years, 23 gene conservation units, including ten forest tree species on a total of 13,890 hectares, have been registered and are included in the EUFGIS database. Some of the units include more than one species. They are shown in Table 5.3. The type of occurrence may differ for different species in the nature reserves and they have different requirements for long-term existence. As an example, *Picea abies* is a highly competitive species that occurs in large stands and with sufficient sexual reproduction and natural regeneration capacity. Other species occur as scattered single trees that may have low competitive ability or may not have sufficient sexual reproduction, e.g. *Tilia cordata*. Management plans for the genetic resource are therefore needed, and in some cases these are already included in the management and conservation plans for the nature reserves.

The genetic resources for a number of Norwegian tree species are considered to be vital, e.g. *Pinus sylvestris, Populus tremula, Betula pendula and Alnus glutinosa* (Table 5.1); they have a wide and continuous occurrence and reproduce easily. It is thus assumed that there is an extensive gene flow among populations and that they maintain genetic diversity on a large scale. For such species, with the exception of *Picea abies*, it has not yet been found necessary to establish specific gene conservation units. However, to follow up the pan-European conservation strategy with at least one conservation unit per country for each climatic zone in that country (de Vries et al. 2015), the establishment of additional conservation units for these species is under consideration.

Several *Sorbus* species are endemic to Norway; they often have marginal geographic ranges and some are considered threatened or vulnerable (Table 5.1). Many species have a large and often unique variation that should be given special protection. Specific conservation activities are needed to manage and conserve these unique genetic resources. Based on former field studies, 43 localities have been identified and described where these species occur with high variability or where rare taxa have been found. Some of these localities are in protected areas, some of which are designated for other purposes. However, most are not yet protected and protection and management have been proposed, in some cases as a combination of in situ and ex situ conservation. Ex situ collections of some of the Sorbus species have been established. Other species for which such strategies should be proposed are *Taxus baccata, Ilex aquifolium* and *Malus sylvestris.* Work is now ongoing to establish in situ areas also for *Malus sylvestris*, based on new information on distribution and genetic characterisation of the species (Tollefsrud et al. 2014).

Table 5.3 Target forest species included within the in situ conservation programme

Species	Number of conservation units	Total area hectares
Picea abies	5	13,151
Ilex aquifolium	3	81.3
Taxus baccata	3	118.4
Ulmus glabra	4	189.3
Fraxinus excelsior	3	74.2
Tilia cordata	3	253.1
Fagus sylvatica	2	25.9
Quercus petraea, Q. robur	3	68.6
Acer platanoides	2	46.7

All conservation units are established as part of the European EUFGIS project

5.4.1 Challenges Related to in Situ Conservation

The climate change scenarios for Norway predict an increase in annual temperature, with warming rates increasing with distance from the coast and with latitude. Higher warming rates are projected in winter than in summer. Increasing annual precipitation is also expected. This offers an opportunity for forest tree populations to expand both to higher altitudes and to the north. The response of individual species will depend on their biology in general and the size and distribution of populations. Phenotypic plasticity, possible epigenetic responses and genetic diversity in adaptive traits are factors that will be crucial for the adaptive potential of species to the climate change conditions. Sufficient knowledge of these factors is available only for *Picea abies* in Norway, indicating that the Norwegian spruce populations have potential for adaptation to the climate change conditions and thus for conserving spruce genetic resources. Several other species have been shown to possess large genetic diversity in adaptive traits which may contribute to the conservation of their genetic resources under in situ conditions.

For some species, a major threat to their genetic resources may be diseases or pests that have been shown to expand their range northward and to higher altitudes. One example is the ash disease (*Hymenoscyphus fraxineus*) that during recent years has spread to ash (*Fraxinus excelsior*) populations across the country as far north as latitude 64°. The mortality of attacked trees is high and the population sizes of the ash stands have been strongly reduced. However, resistance against the disease has been demonstrated and research is under way to characterise the genetic diversity in resistance and explore the possibility for breeding for more healthy ash trees. Seeds are being collected from healthy trees with the intention to plant their offspring in stands with few trees left in order to increase the number of healthy trees in the next generation.

It is important to obtain full acceptance from the nature reserve managers that the genetic resources of selected forest tree species should be conserved in nature reserves and that management may be needed to fulfil this objective of the reserve. Such management should be an integrated part of the management plan of reserves that are selected for this purpose. More gene conservation units and more species should be evaluated to be included in the in situ conservation programme in nature reserves. A better cooperation is needed between local and regional managers of both protected areas and forests.

Greater public awareness is needed about in situ conservation of vulnerable tree species and the role of protected areas for such conservation. The online database of protected areas in forest is updated with information about the gene reserves. It is important that the Norwegian in situ conservation units at the marginal range of several tree species are considered as an integral part of the gene conservation efforts across their whole natural range through EUFGIS.

5.5 The State of Ex situ Genetic Conservation

Ex situ conservation of forest genetic resources in Norway includes collections in arboreta and botanical gardens; long-term tests of clones, families and provenances in research plantations; progeny tests, clonal archives and seed orchards belonging to the national breeding programme; and seed lots stored at The Norwegian Forest Seed Centre and at the Svalbard Global Seed Vault. No in vitro storage for conservation purpose of forest genetic material is performed in Norway. Facilities for such storage are available, however, both at research institutions and private companies.

5.5.1 Arboreta and Botanical Gardens

Several arboreta and botanical gardens possess collections of forest trees, of both native and exotic species. In most cases, these collections contain a small number of individuals of each species and do not have a strategy for long-term regeneration. Therefore, they are not considered as the main elements of the national conservation strategy. Such collections often contribute to the maintenance of unique and rare genotypes, but may also contain locally adapted populations of native species and individuals from transferred provenances of native or exotic species. Some arboreta and botanical gardens possess collections that have the potential to be important for the ex situ conservation of some forest tree species. As an example, nearly 200 accessions of endemic *Sorbus* species have been established in a collection in a cooperative project between arboreta and botanical gardens. A status report of ex situ collections in arboreta and botanical gardens was published in Norwegian during spring 2015 (Grundt and Fjellstad 2015). Collections of trees in arboreta often have a role as public parks and are important for raising public awareness. They are therefore valuable for demonstration and education.

5.5.2 Collections for Landscaping Purposes

A substantial number of trees of both native and exotic origins are planted in the landscape: in parks, along the roadside and in private gardens. Cultivars and clones with specific aesthetic values have been developed, tested and propagated for use in such plantings. For Norwegian conditions, testing for frost hardiness is of specific importance. Collections of such material, intended both for testing and demonstration purpose, contain valuable genetic resources. They offer a large variety of genetic material and contribute towards increasing the diversity of tree plantings in the landscape. The largest collection, located at the Norwegian University of Life

Sciences, contains trees of approximately 120 different species of the genera *Acer, Alnus, Betula, Carpinus, Fraxinus, Prunus, Quercus, Salix, Sorbus, Tilia* and *Ulmus*. However, only a small number of trees of each genetic unit have been planted, in most cases four, and the long-term conservation of the material is not secured. Smaller collections exist in other parts of the country.

5.5.3 Long-term Trials of Clones, Families and Provenances in Research and Breeding Plantations

The first trials planted have given valuable information about the use of species and provenances. However, the early research plantations were often planted in experimental designs that were not suitable for long-term studies, and few of these old trials exist today. During the last 50 years, a large proportion of the field trials have dealt with the most important native conifer, *Picea abies*, and many of these more recent trials still exist and constitute valuable genetic resources containing genetic units from which genetic information on phenotypic traits is available. In years with abundant flowering, seed lots were obtained from several individual trees in a number of natural populations that often were distributed along climatic gradients, and progeny tests were established. Several such collections were made by the Norwegian Forest Research Institute during the period 1951–1997. More recent trials with *Picea abies* were based on families from controlled crosses, and also some with clones, with the objectives of characterising the genetic variability and inheritance patterns of the species, both within natural populations and in breeding populations. The database of research and breeding field experiments lists 160 species or provenance trials and 148 progeny trials for *Picea abies* alone. In addition, several field tests have been conducted on broadleaved species.

5.5.4 Clonal Archives and Seed Orchards

Forest tree breeding with *Picea abies* started in the mid-1950s with the selection of plus trees (superior phenotypes) in natural populations. These plus trees were grafted in clonal archives and seed orchards. Seed lots were collected from the selected trees, either in the forest or in the clone collections, or controlled crosses were made on the grafts. The resulting families were planted in progeny tests at multiple sites. Altogether, 5186 plus trees of *Picea abies* were selected and kept as grafts in clonal archives or seed orchards, and 3832 families are being tested in progeny tests. The Norwegian Forest Seed Centre is responsible for all breeding material.

5.5.5 Seed Lots in Storage

The Svalbard Global Seed Vault (SGSV) was established in 2008 by the Norwegian Government as a backup storage of seed lots of the world's crop genetic resources and contains (in 2015) nearly 800,000 accessions. Recently seed lots of *Picea abies* and *Pinus sylvestris* from the Nordic countries were stored at SGSV. The main objectives of including forest tree seed lots at SGSV were

- Conservation of backup seed of, e.g. threatened populations, gene reserve forests or other in situ conservation units that are managed by the depositor.
- Conservation of seed samples from representative natural populations for backup storage of material that can be used for future monitoring of long-term changes in genetic diversity of the natural forest.
- Conservation of seed samples from different stages and generations of breeding populations or seed orchards to monitor changes in genetic diversity taking place during breeding operations.

The Norwegian Forest Seed Centre is responsible for the procurement, storage and trade of seeds for the forest sector. Seeds of recommended seed sources and of both native and imported species are stored, with main emphasis on a wide selection of native *Picea abies* provenance and seed orchard seed lots and *Pinus sylvestris* provenances. Long-term seed storage is needed as seed years are scarce at northern latitudes and at high altitudes. Seed lots are therefore kept for as long as 20–30 years until new representative seed crops become available. Optimal storage conditions will guarantee a high-germination rate even after several decades of storage. This seed storage is an important component in the management of the forest tree genetic resources in artificial regeneration.

5.5.6 Challenges in Ex situ Conservation

There is no complete catalogue of all the material contained in ex situ collections and no holistic evaluation has been made to assess which material should have highest priority in ex situ conservation. Management is required in order to maintain collections in long-term field trials. At the time field trials must be thinned and finally harvested, decisions must be made for further conservation of the genetic material. There is a particular need for strategies for ex situ conservation of former breeding material in spruce as well as of genetic resources of species threatened by diseases, the present ash decline being a case in point.

5.6 Sustainable Use of Forest Genetic Resources

Forest genetic resources are used in production forestry when forests are regenerated after harvest, in afforestation on treeless land or for the replacement of other tree species. They are also used for Christmas tree and greenery production, for landscaping purposes or for ornamental use in gardens.

The Forestry Act requires that regeneration generally should take place within 3–5 years after harvest, depending on environmental conditions. The local forest authority is mandated to demand that the forest owner takes action to establish a commercially viable stand within a reasonable period of time. Regulations are given for silvicultural and environmental actions in the regeneration, such as change of tree species, the introduction of exotic species, the transfer of provenances and the recommended number of seedlings planted per hectare.

In production forestry, regeneration after harvest is executed differently for the two major commercial species, *Pinus sylvestris* and *Picea abies. Pinus sylvestris* is to a large extent naturally regenerated, using the seed-tree method. In the regeneration fellings, 30–150 seed trees are retained per hectare, depending on site conditions. On sites with difficult regeneration conditions due to, e.g., harsh climate and/or thick humus layers, soil scarification may be used to improve seedling establishment.

Picea abies is regenerated both naturally and artificially by planting. When natural regeneration is planned, the use of patch clear-cuts and shelter wood fellings are common. When using the latter method, 150–400 trees are retained per hectare for seed dispersal and to provide shelter. The shelter trees can also be other tree species than *Picea abies*. At higher elevations, a significant proportion of the spruce forest is also harvested by means of mountain forest selective cutting, where subsequent recruitment is initiated by either natural regeneration or planting, or both. However, clear-cut fellings and subsequent planting of seedlings is most common in *Picea abies* and is considered to be the fastest regeneration method on most forest sites.

National assessments are made annually of the regeneration methods used and the results of the regeneration 3 years after harvest. In the assessment made in 2010 of the areas harvested in 2007, planting was the regeneration method used on 52.5% of the area, while 10.2% was regenerated by a combination of planting and natural regeneration. On 26.6% of the harvested areas, treatments were initiated to favour natural regeneration. No actions were initiated to re-establish the forest on 10.6% of the area, which is a reduction compared to earlier years. Approximately 75% of the planted areas had a seedling density equal to or higher than the recommendations in the legal regulations.

5.6.1 Reproductive Material in Use

The number of seedlings delivered from Norwegian forest nurseries and planted in production forestry has been strongly reduced, from close to 50 million seedlings 20 years ago to an average of 23 million during the last 5-year period and 30 million in 2014. More than 95% of the seedlings are *Picea abies*. The seedlings are produced in 13 forest nurseries, which are located in different parts of the country. In general, each nursery produces seedlings for its local region. Most of the seedlings are 1- or 2-year-old container plants.

During the period 1950–1980, Central European provenances of *Picea abies* were planted to a large extent in southern Norway. Both practical experience and results from surveys showed that this was a bad choice of provenances for south-eastern Norway, resulting in plantations with climatic damage and reduced saw timber qualities (Skrøppa et al. 1993). It was feared that gene flow from such stands would lead to reduced adaptedness in the next generation. Research results have shown that this may not be the case, as there seems to be a rapid change in adaptive performance from one generation to the next in *Picea abies* due to the epigenetic mechanism discussed above (Skrøppa et al. 2009).

Forest reproductive material in Norway is classified according to the categories of the OECD Forest Seed and Plant Scheme: *source-identified, selected, qualified* and *tested* (OECD 2015). Import of seeds of *Picea abies*, which was quite high 50 years ago, is now at a low level. The percentage of seeds from seed orchards (*qualified* or *tested*) has increased considerably during the last 5 years, in particular in the southern part of the country, where seed orchard seed sold during the last 2 years amounted to 75% of the *Picea abies* seed sold, and with a percentage close to 100 in south-eastern Norway.

5.6.1.1 Regions of Provenance and Transfer Rules

Before seeds from seed orchards came into use, most reproductive material of *Picea abies* was of the category *source-identified*, collected in natural stands and characterised by its region of provenance (Fig. 5.2). According to regulations mandated in the Forestry Act, transfers within the country should not be made more than 200 km north or south or more than 300 m in altitude. During the last 20-year period, the vast majority of *Picea abies* seed used has been from sources of native origin. However, in western Norway, where *Picea abies* did not occur naturally, provenances from Central Europe are recommended due to their superior volume growth in coastal areas.

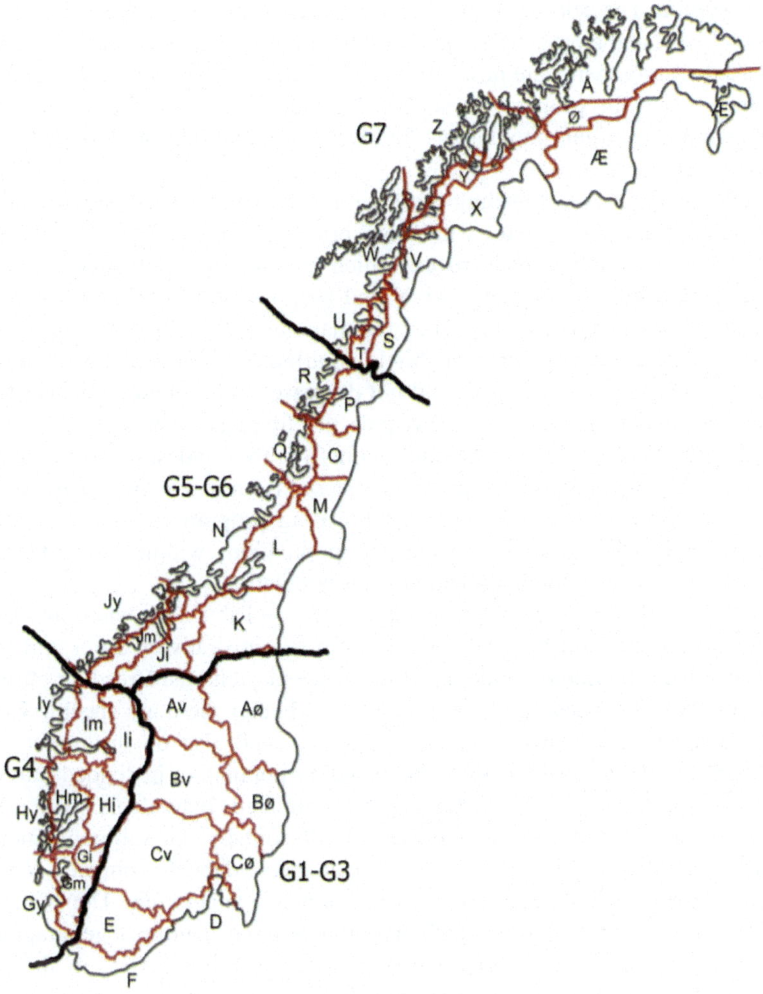

Fig. 5.2 Regions of provenance in Norway (*red* borders) and breeding zones according to the revised breeding strategy from 2010. The regions of provenance are in addition characterised by altitude in 100 m intervals. *Source* The Norwegian Forest Seed Centre

5.6.2 Tree Improvement Programmes and Their Implementation

Tree breeding activities in *Picea abies* started in Norway more than 60 years ago with the selection of plus trees in natural stands, and grafted seed orchards were established in the 1960s and 1970s. Breeding activities were also initiated with other species (*Pinus sylvestris, Picea sitchensis*), but with a lower intensity than for

Norway spruce. The selected plus trees were kept as grafts in the seed orchards or clonal archives, and a progeny testing programme was slowly initiated. The earliest established seed orchards are now terminated or are re-established with parents that are being tested in progeny tests. All Norwegian seed orchards are still of first-generation breeding populations, but with an increasing number of the tested category.

A new national tree breeding strategy for the period 2010–2040 has been developed during recent years with main priority given to *Picea abies* and the introduced species *Abies lasiocarpa,* which is used for Christmas trees. Bred material should provide higher survival and be possible to use over a larger area than material from natural stands. The bred material should also be robust to future climatic changes. Increased growth should contribute to mitigate the effect of CO_2 from the atmosphere. Breeding should not decrease, and preferably increase, wood density, improve form stability and reduce the frequencies of defects that cause reduced value. Except for one breeding zone, for the mildest climatic conditions along the coast, breeding is performed on material from native provenances. A high level of genetic diversity is kept in the breeding programme of *Picea abies* by having several breeding zones, several sub-populations within most zones and a sufficient number of individuals within each population.

Norway is divided into eight breeding zones (Table 5.4) based on latitude, altitude and known climatic gradients, for both administrative reasons and optimal use of adapted reproductive material from the seed orchards. Breeding efforts and objectives differ between zones depending on whether there are specific issues in the wood production that have to be addressed, and also on the importance of forestry in the region. In each zone, the breeding population is divided into one or more sub-populations each containing 50 unrelated individuals. The planned breeding zone G0, not shown in Fig. 5.2, will contain one sub-population with individuals selected from more southern provenances adapted to climatic conditions corresponding to a 2 °C increase in mean annual temperature. The other zones should contain sub-populations with individuals from a limited geographic area within the zone. Hence, the populations should then be adapted to the present climate in the zone, but may also be ranked according to climatic gradients within the zone. They should provide the basic material for reproductive material from seed orchards that could be used in a wide area, but also be flexible for transfer if the climate changes. All individuals in the sub-populations should be tested in progeny tests planted at several sites. The importance of traits in selection will vary among zones, but will generally include annual growth rhythm, height growth and wood quality traits. The timing of flushing in spring is a key trait in regions where spring frosts frequently occur and early flushing will be avoided. The recommendations for the deployment of reproductive material should be revised as more field test information becomes available.

It should be noted that breeding zones and deployment zones for seed orchards are different and that there can be several deployment zones within each breeding zone. The deployment zones are defined by the adaptive properties of the seedlings

Table 5.4 Breeding zones and regions of deployment for *Picea abies* in Norway

Breeding zone	Region of deployment	Altitude	Number of sub-populations
G0	Same as G1 and G4 with a 2 °C increase in mean annual temperature	0–250 m	1
G1	Interior south-eastern Norway Lat. 58°–62°N	0–350 m	5
G2	Interior south-eastern Norway Lat. 58°–62°N	350–650 m	4
G3	Interior south-eastern Norway Lat. 58°–62°N	650–950 m	4
G4	Western Norway Lat. 58°–62°N	0–350 m	2
G5	Central and northern Norway Lat. 62°–66°30'N	0–250 m	3
G6	Central and northern Norway Lat. 62°–66°30'N	250–550 m	3
G7	Northern Norway Lat. 66°30'–70°N	0–250 m	1

In each breeding zone, there are one or more sub-populations each containing 50 unrelated individuals from a limited geographic area within the zone

from each seed orchard, which have to be tested. Their performance may to some extent be influenced by the seed orchard locality due to both pollen contamination from surrounding forests and by epigenetic effects caused by the climatic conditions at the seed orchard site, as discussed in Sect. 2.3.1. If a seed orchard is located further south or at a lower altitude than the region of origin of its clones, then pollen contamination may reduce the adaptability to that region of seedlings from seeds from the orchard. The epigenetic effects due to the milder climate at the seed orchard site may cause similar implications. In both cases, adjustment of the deployment zones may have to be made based on genetic tests.

Adaptation to future climatic conditions is a major concern in the Norway spruce tree breeding programme. An increase in winter and spring temperatures implies an earlier start of growth processes and greater chances for damage caused by frost events that most likely will occur during the spring. A proper timing of phenology with late initiation of shoot growth is therefore important in the selection of improved material. Another concern is to secure a large genetic diversity in the base population, and in later generations both within and among the breeding populations. It is planned that in some zones individuals should be selected in commercial populations established with bulked seed lots from the original base population. DNA fingerprinting should be carried out to avoid relatedness among selected individuals, a so-called breeding without breeding approach (El-Kassaby and Lstiburek 2009).

5.6.2.1 Benefits from Breeding

Results from field trials suggest that seedlings of *Picea abies* from seeds produced in untested seed orchards will have 10–15% better height growth when the stand closes, and similar or better quality, compared to seedlings from stand seed (Jansson et al. 2015). The gain from seed orchards with material tested in progeny tests may be an additional 5–10%. This will contribute to an increase in the value of the forest, and it has been estimated that the planting of 40 million seedlings from seed orchard seed will result in an extra uptake of one million tons of CO_2 (Kvaalen unpublished).

5.7 Concluding Remarks

The conservation strategy for forest genetic resources in Norway has two major objectives: to provide good conditions for future evolution and adaptation of natural populations and to secure that reproductive material with sufficient genetic diversity will be available for sustainable use of the resources in the future.

For the most important commercial species, Norway spruce, the use of sustainable tree breeding to develop plant material in synthetic populations that are adapted to a range of climatic conditions is an important component of the last objective. This is particularly important considering the climate change scenarios for the northern part of Europe, predicting increased mean temperatures, milder winters, a longer growth season and an increase in rainfall, but with local variations in the climatic parameters. The goal is to keep quite a large number of breeding populations, each with a high-effective population size. Considerations of adaptive traits, in particular phenology, in selection, should secure reproductive material with both high genetic diversity and climate change adaptations.

For the non-commercial tree species, most of which have their northern limit in Norway, the climatic change conditions should offer possibilities for migration further north and to higher altitudes. However, a requirement is that the trees have the ability to reproduce sexually and that they have available genetic diversity in the adaptive traits. For some species, these conditions may be fulfilled, but for many of them knowledge is lacking.

In situ conservation of forest genetic resources is carried out in nature reserves in Norway. More gene conservation units and more species should be evaluated to be included in the in situ conservation programme, in order to follow up the pan-European conservation strategy and to secure forest genetic resources for future adaptation and evolution. Greater public awareness is needed about in situ conservation of vulnerable tree species and the role of protected areas for such conservation.

Norway has taken a broad approach in the conservation of forest genetic resources: generation and dissemination of knowledge, activities in situ and ex situ and sustainable use. The active involvement of a large number of institutions and actors, all with a common understanding of the importance of the programme, should be the best guarantee for success.

References

De Vries SMG, Alan M, Bozzano M, Burianek V, Collin E, Cottrell J, Ivanovic M, Kelleher CT, Koskela J, Rotach P, Vietto L, Yrjänä L (2015) Pan-European strategy for gene conservation of forest trees and establishment of a core network of dynamic conservation units. *European Forest Genetic Resources Programme (EUFORGEN)*. Bioversity International, Rome, Italy. *Xii*, 40 p

El-Kassaby YA, Lstiburek M (2009) Breeding without breeding. Genetic Res Camb 91:111–120. doi:10.1017/S001667230900007X

FAO (2014) Global plan of action for the conservation. Sustainable use and development of forest genetic resources. Rome. 31 p

Grundt HH, Fjellstad KB (2015) Ex situ-samlinger av norske skogtrær i arboreter og botaniske hager. Rapport fra Norsk genressurssenter, Skog og landskap 9:1–17

Grundt HH, Salvesen PH (2011) Kjenn din *Sorbus*. Rogn og asal i Norge. Rapport fra Norsk genressurssenter, Skog og landskap 23:1–104

Henriksen S, Hilmo O (red.) (2015). Norwegian Red List for Species 2015. Norwegian Biodiversity Information Centre, Norway

Jansson G, Hanssen JK, Haapanen M, Steffenrem A (2015) Genetic and economic gains from forest tree breeding in the Nordic countries. Submitted

Johnsen Ø, Kvaalen H, Yakovlev I, Dæhlen OG, Fossdal CG, Skrøppa T (2009) An epigenetic memory from time of embryo development affects climatic adaptation in Norway spruce. In: Gusta, I, Wisiewski, M, Tanio, K (eds) Plant cold hardiness—from Laboratory to the Field CABI Publication, pp 99–107

Myking T, Skrøppa T (2001) Bevaring av genetiske ressurser hos norske skogstrær. Aktuelt fra skogforskningen 2(1):1–44

OECD (2015) The OECD forest seed and plant scheme. Rules and Regulations. http://www.oecd.org/tad/code/RuleAndRegulations_EN_2015.pdf

Parducci L, Jørgensen T, Tollefsrud MM, Elverland E, Alm T, Fontana SL, Bennett KD, Haile J, Matetovici I, Suyama Y, Edwards ME, Andersen K, Rasmussen M, Boessenkool S, Coissac E, Brochmann C, Taberlet P, Houmark-Nielsen M, Krog Larsen N, Orlando L, Gilbert MTP, Kjær KH, Alsos IG, Willerslev E (2012) Glacial survival of boreal trees in Northern Scandinavia. Science 335:1083–1085

Skrøppa T, Martinsen DR, Følstad A (1993) Vekst og kvalitet av mellom-europeiske og norske granprovenienser plantet i Østfold. Rapport fra Skogforsk 7

Skrøppa T, Tollefsrud MM, Sperisen C, Johnsen Ø (2009) Rapid change in adaptive performance from one generation to the next in *Picea abies*—central European trees in a Nordic environment. Tree Genetics Genome 6:93–99

Tollefsrud MM, Kissling R, Gugerli F, Johnsen Ø, Skrøppa T, Cheddadi R, van der Knaap WO, Latałowa M, Terhürne-Berson R, Litt T, Geburek T, Brochmann C, Sperisen C (2008) Genetic consequences of glacial survival and postglacial colonization in Norway spruce: combined analysis of mitochondrial DNA and fossil pollen. Mol Ecol 17:4134–4150

Tollefsrud MM, Brochmann C, Johnsen Ø, Skrøppa Vendramin GG (2009) Combined analysis of nuclear and mitochondrial markers provides new insight into the genetic structure of North European *Picea abies*. Heredity 102:549–562

Tollefsrud MM, Sønstebø JH, Åsen, PA (2014) Registrering og genetisk karakterisering av villeple i Norge. Rapport fra Skog og landskap 18

Yakovlev I, Lee YK, Rotter B, Olsen JE, Skrøppa T, Johnsen Ø, Fossdal CG (2014) Temperature-dependent differential transcriptomes during formation of an epigenetic memory in Norway spruce embryogenesis. Tree Genetics Genomes 2014(10):355–366. doi:10.1007/s11295-013-0691-z

Chapter 6
Use of Molecular Markers in the Conservation Management of Tropical Trees

S. Changtragoon, R. Jalonen and A.J. Lowe

Abstract Thousands of tropical tree species provide timber, fuel, food, fodder and other ecosystem services for local and global communities. Yet, knowledge about the status and trends of tropical tree genetic diversity, which provides the basis for conservation and sustainability, remains limited. Nevertheless, the number of molecular studies is rapidly increasing thanks to technological advances and reduced costs. DNA markers provide ways to identify species or describe the genetic diversity of populations that can be used for gene conservation and restoration programs. Increasingly, molecular methods are also being used to identify the origin of timber to help combat illegal logging. Technical developments are also facilitating studies of adaptive traits, with expressed gene analysis offering real potential for detecting adaptive genetic diversity. Combining information from molecular studies with ecological and socioeconomic information related to, for instance, current and predicted species distribution, protected area networks or competing land use demands, enables the identification of the most suitable gene conservation, management or restoration strategies for each species. In this chapter we review advances in the development and application of molecular markers in conservation planning, landscape restoration and DNA forensics of tropical tree species, and discuss future directions.

S. Changtragoon (✉)
Forest and Plant Conservation Research Office, Department of National Parks, Wildlife and Plant Conservation, 61 Phaholyothin, Chatuchak, Bangkok 10900, Thailand
e-mail: suchitra.changtragoon@gmail.com

R. Jalonen
Bioversity International, Post Office, PO Box 236, 43400 Serdang, Selangor Darul Ehsan, Malaysia
e-mail: r.jalonen@cgiar.org

A.J. Lowe
Environment Institute and School of Biological Sciences, University of Adelaide, Level 12 Schulz Building, North Terrace, Adelaide, SA 5005, Australia
e-mail: andrew.lowe@adelaide.edu.au

M.R. Ahuja and S.M. Jain (eds.), *Biodiversity and Conservation of Woody Plants*, Sustainable Development and Biodiversity 17, DOI 10.1007/978-3-319-66426-2_6

Keywords Tropical tree genetic resources · Molecular markers
Gene conservation · DNA forensics · Forest and landscape restoration

6.1 Introduction

Estimates of the number of tropical tree species in the world vary between 40,000 and 53,000 (Slik et al. 2015). Thousands of tree species across Africa, Asia-Pacific, and Latin America and the Caribbean are actively managed for diverse purposes such as timber, fuel, food and fodder, and provide important contributions to local diets and livelihoods (FAO 2014a). Conservation and sustainable use of these species is hampered by continued deforestation and forest degradation across the tropical regions (FAO 2015) that result in habitat fragmentation and genetic erosion (Lowe et al. 2005; Vranckx et al. 2012), and by the progressive climate change that increases selection pressure on tree populations (Alfaro et al. 2014; Christmas et al. 2016). At a time when genetic diversity as the raw material of natural selection is most needed, the world is losing it faster than ever.

Despite the importance of genetic diversity as the foundation for productivity, resistance and adaptive capacity of tree populations (Ellstrand and Elam 1993; Hughes et al. 2008), it is commonly overlooked in conservation planning and biodiversity monitoring (Laikre 2010). In addition, knowledge of the status and trajectory of genetic diversity in tropical tree species remains limited. The TreeGenes database, maintained by the University of California, Davis, USA, compiles data on tree and shrub species that have been subject to genetic, genomic, transcriptome or proteomic studies. Currently, the database includes information on 461 tropical tree and shrub species from 65 genera, which is approximately 0.1% of the estimated number of tropical tree species[1] (Table 6.1). The majority of genetic diversity studies have been carried out on commercially important, widely planted species and genera, such as acacias (145 species), pines (42 species) and eucalypts (34 species) (data from TreeGenes database). Many of the species identified as national conservation priorities, especially for local use, have received little or no attention in research (FAO 2014a). However, the number of species being studied has increased rapidly since the early 2000s, as technological advances make studies easier and cheaper.

Molecular markers have many useful applications for gene conservation management. Neutral DNA markers can provide information about the mating system, gene flow and population history of a species, including determining evolutionary lineages (Lowe et al. 2004a, b). They can support species identification and delimitation of subspecies to help accurately assess the status of genetic diversity of taxa (Ng et al. 2013; Ng and Szmidt 2013). Within species, molecular markers can also be used to analyze patterns of genetic diversity, which are often considered as

---

[1]All species in the database which have at least part of their range within the tropical regions; http://dendrome.ucdavis.edu/treegenes/species/ [February 16, 2016].

Table 6.1 Indicative number of tropical tree and shrub species for which genetic diversity has been generated (figures in parentheses indicate number of genera studied). Data from TreeGenes database (http://dendrome.ucdavis.edu/treegenes/species/ [February 16, 2016])

Region	DNA analyses	Transcriptome studies	Proteomic studies	Total
Asia	161 (48)	6 (6)	151 (46)	161 (48)
Africa	49 (10)	0	44 (10)	49 (10)
Latin America and the Caribbean	111 (24)	4 (4)	104 (22)	111 (24)
Oceania (Australia)	138 (6)	6 (2)	123 (6)	138 (6)

baseline information for developing gene conservation strategies (e.g., Lee et al. 2006; Ayele et al. 2011; Vinceti et al. 2013; Ottewell et al. 2015). Recently, molecular methods have been used increasingly to identify species and origin of timber in combating illegal logging (Finkeldey et al. 2010; Tnah et al. 2010; Degen et al. 2013; Dormontt et al. 2015; Lowe et al. 2016). Planning and monitoring of forest and landscape restoration, a topic of many recent international commitments, would equally benefit from molecular information on the target species, but to date molecular studies in a forest restoration context have been very limited (Bozzano et al. 2014).

Neutral DNA markers are unsuitable for estimating local adaptation and population divergence or understanding the adaptive potential of populations under changing environments, which are of equal importance in terms of information for developing conservation plans (Kramer and Havens 2009). Only a weak correlation between molecular marker diversity and quantitative trait variation has been observed in a meta-analysis of 71 studies, that included both plant and animal species (Reed et al. 2001). For example, a study of the monkey puzzle tree (*Araucaria araucana*) demonstrated that neutral DNA makers (RAPDs) failed to detect genetic divergence in an ecologically important trait related to drought tolerance when neutral and quantitative genetic heterogeneity were compared within and among populations across the natural range of the species in southern South America (Bekessy et al. 2002). Karp et al. (1996) and Lowe et al. (2004b) have emphasized the importance of first focusing on the specific question to be addressed using molecular markers, and whether the choice of markers, sampling strategy and data analysis are appropriate. Porth and El-Kassaby (2014) stressed the value of integrating knowledge on adaptive complex traits as a companion to molecular markers for making informative management and conservation decisions. One option is to use molecular markers that can detect potentially adaptive genetic diversity (Krutovskii and Neale 2001a, b; Szmidt 1995; Szmidt and Wang 2000). Information from molecular studies must also be complemented by ecological and socioeconomic information to design effective strategies for conservation and sustainable use of tree genetic resources.

In this chapter we review advances in molecular marker application to the conservation and management of tropical tree species, and with a particular emphasis on how such methods can contribute to the development of practical conservation and management strategies. Case studies from across the tropics on socioeconomically valuable tree species illustrate the applications of molecular methods.

6.2 Types of Molecular Markers

Studies of genetic diversity should consider the three distinct genomes that underlie the biological function of a plant: the nuclear, chloroplast and mitochondria (Szmidt 1995). All these genomes harbor genes which are vital to growth, photosynthesis, respiration and other key biological processes. Each of these genomes differs in sequence, structure, RNA synthesis, regulatory genes expression and has different modes of inheritance, which may affect the extent of population differentiation for biparentally (nuclear) versus uniparentally (chloroplast and mitochondrial) inherited genes. It is therefore important to take account of the location of molecular markers and their distribution in the genomes, as well as their potential adaptive significance (strongly adaptive or near neutral) (Karp et al. 1996; Lowe et al. 2004a; Szmidt 1995).

Various types of molecular markers have been applied to the analysis of genetic variation of plants and forest trees over the past few decades.

Isoenzymes (or allozymes) are codominant markers that have been used in forest genetics since the 1960s, first on temperate and boreal tree species, and from the mid-1980s also on tropical tree species, especially in Australia and Central and South America (Finkeldey 2005). These markers are still used in some laboratories for applications such as estimation of mating system parameters and genetic diversity. However, the limited availability of allozymes would never allow for a genome-wide scan of variability (Porth and El-Kassaby 2014; Table 6.2).

More recently, DNA markers have been developed, and since the 1980s, restriction fragment length polymorphisms (RFLPs) have been applied in forest genetics and breeding. However, there are many limitations with RFLPs, such as the requirement for large amount of good-quality DNA, radioactive labelling, cloning and characterization of probe (Kumar et al. 2009).

In the early 1990s many new DNA techniques and markers were developed, made possible by the polymerase chain reaction (PCR), including randomly amplified polymorphic DNA (RAPDs) and amplified fragment length polymorphisms (AFLPs) which are dominant markers.

Recently, DNA markers such as PCR-RFLPs, microsatellites or simple sequence repeats (SSRs), AFLPs and RAPD markers have been widely used to determine genetic diversity of forest populations because they can detect more polymorphic loci than isoenzymes (Szmidt 1995; Changtragoon 1998; Table 6.2). Occasionally, other markers have also been used on tropical trees, for example direct amplification of length polymorphism (DALP) in *Gonystylus bancanus* Kurz (Fatma et al. 2011).

Table 6.2 Advantages and disadvantages of different types of molecular markers for plant biotechnology and genetic studies (Krutovsky and Neale 2005b)

Marker type	RFLP	SSR	RAPD	AFLP	Isoenzymes	STS-EST
Origin	Anonymous genic	Anonymous	Anonymous	Anonymous	Genic	Genic
Maximum theoretical number of possible loci	Limited by the restriction site (nucleotide) polymorphism (tens of thousands)	Limited by the size of genome and number of simple repeats in a genome (tens of thousands)	Limited by the size of genome and by nucleotide polymorphism (tens of thousands)	Limited by the restriction site (nucleotide) polymorphism (tens of thousands)	Limited by the number of enzyme genes and histochemical enzyme assays available (30–50)	Limited by the number of expressed genes (10,000–30,000)
Dominance	Codominant	Codominant	Dominant	Dominant	Codominant	Codominant
Null alleles	Rarely to extremely rare	Occasional to common	Not applicable (presence/absence type of detection)	Not applicable (presence/absence type of detection)	Rare	Rare
Transferability	Across genera	Within genus or species	Within species	Within species	Across families and genera	Across related species
Reproducibility	High to very high	Medium to high	Low to medium	Medium to high	Very high	High
Amount of sample required per sample	2–10 mg DNA	10–20 ng DNA	2–10 ng DNA	0.2–1 µg DNA	Several mg of tissue	10–20 ng DNA
	Difficult	Difficult	Easy	Moderate	Moderate	Moderate
Ease of assay	Difficult	Easy to moderate	Easy to moderate	Moderate to difficult	Easy to moderate	Easy to moderate
Automation/multiplexing	Difficult	Possible	Possible	Possible	Difficult	Possible
Genome and QTL mapping potential	Good	Good	Very good	Very good	Limited	Good

(continued)

Table 6.2 (continued)

Marker type	RFLP	SSR	RAPD	AFLP	Isoenzymes	STS-EST
Comparative mapping potential	Good	Limited	Very limited	Very limited	Excellent	Good to very limited
Candidate gene mapping potential	Limited	Useless	Useless	Useless	Limited	Excellent
Potential for studying adaptive genetic variation	Limited	Limited	Limited	Limited	Good	Excellent
Cost of development	Moderate	Expensive	Inexpensive	Moderate	Inexpensive	Expensive
Cost of assay	Moderate	Moderate	Inexpensive	Moderate to expensive	Inexpensive	Moderate
Cost of equipment	Moderate	Moderate to expensive	Moderate	Moderate to expensive	Inexpensive	Moderate to expensive

Using microsatellite markers provides an improved method, compared to isoenzymes, for estimating mating system parameters and gene flow in forest trees since they are also codominant, but can detect higher levels of variation. In a comparative study, Degen et al. (1999) investigated genetic variation and differentiation of two pedunculate oak (*Quercus roburs* L.) stands using microsatellite DNA makers and allozyme loci. The results showed that the number and effective number of alleles were five to six times higher, and the observed heterozygosity was three times higher, for SSR compared to allozyme loci. The results are consistent with the study of Chase et al. (1996) who used SSR markers for an investigation of the population and conservation genetics of a tropical tree species, *Pithecellobium elegans* Ducke. (Mimosoidaeae). The authors estimated the frequency of SSR markers and assessed the level and distribution of variability at these makers, and then contrasted the results with variability found at allozyme makers in the same population. The results indicated that SSR loci are powerful tools for the analysis of population structure and that they provide a means of accurately estimating gene flow and paternity, two important parameters in conservation biology.

One of the shortcomings of SSR markers is that they cannot generally be used for comparisons between species, although in some cases they have been applied to species within a single genus (e.g., *Shorea* spp., Pandey and Geburek 2009; Senakun et al. 2011) and also among closely related genera (e.g., *Shorea* and *Parashorea*, Abasolo et al. 2009). Other disadvantages of SSR markers have been the traditionally high cost and length of time required to develop markers (Lowe et al. 2004a; Squirrell et al. 2003; also see, e.g., Boontong et al. 2008 and Pandey and Changtragoon 2012 for *Azadirachta indica* A. Juss. var. *indica* and *Phyllanthus emblica* L., respectively). With technological advances, these disadvantages have been ameliorated by using next-generation sequencing (NGS) approaches to isolate microsatellite loci, which allows the efficient identification of large numbers of microsatellites at a fraction of the cost and effort of traditional approaches (Gardner et al. 2011). The major advantage of NGS methods is their ability to produce large amounts of sequence data from which to select and develop numerous genome-wide and gene-based SSR loci (Zalapa et al. 2012). Ekblom and Galindo (2011) reviewed how NGS has been, and can be, applied to ecological, population genetic and conservation genetic studies of non-model species, in which there are no (or very limited) genomic resources.

The innovation of automated DNA sequencing combined with the availability of plant and forest tree genes (dendrome.ucdavis.edu/treegenes) and DNA sequence databases on the NCBI database (National Center for Biotechnology Information, www.ncbi.nlm.nih.gov), facilitate the development of new molecular markers for determining genetic variation of forest trees at specific functional and regulatory genes, regions of DNA and (nuclear, chloroplast and mitochondrial) genomes. These new marker classes include specific amplicon polymorphisms (SAPs), expressed sequence tag polymorphisms (ESTPs) and single nucleotide polymorphisms (SNPs) (Changtragoon 2004).

According to Krutovskii and Neale (2001a, 2005b), ESTPs are the most informative markers in terms of gene function among these most recently developed markers and are the first genetic markers that offer real potential for broad detection of adaptive genetic diversity. Many EST sequences are available for several forest tree species, such as *Eucalyptus* spp. and *Populus* spp. Yan et al. (2012) studied the variations of SSRs within genes in different tree species and performed a meta-analysis of 30,000 ESTs in pine (*Pinus* spp.), poplar (*Populus* spp.) and eucalyptus (*Eucalyptus* spp.) downloaded from the NCBI database. The results showed that the proportion of ESTs containing microsatellites was 18.7% in eucalyptus and 15.3% in poplar, whereas only 8.2% occurred in pines. This study not only provides important parameters for understanding the differentiation of genomes of different tree species, but also provides a bioinformatics reference for using sequence resources to develop polymorphic SSR markers (Yan et al. 2012). Külheim et al. (2009) generated comparative SNP diversity among four Eucalyptus species for genes from secondary metabolite biosynthetic pathways. Cheua-ngam and Volkaert (2006) developed a DNA-based catalase gene marker in teak (*Tectona grandis* Linn.f.) for applications such as evolutionary studies, population genetics, outcrossing rate estimation and mating patterns in this tropical tree species.

New DNA technologies such as genome-wide discovery of SSRs and SNPs and the detection of common and rare functional variants through next-generation sequencing approaches have provided opportunities for affordable in-depth characterization of plant genomes (Davey et al. 2011; Tsai et al. 2011; Marroni et al. 2011; Chen et al. 2013; Porth and El-Kassaby 2014) and have allowed the genome to be much more densely sampled to determine the full range of evolutionary processes acting across the genome (Allendorf et al. 2010; Stapley et al. 2010; Li et al. 2012a; Narum et al. 2013). However, these approaches still face some challenges in terms of capacity and support in developing countries if more work is to be done on tropical forest trees.

The details of some of the above-mentioned molecular markers and their application can be found in Amaral (2001), Harry et al. (1998), Karp et al. (1997), Kristensen et al. (2001), Krutovskii and Neale (2001b), Wickneswari (2001), Wang and Szmidt (2001), Szmidt (1995), Szmidt and Wang (2000), Schnabel and Krustovskii (2004), Slavov et al. (2004), Krustovsky and Neale (2005a, b), Petit et al. (2005), Krutovsky (2006), González Martínez et al. (2006), Krustovsky et al. (2012, 2014).

6.3 Application of Molecular Markers to Conservation and Management of Tropical Tree Species

6.3.1 Assessment of Genetic Diversity Using Molecular Markers

Genetic diversity of species is influenced by six evolutionary processes (Lowe et al 2004a, b): (1) random genetic drift—the non-directional changes in genotypes frequencies among generations due to random chance in small populations,

(2) natural selection, due to relative differences among genotypes in viability or reproductive success, (3) migration, the exchange of genes between populations that differ in genotypic frequencies, (4) mating, the process mediating the recombination and assortment of genes between generations, (5) mutation (Wickneswari 2001) and (6) the Wahlund effect, inbreeding and the presence of subpopulation structure, where gene flow is prevented by habitat fragmentation (Hartl and Clark 2007, Lowe et al. 2005; Porth and El-Kassaby 2014). These evolutionary processes may affect the level and distribution of genetic variation either individually or in concert. Migration may increase variation within populations if the populations were originally distinct, but reduce variation between populations and possibly reduce fitness. The size of the recipient population may increase, but the adaptability of the resultant genetic variation is dependent on the forms of selection and migration. However, if the processes that maintain the levels and distribution of genetic variation remain unchanged, then the biological capacity for adaptability is expected to remain at least as good into the immediate future (Wickneswari 2001).

Changtragoon (2004) reviewed papers related to molecular makers as follows: Molecular genetic markers hold great promise for several conservation applications, including measuring fundamental conservation parameters, such as effective population size, past bottlenecks, sex-specific gene flow or founder contributions (Lowe et al. 2004b). Hedrick (2001) stated that they also can be used to infer the historical and geographic relationships between groups (Hedrick 2001). The use of molecular markers has revolutionized studies of mating systems, pollen movement, seed dispersal and genetic processes. Results of such studies are of considerable practical significance in relation to conservation and breeding programs and include population sampling, seed orchard design and management, controlled pollination methods and clonal forestry programs for conservation and breeding (Haines 1994). Besides, Karp et al. (1996) mentioned that molecular markers may be used in four types of measurements required for effective ex situ and in situ conservation of plants: identity, similarity and structure, and detection of genetic background of individuals, accessions, populations and taxa. However, Rajora and Mosseler (2001) and Ottewell et al. (2015) stated that designing an effective gene conservation strategy begins with an assessment of the genetic status (genetic diversity, population structure, mating system, etc.) of a species across its range. While an assessment based on both adaptive traits and molecular or biochemical genetic markers is preferable, biochemical and molecular genetic variation alone can now be used to provide an initial assessment.

Changtragoon (2004) stated that most of tropical countries have in situ and ex situ forest gene conservation measures established based on silvicultural practices. The application of molecular markers to genetic conservation of these programs should involve two steps. Firstly, markers can be used to evaluate the status of genetic background of ex situ plantations and in situ sites that are established based on conventional silvicultural practices. Secondly, they can be used to check whether the populations contain correctly identified clones and ramets and have sufficient genetic diversity for conservation as a representative sample of the species' gene pool. If not, remedial measures could then be taken according to the information guidelines provided by molecular genetic investigations.

Table 6.3 shows an example of the application of molecular markers to a study of genetic variation in some tropical forest trees and wild plants, including ecologically (e.g., mangrove species) and economically important species (e.g., teak, *Tectona grandis* Linn.f, *Dipterocarp* spp., *Shorea* spp.). Ng et al. (2013) and Ng and Szmidt (2013) used SNPs from the nuclear genome to identify natural hybrids of the mangrove *Rhizophora apiculata* Blume, *R. mucronata* Poir. and *R. stylosa* Griff., since it is difficult to identify those species and their natural hybrids based on morphological features alone. The results have significance for conservation and restoration approaches. Ng et al. (2015) also investigated genetic variation of Rhizophora mangroves across the Malay Peninsula using one chloroplast (cpDNA) region and five partial nuclear gene (nDNA) regions. They found that *R. apiculata* Blume, *R. mucronata* Poir. and *R. stylosa* Griff. were characterized by low intraspecific genetic variation and a deficiency of heterozygotes within populations. These species also have different demographic histories, despite being closely related and having sympatric distributions today. Furthermore, all three species exhibited high levels of inbreeding, apparently as a result of limited pollen and propagule dispersal, which also contributed to population differentiation. These results are consistent with the study of Inomata et al. (2009) on *R. apiculata* and *R. mucronata* from Thailand who used two cpDNA regions and five nDNA genes. However, the study of Changtragoon (2007) on genetic diversity of R. *apiculata* and *R. mucronata* in Thailand using AFLP makers revealed higher genetic diversity and differentiation among populations (Table 6.3).

Minobe et al. (2010) evaluated the genetic diversity of another mangrove species, *Bruguiera gymnorrhiza* (L.) from the southern islands of Japan, Thailand, Malaysia, Indonesia, Macronesia and India, using one nDNA region and two cpDNA regions. Low levels of polymorphism were found within each of the three geographic regions, Pacific Ocean, Bay of Bengal and Arabian Sea. These results are consistent with the hypothesis that low genetic diversity within any local population and differentiation between the different regions are caused by very low gene flow between the regions, coupled with frequent fluctuation of population sizes due to changes in sea level.

Numerous studies have assessed genetic diversity of teak (*Tectona grandsis* Linn.f.), which is a valuable timber species across tropical regions, even though its natural distribution is limited to Asia. Allozymes, SSR, AFLPs, ISSR and cpSNP have all been used to investigate genetic diversity of natural populations in India (Ansari et al. 2012), Thailand (Changtragoon and Szmidt 1999, Changtragoon 2001a) and Myanmar (Thwe-Thwe-Win and Watanabe et al. 2015, Thwe-Thwe-Win and Hirao et al. 2015), and in plantations within (Thwe-Thwe-Win and Watanabe et al. 2015), and outside of species' natural range, in Brazil (Alcântara and Veasey 2013). The genetic diversity and differentiation among populations as well as outcrossing rates in some of the studies are shown in Tables 6.3 and 6.4, respectively. Watanabe et al. (2004) discriminated teak plus trees in Indonesia using RAPD markers, while Narayanan et al. (2007)

Table 6.3 Examples of application of molecular markers to study genetic variation in tropical forest trees and wild plants

Species	Type of molecular markers		Genetic diversity (He: expected heterozygosity/Ht: gene diversity	Genetic differentiation		References
	Isoenzyme gene markers	DNA markers		F_{ST}	G_{ST}	
Acacia auriculiformis A. Cunn. ex Benth.	√		0.081		0.27	Wickneswari and Norwati (1993)
Acacia mangium Willd.		RFLP	0.130	0.311		Butcher et al. (1998), Moran et al. (1989)
Adansonia digitata L.		AFLP	0.355	0.127		Assogbadjo et al. (2006)
Bambusa bambos(L.) Voss		SSR (microsatellite)	0.369	0.243		Laphom and Changtragoon (2005)
Dipterocarpus alatus Roxb. ex G.Don	√		0.0924	0.182		Changtragoon and Boontawee (1999)
				0.128	0.067	Changtragoon (2001b)
Dryobalanops aromatic Gaertn.		SSR	0.709		0.193	Lim et al. (2002)
Dryobalanops aromatic Gaertn.		SSR	0.635	0.194		Dwiyanti et al. (2015)
Dryobalanops beccarii Gaertn.		SSR	0.430	0.283		Dwiyanti et al. (2015)
Dryobalanops lancerolata Gaertn.		SSR	0.471	0.246		Dwiyanti et al. (2015)
Dryobalanops rappa Gaertn.		SSR	0.392	0.159		Dwiyanti et al. (2015)
Dryobalanops keithii Gaertn.		SSR	0.420	0.156		Dwiyanti et al. (2015)
Dryobalanops oblongifolia subsp. *oblongifolia* Gaertn.		SSR	0.583	0.194		Dwiyanti et al. (2015)

(continued)

Table 6.3 (continued)

Species	Type of molecular markers		Genetic diversity (He: expected heterozygosity/Ht: gene diversity	Genetic differentiation		References
	Isoenzyme gene markers	DNA markers		F_{ST}	G_{ST}	
Dryobalanops oblongifolia subsp. *occidentails* Gaertn.		SSR	0.464			Dwiyanti et al. (2015)
Hopea odorata Roxb.		SSR	0.356	0.251		Nguyen et al. (2014)
Paphiopedilum exul (Ridl.) Rolfe		AFLP	0.301	0.082		Wanichkul and Changtragoon (2005)
Pinus merkusii Jungh. & de Vries	√		0.058	0.104		Changtragoon and Finkeldey (1995a)
Prunus cerasoides D.Don		SSR	0.697	0.115	0.12 (nSSR) 0.92 (cpSSR)	Pakkad et al. (2003)
Quercus semiserrata Roxb.		SSR (microsatellite)	0.68 (nSSR) 0.16 (cpSSR)	0.12 (nSSR) 0.83 (cpSSR)		Pakkad et al. (2003)
Rhizophora apiculata Blume		AFLP	0.316	0.250		Changtragoon (2007)
Rhizophora mucronata Poir.		AFLP	0.385	0.212		Changtragoon (2007)
Shorea obtusa Wall. ex Blume		SSR (microsatellite)	0.664	0.030	0.117	Senakun et al. (2011)
Shorea leprosula Miq.	√		0.406			Lee et al. (2000a)
Tectona grandis Linn.f.	√	RAPD	0.310	0.217	0.632	Changtragoon (2001a),

(continued)

Table 6.3 (continued)

Species	Type of molecular markers		Genetic diversity (He: expected heterozygosity/Ht: gene diversity	Genetic differentiation		References
	Isoenzyme gene markers	DNA markers		F_{ST}	G_{ST}	
						Changtragoon and Szmidt (1999, 2000)
		SSR	0.63	0.227		Hansen et al. (2015)
		SSR	0.28–0.78	0.22		Fofana et al. (2009)
	√		0.347			Kertdikara and Prad Prat (1995a, b)
		SSR	0.489	0.21		Fofana et al. (2008)
		SSR	0.599	0.074		Thwe-Thwe-Win, Watanabe et al. (2015)
		SSR	0.652	0.094	0.346	Thwe-Thwe-Win, Hirao et al. (2015)
		cp SNP		0.293	0.09	Thwe-Thwe-Win, HIirao et al. (2015)
	√			0.12	0.1533	Kertadikara and Prat (1995a)
		ISSR	0.40			Ansari et al. (2012)

Table 6.4 Estimate of outcrossing rate (t_m) of some tropical trees and wild plants, obtained using isoenzyme gene and mixed mating model

Species	Estimate of outcrossing rate (t_m)	References
Shorea megistophylla Ashton	0.71–0.87	Murawski et al. (1994)
Stemonoporus oblongifolius Thwaites	0.84	Murawski and Bawa (1994)
Shorea congestiflora (Thw.) P. Ashton	0.874	Murawski et al. (1994)
Shorea trapezifolia (Thw.) P. Ashton	0.519–0.602	Murawski et al. (1994)
Pterocarpus indicus Willd.	0.908	Finkeldey et al. (1998, 1999)
Pterocarpus macrocarpus Kurz	0.719–0.959	Liengsiri et al. (1995)
Pinus merkusii Jungh. & de Vries	0.017–0.843	Changtragoon and Finkeldey (1995a)
Tectona grandis Linn.f.	0.95	Kjær and Suangtho (1995)
Tectona grandis Linn.f.	0.983	Kertadikara and Prat (1995b)
Tectona grandis Linn.f.	0.872–0.995	Changtragoon and Szmidt (1999)
Calamus palustris Griff.	0.862–0.980	Changtragoon, unpublished
Rhizophora apiculata Blume	0.2–0.97	Changtragoon, unpublished
Shorea leprosula Miq.	82.7	Lee et al. (2000b)

characterized teak plus trees in India using RAPD and ISSR makers. Thongthawee and Volkaert (2014) analyzed the teak genome and its diversity using a complete chloroplast genome assembly and partial mitochondrial genome. Three different chloroplast haplotypes were detected among 10 trees collected from different sources. This effort could be expanded and applied for the identification of origin in future (e.g., Dormontt et al. 2015).

In a regional study of teak, Kertadikara and Prat (1995a, b) investigated genetic structure and mating system in teak provenances within (India and Thailand) and outside of the species' natural distribution range (Indonesia, Ivory Coast and Tanzania) using allozymes. They found a general lack of heterozygosity (average fixation index 0.11), and the multilocus outcrossing rate was approximately 0.98. More recently, Hansen et al. (2015) investigated genetic diversity and genetic structure of twenty-nine provenances of teak (*Tectona grandis* Linn.f) representing the full distribution range of the species in Asia, including India, Myanmar, Thailand and Laos, using six SSR markers. The provenances originating from the semi-moist east coast of India were the most diverse. On average, these provenances had 34% more alleles than the overall mean across all 29 provenances. At the opposite end of the spectrum, the two provenances from Laos clearly showed the lowest genetic diversity, having approx. 60% fewer alleles than the overall mean of the study. In the eastern part of the natural distribution area, comprising Myanmar, Thailand and Laos, there was a strong clinal decrease in genetic diversity the further east the provenance was located.

Overall, the pattern of genetic diversity supports the hypothesis that teak has its center of diversity in India, from where it spread eastwards. This finding is consistent with the study of Fofana et al. (2009) of Indian, Lao and Thai provenances, using fifteen SSR makers. Hansen et al. (2015) analyzed molecular variance (AMOVA) which gave an overall highly significant F_{ST} value of 0.227. The estimated overall differentiation (G_{ST}) was 0.632, implying a strong genetic structure among populations. A neighbor-joining (NJ) tree contained three distinct groups: (1) the eight provenances from Thailand and Laos, (2) Indian provenances from the dry interior and moist west coast and (3) provenances from northern Myanmar. Knowledge of the genetic patterns within and between major regions, as revealed in this study, can support local efforts to develop and refine effective gene conservation measures for this species. The results of this study can also serve as a reference to test the origin of landraces which could support sustainable domestication of teak genetic resources in several countries.

Examples of the application of molecular markers to study genetic variation of tropical tree species using genotype data to infer (a) genetic variation within and among populations and (b) mating system are given in Tables 6.3 and 6.4, respectively. The results could be used as a basis for conservation management in the future.

6.3.2 Development of Gene Conservation Strategies

Up to half of the world's tree species may be threatened, according to the Food and Agriculture Organization of the United Nations, based on reports from 86 countries (FAO 2014a). This proportion is considerably higher than that from a global assessment a decade and a half earlier, which had estimated that 10% of all tree species were threatened (Oldfield et al. 1998). The differences in the estimates are partly explained by different assessment methods, but they also suggest that many species estimated as not threatened at a global scale may be threatened within parts of their distribution range. Furthermore, more than 1.5 million km^2 of forests have been lost between the two assessments, 32% of that loss being in the tropical rainforest ecozone alone (2000–2012; Hansen et al. 2013), thus worsening the conservation status for tropical tree species and limiting their potential to contribute to socioeconomic development of the countries in which they occur. Genetic diversity of offspring in fragmented tree populations is generally less than that of adult trees, suggesting an extinction debt resulting from historical deforestation (Lowe et al. 2015; Vranckx et al. 2012). The majority of the world's tropical tree species are assumed to have small population sizes or to be rare (Slik et al. 2015; Hubbell 2013), which makes them vulnerable to overexploitation, habitat fragmentation and environmental changes. Yet, common plant species may be just as susceptible to the population genetic consequences of habitat fragmentation as rare species (Honnay and Jacquemyn 2007). Intraspecific diversity is commonly overlooked in conservation assessments and planning and monitoring that are based on generic indicators such as species richness (Laikre 2010).

Gene conservation objectives vary from preserving the potential for particular trait expressions to preserving maximum genetic diversity, and maintaining adaptability (Finkeldey and Hattemer 2007). Maintaining adaptability is often considered as the main objective for gene conservation (Christmas et al. 2016) as it allows tree populations to respond to a range of potential selection pressures under current and largely unknown future environments (Gregorius 1991). In practice, genetic information for developing conservation strategies usually comes from inventories of a small number of gene loci with neutral DNA markers. Even though such markers cannot detect adaptive capacity of tree populations, they can importantly complement conservation assessments and planning for example by helping to identify centers of diversity (e.g., van Zonneveld et al. 2012; Vinceti et al. 2013) and refugia (Thomas et al. 2015), genetic bottlenecks (Kamiya et al. 2012), connectedness of populations (Finger et al. 2012), and evaluating human impacts, for instance from logging or habitat fragmentation (Lowe et al. 2005, 2015; Jalonen et al. 2014).

Independent of the specific conservation objectives, all genetic conservation strategies need to consider (i) the number and location of conserved or sampled populations to capture useful genetic diversity, (ii) the different types of populations or collections to be conserved and management options for them, and (iii) in the case of in situ or on-farm conservation, the number of individuals within a population needed to maintain genetic diversity and withstand inbreeding depression (White et al. 2007). These parameters vary depending on the conservation objectives, the characteristics of the target species and environmental processes, including changing climate—and even on the genetic parameters used for defining priorities. Proposed approaches for identifying the number of populations and individuals within populations to be conserved, by using molecular methods, are briefly discussed below, and illustrated through Case study 1.

Conducting an inventory of the extent and distribution of genetic resources of target tree species, to identify the number and location of populations that are genetically particularly diverse or distinct, is commonly the first step in developing practical gene conservation strategies for species. Populations may be clustered according to their genetic similarity (Vinceti et al. 2013). Gene inventories can be followed by the use of geographic information systems to overlay genetic information with, for instance, species distribution data, protected area coverage and occurrence of threats such as climate change or deforestation to identify gaps in current conservation efforts and, subsequently, priority areas or populations for conservation that best complement existing efforts (Lipow et al. 2003; van Zonneveld et al. 2014). However, large datasets are required to analyze genetic diversity across species distribution ranges using such approaches.

Practical approaches for designing a network of gene conservation areas combining molecular data with ecological and socioeconomic information have been developed for some threatened and socioeconomically important tropical species (Ottewell et al. 2015). Changtragoon and Finkeldey (1995a, b) investigated genetic diversity and mating system of *Pinus merkusii* using allozymes genes and identified priority populations for conservation, based on highest diversity and complementarity of the populations. Very high selfing rates were discovered in 9 of the 10 studied populations and associated with low population densities and overmature stands, resulting in limited flowering and pollen production, poor synchronization of flowering and scarcity of foreign pollen available for fertilization of ovules. Hence, the authors recommended establishing complementary ex situ conservation stands with high population densities to facilitate outcrossing, and seed collection strategies to capture adequate diversity for establishing these stands.

Lee et al. (2006) studied the ecology and genetic diversity of a rare dipterocarp, *Shorea lumutensis*, using SSR markers, and recommended specific attention be given to one of the five existing populations, which did not cluster with the others and also displayed the highest allelic richness of all populations. Ayele et al. (2011) developed a weighted-score population prioritization matrix based on genetic (microsatellite and AFLP markers) and morphological characteristics (leaf traits) and demographic criteria to identify priority populations of endangered *Hagenia abyssinica* for conservation and domestication. Vinceti et al. (2013) prioritized the genetically most diverse populations (measured as allelic or haplotype richness) of an Afromontane tree species, *Prunus africana*, but also selected populations for conservation based on the highest presence of locally common alleles, climatic clusters, and the minimum number of populations required to capture all genetic diversity through the species' range based on chloroplast and nuclear markers (Case study 1). Analyzing the presence of locally common alleles and climatic clustering can help capture variation in adaptive capacity and can complement information obtained with neutral DNA markers. Including not only the most genetically diverse populations, but also those that are peripheral could also help conserve adaptive capacity, because peripheral populations may often harbor valuable adaptive variation that has developed in response to the specific marginal environments (Pandey and Geburek 2010; Fady et al. 2016).

After the priority populations have been identified, appropriate conservation methods need to be selected. In situ conservation is often the only practical approach for gene conservation of tropical tree species, the taxonomy, silvicultural features, and potential uses of which are generally poorly known (Finkeldey and Hattemer 2007). Moreover, many tropical tree species have recalcitrant seeds, which do not tolerate desiccation and cannot be easily conserved in genebank collections (Sacandé et al. 2004). Since forest trees live longer than annual or crop plants, high genetic diversity and outcrossing rate in priority populations should improve survival, viability, longevity, disease and insect resistance in an unpredictable and changing environment (Changtragoon and Szmidt 1997). However, if genetic and ecological studies reveal that genetically distinct priority populations, for example, have too few reproductive individuals to be viable, have poor regeneration, or are at risk because of climate change or deforestation, ex situ conservation may be necessary (Lee et al. 2006; Vinceti et al. 2013). Van Zonneveld et al. (2012) identified priority populations for ex situ collections of the neotropical fruit tree *Annona cherimola* based on high frequencies of alleles not included in existing genebank collections and high observed levels of inbreeding in these populations, suggesting a risk of genetic erosion.

Methods used to estimate effective population sizes for maintaining genetic diversity were reviewed by Nunney and Elam (1994). Estimates of effective population sizes in natural populations have often been based on indirect information, either demographic (e.g., mean and variance of the number of progeny per parent) or genetic data (e.g., temporal changes in allele frequencies) that are generally difficult to obtain (Lee et al. 2006).

Lee et al. (2006) concluded that estimates of minimum population sizes vary based on the genetic and ecological assumptions, breeding systems of the target species, proportion of genetic variation to be conserved, and levels of uncertainty, necessitating species-specific estimates. Using microsatellite markers and simulation methods, they estimated the minimum population size for maintaining 95% of the genetic variation of the rare, locally common and predominantly outcrossing *Shorea lumutensis* to be 270 individuals. The breeding unit was estimated to be 52 trees (12 ha with 4.3 trees ha^{-1}). Similar methods were previously used to estimate minimum population sizes for the common, outcrossing tree species *Intsia palembanica* (Lee et al. 2002). Based on estimates of pollen dispersal within and between populations, Konuma et al. (2000) estimated the breeding unit area for the widespread, semi-gregarious and predominantly outcrossing dipterocarp *Neobalanocarpus heimii* to include 62 trees, corresponding to an area of 68 ha and population density of 0.7 reproductive trees ha^{-1}.

For most species in need of conservation action, information on minimum population sizes or even the genetic factors influencing them is simply not available, and rules of thumb are needed. Franklin (1980) suggested that a minimum population size of 50 individuals is needed to avoid inbreeding depression and 500 individuals are needed to prevent genetic erosion as a result of genetic drift, while subsequent estimates of minimum population sizes range from 150 to 5000 or more individuals (Graudal et al. 2004; Lawrence and Marshall 1997; Lowe et al. 2005). Based on evidence accumulated since the 1980s, Frankham et al. (2014) published revised recommendations on minimum population sizes (to apply to both plant and animal species), recommending effective population sizes of at least 100 individuals to avoid inbreeding depression and at least 1000 individuals to maintain diversity and evolutionary potential. They also highlighted that population viability assessments typically overlook the impacts of inbreeding depression and evolutionary potential which is particularly important with a changing environment.

While the above examples focus on the identification of populations for the purpose of establishing in situ conservation sites or germplasm collections for ex situ conservation, similar considerations can also be applied to production forests or agroforestry species for conserving useful genetic diversity to maintain productivity and adaptive capacity of the populations (Dawson et al. 2009; Jamnadass et al. 2009; Jalonen et al. 2014; Wickneswari 2016, this volume).

Case study 1 Identifying the number and location of populations to conserve: *Prunus africana*

Prunus africana is an indigenous African tree species widely collected and exported for its medicinal value. Occurrence data from herbaria collections show a broad but disjunct distribution of the species across Afromontane regions. The species is important as a source of income for collectors and local processors, but is threatened by overexploitation, agricultural expansion and environmental change. Vinceti et al. (2013) identified priority populations for the conservation of genetic diversity (measured as allelic and haplotype diversity) of *P. africana*, using two alternative methods. The first method (S1) was to maximize genetic diversity and distinctiveness of conservation units based on genetic and climatic criteria; the second method (S2) was to optimize representativeness of the genetic diversity found throughout the species' distribution range. Both approaches aim at maximizing the evolutionary potential of the species by identifying and conserving populations with the highest genetic diversity.

The analysis was based on occurrence data of the species from 1500 georeferenced observations and genetic information from a recent continent-wide genetic study that sampled trees from 32 natural populations in nine countries (Kadu et al. 2011, 2013). Genetic data were analyzed using microsatellite markers at seven chloroplast DNA loci (cpSSR) and six nuclear loci (nSSR). Molecular marker data were formatted so as to attribute coordinates to each allele or haplotype, to enable spatial analysis.

In the first method (S1) for identifying priority populations, populations were grouped into five clusters according to their genetic (chloroplast- and nuclear-based) and climatic similarity. In each genetic cluster, the population with either the highest allelic or haplotype richness, and the highest presence of locally common alleles was selected as priority sites. Only one of the populations selected based on allelic diversity overlapped with those selected based on haplotype diversity, resulting in a combined list of eight priority populations across five countries (Cameroon, Tanzania, Kenya, Madagascar and Zimbabwe). Subsequently, the results were compared with the four identified climatic clusters. Populations within two climate clusters had not been selected based on genetic data, and in each of these, the population with the highest rank in both allelic or haplotype richness and locally common alleles was added to the list of priority populations, resulting in a total of 10 populations in 6 countries (above-listed countries and Equatorial Guinea).

The second method (S2) was based on the identification of a minimum number of populations needed to capture all genetic diversity based on chloroplast and nuclear markers. The first population chosen in this method was the one with the highest allelic richness, and successive populations were selected based on how they best complemented the intraspecific diversity already represented in the previously selected populations (Rebelo and Siegfried 1992). This method allows identification of priority populations based not only on their intraspecific diversity, but also on their complementarity. The method resulted in a list of 18 priority populations to conserve the diversity found in the 31 studied populations (after combining two populations that fell in the same grid cell in the spatial analysis because of their proximity). This number was higher than that obtained using the first method because the second method focused on the representativeness of diversity across the species distribution range. It also identified one of the populations in a distinct climatic cluster as a priority, in contrast to the first method.

Six populations in five countries (Tanzania, Kenya, Madagascar, Zimbabwe and Equatorial Guinea) were identified as priority populations by both methods S1 and S2 and constitute a core set of proposed conservation areas. The authors recommended to inventory the populations in terms of area, number of individuals, effective population sizes and regeneration success, to assess the viability of and threats to these priority populations. Results of the molecular study were also compared with the protection status of populations and predicted climate changes to further prioritize populations for action. The modelled distribution of the species was compared with the location of protected areas, which showed that two of the six priority populations were not within the boundaries of protected areas and should be protected as a priority. The impact of climate change was assessed by modelling the species' potential distribution under future climate conditions, using climate projections for the year 2050. Results suggested that another two of the six populations would become marginal for the species distribution range by 2050 under progressive climate change and deserve particular attention to understand and help conserve their diversity to maintain evolutionary potential. Genetically less diverse populations away from the core distribution area might be more vulnerable to climate change than highly diverse populations. However, marginal populations may also harbor useful traits for climate adaptation, which are not reflected in results obtained using neutral markers.

6.3.3 DNA Forensics

Tropical forest trees have been illegally logged over decades as a result of poverty and unsustainable management of forest and land use. Even though lots of effort has been made to improve legislation and strengthen law enforcement in some tropical countries, the problem remains. Therefore, timber consumer countries have developed regulations such as the European Union Timber Regulation (EUTR) (www.euflegt.efi.int) and the Lacey Act of the United States (forestlegality.org) to ensure timber imported to their markets has not been illegally logged (Lowe et al. 2016).

DNA forensics has been long used on humans for legal matters such as paternity tests (Butler 2005 and Kobilinsky et al. 2005), and more recently for plants and wildlife (Craft et al. 2007; Zaya and Ashley 2012), to identify species or subspecies and origin (Changtragoon 2012; Deguilloux et al. 2004; Dormontt et al. 2015; Manel et al. 2002; Withler et al. 2004; Wasser et al. 2004, 2007, 2015).

Molecular genetic markers have been suggested as potential suitable tools for identifying the origin of wood and wood products, and to identify false declarations, due to the great stability of DNA and the impossibility to manipulate the DNA contained in dead plant tissue (Finkeldey et al. 2007, 2010; Lowe 2007;

Lowe and Cross 2011). Two basic requirements are needed to use DNA markers for tracing the origin of wood and wood products. DNA needs to be isolated from woody tissues, and informative markers need to be identified and investigated (Finkeldey et al. 2010; Lowe and Cross 2011). In this regard, DNA markers and DNA profiling have been acknowledged as an efficient tool to identify and verify confiscated wood, whether it is illegally logged from protected species or from individual trees and populations in National Parks or other protected areas. There have been a number of efforts to develop DNA tools and markers as well as methodologies for DNA extraction from wood samples in order to be used in databases and as standards to compare with confiscated wood samples, and to investigate potential illegal logging. DNA extraction from wood has been developed for various tropical tree species such as species of the Dipterocarpaceae (Rachmayanti et al. 2006, 2009; Tnah et al. 2012; *Gonystylus bancanus* (Asif and Cannon 2005); *Tectona grandis* Linn.f, *Dalbergia cochinchinensis* Pierre., and *Mangifera* spp. (Changtragoon, unpublished).

Lowe (2007) and Lowe and Cross (2011) stated that the development of methodologies in each of the areas of DNA extraction from wood, identification of genetic variation across the natural range of tropical tree species and statistical assignment frameworks, offers exciting possibilities for using DNA to verify timber sources. Finkeldey et al. (2010) reviewed molecular genetic tools to infer the origin of forest plants and wood and provided suggestions on DNA extraction practices and protocols, as well as examples of the identification of species and origin of wood of Dipterocarpaceae by using cpDNA (Indriogo et al. 2006) and AFLP and sequence characterized amplified regions (SCAR) (Nuroniah 2009).

Verhaegen et al. (2010) used fifteen microsatellite DNA markers to determine the genetic origin of teak introduced in Africa and Indonesia and found that nearly 95% of teak landraces in Benin, Cameroon, Ivory coast, Tanzania, Togo and Senegal came from India, and 96% of Indonesian and Ghanaian teak appeared to be very closely linked to Central Laos. The genetic origin of introduced teak was confirmed by the main traits of interest of provenances observed in international trials. These results could be applied for verification of origin of confiscated teak timber and products in the future. Changtragoon and Singthong (unpublished) developed SNPs to identify *Dalbergia* spp. for use in verifying whether confiscated logged trees or products are *Dalbergia* spp., of which *Dalbergia cochinchinensis* Pierre is a protected species and listed in CITES appendix II. Table 6.5 shows areas of genetic research on DNA barcoding, phylogeography, genetic diversity, structure and differentiation of tropical trees which could be applied for identification of endangered and protected species and verification of timber origin. Lowe (2007) mentioned that the main challenge is usually to establish whether timber has come from sustainably managed, or at least legal, sources rather than illegal sources. To verify such a claim using DNA-based methods, genetic variation needs to exist within the species of interest and ideally needs to be geographically structured. Suitable variation is established within natural populations of nearly all species due to a combination of mutational and population processes (i.e., gene flow and drift).

Table 6.5 Some areas of genetic researches in forest tropical trees which could be used as the basis for potential applications in DNA forensics

Investigated species/taxa/families	Country/region in the tropics	Molecular techniques/markers used	References
1. DNA barcodes/molecular assay and phylogeny (application: species identification)			
Many species and families	Panama/Latin America	matK, trnH-psbA, rbcLa nucleotide sequences in cpDNA	Kress et al. (2009)
Many species and families	Puerto Rico	matK, trnH-psbA, rbcLa nucleotide sequences in cpDNA	Kress et al. (2010)
149 species, 82 genera and 38 families	India/South Asia	matK, trnH-psbA, rbcL, ITS, ITS2	Tripathi et al. (2013)
Pinaceae	Asia	cp (RFLP)	Wang and Szmidt (1993)
Many species and families	French Guiana	*r*bcLa, rpoB, matK, ycf5, trnl,psbA-trnH, ITS nucleotide sequences in cpDNA	Gonzalez et al. (2009)
Hopea spp.	Malaysia/Southeast Asia	cp DNA and Nu PgiC sequences	Choong et al. (2008)
Dipterocapaceae	Southeast Asia	PCR-RFLP	Tsumura et al. (1996)
Diperocarpaceae	Southeast Asia	matK, trnL Intron, trnL-trnF nucleotide sequences in cpDNA	Kajita et al. (1998)
Rhizophora apiculata Blume *R. mucronata* Poir *R. stylosa* Griff.	Indo-west Pacific/Southeast Asia	Multiplex PCR	Ng and Szmidt (2013)
Rhizophora apiculata Blume *R. mucronata* Poir R. stylosa Griff. and their hybrid	Malaysia/Southeast Asia	Nucleotide sequences in nDNA	Ng et al. (2013)
2. Phylogeography (application: origin verification)			
Cedrela odorata L.	Mesoamerica	cpDNA	Cavers et al. (2003)
Lithocarpus spp.	Southeast Asia	cpDNA sequence, ITS sequence	Cannon and Manos (2003)
Swietenia macrophylla King	Amazonia and Central America	cpDNA (SSR)	Lemes et al. (2010)
Erythrophleum ivorense A. Chev.	Africa	cpDNA	Duminil et al. (2010)

(continued)

Table 6.5 (continued)

Investigated species/taxa/families	Country/region in the tropics	Molecular techniques/markers used	References
E. suaveolens (Guill. & Perr.) Brenan			
3. Genetic diversity, structure and differentiation (application: origin verification)			
Swietenia macrophylla King.	America Central, Southern Brazil	SSRs and RAPD	Novick et al. (2003), Lemes et al. (2003), Gillies et al. (1999)
Ceiba pentandra (L.) Gaertn	Neotropics and Africa	ITS nucleotide sequences in cpDNA	Dick et al. (2007)
Vouacapoua Americana Aubl	French Guiana and the Amazon Basin/Latin America	Nucleotide sequences in cpDNA	Dutech et al. (2000, 2003, 2004)
Carapa guianensis Aubl	Amazon Basin/Latin America	Nucleotide sequences in cpDNA	Cloutier et al. (2005)
Pterocarpus officinalis Jacq.	Caribbean Basin	AFLP	Rivera-Ocasio et al. (2002)
Vochysia ferruginea Mart.	Costa Rica/Latin America	PCR-RFLP and AFLP	Cavers et al. (2005)
Lonchocarpus costaricensis (J.D. Smith)Pittier	Guanacaste, Costa Rica/Latin America	PCR-RFLP and AFLP	Navarro et al. (2005)
Irvingia wombulu Vermoesen and *I.gabonensis* Aubry-Lecomte ex O'Rorke	Nigeria, Cameroon Gabon/Africa	RAPD	Lowe et al. (2000)
Aucoumea Klaineana Pierre	Gabon	Nucleotide sequences in cpDNA	Muloko-Ntoume et al. (2000)
Gonystylus bancanus (Miq.) Kurz	Malaysia/Southeast Asia	DALP fingerprinting	Fatma et al. (2011)
Tectona grandis Linn.f	India, Myanmar, Laos, Thailand/Asia	SSR	Hansen et al. (2015)

Over time, genetic structure can develop at different spatial scales due to limited pollen and seed dispersal within a population, limited gene flow between populations or historic disruption of a once continuous range of a species by climatic or geographic changes. Each of these changes will produce genetic discontinuities and clines within a species that can be identified using a range of DNA-based markers (Lowe 2007).

A new European Union (EU) regulation came into force in 2010 that made it illegal to place illegally harvested timber and timber products on the European market. However, in the past there has been a lack of practical mechanisms for identifying the origin of timber and wood products. Degen et al. (2013) developed DNA tools and databases to verify the geographic origin of mahogany (*Swietenia macrophylla* King.) in Latin America with DNA fingerprinting, and used six nuclear microsatellite markers to generate DNA fingerprints for a generic reference database characterizing the populations of origin of large set of mahogany samples. For the databases, leaves and cambium from 1971 tree samples from 31 stands in Bolivia, Costa Rica, Guatemala, Honduras, Mexico, Nicaragua, Panama, Belize and Brazil were genotyped. A total of 145 different alleles were found, showing strong genetic differentiation ($g = 0.52$, $F_{st} = 0.18$, $G_{st} = 0.65$) and clear correlation between genetic and spatial distances among stands ($r = 0.82$, $p < 0.05$). The authors used the genetic reference database and Bayesian assignment testing to determine the geographic origins of two sets of mahogany wood samples, based on their multilocus genotypes. In both cases the wood samples were assigned to the correct country of origin. They also discussed the overall applicability of this methodology to tropical timber trading.

Illegal logging of tropical trees of especially high economic value in natural forests and protected areas is still common due to high demands in international trade. DNA forensics is one tool to efficiently facilitate and enhance law enforcement. Tnah et al. (2010) developed a feasibility study by using short tandem repeats (STRs) as a tool to identify the sources of illegally logged timber in Malaysia, using *Neobalanocarpus heimii* (King) (Dipterocarpaceae) as a model species. Thirty natural populations of *N. heimii* were profiled using 12 STRs to develop DNA profile databases. Cluster analysis divided the 30 populations into three genetic clusters, corresponding to three subregions within Peninsular Malaysia. The DNA databases were characterized at the levels of population, subregion and Peninsular Malaysia. Independence tests within and among loci were violated in all the databases. Yooyuen et al. (2011) have developed and used single nucleotide polymorphisms (SNPs) in chloroplast genomes to identify populations of endangered and CITES-listed species *Dalbergia cochinchinensis* (Pierre.) in Thailand, which could be useful for identification of origin of illegally logged timber in the future. Changtragoon (unpublished) used DNA tools to identify whether confiscated wood was from protected forest tree species (Case study 2).

DNA markers can also be used by those in the timber trade to identify risk in their own supply chains. In a study of merbau, *Intsia palembanica*, a high-value timber species from Southeast Asia, samples were taken from 2627 logs harvested from a target logging concession (Lowe et al. 2010). These logs were delivered to a sawmill in Java, Indonesia, and wood samples were taken from a total of 741 logs at the sawmill. It is this stage in the supply chain, between the concession and mill, which is most prone to the substitution of illegally harvested timber and falsification of documents certifying origin is rampant. For a random sample of 32 pairs of samples (wood from the log collected at the logging concession and samples from the same putative log collected at the saw mill), a unique genetic fingerprint was generated using 14 microsatellite markers. An exact genotype match was provided between forest and sawmill samples for 27 out of 32 random paired log samples. Of the five samples where this test was not possible to complete, four of the cases involved the sawmill samples failing to amplify and for one case, non-overlapping loci were amplified between forest and sawmill samples. This DNA tracking methodology is actively being applied in the forestry industry to check for illegal substitutions along supply chains, and complements paper certification methods and DNA source identification methods based on genetic structure.

Case study 2 Species identification of confiscated wood in Thailand

A law enforcement case involved investigation of whether confiscated wood was wild mango from protected tree species in the area of Ramkhamhaeng National Park in the northern part of Thailand. Wild mango (*Mangifera* spp.) is a protected species in Thailand. In order to investigate this case, cpDNA sequences at the trnH-psbA region of the confiscated sample were compared with domesticated mango varieties and wild mango species in Ramkhamheang National Park. The results showed that expected size of the PCR products of DNA from the confiscated wood had the same size as domesticated mango varieties, as shown in Fig. 6.1.

Furthermore, variable nucleotide sites in a noncoding trnH-psbA spacer region in chloroplast DNA of the confiscated wood were the same as domesticated mango varieties as shown in Fig. 6.2. The phylogenetic analysis in Fig. 6.3 shows that this

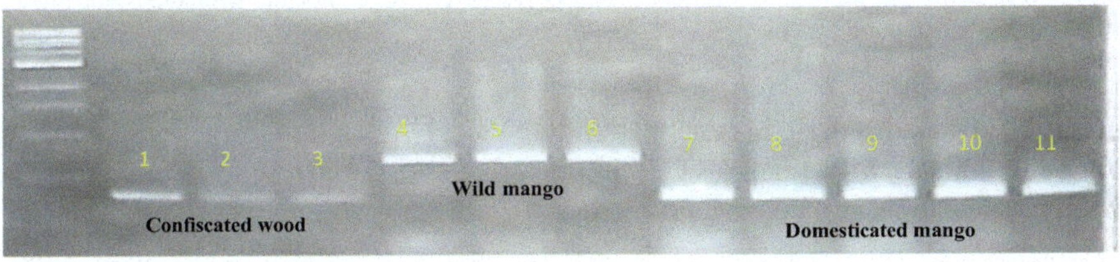

Fig. 6.1 PCR product from the DNA of confiscated wood (380 bp expected size), wild mango (650 bp expected size) and domesticated mango (380 bp expected size) at noncoding trnH-psbA spacer region in Chloroplast DNA (Changtragoon, unpublished). **Remarks** *No. 1, 2, 3* confiscated wood; *No. 4, 5, 6* wild mango; *No. 7, 8, 9, 10, 11* domesticated mango varieties; *No. 7* Namdokmai; *No. 8* Khieosawoey; *No.9* Phimsen; *No.10* Chok Anan; *No.11* Khieo Yai

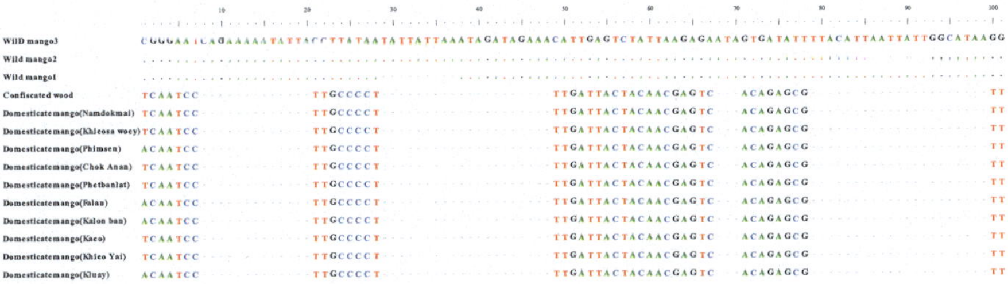

Fig. 6.2 PCR product from noncoding trnH-psbA spacer region in chloroplast DNA. The expected size wild of mango is 650 bp; expected size of confiscated and domesticated mango is 380 bp. **Remarks** *No.. 1, 2, 3* confiscated wood; *No. 4, 5, 6* wild mango; *No. 7, 8, 9, 10, 11* domesticated mango; *No. 7* Namdokmai; *No. 8* Khieosawoey; *No. 9* Phimsen; *No. 10* Chok Anan; *No. 11* Khieo Yai

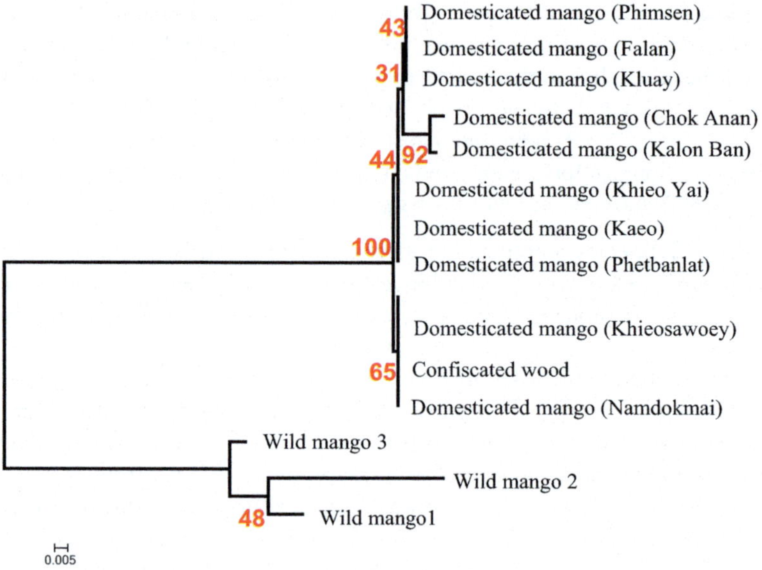

Fig. 6.3 The phylogenetic analysis at *trnH-psbA* shows that the confiscated wood sample has genetically closely related to domesticated mango varieties than wild mango (*Mangifera* spp). This means it is from one among domesticated mango relatives and is not the protected wild mango (Changtragoon, unpublised)

confiscated wood sample is genetically more closely related to domesticated mango varieties than to wild mango. This implies that the sample is from domesticated mango rather than protected wild mango species.

6.3.4 Forest and Landscape Restoration

Global efforts to restore hundreds of millions of hectares of degraded and deforested lands, in line with the 15th Sustainable Development Goal (SDG),[2] are hampered by the lack of information on how to ensure that restored sites will indeed develop into functional and resilient forests able to adapt to changing environments and deliver ecosystem services needed to meet the other SDGs, including those on eliminating poverty and hunger and combating climate change. Typical survival rates in restoration projects over the first years are as low as 50%, and it is estimated that only a quarter of restoration projects succeed in establishing self-sustaining

[2]Goal 15.3, Life on Land: By 2020, promote the implementation of sustainable management of all types of forests, halt deforestation, restore degraded forests and substantially increase afforestation and reforestation globally.

populations that are capable of regenerating on their own, over time (Godefroid et al. 2011).

Restoration success is commonly defined as the capacity of the reintroduced populations to survive, reproduce and adapt to changing environmental conditions (Godefroid et al. 2011). These outcomes are in part determined by the genetic diversity of the propagation material and its match to the environmental conditions on the restoration site (Ellstrand and Elam 1993; Hughes et al. 2008). A survey among restoration practitioners showed that having some knowledge of the genetic diversity of the target plant species and using propagation material from stable rather than declining populations improved plant survival rates, which had an increasing impact over time (Godefroid et al. 2011). However, genetic studies in restored tree populations have repeatedly revealed notable founder effects resulting from collecting germplasm from very few parent trees (Burgarella et al. 2007; Navascués and Emerson 2007; Kettle et al. 2008; Salas-Leiva et al. 2009; Li et al. 2012b), and concerns about the viability of the seed, which often exhibits high levels of inbreeding (Broadhurst et al. 2006; Broadhurst 2011; Salas-Leiva et al. 2009). For example, a seed source population used for restoring a mangrove species, *Avicennia germinans,* was found to have been significantly reduced in size because of habitat conversion, resulting in a higher inbreeding coefficient among younger (<40 yrs) individuals, compared with more mature (>40 yrs) individuals (Salas-Leiva et al. 2009). Moreover, few restoration practitioners are aware of, or use specific strategies in sourcing germplasm to help the restored tree populations to adapt to a changing climate (Bozzano et al. 2014). Tree populations that have limited viability and capacity to adapt to environmental variation are unlikely to fulfill the much anticipated role of forests in mitigating climate change. The identification and selection of appropriate sources of tree germplasm for a given restoration site would ideally be guided by the strength of the interaction between genotype performance and current and future environmental conditions (genotype-by-environment, G x E interactions; Breed et al. 2013). Such interactions are studied with multilocation progeny or provenance trials and climate modelling, respectively (Sgrò et al. 2011), which help identify planting sources that are adapted to particular sites and the range within which reproductive material of a species can be moved.

Restoration efforts could make much better use of native, including threatened, tree species than they have done in the past. For example, the proportion of native species in afforestation or reforestation initiatives in South and Southeast Asia between 2003 and 2007 was reported at six percent (FAO 2010: 92). However, restoring self-sustaining forests with threatened tree species, which may already suffer from the negative genetic impacts of disturbance, requires a careful choice of approaches. Habitat fragmentation and exposure to other disturbances limit the quality of remnant forest patches as seed sources in many of the landscapes that are currently in need of restoration (Vranckx et al. 2012; Lowe et al. 2015). This situation has been aggravated by a fairly common emphasis on using only 'local' germplasm for restoration purposes sourced close to the restoration site (Broadhurst 2013; Breed et al. 2013). Although such an approach can help to maintain the genetic integrity of the species, it can result in using planting material that is inbred and having a narrow genetic base, especially when the targeted species have limited pollen and seed dispersal. Molecular studies can help to assess inbreeding in remnant forests and potential seed source populations (Broadhurst et al. 2006, Salas-Leiva et al. 2009; Broadhurst 2011), genetic distances between planned seed source populations and remnant natural populations near the site to identify the range within which germplasm could be transferred (Pakkad et al. 2008; Ritchie and Krauss 2012; Krishnan et al. 2012), or the potential of gene flow to restore genetic diversity in restored forests (Céspedes et al. 2003; Ritchie and Krauss 2012; Cruz Neto et al. 2014). Such information will help adapt restoration methods and approaches for species and site contexts depending on the status of genetic diversity of the target species. For species with very few remaining individuals, restoration efforts should start with gene inventories across the remaining populations, followed by tailored approaches to enhance cross-pollination and genetic diversity (Ducci 2014). Despite the enormous demand for forest and landscape restoration in tropical countries, there have been only a handful of studies to date assessing the success of restoring genetic diversity of tropical tree species and populations (Krishnan et al. 2012; Cruz Neto et al. 2014; Kettle et al. 2008; Pakkad et al. 2008; Salas-Leiva et al. 2009).

Case study 3 Genetic diversity and gene flow between natural and planted populations of *Inga vera*

To study how genetic diversity is retained in fragmented landscapes and transmitted between remnant natural and planted populations, Cruz Neto et al. (2014) genotyped five natural and five planted populations of *Inga vera* ssp. *affinis* in the Atlantic forest of northeastern Brazil. Ten reproductive trees and approximately 50 seedlings were sampled in each population. All individuals were genotyped at four microsatellite loci. Genetic diversity of adult trees in planted populations, measured as allelic richness, was only half ($A_R = 7.599$) that found in the remnant natural populations ($A_R = 14.128$) (Table 6.6). Twelve alleles were found exclusively in natural populations, whereas no alleles were found exclusively in planted populations. Gene diversity was also lower in planted (Hs = 0.696) compared with natural (Hs = 0.885) populations.

Of the 63 alleles found in natural populations, 62 were also found in seedlings. In planted populations the number of alleles found in seedlings was 38% higher (43 alleles) than that found in adults (31 alleles). All of the alleles found in seedlings but absent in adult trees of planted populations were present adults in natural populations. The authors concluded that for self-incompatible species with extensive pollen flow, plantations established with a narrow genetic base may regain genetic diversity by outcrossing with remnant natural populations within the landscape.

Table 6.6 Genetic diversity of adults and seedlings of *Inga vera* in five planted and five natural stands in the Atlantic forest of Brazil. Source: Cruz Neto et al. (2014)

Population	Adults		Seedlings	
	A	A_R	A	A_R
Planted				
P1	4.3	3.62	6	3.96
P2	4.8	3.98	5.75	5.35
P3	4.0	3.42	5.25	4.01
P4	4.3	3.45	6.25	5.58
P5	4.8	4.26	6.5	5.66
Mean	4.4 ± 0.33	3.7 ± 0.36	6.01 ± 0.48	4.91 ± 0.85
Natural				
N1	8.3	8	7.5	6.64
N2	7.8	7.29	8	7.06
N3	8.0	7.29	7.25	6.37
N4	9.0	8.1	8.25	7.26
N5	8.5	7.89	9.5	7.95
Mean	8.3 ± 0.48	7.7 ± 0.4	8.1 ± 0.87	7.06 ± 0.6

A average number of alleles per locus
A_R allelic richness rarefied to nine individuals

6.4 Conclusions

The studies reviewed in this chapter demonstrate how different types of molecular markers can be applied to yield information on genetic diversity patterns of tropical tree species, in support of effective, targeted conservation strategies and planning, DNA forensics, and forest and landscape restoration. Nevertheless, application of molecular tools and, in particular, of the most recent technologies, still remains woefully limited in developing countries, which harbor most of the world's tree diversity, and this is despite rapid technological advances and declining costs of genetic and genomic analyses. In these countries, technical capacities, facilities and financial resources are often inadequate for conducting molecular research on tree species, especially on a systematic, programmatic basis. Existing projects often focus narrowly on species of commercial importance. Recognizing these limitations, the Global Plan of Action on Forest Genetic Resources (FAO 2014b) called for increased efforts in north–south as well as south–south cooperation to strengthen both education and research on forest genetic resources. It is our hope that such collaborative initiatives multiply and benefit developing country researchers and research organizations working to help conserve and sustainably manage these species, and improve conservation management for a broader group of tropical tree species.

Acknowledgements We would like express our sincere appreciation to Prof. Dr. S. Mohan Jain, Department Agricultural Sciences, University of Helsinki, Finland; Dr. Timothy Bole, regional technical advisor, UN-REED Programme, UNDP; Associate Prof. Dr. A.E Szmidt, Department of Biology, Graduate School of Science, Kyushu University, Japan; Dr. Pandey Manhav, senior scientist, Molecular Research, LP; and Dr. Ng Wei Lun, postdoctoral researcher, Faculty of Biomolecular Sciences, Putra University, Malaysia, for their editing, comments and suggestion for this chapter.

References

Abasolo MA, Fernando ES, Borromeo TH, Hautea DM (2009) Cross-species amplification of *Shorea* microsatellite DNA markers in *Parashorea malaanonan* (Dipterocarpaceae). Philippine J Sci 138(1):23–28

Alcântara BK, Veasey EA (2013) Genetic diversity of teak (*Tectona grandis* L.F.) from different provenances using microsatellite markers. Rev Árvore 37(4):747–758

Alfaro RI, Fady B, Vendramin GG et al (2014) The role of forest genetic resources in responding to biotic and abiotic factors in the context of anthropogenic climate change. For Ecol Manage 333:76–87

Allendorf FW, Hohenlohe PA, Luikart G (2010) Genomics and the future of conservation genetics. Nat Rev Genet 11:697–709

Amaral W (2001) Characterization, evaluation and conservation of forest genetic resources. The potential and limitation of new biotechnology tools. In: Uma Shaanker R, Ganeshaiah KN, Bawa KS (eds) Forest genetic resources: status, threats and conservation strategies. Oxford and IBH Publishing, New Delhi/Calcutta, pp 115–125

Ansari SA, Narayanan C, Wali SA, Kumar R, Shukla N, Rahangdale SK (2012) ISSR markers for analysis of molecular diversity and genetic structure of Indian teak (*Tectona grandis* L.f.) populations. Ann For Res 55(1):11–23

Asif MJ, Cannon C (2005) DNA extraction from processed wood: A case study for the identification of a endangered timber species (Gonystylusbancanus). Plant Mol Biol Reporter 23:180–187

Assogbadjo AE, Kyndt T, Sinsin B, Gheysen G, Van Damme P (2006) Patterns of genetic and morphometric diversity in baobab (*Adansonia digitata*) populations across different climatic zones of Benin (West Africa). Ann Bot 97:819–830

Ayele TB, Gailing O, Finkeldey R (2011) Assessment and integration of genetic, morphological and demographic variation in *Hagenia abyssinica* (Bruce) J.F. Gmel to guide its conservation. J Nat Conserv 19:8–17

Bekessy SA, Allnutt TR, Premoli AC, Lara A, Ennos RA, Burgman MA, Cortes M, Newton AC (2002) Genetic variation in the vulnerable and endemic monkey puzzle tree, detected using RAPDs. Heredity 88:243–249

Boontong C, Pandey M, Changtragoon S (2008) Isolation and characterization of microsatellite markers in Indian Neem (*Azadirachta Indica* Var. *Indica* A. Juss) and cross-amplification in Thai Neem (A. *Indica* Var. Siamensis Valenton). Conserv Genet. doi:10.1007/s10592-008-9610-5

Bozzano M, Jalonen R, Evert T et al (eds) (2014) Genetic considerations in ecosystem restoration using native tree species. A thematic study for the state of the World's forest genetic resources report. Food and Agriculture Organization of the United Nations, Rome, p 251

Breed MF, Stead M, Ottewell K, Gardner MG, Lowe AJ (2013) Which provenance and where? Seed sourcing strategies for revegetation in a changing environment. Conserv Genet 14:1–10

Broadhurst LM (2011) Genetic diversity and population genetic structure in fragmented *Allocasuarina verticillata* (Allocasuarinaceae)—implications for restoration. Aust J Bot 59:770–780

Broadhurst LM (2013) A genetic analysis of scattered yellow box trees (*Eucalyptus melliodora* A. Cunn. ex Schauer, Myrtaceae) and their restored cohorts. Biol Conserv 161:48–57

Broadhurst LM, North T, Young AG (2006) Should we be more critical of remnant seed sources being used for revegetation? Ecol Manage Restor 7(3):211–217

Burgarella C, Navascués M, Soto Á et al (2007) Narrow genetic base in forest restoration with holm oak (*Quercus ilex* L.) in Sicily. Ann For Sci 64:757–763

Butcher PA, Moran GF, Perkins HD (1998) RFLP diversity in the nuclear genome of *Acacia mangium*. Heredity 81:205–213

Butler JM (2005) Forensic DNA typing: biology, technology, and genetics of STR marker, 2nd edn. Elsevier Academic Press, Burlington, Massachusetts, p 660

Cannon CH, Manos PS (2003) Phylogeography of the Southeast Asian stone oaks (*Lithocarpus*). J Biogeography 30:211–226

Cavers S, Navarro C, Lowe AJ (2003) Chloroplast DNA phylogeography reveals colonisation history of a Neotropical tree, *Cedrela odorata* L., in Mesoamerica. Mol Ecol 12:1451–1460

Cavers S, Navarro C, Hopkins P, Ennos RA, Lowe AJ (2005) Regional and population-scale influences on genetic diversity partitioning within Costa Rican populations of the pioneer tree Vochysia ferruginea Mart. Silvae Genet 54:258–264

Céspedes M, Gutierrez MV, Holbrook NM et al (2003) Restoration of genetic diversity in the dry forest tree *Swietenia macrophylla* (Meliaceae) after pasture abandonment in Costa Rica. Mol Ecol 12:3201–3212

Changtragoon S (1998) Application of Molecular Markers in Forest trees for developing countries. In: The 7th Annual international conference of BIOREFOR, Manila, Phillipins, 3–5 Nov 1998, pp 111–114

Changtragoon S (2001a) Forest genetic resources of Thailand: status and conservation. In: Uma Shaanker R, Ganesaiah KN, Bawa KS (eds) Forest Genetic Resources: Status, Threats and Conservation Strategies. Oxford and IBH Publishing Co. Pvt. L.t.d., pp 141–151

Changtragoon S (2001b) Evaluation of genetic diversity of *Dipterocarpus alatus* genetic resources in Thailand using isozyme gene markers. In: Ex Situ and In Situ conservation of commercial tropical trees: the contribution of genetic resource conservation to tree breeding, biotechnology, and future commercial plantation programs, Yogyakarta, Indonesia, 11–13 June 2001 pp 351–356

Changtragoon S (2004) Thepotential for using molecular markers to facilitate gene management and the in situ and ex situ conservation of tropical forest trees. In: Proceedings of the Asia Pacific Forest Genetic Resources Programme (APFORGEN) inception workshop, Kepong, Kuala Lumpur, Malaysia, 15–18 July 2003, pp 305–315

Changtragoon S (2007) Evaluating the status of genetic resources of *Rhizophora apiculata* and *R. mucronata* in Thailand by using DNA and isoenzyme gene markers for conservation and reforestation guideline. In: Mangrove Seminar, Holiday Inn Hotel, Petchaburi, Thailand, 12–14 September 2007 pp 201–210

Changtragoon S (2012) Wildlife Forensics in Thailand. In Huffman JE, Wallace JR (eds) Wildlife forensics: methods and applications. John Wiley and Sons, Chichester, UK. doi: 10.1002/9781119953142.ch16

Changtragoon S, Boontawee B (1999) The study of genetic diversity of *Dipterocarpus alatus* by isoenzyme gene markers. In: Seminars on Dipterocarpus alatus and Dipterocapaceae, Kasetsart University, Bangkok, Thailand, 17–18 November 1999, pp 107–114

Changtragoon S, Finkeldey R (1995a) Genetic variation of Pinus merkusii in Thailand. I: genetic analysis of isozyme phenotypes. J Tropical Forest Sci 8(2):167–177

Changtragoon S, Finkeldey R (1995b) Patterns of genetic variation and characterization of the mating system of Pinus merkusii in Thailand. Forest Genetics 2(2):87–97

Changtragoon S, Szmidt AE (1997) The evaluation of genetic diversity and resources of tropical forest trees in Thailand by using molecular markers. In: The 6th annual international workshop of BIO-REFOR, the university of Queensland, Brisbane, Australia, 2–9 December 1997, pp 169–171

Changtragoon S, Szmidt AE (1999) Genetic diversity of Teak by means isoenzyme gene markers. In: International teak conference: teak beyond year 2000, Empress Hotel, Chiang Mai, Thailand, 23–25 August 1999, p 13

Changtragoon S, Szmidt AE (2000) Genetic diversity of teak (*Tectona grandis* L.f.) in Thailand revealed by Randam Amplified Polymorphic DNA (RAPD. In: IUFRO working party 2.08.01 Tropical species breeding and genetic resources: forest genetics for the next millennium, international conferences center, Durban, South Africa, 8–13 October 2000, pp 82–83
Chase M, Kesseli R, Bawa K (1996) Microsatellite markers for population and conservation genetics of tropical trees. Am J Bot 83(1):51–57
Chen C, Mitchell S, Elshire R, Buckler E, El-Kassaby Y (2013) Mining conifers' mega-genome using rapid and efficient multiplexed high-throughput genotyping-by-sequencing (GBS) SNP discovery platform. Tree Genet Genomes 9:1537–1544
Cheua-ngam J, Volkaerf H (2006) Development of catalase gene nuclear DNA-based marker for population genetic analysis in Thai teak (*Tectona grandis* L.f.). Kasetsart J (Nat Sci) 40:91–98
Choong CY, Wickneswari R, Norwati M, Abbott RJ (2008) Phylogeny of *Hopea* (Dipterocarpaceae) inferred from chloroplast DNA and nuclear *PgiC* sequences. Mol Phylogenet Evol 48:1238–1243
Christmas M, Breed M, Lowe AJ (2016) Constraints to and conservation implications for climate change adaptation in plants. Conserv Genet 17:305–320
Cloutier D, Povoa JSR, Procopio LC, Leao NVM, Wadt LH, Ciampi AY, Schoen DJ (2005) Chloroplast DNA variation of Carapa guianensis in the Amazon basin. Silvae Genet 54:270–274
Craft KJ, Owens JD, Ashley MV (2007) Applications of plant DNA markers in forensic botany: genetic comparison of Quercus evidence leaves to crime scene trees using microsatellites. Forensic SciInt 165(1):64–70
Cruz Neto O, Aguiar AV, Twyford AD et al (2014) Genetic and ecological outcomes of *Inga vera* subsp. *affinis* (Leguminosae) tree plantations in a fragmented tropical landscape. PLoS ONE 9(6):e99903. doi:10.1371/journal.pone.0099903
Davey JW, Hohenlohe PA, Etter PD, Boone JQ, Catchen JM, Blaxter ML (2011) Genome-wide genetic marker discovery and genotyping using next-generation sequencing. Nat Rev Genet 12:499–510
Dawson IK, Lengkeek A, Weber JC, Jamnadass R (2009) Managing genetic variation in tropical trees: linking knowledge with action in agroforestry ecosystems for improved conservation and enhanced livelihoods. Biodivers Conserv 18(4):969–986
Degen B, Streiff R, Ziegenhagen B (1999) Comparative study of genetic variation and differentiation of two pedunculate oak (*Quercus robur*) stands using microsatellite and allozyme loci. Heredity 83:597–603
Degen B, Ward SE, Lemes MR, Navarro C, Cavers S, Sebbenn AM (2013) Verifying the geographic origin of mahogany (*Swieteniamacrophylla* King) with DNA-fingerprints. Forensic Sci Int: Genet 7(1):55–62
Deguilloux MF, Pemonge MH, Petit RJ (2004) DNA-based control of oak wood geographic origin in the context of the cooperage industry. Ann For Sci 61:97–104
Dick CW, Bermingham E, Lemes MR, Gribel R (2007) Extreme long-distance dispersal of the lowland tropical rainforest tree *Ceiba pentandra* L. (Malvaceae) in Africa and the Neotropics. Mol Ecol 16:3039–3049
Dormontt E, Bonor M, Degen B, Breulmann G, Guillery P, Espinoza E, Gardner S, Koch G, Lee SL, Rimbawanto A, Thomas D, Wiedenhoft A, Yafang Zahnen J, Lowe AJ (2015) Forensic wood identification: it's time to integrate disciplines to combat illegal logging. Biol Cons 191:790–798
Ducci F (2014) Species restoration through dynamic ex situ conservation: *Abies nebrodensis* as a model. In: Bozzano M, Jalonen R, Evert T et al (eds) Genetic considerations in ecosystem restoration using native tree species. A thematic study for the state of the World's forest genetic resources report. Food and Agriculture Organization of the United Nations, Rome, pp 225–232
Duminil J, Heuertz M, Doucet J-L, Bourland N, Cruaud C, Gavory F, Doumenge C, Navascués M, Hardy OJ (2010) CpDNA-based species identification and phylogeography: application to African tropical tree species. Mol Ecol. doi:10.1111/j.1365-294X.2010.04917.x

Dutech C, Maggia L, Joly HI (2000) Chloroplast diversity in Vouacapoua americana (Caesalpiniaceae), a neotropical forest tree. Mol Ecol 9:1427–1432

Dutech C, Maggia L, Tardy C, Joly HI, Jarne P (2003) Tracking a genetic signal of extinction-recolonization events in a neotropical tree species: Vouacapoua americana Aublet in French Guiana. Evolution 57:2753–2764

Dutech C, Joly HI, Jarne P (2004) Gene flow, historical population dynamics and genetic diversity within French Guianan populations of a rainforest tree species, Vouacapoua americana. Heredity 92:69–77

Dwiyanti FG, Chong L, Diway B, Fah LY, Siregar IZ, Subiakto A, Kamiya K, Ninomiya I, Harada K (2015) Population genetic diversity in the genus *Dryobalanops Gaertn*. f. (Dipterocarpaceae) Based on nuclear microsatellite markers. J Sustain 3(1):12–20

Ekblom R, Galindo J (2011) Applications of next generation sequencing in molecular ecology of non-model organisms. Heredity 107:1–15

Ellstrand NC, Elam DR (1993) Population genetic consequences of small population-size: implications for plant conservation. Annu Rev Ecol Syst 24:217–242

Fady B, Aravanopoulos FA, Alizoti P, Mátyás C, von Wühlisch G, Westergren M, Belletti P, Cvjetkovic B, Ducci F, Huber G, Kelleher CT, Khaldi A, Dagher Kharrat MB, Kraigher H, Kramer K, Mühlethaler U, Peric S, Perry A, Rousi M, Sbay H, Stoinic S, Tijardovic M, Tsvetkov I, Varela MC, Vendramin GG, Zlatanov T (2016) Evolution-based approach needed for the conservation and silviculture of peripheral forest tree populations. Forest Ecol Manage 375:66–75

FAO (2010) Global forest resources assessment 2010. Main report. Food and Agriculture Organization of the United Nations, Rome

FAO (2014a) State of the World's Forest Genetic Resources. Commission of the Genetic Resources for Food and Agriculture, Food and Agriculture Organization of the United Nations. Rome

FAO (2014b) Global plan of action for the conservation, sustainable use and development of forest genetic resources. Commission of the genetic resources for food and agriculture, food and agriculture organization of the United Nations. Rome

FAO (2015) Global forest resources assessment 2015. Desk reference. Food and Agriculture Organization of the United Nations, Rome

Fatma NAH, Wickneswari R, Cannon CH (2011) Detection of genetic structure among ramin (*Gonystylusbancanus* (Miq.) Kurz) populations in Peninsular Malaysia using a rapid DNA fingerprinting technique. Malays Appl Biol 40(2):43–50

Finger A, Kettle CJ, Kaiser-Bunbury CN et al (2012) Forest fragmentation genetics in a formerly widespread island endemic tree: *Vateriopsis seychellarum* (Dipterocarpaceae). Mol Ecol 21(10):2369–2382

Finkeldey R (2005) An introduction to tropical forest genetics. Georg-August-University, Göttingen, Germany, Institute of forest genetics and forest tree breeding

Finkeldey R, Hattemer HH (2007) Tropical forest genetics. Springer, Berlin

Finkeldey R, Guzman ND, Changtragoon S (1998) Inheritance of isozyme phenotypes in *Pterocrpus indicus* willd. Forest Genet 5(3):165–170

Finkeldey R, Guzman ND, Changtragoon S (1999) The mating system of *Pterocarpus Indicus* Willd. at Mt. Makiling, Philippines. Biotropica 31(3):525–530

Finkeldey R, Rachmayanti Y, Gailing O (2007) Molecular genetic tools for the identification of the origin of wood. In: Kües U (ed) Wood production, wood technology and biotechnological impacts. Universitätsverlag Göttingen, Göttingen, pp 143–158

Finkeldey R, Leinemann L, Gailing O (2010) Molecular genetic tools to infer the origin of forest plants and wood. Appl Microbiol Biotechnol 85:1251–1258. doi:10.1007/s00253-009-2328-6

Fofana IJ, Lidah YJ, Diarrassouba N, N'guetta SPA, Sangare A, Verhaegen D (2008) Genetic structure and conservation of Teak (*Tectona grandis*) plantations in Côte d'Ivoire, revealed by site specific recombinase (SSR). Trop Conserv Sci 1(3):279–292

Fofana IJ, Ofori D, Poitel M, Verhaegen D (2009) Diversity and genetic structure of teak (*Tectona grandis* L.f) in its natural range using DNA microsatellite markers. New Forest 37:175–195

Frankham R, Bradshaw CAJ, Brook BW (2014) Genetics in conservation management: revised recommendations for the 50/500 rules, Red ist criteria and population viability analyses. Biol Conserv 170:56–63

Franklin IR (1980) Evolutionary change in small populations. In: Soulé ME, Wilcox BA (eds) Conservation biology: an evolutionary-ecological perspective. Sinauer Associates, Sunderland, MA, pp 135–150

Gardner MG, Fitch AJ, Bertozzi T, Lowe AJ (2011) Rise of the machines—recommendations for ecologists when using second generation sequencing for microsatellite development. Mol Ecol Res 11:1093–1101

Gillies ACM, Navarro C, Lowe AJ, Newton AC, Hernández M, Wilson J, Cornelius JP (1999) Genetic diversity in Mesoamerican populations of mahogany (*Swietenia macrophylla*), assessed using RAPDs. Heredity 83:722–732

Godefroid S, Piazza C, Rossi G et al (2011) How successful are plant species reintroductions? Biol Conserv 144(2):672–682

González Martínez SC, Krutovsky KV, Neale DB (2006) Forest tree population genomics and adaptive evolution. New Phytol 170(2):227–238

Gonzalez MA, Baraloto C, Engel J, Mori SA, Pétronelli P, Riéra B, Roger A, Thébaud C, Chave J (2009) Identification of amazonian trees with DNA barcodes. PLoS ONE 4(10):e7483. doi:10.1371/journal.pone.0007483

Graudal L, Yanchuk A, Kjær E (2004) National planning. In FAO, DFSC, IPGRI, forest genetic resources conservation and management. overview, concepts and some systematic approaches, vol 1. International Plant Genetic Resources Institute, Rome, pp 25–36

Gregorius H-R (1991) Gene conservation and the preservation of adaptability. In: Seitz A, Loeschcke V (eds) Species conservation: a population-biological approach. Birkhäuser, Basel, pp 31–47

Haines R (1994) Biotechnology in forest tree improvement. FAO Forestry Paper 118:230

Hansen MC, Potapov PV, Moore R et al (2013) High-resolution global maps of 21st-century forest cover change. Science 15:850–853

Hansen OK, Changtragoon S, Ponoy B, Kjær ED, Minn Y, Finkeldey R, Nielsen KB, Graudal L (2015) Genetic resources of teak (*Tectona grandis* Linn.f.)—strong genetic structure among natural populations. Tree Genet Genomes 11:802. doi:10.1007/s11295-014-0802-5

Harry DE, Temesgen B, Neale DB (1998) Codominant PCR-based marker for *Pinus taeda* developed from mapped cDNA clones. Theor Appl Genet 97:327–336

Hartl DL, Clark AG (2007) Principles of population genetics. Sinauer Associates Inc Publishers, USA

Hedrick PW (2001) Conservation genetics: where are we now? Trends Ecol Evol 16(11):629–636

Honnay O, Jacquemyn H (2007) Susceptibility of common and rare plant species to the genetic consequences of habitat fragmentation. Conserv Biol 21(3):823–831

Hubbell SP (2013) Tropical rain forest conservation and the twin challenges of diversity and rarity. Ecol Evol 3(10):3263–3274

Hughes AR, Inouye BD, Johnson MTJ et al (2008) Ecological consequences of genetic diversity. Ecol Lett 11(6):609–623

Indriogo S, Gailing O, Finkeldey R (2006) Moleculra phylogeny of diperocarpaceae in Indonesia based on choroplst DNA. Plant Syst Evol 261:99–115

Inomata N, Wang X-R, Changtragoon S, Szmidt AE (2009) Levels and patterns of DNA variation in two sympatric mangrove species, *Rhizophoraapiculata* and *R. mucronata* from Thailand. Genes Genet Syst 84:277–286

Jalonen R, Hong LT, Lee SL, Loo J, Snook L (2014) Integrating genetic factors into management of tropical Asian production forests: a review of current knowledge. For Ecol Manage 315:191–201

Jamnadass R, Lowe A, Dawson IK (2009) Molecular markers and the management of tropical trees: the case of indigenous fruits. Trop Plant Biol 2(1):1

Kadu CAC, Schueler S, Konrad H et al (2011) Phylogeography of the Afromontane *Prunus africana* reveals a former migration corridor between East and West African highlands. Mol Ecol 20:165–178

Kadu CAC, Konrad H, Schueler S et al (2013) Divergent pattern of nuclear genetic diversity across the range of the Afromontane *Prunus africana* mirrors variable climate of African highlands. Ann Bot 111(1):47–60

Kajita T, Kamiya K, Nakamura K, Tachida H, Wickneswari R, Tsumura Y, Yoshimaru H, Yamazaki T (1998) Molecular phylogeny of dipetrocarpaceae in southeast Asia based on nucleotide sequences of *mat*K, *trn*L Intron, and *trn*L-*trn*F Intergenic spacer region in chloroplast DNA. Mol Phylogenet Evol 10(2):202–209

Kamiya K, Nanami S, Kenzo T, Yoneda R, Diway B, Chong L, Azani MA, Majid NM, Lum SKY, Wong K, Harada K (2012) Demographic history of *Shorea curtisii* (Dipterocarpaceae) inferred from chloroplast DNA sequence variations. Biotropica 44:577–585

Karp A, Seberg O, Buiatti M (1996) Molecular techniques in the assessment of botanical diversity. Ann Bot 78:143–149

Karp A, Kresovich S, Bhat KV, Ayad WG, Hodgkin T (1997) Molecular tools in plant genetic resources conservation: a guide to the technologies. IPGRI Tech Bull 2:47

Kertadikara AWS, Prat D (1995a) Isozyme variation among teak (*Tectona grandis* L.f.) provenances. Theor Appl Genet 90:803–810. doi:10.1007/BF00222015

Kertadikara AWS, Prat D (1995b) Genetic structure and mating system in teak (*Tectona grandis* L.f.) provenances. Silvae Genet 44:104–110

Kettle CJ, Ennos RA, Jaffre T et al (2008) Cryptic genetic bottlenecks during restoration of an endangered tropical conifer. Biol Conserv 141(8):1953–1961

Kjær ED, Suangtho V (1995) Outcrossing rate in *Tectona grandis* L. Silvae Genetica 44:175–177

Kobilinsky L, Liotti T, Oeser-Sweat J (2005) DNA: forensic and legal applications. Wiley, Hoboken, New Jersey, p 364

Konuma A, Tsumura Y, Lee CT et al (2000) Estimation of gene flow in the tropical-rainforest tree *Neobalanocarpus heimii* (Dipterocarpaceae), inferred from paternity analysis. Mol Ecol 9(11):1843–1852

Kramer AT, Havens K (2009) Plant conservation genetics in a chnaging world. Trends Plant Sci 14:599–607

Kress WJ, Erickson DL, Jones FA, Swenson NG, Perez R, Sanjur O, Bermingham E (2009) Plant DNA barcodes and a community phylogeny of a tropical forest dynamics plot in Panama. PNAS 106(44):18621–18626. doi:10.1073/pnas.0909820106

Kress WJ, Erickson DL, Swenson NG, Thompson J, Uriarte M, Zimmerman JK (2010) Advances in the use of DNA barcodes to build a community phylogeny for tropical trees in a puerto rican forest dynamics plot. PLoS ONE 5(11):e15409. doi:10.1371/journal.pone.0015409

Krishnan SA, Ranker TP, Davis A et al (2012) The study of genetic diversity patterns of *Coffea commersoniana*, an endangered coffee species from Madagascar: a model for conservation of other littoral forest species. Tree Genet Genom 9(1):179–187

Kristensen VN, Kelefiotis D, Kristensen T, Borrensen-Dale AL (2001) High-throughput methods for detection of genetics variation. Bio Techniques 30:318–332

Krutovskii KV, Neale DB (2001a) Forest genomics for conserving adaptive genetic diversity. In: Forest genetic resources working papers, Working Paper FGR/3. Forest Resources Development Service, Forest Resources Division, FAO, Rome

Krutovskii KV, Neale DB (2001b) Forest genomics for conserving adaptive genetic diversity. Forest Genet Resour 29:7–9

Krutovsky KV (2006) From population genetics to population genomics of forest trees: integrated population genomics approach. Russ J Genet 42(10):1088–1100

Krutovsky KV, Neale DB (2005a) Nucleotide diversity and linkage disequilibrium in cold-hardiness and wood quality-related candidate genes in Douglas fir. Genetics 171:2029–2041

Krutovsky KV, Neale DB (2005b) Forest genomics and new molecular genetic approaches to measuring and conserving adaptive genetic diversity in forest trees. In: Geburek Th, Turok J

(eds) Conservation and management of forest genetic resources in Europe. Arbora Publishers, Zvolen, pp 369–390
Krutovsky KV, Burczyk J, Chybicki I, Finkeldey R, Pyhäjärvi T, Robledo-Arnuncio JJ (2012) Gene flow, spatial structure, local adaptation and assisted migration in trees. In: Schnell RJ, Priyadarshan PM (eds) Genomics of tree crops. Springer, New York, pp 71–116. doi:10.1007/978-1-4614-0920-5_4
Krutovsky KV, Tretyakova IN, Oreshkova NV, Pak ME, Kvitko OV, Vaganov EA (2014) Somaclonal variation of haploid in vitro tissue culture obtained from Siberian larch (*Larix sibirica* Ledeb.) megagametophytes for whole genome de novo sequencing. In: Vitro cellular and developmental biology—Plant 50:655–664
Külheim C, Yeoh S, Maintz J, Foley WJ, Moran GF (2009) Comparative SNP diversity among four *Eucalyptus* species for genes from secondary metabolite biosynthetic pathways. BMC Genom 10(452):1–11
Kumar P, Gupta VK, Misra AK, Modi DR, Pandey BK (2009) Potential of molecular markers in plant biotechnology. Plant Omics J 2(4):141–162
Laikre L (2010) Genetic diversity is overlooked in international conservation policy implementation. Conserv Genet 11:349–354
Laphom R, Changtragoon S (2005) Evaluation of genetic diversity of some bamboos in Thailand using AFLP(Amplified Fragment Length Polymorphism) and microsatellite markers. In: The Proceedings of biodiversity of forest and wildlife seminar: research and activities in 2005, Regent Cha-am Hotel, Petchaburi, Thailand, 21–24 Aug 2005, pp 352–370
Lawrence MJ, Marshall DF (1997) Plant population genetics. In: Maxted N, Ford-Lloyd BV, Hwakes JG (eds) Plant genetic conservation: the in situ approach. Kluwer Academic Publishers, Dordrecht, pp 99–113
Lee SL, Wickneswari R, Mahani MC, Zakri AH (2000a) Genetic diversity of a tropical tree species, *Shorealeprosula* Miq. (Dipterocarpaceae), in Malaysia: implications for conservation of genetic resources and tree improvement. Biotropica 32(2):213–224
Lee SL, Wickneswari R, Mahani MC, Zakri AH (2000b) Mating system parameters in a tropical tree species, *Shorealeprosula* Miq. (Dipterocarpaceae), from Malaysian lowland dipterocarp forest. Biotropica 32(4a):693–702
Lee SL, Ng KKS, Saw LG et al (2002) Population genetics of *Intsia palembanica* (Leguminosae) and genetic conservation of virgin jungle reserves in peninsular Malaysia. Am J Bot 89(3):447–459
Lee SL, Ng KKS, Saw LG et al (2006) Linking the gaps between conservation research and conservation management of rare dipterocarps: a case study of *Shorea lumutensis*. Biol Conserv 131(1):72–92
Lemes MR, Gribel R, Proctor J, Grattapaglia D (2003) Population genetic structure of mahogany (Swietenia macrophylla King, Meliaceae) across the Brazilian Amazon, based on variation at microsatellite loci: implications for conservation. Mol Ecol 12(11):2875–2883
Lemes MR, Dick CW, Navarro C, Lowe AJ, Cavers S, Gribel R (2010) Chloroplast DNA microsatellites reveal contrasting phylogeographic structure in mahogany (*Swieteniamacrophylla* King, Meliaceae) from Amazonia and Central America. Trop Plant Biol 3:40–49. doi:10.1007/s12042-010-9042-5
Li J, Haipeng L, Jakobsson M, Sjodin P, Lascoux M (2012a) Joint analysis of demography and selection in population genetics: where do we stand and where could we go? Mol Ecol 21:28–44
Li Y, Tsang EPK, Cui M et al (2012b) Too early to call it success: an evaluation of the natural regeneration of the endangered *Metasequoia glyptostroboides*. Biol Conserv 150(1):1–4
Liengsiri C, Yeh FC, Boyle TJB (1995) Isozyme analysis of a tropical forest tree, Pterocarpus macrocarpus Kurz, in Thailand. For Ecol Manage 74:13–22
Lim LS, Wickneswari R, Lee SL, Latiff A (2002) Genetic variation of *Dryobalanopsaromatica Gaertn*. F. (Dipterocarpaceae) in Peninsular Malaysia using microsatellite DNA markers. Forest Genet 9(2):127–138

Lipow SR, Vance-Borland K, St. Clair JB et al (2004) Gap analysis of conserved genetic resources for forest trees. Conserv Biol 18:412–423

Lowe A (2007) Can we use to identify the geographic origin of tropical timber? In: Proceedings of the international workshop Fingerprinting methods for the identification of timber origins, Bonn, Germany, Oct 8–9 2007

Lowe AJ, Cross HB (2011) The application of DNA to timber tracking and origin verification. J Int Assoc Wood Anatomists 32(2):251–262

Lowe AJ, Wilson J, Gillies ACM, Dawson I (2000) Conservation genetics of bush mango from central/west Africa, implications from RAPD analysis. Mol Ecol 9:831–841

Lowe AJ, Moule C, Trick M, Edwards KJ (2004a) Efficient large-scale development of microsatellites for marker and mapping applications in Brassica crop species. Theor Appl Genet 108:1103–1112

Lowe AJ, Harris SA, Ashton P (2004b) Ecological genetics: design. Anal Appl, Blackwells, Oxford, p 326

Lowe AJ, Boshier D, Ward M et al (2005) Genetic resource impacts of habitat loss and degradation; reconciling empirical evidence and predicted theory for neotropical trees. Heredity 95(4):255–273

Lowe AJ, Wong KN, Tiong Y-S, Iyerh S, Chew F-T (2010) A DNA method to verify the integrity of timber supply chains; confirming the legal sourcing of merbau timber from logging concession to Sawmill. Silvae Genetica 59:263–268

Lowe AJ, Cavers S, Boshier D, Breed MF, Hollingsworth P (2015) The resilience of forest fragmentation genetics—no longer a paradox—we were just looking in the wrong place. Heredity 115:97–99

Lowe AJ, Dormontt EE, Bowie MJ, Degen B, Gardner S, Thomas D, Clarke C, Espinoza E, Rimbawanto A, Wiedenhoeft A, Yin Y, Sasaki N (2016) Technologies for transparency in the timber trade. BioScience 66(11):990–998

Manel S, Berthier P, Luikart G (2002) Detecting wildlife poaching: identifying the origin of individuals with bayesian assignment tests and multilocus genotypes. Conserv Biol 16(3):650–659

Marroni F, Pinosio S, di Centa E, Jurman I, Boerjan W, Felice N, Cattonaro F, Morgante M (2011) Large-scale detection of rare variants via pooled multiplexed next-generation sequencing: towards next-generation Ecotilling. Plant J 67:736–745

Minobe S, Fukui S, Saiki R, Kajita T, Changtragoon S, Ab Shukor NA, Latiff A, Ramesh BR, Koizumi O, Yamazaki T (2010) Highly differentiated population structure of a Mangrove species, *Bruguiera gymnorhiza (Rhizophoraceae)* revealed by one nuclear *GapCp* and one chloroplast intergenic spacer *trnF-trnL*. Conserv Genet 11:301–310

Moran GF, Muona O, Bell JC (1989) Acacia mangium: A tropical forest tree of the coastal lowlands with low genetic diversity. Evolution 43(1):231–235

Muloko-Ntoutoume N, Petit RJ, White L, Abernethy K (2000) Chloroplast DNA variation in a rainforest tree (Aucoumea klaineana, Burseraceae) in Gabon. Mol Ecol 9:359–363

Murawski DA, Bawa KS (1994) Genetic structure and mating system of Stemonoporus oblongifolius (Dipterocarpaceae) in Sri Lanka. Am J Bot 81:155–160

Murawski DA, Dayanandan B, Bawa KS (1994) Outcrossing rates of two endemic Shorea species from Sri Lanka tropical rain forests. Biotropica 26:23–29

Narayanan C, Wali SA, Shukla N, Kumar R, Mandal AK, Ansari SA (2007) RAPD and ISSR markers for molecular characterization of teak (*Tectona grandis*) plus trees. J Trop Forest Sci 19(4):218–225

Narum SR, Buerkle CA, Davey JW, Miller MR, Hohenlohe PA (2013) Genotyping-by-sequencing in ecological and conservation genomics. Mol Ecol 22:2841–2847

Navarro C, Cavers S, Colpaert N, Hernandez G, Breyne P, Lowe AJ (2005) Chloroplast and total genomic diversity in the endemic Costa Rican tree Lonchocarpus costaricensis (J.D.Smith) Pittier (Papilionaceae). Silvae Genet 54:293–300

Navascués M, Emerson BC (2007) Natural recovery of genetic diversity by gene flow in reforested areas of the endemic Canary Island pine. Pinus canariensis. For Ecol Manage 244 (1–3):122–128

Ng WL, Szmidt AE (2013) A simple and inexpensive molecular assay for species identification of Indo-West Pacific Rhizophora mangroves for conservation and management. Conserv Genet Resour. doi:10.1007/s12686-013-9973-x

Ng WL, Chan HT, Szmidt AE (2013) Molecular identification of natural mangrove hybrids of Rhizophora in Peninsular Malaysia. Tree Genet Genomes. doi:10.1007/s11295-013-0619-7

Ng WL, Onishi Y, Inomata N, Teshima KM, Chan HT, Baba S, Changtragoon S, Siregar IZ, Szmidt AE (2015) Closely related and sympatric but not all the same: genetic variation of Indo-West Pacific Rhizophora mangrove across the Malay Peninsula. Conserv Genet 16 (1):137–150

Nguyen TPT, Tran TH, Nguyen MD, Sierens T, Triest L (2014) Genetic population of threatened *Hopea odorata* Roxb. in the protected areas of Vietnam. J Viet Env 6(1):69–76. doi:10.13141/jve.vol6.no1

Novick RR, Dick CW, Lemes MR, Navarro C, Caccone A, Bermingham E (2003) Genetic structure of Mesoamerican populations of Big-leaf mahogany (*Swieteniamacrophylla*) inferred from microsatellite analysis. Mol Ecol 12:2885–2893

Nunney L, Elam DR (1994) Estimating the effective population size of conserved populations. Conserv Biol 8(1):175–184

Nuroniah HS (2009) Diagnostic markers for the identification of the tree species *Shorea leprosula* Miq. and *S. parvifolia* dyer and the geographic origin of *S. leprosula* Miq. Ph.D. Thesis, Göttingen University, Göttingen

Oldfield SF, Lusty C, MacKinven A (1998) The world list of threatened trees. World Conservation Press, Cambridge

Ottewell KM, Bickerton DC, Byrne M, Lowe AJ (2015) Bridging the gap: A genetic assessment framework for population-level threatened plant conservation prioritization and decision-making. Divers Distrib 22:174–188

Pakkad G, James C, Torre F, Elliott S, Blakesley D (2003) Genetic variation of Prunus cerasoides D. Don, a framework tree species in northern Thailand. New Forest 27:189–200

Pakkad G, Mazrooei SA, Blakesley D et al (2008) Genetic variation and gene flow among *Prunus cerasoides* D. Don populations in northern Thailand: analysis of a rehabilitated site and adjacent intact forest. New For 35:33–43

Pandey M, Changtragoon S (2012) Isolation and characterization of microsatellites in a medicinal plant, Phyllanthusemblica (Euphorbiaceae). Am J Bot 99(12):e468–e469

Pandey M, Geburek T (2009) Successful cross-amplification of Shorea microsatellites reveals genetic variation in the tropical tree, Shorea robusta Gaertn. Hereditas 146:29–32

Pandey M, Geburek T (2010) Genetic differences between continuous and disjunct populations: some insights from sal (*Shorea robusta* Roxb.) in Nepal. Conserv Genet 11:977–984

Petit RJ, Duminil J, Fineschi S, Hampe A, Salvini D, Vendramin GG (2005) Comparative organization of chloroplast, mitochondrial and nuclear divesity in plant populations. Mol Ecol 14:689–701

Porth I, El-Kassaby YA (2014) Assessment of the genetic diversity in forest tree populations using molecular markers. Diversity 6:283–295. doi:10.3390/d6020283

Rachmayanti Y, Leinemann L, Gailing O, Finkeldey R (2006) Extraction, amplification and characterization of wood DNA from Dipterocarpaceae. Plant Mol Biol Report 24:45–55

Rachmayanti Y, Leinemann L, Gailing O, Finkeldey R (2009) DNA from processed and unprocessed wood: factors influencing the isolation success. Forensic Sci Int Genet 3:185–192

Rajora OP, Mosseler A (2001) Challenges and opportunities for conservation of forest genetic resources. Euphytica 118:197–212

Rebelo AG, Siegfried WR (1992) Where should nature reserves be located in the Cape Floristic Region, South Africa? Models for the spatial configuration of a reserve network aimed at maximizing the protection of floral diversity. Conserv Biol 6:243–252

Reed DH, Frankham R, Roff D (2001) How closely correlated are molecular and quantitative measures of genetic variation? A meta-analysis. Evol 55(6):1095

Ritchie AL, Krauss SL (2012) A genetic assessment of ecological restoration success in *Banksia attenuata*. Restor Ecol 20(4):441–449

Rivera-Ocasio E, Aide TM, McMillan WO (2002) Patterns of genetic diversity and biogeographical history of the tropical wetland tree, pterocarpus officinalis (Jacq.), in the Caribbean basin. Mol Ecol 11:675–683

Sacandé M, Jøker D, Dulloo ME et al (eds) (2004) Comparative storage biology of tropical tree seed. IPGRI, Rome

Salas-Leiva DE, Mayor-Durán VM, Toro-Perea N (2009) Genetic diversity of black mangrove (*Avicennia germinans*) in natural and reforested areas of Salamanca Island Parkway, Colombian Caribbean. Hydrobiol 620:17–24

Schnabel A, Krutovskii KV (2004) Conservation genetics and evolutionary history of Gleditsia caspica: inferences from allozyme diversity in populations from Azerbaijan. Conserv Genet 5:195–204

Senakun C, Changtragoon S, Pramual P, Prathepha P (2011) Genetic structure and diversity of *Shorea obtusa* (Dipterocarpaceae) in Thailand. J Syst Evol 49(2):120–125

Sgrò CM, Lowe AJ, Hoffmann AA (2011) Building evolutionary resilience for conserving biodiversity under climate change. Evol Appl 4(2):326–337

Slavov GT, Howe GT, Yakovlev I, Edwards KJ, Krutovskii KV, Tuskan GA, Carlson JE, Strauss SH, Adams WT (2004) Highly variable SSR markers in Douglas-fir: mendelian inheritance and map locations. Theor Appl Genet 108:873–880. doi:10.1007/s00122-003-1490-y

Slik JWF, Arroyo-Rodríguez V, Aiba S et al (2015) An estimate of the number of tropical tree species. PNAS 112(24):7472–7477

Squirrell J, Hollingsworth PM, Woodhead M, Russell J, Lowe AJ, Gibby M, Powell W (2003) How much effort is required to isolate nuclear microsatellites from plants? Mol Ecol 12:1339–1348

Stapley J, Reger J, Feulner PGD, Smadja C, Galindo J, Ekblom R, Bennison C, Ball AD, Beckerman AP, Slate J (2010) Adaptation genomics: the next generation. Trends Ecol Evol 25:705–712

Szmidt AE (1995) Molecular population genetics and evolution: two missing elements in studies of biodiversity. In: Boyle TJ, Boontawee B (eds) Measuring and monitoring biodiversity in tropical and temperate forest. CIFOR, Bogor, Indonesia, pp 177–193

Szmidt AE, Wang XR (2000) Genetics markers in forest genetics and breeding—the tunnel remains dark. In: Matyas C (ed) Forest Genetics and Sustainability, vol 63. Kluwer Acadermic Publishers. Dordrecht, The Netherlands, pp 31–48

Thomas E, Alcázar Caicedo C, McMichael CH et al (2015) Uncovering spatial patterns in the natural and human history of Brazil nut (*Bertholletia excelsa*) across the Amazon Basin. J Biogeogr 42(8):1367–1382

Thongthawee S, Volkaert H (2014) Analysis of teak (*Tectona grandis*) genome and its diversity. In: The 26th annual meeting of the Thai Society for Biotechnology and international conference, pp 508–514

THWE-THWE-WIN, Hirao T, Watanabe A, Goto S (2015) Current genetic structure of teak (*Tectona grandis*) in Myanmar based on newly developed chloroplast single nucleotide polymorphism and nuclear simple sequence repeat markers. Trop Conserv Sci 8(1):235–256

THWE-THWE-WIN, Watanabe A, Hirao T, Isoda K, Ishizuka W, Goto S (2015) Genetic diversity of teak populations in native regions and plantations in Myanmar detected by microsatellite markers. Bull Univ of Tokyo For 132:1–15

Tnah LH, Lee SL, Ng KKS, Faridah QZ, Faridah-Hanum I (2010) Highly variable STR markers of Neobalanocarpus heimii (Dipterocarpaceae) for forensic DNA profiling. J trop forest sci 22(2):214–226

Tnah LH, Lee SL, Ng KKS, Bhassu S, Othman RY (2012) DNA extraction from dry wood of Neobalanocarpus heimii (Dipterocarpaceae) for forensic DNA profiling and timber tracking. Wood Sci Technol 46(5):813–825

Tripathi AM, Tyagi A, Kumar A, Singh A, Singh S, Chaudhary LB, Roy S (2013) The internal transcribed spacer (ITS) region and trnhH-psbA are suitable candidate loci for DNA barcoding of tropical tree species of India. PLoS ONE 8(2):e57934. doi:10.1371/journal.pone.0057934

Tsai H, Howell T, Nitcher R, Missirian V, Watson B, Ngo KJ, Lieberman M, Fass J, Uauy C, Tran RK, Khan AA, Filkov V, Tai TH, Dubcovsky J, Comai L (2011) Discovery of rare mutations in populations: TILLING by sequencing. Plant Physiol 156:1257–1268

Tsumura Y, Kawahara T, Wickneswari R, Yoshimura K (1996) Molecular phylogeny of dipterocarpaceae in Southeast Asia using RFLP of PCR-amplified chloroplast genes. Theor Appl Genet 93:22–29

van Zonneveld M, Scheldeman X, Escribano P et al (2012) Mapping genetic diversity of cherimoya (*Annona cherimola* Mill.): application of spatial analysis for conservation and use of plant genetic resources. PLoS ONE 7(1):e29845

van Zonneveld M, Dawson I, Thomas E et al (2014) Application of molecular markers in spatial analysis to optimize in situ conservation of plant genetic resources. Genomics of plant genetic resources. Springer, Netherlands, pp 67–91

Verhaegen D, Fofana IJ, Logossa ZA, Ofori D (2010) What is the genetic origin of teak (*Tectona grandis* L.) introduced in Africa and in Indonesia? Tree Genet Genomes 6:717–733

Vinceti B, Loo J, Gaisberger H, van Zonneveld MJ et al (2013) conservation priorities for *Prunus africana* defined with the aid of spatial analysis of genetic data and climatic variables. PLoS ONE 8(3):e59987. doi:10.1371/journal.pone.0059987

Vranckx G, Jacquemyn H, Muys B et al (2012) Meta-analysis of susceptibility of woody plants to loss of genetic diversity through habitat fragmentation. Conserv Biol 26(2):228–237

Wang XR, Szmidt AE (1993) Chloroplast DNA-based phylogeny of Asian *Pinus* species (*Pinaceae*). Pl Syst Evol 188:197–211

Wang XR, Szmidt AE (2001) Molecular markers in population genetics of forest trees. Scand J For Res 16:199–220

Wanichkul J, Changtragoon S (2005) Investigation of genetic diversity of *Paphiopedilum exul* by AFLP (Amplified Fragment Length Polymorphism) markers. In: The proceedings of biodiversity of forest and wildlife seminar: research and activities in 2005, Regent Cha-am Hotel, Petchaburi, Thailand, 21–24 Aug 2005, pp 342–36

Wasser SK, Shedlock A, Comstock K, Ostrander E, Mutayoba B, Stephens M (2004) Assigning African elephant DNA to geographic region of origin. Applications to the ivory trade. Proc Natl Acad Sci 101:14847–14852

Wasser SK, Mailand C, Booth R, Mutayoba B, Kisamo ES, Clark B, Stephens M (2007) Using DNA to track the origin of the largest ivory seizure since the 1989 trade ban. Proc Natl Acad Sci 104(10):4228–4236

Wasser SK, Brown L, Mailand C, Mondol S, Clark W, Laurie C, Weir BS (2015) Genetic assignment of large seizures of elephant ivory reveals Africa's major poaching hotspots. Science 349(6243):84–87. doi:10.1126/science.aaa2457

Watanabe A, Widyatmoko A, Rimbawanto A, Shiraishi S (2004) Discrimination of teak (*Tectona grandis*) plus trees using selected random amplified polymorphic DNA (RAPD) markers. J Tropical Forest Sci 16(1):17–24

White TL, Adams WT, Neale DB (2007) Forest genetics. CABI

Wickneswari R (2001) Use of molecular markers to quantify genetic diversity parameters of forest trees. In: Uma Shaanker R, Ganeshaiah KN, Bawa KS (eds) Forest genetic resources: status, threats and conservation strategies. Oxford and IBH Publishing, New Delhi/Calcutta, pp 127–139

Wickneswari R (2016) Impact of logging and disturbance on genetic diversity of tropical forest species. In: Ahuja MR, Jain SM (eds) Biodiversity and conservation of woody plants. Place of Publication, Springer, pp 37–67

Wickneswari R, Norwati M (1993) Genetic diversity of natural populations of *Acacia auriculiformis*. Aust J Bot 41:65–77

Withler RE, Candy JR, Beacham TD, Miller KM (2004) Forensic DNA analysis of Pacific salmonid samples for species and stock identification. Environ Biol Fishes 69:275–285

Yan M, Dai X, Li S, Yin T (2012) A meta-analysis of EST-SSR sequences in the genomes of pine, poplar and eucalyptus. Tree Genet Mol Breed 2(1):1–7. doi:10.5376/tgmb.2012.02.0001

Yooyuen R, Duangjaj S, Changtragoon S (2011) Chloroplast DNA Variation of *Dalbergiacochinchinensis* Pierre in Thailand and Laos. In Heok-Choh S, Hamid SA, Mei L (eds) Asia and the pacific workshop—multinational and transboundary conservation of valuable and endangered forest tree species, Guangzhou, China, 5–7 Dec 2011, pp 84–87

Zalapa JE, Cuevas H, Zhu HY, Steffan S, Senalik D, Zeldin E, McCown B, Harbut R, Simon P (2012) Using next-generation sequencing approaches to isolate simple sequence repeat (SSR) loci in the plant sciences. Am J Bot 99:193–208

Zaya DN, Ashley MV (2012) Plant genetics for forensic applications. Methods Mol Biol 862:35–52. doi:10.1007/978-1-61779-609-8_4

Chapter 7
Genetic Diversity Utilization and Conservation of Neem (*Azadirachta indica* A. Juss.)

A.V.C. Silva, A.S. Jesus, A.N.R. Soares and A.S. Ledo

Abstract This chapter deals with recent studies on the genetic diversity of neem (*Azadirachta indica* A. Juss.), its forms of use and conservation of genetic resources, focusing on the applications of this species for livestock rearing, agriculture, medicine and industry. Information about germplasm banks and collections of the species worldwide is also included. In addition, as a case study, we have analyzed neem's wide distribution in Brazil and the maintenance of an active neem germplasm bank operated by Embrapa (Brazilian Agricultural Research Corporation).

Keywords Azadirachtin · Bioinsecticide · Genetic diversity · In vitro conservation · Cryopreservation · RAPD markers · Micropropagation

7.1 Introduction

The neem tree (*Azadirachta indica*, Meliaceae) was discovered in India approximately 4500 years ago (Patel et al. 2016) and is widely distributed today, including in some parts of the Americas and other tropical and subtropical climatic regions (Azevedo et al. 2015). In recent years, this species has become a focus of interest in the scientific community from several areas, as its biocompounds have a myriad of applications in agriculture, livestock rearing and medicine (Brasil 2013).

A.V.C. Silva (✉) · A.S. Jesus · A.N.R. Soares · A.S. Ledo
Embrapa Coastal Tablelands, Av. Beira Mar, 3250, Aracaju, SE 49025-040, Brazil
e-mail: ana.veruska@embrapa.br

A.S. Jesus
e-mail: alexdabio@hotmail.com

A.N.R. Soares
e-mail: adrielle.naiana@hotmail.com

A.S. Ledo
e-mail: ana.ledo@embrapa.br

M.R. Ahuja and S.M. Jain (eds.), *Biodiversity and Conservation of Woody Plants*,
Sustainable Development and Biodiversity 17, DOI 10.1007/978-3-319-66426-2_7

A large number of biologically active compounds have been isolated from various parts of the plant, making them commercially exploitable. More than 135 compounds of the tree have been isolated and divided into two main classes, the isoprenoids and others (Brasil 2013). This tree adapts very well in hot climates, including in semiarid regions. The neem tree grows rapidly and can reach heights of 15–20 m tall, even in areas with little precipitation (Moura and Soares 2006). The fruits are drupe-like, variable in shape, from elongated to rounded, and can reach 1.4–2.8 cm x 1.0–1.5 cm when mature. The drupes are green in color when young and yellow-green to yellow when ripe (Neem Foundation 2016).

The applications of neem are as follows: wood production for fences, furniture and car bodies; plant insecticide against more than 100 types of insects (whitefly, stable fly, cochineals, aphids, caterpillars); use against ticks and lice; vermicide for animals (goats and sheep); and use in the cosmetics industry, including in creams, toothpaste, soap and drugs for chilblains and head lice (Moura and Soares 2006).

Given its intrinsic nature as a medicinal plant and the fact it has “a thousand and one uses,” this chapter will address recent studies on the genetic diversity of the neem tree, its forms of use and conservation of genetic resources, focusing on the applications of this species for livestock rearing, agriculture, medicine and industry.

7.2 Source and Geographical Distribution

Of Asian origin, from the arid regions of the Indian subcontinent, the neem tree is widely distributed, especially in dry tropical and subtropical areas of Asia, Africa, the Americas, Australia and South Pacific islands. The neem tree has an excellent ability to adapt to climate change. It is cultivated successfully in arid, semiarid, wet tropical and subtropical areas and tolerates temperatures of up to 49 °C; however, it is susceptible to extreme cold (Hegde 1993).

In India, it is widely distributed in many states. In Myanmar, it is very common in central parts of the country. There, studies with neem have intensified, focusing on its potential for agriculture and medicinal purposes. Currently, the species plays an important role in Myanmar’s rural development, and the deployment of projects for commercial use has successfully been carried out in several countries such as Kenya. This plant is also found in Fiji, in the South Pacific. In Australia, it was introduced in the twentieth century. In Indonesia, the neem tree primarily occurs in the northern and eastern lowland areas of Java. In Nepal, the trees are found in the south, in the region of Tarai. In Sri Lanka, it is quite common in the driest parts of the northern region of the island (Neem Foundation 2016).

The neem tree is found in Malaysia, especially in Kedah, Penang, Langkawi and Perlis (Kumar and Navaratnam 2013). In China, the species was introduced in Pan Zhi Hua, Sichuan Province. In Yunnan, 400,000 established trees contribute to the city being the largest artificial area on the planet and the center for raw materials of neem products in China. The species is also found in Iran, Iraq and the Arabian Peninsula. In Qatar and Abu Dhabi, the neem tree was introduced with the aid of an irrigation system using desalinated seawater. This species was also introduced in the plains of Arafat, near Mecca, to provide shade for pilgrims (Zhang et al. 2007).

In Brazil, the species was introduced for study as an insecticide plant by the Paraná Agronomic Institute (Instituto Agronômico do Paraná—IAPAR) and has been well received in the Brazilian North, Midwest, Southeast and Northeast regions. Neem plantations have provided sizeable financial revenue to producers, given the tree's varied uses of dried leaves, wood and, especially, pesticide oil extracts (Mourão et al. 2004).

7.3 Neem Uses

The neem tree (Fig. 7.1) is a source of a large number of natural products, including azadirachtin, a powerful bioinsecticide (Rodrigues et al. 2014). In some countries, it is highly valued for its medicinal value and other bioactive properties (Narnoliva et al. 2014).

This species has been the target of research to better understand its properties to provide alternatives to chemical products, such as fruit, seed, branch and leaf extracts, as well as for the control of pests in crops where pesticide use is prohibited, such as in the case of organic crops.

Fig. 7.1 Tree, leaves, flowers, fruits and seeds of neem

A common characteristic of Meliaceae species is the presence of oxygen-containing triterpenes, known as meliacines. Among them is the most promising antifeedant discovered to date, azadirachtin, which is found in the leaves, fruits and seeds and was initially isolated from neem. Other compounds, such as triterpenoids, gedunin, nimbin and limonoids, among other substances, act together to increase the insecticide activity (Brasil 2013). The neem tree is the most widely known and used insecticide plant, and some formulations based on organic extracts, raw oil or limonoids purified from its seeds are marketed in different countries (Carvalho et al. 2015).

There are a vast number of published studies about neem use, and this chapter will address recent studies on the use and application of oils and extracts, primarily with respect to acaricide, insecticide, antimicrobial and antitumoral activity, as well as the potential of the species to restore degraded areas.

The search for new acaricides from plant extracts comes from the need for safer and less aggressive practices that impact humans and the environment, and it is believed that the use of plant extracts alone or in combination can slow down the rate of developed resistance (Brasil 2013). The use of an emulsion containing 10% neem oil was effective in the treatment of sheep scab (Fernandes et al. 2012) and against *Sarcoptes scabiei* var. in infested rabbits, presenting a promising acaricide for veterinary use (Seddiek et al. 2013).

The oil and extracts of neem bark and leaves have therapeutic effectiveness against leprosy, intestinal helminthiases and respiratory diseases in children (Chouhan et al. 2015). The ingestion of neem oil affects the replacement of the midgut epithelium, causing cytotoxic effects that can change the physiology of the organism due to extensive cell injuries during metamorphosis of *Ceraeochrysa claveri* (Scudeler et al. 2013).

The aqueous extract of the leaves inhibits the growth of the alga *Scenedesmus quadricauda* because the extract contains bioactive compounds with activity against microalgae, making it a cheap and non-toxic alternative for microalgal control in aquatic ecosystems (Chia et al. 2016).

Azadirachtin is a tetraterpenoid that is more abundant in seeds compared with other limonoid molecules. These neem oil compounds are responsible for the toxic effects that affect growth and molting, as well as repellency, sterility and mortality, resulting in cytotoxic effects in different organs and tissues with dysfunction of the endocrine and neuroendocrine systems. The ingestion of oil by *Ceraeochrysa claveri* causes morphological and ultrastructural changes in their cocoons (Scudeler et al. 2014).

The sulfoquinovosyl diacylglycerol (SQDG) isolated from the methanol extract of mature neem leaves showed antimicrobial activity against Gram-positive and Gram-negative bacteria and the herpes simplex virus (Bharitkar et al. 2014). Ethanol fractions of leaves and seeds have a leishmanicidal effect against *Leishmania donovani,* the causative agent of visceral leishmaniasis (Chouhan et al. 2015). The azadiradione extracted from the ethanol extracts of the seeds exhibited a potent antiulcer effect and displayed cytoprotective and anti-secretor effects (Singh et al. 2015). The analysis by HPLC (high performance liquid chromatography) revealed the presence of terpenoids in the methanol extract with antifungal action for the control of dermatophytosis (Ospina Salazar et al. 2015).

The methanol extract of neem has been used in vivo because of its pharmacological, antioxidant and anticoccidial properties, presenting effects against the protozoan *Eimeria papillata,* the causative agent of coccidiosis (Dkhil et al. 2013). Because the fractions of the leaf extracts contain a wide variety of limonoids that inhibit cell proliferation and induce apoptosis, this species has become a promising source of cancer chemopreventive agents (Manikandan et al. 2012).

Several studies indicate that the neem tree has the potential for the prevention and treatment of multiple human cancers (breast, cervix, colon, bowel, liver, lung, mouth, ovarian, prostate, skin, stomach and uterus) through cellular and molecular mechanisms, including the elimination of free radicals, DNA repair, cell cycle change, programmed cell death (apoptosis) and autophagy. However, quality control of the extraction process is required, as is control of the environmental factors that can vary and influence the active compounds (Patel et al. 2016).

Neem cake has activity against the eggs and larvae of *Aedes albopictus* (*Stegomyia albopicta*) (Diptera: Culicidae), which is attributed to the composition of nor-triterpenes, and is therefore a potential household insecticide (Nicoletti et al. 2012). The tree and its derivatives can be used in the control of the *Aedes aegypti* mosquito population, responsible for Dengue and recently for Chikungunya. In addition to being a natural insecticide, neem oil presents the advantage of being biodegradable and cheaper than synthetic insecticides. The action mechanism of the azadirachtin molecule present in the oil comprises the inhibition of a molting protein, whose function is the development of arthropod eggs, influencing also the larval development and overall growth of the insect (Oliveira 2015). The aqueous extract of the leaves and seeds has anthelmintic activity in vitro (Barrabí-Puerta and Arece-García 2013).

In agriculture, the oil extract exhibits repellency against the aphid *Aphis gossypii* and its natural enemy *Cycloneda sanguinea,* which attack the watermelon (Souza et al. 2015). The encapsulation and nanoencapsulation of neem oil using sodium alginate, in addition to being stable, proved to be toxic in lymphocyte cells, which has applications in medicine, aquaculture and soil agriculture (Jerobin et al. 2012). Another study evaluated the systemic insecticide action and residual effect of neem-based nanoformulations (azadirachtin and 3-tigloylazadirachtol) on nymphs of *Bemisia tabaci* Biotype B (Hemiptera: Aleyrodidae), a pest of various crops of economic importance (Carvalho et al. 2015). The application of neem-based nanoformulations as an alternative to control the fall armyworm was also reported by Maria et al. (2016).

Aqueous neem extracts have an effect on oocyte maturation, oviposition, embryonic development and egg hatching of the fish ectoparasite *Argulus bengalensis,* suggesting population control effects for the parasites (Banerjee et al. 2014). The methanol extract of neem seeds inhibit mycelial growth in vitro of the fungus *Colletotrichum scovillei,* which causes anthracnose in bell peppers, demonstrating better efficiency than the fungicide mancozeb (Araujo et al. 2014).

Similarly, neem oil can be used for the control of fungi associated with cowpea seeds (Silva et al. 2014).

In addition to its pharmacological and agricultural uses, the species is increasingly used for restoring arid and degraded lands in tropical and subtropical regions worldwide, as it is highly adaptive (Kuravadi et al. 2013).

7.4 Urban and Rural Afforestation

In urban environments, thermal comfort has been gradually threatened by climate change.

Due to multiple factors, a large percentage of urban populations increasingly require methods to improve life within an often-hostile environment (Pivetta and Silva-Filho 2002). It is precisely in such situations that plants can play a role by bordering streets and avenue, thus reducing the aggressive effect of construction that dominates landscapes in urban centers.

In India, neem is widely cultivated along avenues, within villages and within rural areas in a very dispersed way (Syed et al. 2013).

In Brazil, the neem tree is one of the most commonly used species in urban afforestation, especially in cities in the Northeast region (Fig. 7.2).

Studies conducted on urban afforestation in Rio Grande do Norte (Batista et al. 2013), Paraíba (Klippel et al. 2013) and Ceará (Souza et al. 2013) show this tree species to be the most commonly used for this purpose.

In northeastern Brazil, neem trees have been widely used as an alternative for urban and rural afforestation, being used in reforestation efforts and as windbreaks in semiarid regions. This is because the species is extremely versatile, i.e., its natural regeneration occurs in the rainfall range between 400 and 700 mm (Delwaulle 1979), and it adapts to degraded soil conditions in arid regions

Fig. 7.2 Use of neem in urban afforestation. Cascavel, Ceará, Brazil

throughout the world (Koul 2004). In addition, neem's ability to adapt to the climatic conditions of the Northeast region of Brazil enables the satisfactory growth of its canopy and radicular system, in addition to providing pleasant aesthetics (Crispim et al. 2014). Despite its widespread use, the neem tree can cause imbalance in Brazilian ecosystems. Because it is an exotic species, it may cause ecological changes in the environment and modify the local flora (Vale et al. 2011).

7.5 Genetic Diversity

Genetic diversity is one of the most important parameters evaluated by plant breeders in the initial phase of a breeding program. Many methods are available to evaluate plant populations, which vary in their ability to detect differences between genotypes (Morales et al. 2011). The quantification of genetic diversity can be accomplished through agronomic, morphological and molecular traits, among others.

In plant breeding programs, information related to diversity is key because it enables the identification of hybrid combinations that can produce high heterotic effects, as well as provide greater genetic variability in segregating generations (Gonçalves et al. 2009). In addition, it enables the identification of duplicates, thereby reducing the maintenance costs of germplasm banks.

One way to study the genetic diversity of plants is through molecular markers. In general, these methods are based on the amplification of DNA fragments by polymerase chain reaction (PCR) and have been widely used in plant breeding programs. They can be used, for example, in the study of the genetic diversity of populations and in the evaluation of the potential of genetic resources available and their associations with agronomic characteristics (Benko-Iseppon et al. 2003).

Using the molecular markers RAPD to estimate the variation and genetic relationships between neem accessions, it was possible to confirm the formation of four genetic independent groups in the Genebank. These data enable the formulation of appropriate strategies for conservation and breeding programs (Silva et al. 2013) (Fig. 7.10). The genetic diversity of 12 neem populations collected in Nigeria was evaluated using RAPD markers. Using this marker, no variation was detected between the 12 populations in terms of genomic composition (Awodele et al. 2011).

Studies by Syed et al. (2013) of populations distributed in 40 locations in nine states of India indicate a large genetic variability among individuals. Comparing 37 Indian genotypes with four Thai genotypes, Mishra (2005) observed a low coefficient of genetic similarity between the Indian and Thai genotypes.

For neem, Rajakani et al. (2013) developed a modified protocol for high-quality RNA extraction that will contribute to the biotechnological study of medicinal plants rich in secondary metabolites such as polyphenols, polysaccharides, flavonoids, alkaloids, terpenoids and quinones that hinder the RNA extraction process.

Experiments to identify genes and metabolic pathways in species were described by Kuravadi et al. (2013). In that study, the genome and transcriptome were sequenced using next-generation sequencing technologies. The sequencing platform Illumina 454 resulted in a 267-Mb sequence, which corresponds to 70% of the estimated size of the neem genome. It was possible to identify candidate genes related to the biosynthesis of azadirachtin, as well as SSRs (simple sequence repeats) and SNPs (single nucleotide polymorphism), and to establish phylogenetic relationships between the neem tree and other species, such as citrus. The results of this study help in the understanding of the biopesticide properties and biochemical pathways of neem and other species of the Meliaceae family.

Krishnan et al. (2012) sequenced the genome and four transcriptomes of *A. indica*, contextualizing the sequences obtained with molecular phylogeny, transcript expression and the terpenoid biosynthesis pathway. Narnoliva et al. (2014) isolated and characterized a total of 387 and 512 ESTs (expressed sequence tags), respectively, from the endocarp and mesocarp of neem fruits, in which genes were related to biological processes and molecular function and cell components. The cDNA library and the EST database represent a valuable resource for breeding this species, which has economic and medicinal value.

7.6 Propagation and Conservation

Despite being the focus of recent studies, the species has been used for a long time in traditional medicine by Ayurveda, Unani and homeopathic systems (Chouhan et al. 2015; Nicoletti et al. 2012; Singh et al. 2015). Every part of the neem tree has a wide variety of biologically active compounds containing medicinal potential.

Due to its importance, high concentrations of azadirachtin were produced in vitro using cell culture techniques (Rodrigues et al. 2014). The clonal propagation of neem was conducted through direct and indirect regeneration in vitro. The formation of calluses in the stem base was observed, and the regenerated stem tips were micropropagated and rooted in a medium supplemented with IBA (indolebutyric acid), which allows entire plants to be obtained (Houllou et al. 2015).

The neem propagates through seeds, but its recalcitrance decreases its viability. In addition, the natural heterozygosis of the plant hinders the selection of plants with a high azadirachtin yield and rapid growth from the seedling stage (Houllou et al. 2015). The degree of maturation at the time of harvest can influence the germination of seeds because when immature seeds are harvested, they can express low vigor and germinating power. This problem is worse in fleshy fruits due to the difficulty in determining the time of physiological maturation of the seeds, as there is not always a need to wait for complete visual maturation (Alvarenga et al. 1984).

In this context, studies were developed with the goal of studying the neem seed germination in different stages of maturation, in addition to factors such as the substrate, fruit maturation stage and temperature. Soares et al. (2016) assessed the influence of two maturation stages in different substrates on the germination of the neem seeds (Figs. 7.3 and 7.4; Tables 7.1 and 7.2). These authors concluded that the seeds from mature fruits in interaction with the vermiculite substrate expressed

Fig. 7.3 Germination of neem seeds

Fig. 7.4 Emission of radicle (A) and seedling (B) of neem

Table 7.1 Germination (%) and germination speed index (GSI) of neem seeds in different substrates and stages of maturity

Substrates	Germination (%)		GSI	
	Mature fruits	Green fruits	Mature fruits	Green fruits
Vermiculite	47 a A	30 b A	2.13 a A	1.1 b A
Paper	0.0 b B	0.0 b B	0.0 b B	0.0 b B

Mean values followed by the same lowercase letter in the row and uppercase letter in the column do not differ among themselves, according to Tukey's test at 5% probability

Table 7.2 Hard seeds (%), deteriorated seeds (%), and abnormal seedlings (%) of neem seed germination in different stages of maturation and substrate

Substrate	Hard seeds (%)		Deteriorated seeds (%)		Abnormal seedlings (%)	
	Mature fruits	Green fruits	Mature fruits	Green fruits	Mature fruits	Green fruits
Vermiculite	46 a B	54 a B	2 a A	14 a A	5 a A	2 a A
Paper	15 b B	0 a A	15a A	62 b B	70 b B	38 a A

Mean values by the same lowercase letter in the row and uppercase letter the column do not differ among themselves, according to Tukey's test at 5% probability

the best results. The largest percentages of hard, deteriorated seeds and abnormal seedlings were observed for seeds subjected to the substrate paper and that originated from green fruits.

In a study that included the germination of seeds, Pedroso et al. (2008) concluded that seeds from fruits collected on the ground exhibited better physiological qualities. Vidigal et al. (2007) noted that the substrates vermiculite and germination (germitest) paper provided the best results for the germination test under a constant temperature of 35 °C. However, the results observed by Paula et al. (2009) demonstrated that seedlings of yellow and green fruits did not differ among themselves in terms of height, root length and dry mass.

7.6.1 Micropropagation

To be used in agroforestry, neem mother plants must have superior characteristics, evaluated based on certain characteristics, such as the age of the tree, height-to-diameter ratio, yield and seed chemical content. This is essential to promote the rapid production of a large number of clonal plants from individuals with desirable characteristics (Arora et al. 2010).

Tissue culture is a tool that can promote the systematic multiplication of plants via a micropropagation process. This technique can be employed in the production of secondary metabolites with pharmaceutical applications (Pereira et al. 2009). In micropropagation, small portions of tissue are used, which can be mature, such as leaf and stem, or meristematic, consisting of cells in the process of division, tissues. Through this technique, it is possible to supply the market plants with long life spans, as well as use them for species conservation purposes (Kerbauy 1997).

The micropropagation of neem can help with the homogeneous production of azadirachtin to enable the cloning of mother plants. Thus, the tissue culture method has broad potential for rapid clonal propagation of high levels of azadirachtin. Some studies have been conducted with the goal of developing tissue culture protocols for the neem species. Neem regeneration was observed in leaf-derived calluses (Narayan and Jaiswal 1985), anthers (Goutam et al. 1993) and cotyledons (Rodrigues et al. 2009). Micropropagation was also performed using stem nodal segments (Arora et al. 2010) and segments of roots cultivated in a liquid medium (Arora et al. 2011).

The in vitro culture of embryos is a tool that can be used to control embryogenesis, the rescue of interspecific embryos, the production of haploids and breaking dormancy, as well as provide insight on issues such as embryo nutrition and other applications (Collins and Grosser 1984; Hu and Ferreira 1998). Studies on organogenesis and somatic embryogenesis of neem were conducted by Parveen et al. (1995) and Wewetzer (1998).

Assays conducted to evaluate the efficiency of extracts from leaves, calluses and cell suspensions in vitro, employing the tissue culture method for the production of bioinsecticides from neem, demonstrated that all extracts were effective in controlling *Schistocerca gregaria* (desert locust), suggesting the effective use of this technique for producing neem extracts (Zypman et al. 1997). Studies involving neem germination in different culture conditions in vitro (Fig. 7.5) were conducted by Lédo et al. (2008). In that study, zygotic embryos and seeds under different culture conditions in vitro were used. Houllou et al. (2015) concluded that it is possible to induce differentiation and development of whole neem plants through tissue culture.

Fig. 7.5 Phases of in vitro germination of neem; abnormal and normal seedlings

7.6.2 Germplasm Databases, Collections, Enrichment and Characterization

There are few reports in the literature about the development of collections and databases of neem germplasm. The first attempts to test and improve the procedures for seed collection of neem accessions, handling and exchange date back to 1993–1994. In 1993, a total of 23 accessions were collected in India, Nepal, Pakistan, Sri Lanka and Thailand and dispatched to 10 countries in Africa, Asia and Latin America. In 1994, a neem evaluation program was started by an international neem network that was developed by several countries and international institutions such as CIRAD, DANIDA (Danish International Development Agency) Forest Seed Centre, FAO (Food and Agriculture Organization of the United Nations), FORSPA (Forestry Research Support Program for Asia) and IPGRI (International Plant Genetic Resources Institute), now known as Bioversity International (FAO 2007).

The National Research Center for Agroforestry, located in Jhansi (in Uttar Pradesh), in India, has a collection with accessions from 40 provinces created in the year 2000 (Syed et al. 2013), where multiple phenotypic and genetic variation studies have been conducted.

In Brazil, many exotic species, such as neem, are introduced for economic purposes. However, studies on genetic diversity for future breeding programs and the selection of superior individuals are needed (Vitória et al. 2013).

In this context, the Active Neem Germplasm Bank of Embrapa Coastal Tablelands was created in 2009, in the municipality of Aracaju, Sergipe state, in Brazil's Northeast region (10°57'01" S and 37°03'06" W longitude). The Embrapa is working to increase the genetic variability of *A. indica* A. Juss. through the introduction of germplasms from multiple origins (Figs. 7.6 and 7.7).

The Active Germplasm Banks (Bancos Ativos de Germoplasma—BAGs) are an alternative for the conservation of plant genetic resources. Additionally, the

Fig. 7.6 Neem Germplasm Bank of Embrapa Coastal Tablelands. Aracaju, Brazil, 2015

Fig. 7.7 Neem Germplasm Bank of Embrapa Coastal Tablelands. Aracaju, Brazil, 2016

development of germplasm collections is the first step for identifying duplicates and genitors. This will allow hybrids with greater commercial potential (Freitas et al. 2012) to be obtained and is a way to reconcile the agrobiodiversity conservation efforts with sustainable development.

The characterization of germplasm databases is important for identifying neem's accessions and to extend its genetic basis, offering multiple materials with high genetic quality. Knowledge of the extension and organization of neem's genetic diversity would be important for the effective assessment, identification, documentation and use of genetic resources for conservation and efficient use programs (Deshwal et al. 2005). For this purpose, the germplasm characterization can be made using different methods, including morphological and agronomic descriptors and molecular markers (Vicente et al. 2005).

Morphological characterization consists of providing an identity for each entry by utilizing knowledge of a series of data that enables the genetic variability of each sample to be studied (Ramos and Queiroz 1999), thus excluding the possibility of duplicates and assessing the influence of morphology in clustering through a dendrogram (Neiva et al. 2011) or other clustering methods.

The neem BAG has 54 genotypes (Table 7.3), periodically evaluated in terms of morphological, agronomic and molecular traits. In 2013, Vitória et al. (2013) evaluated 18 genotypes in terms of total height, circumference, diameter at breast height (DBH) and stem diameter.

The knowledge of the degree of genetic variability, through divergence studies, is fundamental in the process of identifying new sources of genes of interest. Molecular characterization represents a way to evaluate this diversity, allowing the degree of diversity between individuals and between populations to be inferred using genetic markers, unaffected by environmental conditions. For the formation of cultivated varieties, it is essential that there be sufficient genetic diversity to allow

Table 7.3 Neem (*Azadirachta* sp.) accessions analyzed by RAPD markers from Germplasm Bank of Embrapa Coastal Tablelands, Aracaju, Brazil

Accessions	Original location	Accessions	Original location
Ae1	Cenargen	Ai28	CPAC
Ae2	Cenargen	Ai29	Cruangi sugar mill
Ae3	Cenargen	Ai30	Cruangi sugar mill
Ae4	Cenargen	Ai31	Cruangi sugar mill
Ae5	Cenargen	Ai32	Cruangi sugar mill
Ae6	Cenargen	Ai33	Cruangi sugar mill
Ae7	Cenargen	Ai34	Cruangi sugar mill
Ae8	Cenargen	Ai35	Cruangi sugar mill
Ae9	Cenargen	Ai36	Cruangi sugar mill
Ae10	Cenargen	Ai37	Cruangi sugar mill
Ae11	Cenargen	Ai38	Cruangi sugar mill
Ae12	Cenargen	Ai39	Cruangi sugar mill
Ae13	Cenargen	Ai40	Cruangi sugar mill
Ae14	Cenargen	Ai41	Cruangi sugar mill
Ae15	Cenargen	Ai42	Petrobras
Ae16	Cenargen	Ai43	Petrobras
Ae17	Cenargen	Ai44	Petrobras
Ae18	Cenargen	Ai45	Petrobras
Ae19	Cenargen	Ai46	Petrobras
Ae20	Cenargen	Ai47	Petrobras
Ae21	Cenargen	Ai48	Petrobras
Ae22	Cenargen	Ai49	Petrobras
Ai23	CPAC	Ai50	Petrobras
Ai24	CPAC	Ai51	Petrobras
Ai25	CPAC	Ai52	Petrobras
Ai26	CPAC	Ai53	Petrobras
Ai27	CPAC	Ai54	Petrobras

for the selection of individuals that can be used in breeding programs. Therefore, the study of the variability components of the species is fundamental, especially for native species that are minimally studied, whose magnitude of diversity is not yet fully known (Costa et al. 2011).

The genetic diversity of the accessions of the neem BAG of Embrapa Coastal Tablelands was estimated by RAPD (random amplified polymorphic DNA) (Silva et al. 2013) and ISSR (inter-simple sequence repeats) (Vitória et al. 2013) and revealed 100% polymorphism between the genotypes. New collections (Fig. 7.8) and introductions (Fig. 7.9) are still being created with the goal of extending the existing diversity and conserving these genetic resources (Fig. 7.10).

Fig. 7.8 Harvest of neem for introduction of new access in germplasm bank

Fig. 7.9 Neem seedlings production

7.6.3 In Vitro *Conservation of Genetic Resources*

In a global change scenario, where ecosystems face major climate change and strong anthropogenic pressures, genetic erosion is an increasingly present reality. Thus, the conservation of genetic resources represents a priority established and recognized at a global level (Lopes et al. 2013).

The germplasm conservation of plant species has been employed as a way to avoid the risk of extinction, as well as allow multiple genotypes to be available for future use. Among the different methods of the preservation of germplasm, conservation in vitro is well established, where plant collections are kept under laboratory conditions in the medium and long term (Sarasan et al. 2006).

The advantages of conservation in vitro include the maintenance of genotypes in aseptic conditions, reductions in costs and labor and physical space optimization, in

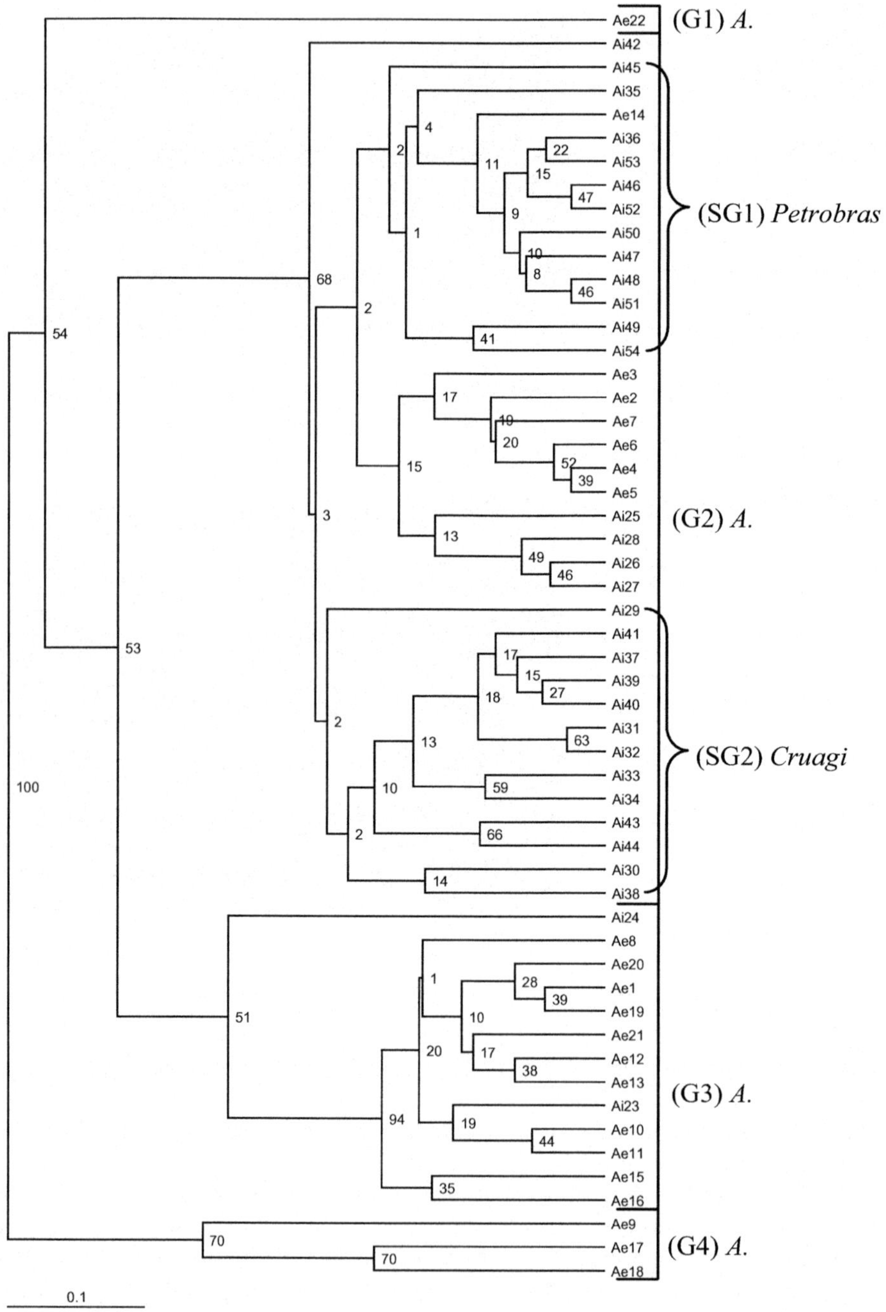

Fig. 7.10 Dendrogram of genetic similarity generated using the Jaccard coefficient and the UPGMA method with analysis bootstraps for 54 Neem accessions (*Azadirachta* sp.) belonging to Embrapa Coastal Tablelands, Aracaju, Sergipe, Brazil

addition to the ease of plant material exchange (Engelmann 2011). In general, two methods have been adopted for plant conservation in vitro: cryopreservation and slow growth. In cryopreservation, the plants are kept at ultra-low temperatures, and therefore, there is a complete suppression of growth (Flores et al. 2013).

In cryopreservation, liquid nitrogen (LN) (−196 °C) or its vapor phase (−150 °C) are used for storing biological material at ultra-low temperatures, where all cell divisions and metabolic activities are stopped. Therefore, the plant material can be stored without change or modification for a theoretically unlimited period of time (Berjak et al. 2011). For Zeliang and Pattanayak (2012), cryopreservation is the most suitable method for the conservation of germplasm in a small space, requiring little maintenance and maintaining the characteristics of the cryopreserved material.

This technique has achieved good results, especially because it highly reduces any cell activity, minimizing biological deterioration during storage (Tresena et al. 2009). It is an alternative to preserving species that are brought in from fields, botanical gardens, natural reserves or in vitro collections in slow-growth conservation systems (Pérez-Molphe-Balch et al. 2012).

The first report on cryopreservation of neem was documented by Chaudhury and Chandel (1991); these authors were successful in the cryopreservation of embryonic axes and desiccated neem seeds from Mombasa, Kenya. Dumet and Berjak (2002), in their studies on the cryopreservation of embryonic axes and neem seeds not desiccated and desiccated for one or two days, observed a decrease in the water content of seeds. The survival rate achieved better results with one-day desiccation, enabling up to four months of storage under cryopreservation conditions. Seeds were cryopreserved successfully for 12 months, with a survival rate of 45% (Varghese and Naithani 2008).

7.7 Concluding Remarks

Despite its large worldwide distribution and applications in various areas, actions on ex situ conservation, characterization and breeding need to be increased. Directed investigations for the selection of genotypes based on their application in the pharmaceutical, cosmetics, pesticide and others should be prioritized in research institutes.

References

Alvarenga EM, Silva RF, Araújo EF (1984) Influence of age and post-harvest storage of fruits on quality of watermelon seeds (*Citrullus vulgaris*). Hortic Bras 2(2):5–8

Araújo ER, Harand W, Lima IC, Dias FCR, Adelmo ADS, Costa RRda, Carvalho DL (2014) Extracts of *Piper marginatum* and *Azadirachta indica* for the control of *Colletotrichum scovillei* in bell pepper. Pesquisa Agropecuária Bras 49(2):88–94

Arora K, Sharma M, Srivastava J, Ranade JSA, Sharma AK (2010) Rapid in vitro cloning of a 40-year-old tree of *Azadirachta indica* A. Juss. (Neem) employing nodal stem segments. Agrofor Syst 78(53):53–63

Arora K, Sharma M, Srivastava J, Ranade AS, Sharma AK (2011) *In vitro* cloning of *Azadirachta indica* from root explants. Biol Plant 55(1):164–168

Awodele O, Ogunkanmi LA, Oyelakin OO, Kuti OO, Olofinnade AT, Akintonwa A (2011) Assessment of genetic diversity of 12 populations of *Azadirachta indica* A. Juss from three different locations in Lagos State, Nigeria using RAPD markers. Niger Q J Hosp Med 21 (1):41–44

Azevedo GTdeOS, Novaes AB, Azevedo GB, Silva HF (2015) Development of indian neem seedlings under different levels of shading. Floresta e Ambiente 22(2):249–255

Banerjee A, Manna S, Saha SK (2014) Effect of aqueous extract of *Azadirachta indica* A. Juss (neem) leaf on oocyte maturation, oviposition, reproductive potentials and embryonic development of a freshwater fish ectoparasite Argulus bengalensis Ramakrishna, 1951 (Crustacea : Branchiura). Parasitol Res 113(12):4641–4650

Barrabí-Puerta M, Arece-García J (2013) *In vitro* antihelmintic activity of an accuos extract of Neem (*Azadirachta indica* A. Juss) leaves and seeds. I. Inhibition of eggs hatching and larval development. Rev Salud Anim 35(2):103–108

Batista FA, Chaves TP, Felismino DC, Dantas IC (2013) Quantitative-qualitative survey of the city of urban afforestation Remígio, Paraíba. Rev Biol e Farmácia. Campina Grande/PB 9(1):70–83

Benko-Iseppon AM, Winter P, Hüttel B, Stagginus C, Mühlbauer F, Kahl G (2003) Molecular markers closely linked to *Fusarium* resistance genes in chickpea show significant alignments to pathogenesis related genes located on *Arabidopsis* chromosomes 1 and 5. Theor Appl Genet 103(2):379–386

Berjak P, Bartels P, Benson EE, Harding K, Mycock DJ, Pammenter NW, Wesley-Smith SJ (2011) Cryoconservation of South African plant genetic diversity. In Vitro Cell Dev Biol-Plant 47:65–81

Bharitkar YP, Bathini S, Ojha D, Ghosh S, Mukherjee H, Kuotsu K, Chattopadhyay D, Mondal NB (2014) Antibacterial and antiviral evaluation of sulfonoquinovosyldiacylglyceride: a glycolipid isolated from *Azadirachta indica* leaves. Lett Appl Microbiol 58(2):184–189

Brasil RB (2013) Aspectos botânicos, usos tradicionais e potencialidades de *Azadirachta indica* (NEEM). 470 Enciclopédia biosfera 9(17):3252–3268. Avaliable in:http://www.conhecer.org.br/enciclop/2013b/MULTIDISCIPLINAR/Aspectos.pdf. Accessed 10 Sep 2016

Carvalho SS, Vendramim JD, Sá ICG, Silva MFGF, Ribeiro LP, Forim MR (2015) Systemic insecticidal effect of neem-based nanoformulations against *Bemisia tabaci* (Hemiptera: Aleyrodidae) biotype B in tomato. Bragantia 74(3):298–306

Chaudhury R, Chandel KPS (1991) Cryopreservation of desiccated seeds of neem(*Azadirachta indica* A. Juss) For germplasm conservation. Indian J Genet Resour 4(2):67–72

Chia MA, Akinsanmi JT, Tanimu Y, Ladan Z (2016) Algicidal effects of aqueous leaf extracts of neem (*Azadirachta indica*) on *Scenedesmus quadricauda* (Turp.) de Brébission. Acta Bot Brasilica 30(1):1–8

Chouhan G, Islamuddin M, Want MY, Abdin MZ, Ozbak HA, Hemeg HA, Sahal D, Afrin F (2015) Apoptosis mediated leishmanicidal activity of *Azadirachta indica* bioactive fractions is accompanied by Th1 immunostimulatory potential and therapeutic cure *in vivo*. Parasites Vectors 26(8):183

Collins GB, Grosser JW (1984) Culture of embryos. In: Vasil IK (ed) Cell culture and somatic cell genetics of plants, Academic Press, New York, 1:241–257

Costa TS, Silva AVC, Lédo AS, Santos ARF, Silva Júnior JFS (2011) Genetic diversity of accessions of the mangaba germplasm bank in Sergipe, Brazil. Pesquisa Agropecuária Trop 46(5):499–508

Crispim DL, Silva MA, Chavess ADC, Almeida RRP, Freitas AJF (2014) Diagnosis of urban afforest center of Pombal-PB. Rev Verde Agroecologia e Desenvolvimento Sustentável 9 (1):191–196

Delwaulle JC (1979) Plantations forestières en Afrique Tropicale Sèche. Revue Bois el Forêst des Tropics 184:3–18

Deshwal RPS, Singh R, Malik K, Randhawa GJ (2005) Assessment of genetic diversity and genetic relationships among 29 populations of *Azadirachta indica* A. Juss. using RAPD markers. Genet Resour Crop Evol 52:285–292

Dkhil MA, Al-Quraishy S, Moneim AEA, Delic D (2013) Protective effect of *Azadirachta indica* extract against Eimeria papillata -induced coccidiosis. Parasitol Res 212(1):101–106

Dumet D, Berjak P (2002) Cryopreservation of neem (*Azadirachta indica* A. Juss.) seeds. Biotechnol Agric Forestry 50:213–219

Engelmann F (2011) Use of biotechnologies for the conservation of plant biodiversity. In vitro Dev Biol 47(4):5–16

Fao. International Neem Network (2007) Available http://www.fao.org/forestry/neem/5309/en/. Accessed 12 sep 2016

Fernandes JI, Verocai GG, Ribeiro FA, Melo RMPS, Correia TR, Veiga CCP, Veira VPC, Scott FB (2012) Miticidal eficacy of an emulsion containing 10% neem (*Azadirachta indica*) oil for control of Psoroptes ovis in naturally infested rabbits. Pesquisa Veterinária Bras 32 (12):1253–1256

Flores R, Uliana SC, Pimentel N, Garlet TMB (2013) Sucrose and sorbitol on the in vitro conservation of *Pfaffia tuberosa* (Spreng.) Hicken (Amaranthaceae). J Biotechnol Biodivers 4 (3):192–199

Freitas MKCde, Coimbra R, Aguiar GB, Aguiar CBN, Chagas DBdas, Ferreira Wde M, Oliveira RJde (2012) Phenotypic variability and morphologic characterization of a natural population of *Hancornia speciosa* Gomes. Biosci J Uberlândia 28(5):833–841

Gonçalves LSA, Rodrigues R, Amaral Júnior AT, Karasawa M, Sudré CP (2009) Heirloom tomato gene bank: assessing genetic divergence based on morphological, agronomic and molecular data using a Ward-modified location model. Genet Mol Res 8(1):364–374

Goutam VK, Nanda N, Gupta SC (1993) Development of shoots and roots in anther derives callus of *Azadirachta indica* A. Juss. Plant Cell Tissue Organ Cult 34(1):13–18

Hedge NG (1993) Neem production and development—constraints at grass root level. In: Souvenir of World Neem Conference. Indian Society of Tobacco Science CTRI Rajamundry (A.P.) India. pp 1–7

Houllou LM, Souza RA, Santos ECPdos, Silva JJP, Barbosa MR, Sauvé JPG, Harand W (2015) Clonal propagation of neem (*Azadirachta indica* A. Juss.) via direct and indirect in vitro regeneration. Rev Árvore 39(3):439–445

Hu CY, Ferreira AG (1998) Cultura de embriões. In: Torres AC, Caldas LS, Buso JÁ (eds) Cultura de tecidos e transformação genética de plantas. EMBRAPA-SPI/ EMBRAPA-CNPH, Brasília 1:93–371

Jerobin J, Sureshkumar RS, Anjali CH, Mukherjee A, Chandrasekaran N (2012) Biodegradable polymer based encapsulation of neem oil nanoemulsion for controlled release of Aza-A. Carbohyd Polym 90(4):1750–1756

Kerbauy GB (1997) Clonagem de plantas in vitro. Revista de Biotecnologia Ciência e Desenvolvimento 1(1):30–33

Klippel VH, Pezzopane JEM, Pezzopane JRM, Cecílio RA, Castro FS, Pimenta LR (2013) Climatic zoning for Teca, australian Cedar, neem and pupunha in state of Espirito Santo. Floresta 43(4):671–680

Koul O (2004) Neem: a global perspective. In: Koul O, Wahab S (eds) Neem: today and in the new millennium. Kluwer Academic Publishers, Dordrecht, pp 1–19

Krishnan NM, Swetansu P, Jain P, Gaur P, Choudhary R, Vaidyanathan S, Deepak S, Hariharan AK, Krishna PGB, Nair J, Varghese L, Valivarthi NK, Dhas K, Ramaswamy K, Panda B (2012) A draft of the genome and four transcriptomes of a medicinal and pesticidal angiosperm *Azadirachta indica*. BMC Genom 464:1–13

Kumar SV, Navaratnam V (2013) Neem (*Azadirachta indica*): Prehistory to contemporary medicinal uses to humankind. Asian Pac J Trop Biomed 3(7):505–514

Kuravadi NA, Yenagi V, Rangiah K, Mahesh HB, Rajamani A, Shirke MD, Russiachand H, Loganathan RM, Lingu CS, Siddappa S, Ramamurthy A, Sathyanarayana BN, Gowda M (2013) Comprehensive analyses of genomes, transcriptomes and metabolites of neem tree. Rev Biología Trop 61(2):515–529

Lédo AS, Blank AF, Barboza SBSC, Rangel MAS, Lédo CAS (2008) In vitro germination of zygotic embryos and seeds of neem (*Azadirachta indica* A. Juss.). Rev Bras Plant Medicinais 10(3):1–5

Lopes KP, Almeida FAC, Carvalho JMFC, Bruno RLA (2013) Cryopreservation of zygotic embryos axes of cotton. Rev Bras Engenharia Agrícola e Ambiental 17(3):291–298

Manikandan P, Ramalingam SM, Vinothini G, Ramamuthi VP, Singh IP, Anandan R, Gopalakrishnan M, Nagini S (2012) Investigation of the chemopreventive potential of neem leaf subfractions in the hamster buccal pouch model and phytochemical characterization. Eur J Med Chem 56:271–281

Maria A, Giongo M, Forim MR (2016) Evaluation of neem-based nanoformulations as alternative to control fall armyworm. Ciência e Agrotecnol 40(1):26–36

Mishra DK (2005) Germplasm variability in neem *(Azadirachta indica)*. J Trop Forest Sci 17 (2):224–227

Morales RGF, Resende JTVde, Faria MV, Silva PRda, Figueiredo AST, Carminatti R (2011) Genetic diversity in strawberry cultivars based on morphological characteristics. Rev Ceres 58 (3):323–329

Moura M, Soares T (2006) Uso do nim: uma árvore de muitas utilidades. Ouricuri-PE: CAATINGA (Centro de Assessoria e Apoio aos Trabalhadores e Instituições Não Governamentais Alternativas). 18 p. Avaliable in: http://www.caatinga.org.br/wpcontent/uploads/2014/03/NIM-IX.pdf-cartilha-do-caatinga.pdf. Accessed 10 Sep 2016

Mourão AS, Silva JCT, Guedes RNC, Venzon M, Jham GN, Oliveira CL, Zanuncio JC (2004) Selectivity of neem extracts (*Azadirachta indica* A. Juss.) to the predatory Mite *Iphiseiodes zuluagai*(Denmark & Muma) (Acari: Phytoseiidae). Neotropical. Entomol 33(5):613–617

Narayan P, Jaiswal VS (1985) Plantlet regeneration from leaflet callus of *Azadirachta indica* A. Juss. J Tree Sci 4(1):65–68

Narnoliva LK, Rajakani R, Sangwan N, Gupta V, Sangwam RS (2014) Comparative transcripts profiling of fruit mesocarp and endocarp relevant to secondary metabolism by suppression subtractive hybridization in *Azadirachta indica* (neem) Expressed sequence Tag National Centre for Biotechnology Information. Mol Biol Rep 41(5):3147–3162

Neem Foundation (2016) All about neem. Mumbai: Neem Foundation. Available http://www.neemfoundation.org/about-neem/introduction-to-neem-tree/. Accessed 12 Aug 2016

Neiva PI, Andrade Júnior VCde, Viana DJS, Figueiredo JA, Mendonça Filho CV, Parrella RAdaC, Santos JB (2011) Morphologic characterization of sweet potato accesses from the germplasm bank of the Universidade Federal dos Vales do Jequitinhonha e Mucuri Brazil. Hortic bras 29 (4):537–541

Nicoletti M, Mariani S, Maccioni O, Coccioletti T, Murugan K (2012) Neem cake: chemical composition and larvicidal activity on Asian tiger mosquito. Parasitol Res 111(1):205–213

Oliveira DABde (2015) Use of neem and its molecular components in mosquito control *Aedes aegypti*. Rev Científica do ITPAC 8(2):1–5

Ospina Salazar DI, Hoyos Sánchez RA, Orozco Sánchez F, Arango Artega M, Gómex Londoño LF (2015) Antifungal activity of neem (*Azadirachta indica*: Meliacea) extracts against dermatophytes. Acta Biológica Colomb 20(3):201–207

Patel SM, Venkata KCN, Bhattacharyya P, Sethi G, Bishayee A (2016) Potential of neem (*Azadirachta indica* L.) for prevention and treatment of oncologic diseases. Semin Cancer Biol 40:100–115

Paula YCM, Ribeiro MCC, Aquino AFMAG, Benedito CPB (2009) Influência da maturação do fruto na germinação de Nim (*Azadirachta indica* A. juss.). Rev Verde de Agroecologia e Desenvolvimento Sustentável 4(1):64–68

Parveen RK, Kakani CS, Tomar UK (1995) Tissue culture studies on neem (*Azadirachta indica* A. Juss.). News lett Int Neem Netw 2(3):15–17

Pedroso L, Azeredo GA, Vanzolini S (2008) Neem seed germination (*Azadirachta indica* A. Juss) depending on the type of collection and of fruits coloring. Nucleus 5(2):197–208

Pereira MAS (2009) Cultura de Tecidos de plantas medicinais. Available http://www.ufmt.br/…/Cultura%20de%20tecidos%20de%20plantas%20medicinais.pdf. Accessed10 Aug 2016

Pérez-Molphe-Balch E, Pérez-Reyes ME, Rosa-Carrillo MLL (2012) *In vitro* conservation of Turbinicarpus (Cactaceae) under slow growth conditions. Haseltonia 17(1):51–57

Pivetta KFL, Silva-Filho DF (2002) Arborização urbana. Boletim acadêmico série arborização urbana. UNESP/FCAV/FUNEP, Jaboticabal, SP, 69p

Rajakani R, Narnoliya L, Sangwan NS, Sangwan RS, Gupta V (2013) Activated charcoal-mediated RNA extraction method for *Azadirachta indica* and plants highly rich in polyphenolics, polysaccharides and other complex secondary compounds. BMC Res Notes 28 (6):125

Ramos SRR, Queiroz MA (1999) Morphological characterization: cucurbit BAG experience of Embrapa Semi-Arid, with pumpkin and squash access. Hortic bras 17(suplemento):9–12

Rodrigues M, Festucci-Buselli RA, Silva LC, Otoni WC (2014) Azadirachtin biosynthesis induction in *azadirachta indica* A. Juss cotyledonary calli with elicitor agents. Braz Arch Biol Technol 57(2):155–162

Rodrigues M, Paiva R, Nogueira RC, Martinotto C, Silva Júnior JM (2009) *In vitro* morphogenesis of neem from cotyledonary-derived explants. Rev Árvore 33(1):21–26

Sarasan V, Cripps R, Ramsay MM, Atherton C, McMichen M, Prendergast G, Rowntree JK (2006) Conservation in vitro of threatened plants—progress in the past decade. In vitro cell dev biol-plant 42(3):206–214

Scudeler EL, Garcia ASG, Padovani CR, Santos DC (2013) Action of neem oil (*Azadirachta indica* A. Juss) on cocoon spinning in Ceraeochrysa claveri (Neuroptera: Chrysopidae). Ecotoxicol Environ Saf 97(1):176–182

Scudeler EL, Padovani CR, Santos DC (2014) Effects of neem oil (*Azadirachta indica* A. Juss) on the replacement of the midgut epithelium in the lacewing Ceraeochrysa claveri during larval-pupal metamorphosis. Acta Histochem 116(5):771–780

Seddiek SA, Khater HF, El-Shorbagy MM, Ali AM (2013) The acaricidal efficacy of aqueous neem extract and ivermectin against Sarcoptes scabiei var. cuniculi in experimentally infested rabbits. Parasites and Vectors 112(1):2319–2330

Silva AVC, Rabbani ARC, Almeida CS, Clivati D (2013) Genetic structure and diversity of the neem germplasm bank from Brazil Northeast. Afr J Biotech 12(20):2822–2829

Silva GC, Santos CC, Gomes DP (2014) Incidence of fungi and germination of cowpea seeds (*Vigna unguiculata* L. (Walp) treated with neem oil (*Azadirachta indica* A. Juss). Rev Bras Plantas Medicinais 16(4):850–855

Singh R, Mishra V, Pandeti S, Palit G, Barthwal MK, Pandey HP, Narender T (2015) Cytoprotective and anti-secretory effects of Azadiradione isolated from the seeds of *Azadirachta indica* (neem) on gastric ulcers in rat models. Phytotharapy Res 1(29):910–916

Syed M, Gupta VK, Pandey HC (2013) Studies of phenotypic and genetic variation in various growth characters in Neem (*Azadirachata indica* A. Juss) germplasm. Agric Sci Res J 33 (3):72–78

Soares ANR, Rocha Júnior VF, Vitória MF, Silva AVC (2016) Neem seed germination in physiological maturity and substrate function. Nucleus 13(1):215–222

Souza DIM, Costa LR, Gurgel MT, Soares SRF (2013) Afforestation neighborhood fish fat in the city of Icapui-Ce. Rev Verde de Agroecologia e Desenvolvimento Sustentável 8(4):238–242

Souza CR, Sarmento RA, Venzon M, Santos GRdos, Silveira MCAC, PH Tschoeke (2015) Lethal and sublethal effects of neem on *Aphis gossypii* and *Cycloneda sanguinea* in watermelon. Acta Scientiarum 37(2):233–239

Tresena NL, Cavalcanti-Mata MERM, Duarte MEM, Moraes AMde, Dias VS (2009) Physiologic quality of pink ipê seed (*Tatebuia chrysotrica* Mart. ex. DC.) Standl.) submitted to cryoconservation. Rev Bras Produtos Agroindustriais 11(1):87–93

Vale NFL, Sousa GS, Mata MF, Braga PET (2011) The urban afforestation in the municipality of Timon-MA: inventory, diversity and qualitative and quantitative diagnosis. Rev da Soc Bras Arborização Urbana 6(4):145–157

Varghese B, Naithani SC (2008) Oxidative metabolism-related changes in cryogenically stored neem (*Azadirachta indica* A. Juss) seeds. J Plant Physiol 165(7):755–765

Vicente MCde, Guzmán FA, Engels J, Ramanatha RAOV (2005) Genetic characterization and its use in decision marking for the conservation of crop germplasm. In: The role of biotechnology. Preocedding. pp 121–128

Vidigal DdeS, Brasileiro BG, Dias DCF, Alvarenga EM, Bhering MC (2007) Germination and post-seminal development of neem (*Azadirachta indica* A. Juss.—MELIACEAE) seeds. Rev Bras de Sementes 29(3):39–46

Vitória MF, Silveira RG, Silva AVC (2013) Morphological and molecular characterization of Indian neem access. Magistra 11(1):228–229

Wewetzer A (1998) Callus cultures of *Azadirachta indica* and their potential for the production of Azadirachtin. Phytoparasitica 26(1):47–52

Zeliang PK, Pattanayak A (2012) Fundamental cryobiology and basic physical, thermodynamical and chemical aspects of plant tissue cryopreservation. In: Abdurakmonov I (ed) Plant breeding, pp 41–56

Zhang K, Lei C, Tan Z (2007) Preliminary report of *Azadirachta indica* A. Juss. introduction and cultivation in Panzhihua. J Sichuan Agric Uni 25(3):282–287

Zypman S, Ziv M, Applebaum S (1997) Tissue culture methods and cloning of the neem tree (*Azadirachata indica*) for bioinsective production. Acta Hort 447:235–236

Part II
Biodiversity and Conservation

Chapter 8
Biotechnological Approaches for Conservation of the Genus *Pistacia*

Hülya Akdemir and Ahmet Onay

Abstract Climate change with the combination of increased population growth, unplanned urbanization, habitat loss and degradation, pollution and diseases, overexploitation of valuable species are the major causes of the loss of plant biodiversity. Moreover, since environment and environmental factors will change faster than the most of the plant's adaptation, conservation strategies become critically essential for preservation of plant species. In this chapter, we described the current status of biotechnological conservation approaches for *Pistacia* species. Since in vitro conservation and most of cryopreservation methods rely on tissue culture-based methods, we described here micropropagation, micrografting, and somatic embryogenesis procedures in addition to in vitro conservation and cryopreservation in *Pistacia* species. Because there is no one 'universal protocol,' biotechnological conservation studies for *Pistacia* are still limited with a few species of the genus as described in this chapter. Development and optimization of appropriate techniques especially for long-term preservation of different *Pistacia* species can lead to establish cryobanks to conserve germplasm. Furthermore, the developed technologies will provide to prevent possible genetic erosion of *Pistacia* species.

Keywords Tissue culture · Ex situ conservation · Genetic erosion
Pistacia · Pistachio · Mastic tree · Cryopreservation

H. Akdemir
Faculty of Science, Department of Molecular Biology and Genetics, Gebze Technical University, 41400 Gebze, Kocaeli, Turkey
e-mail: pinarakdemir@gmail.com

A. Onay (✉)
Faculty of Science, Department of Biology, Dicle University, 21280 Diyarbakır, Turkey
e-mail: ahmetonay45@gmail.com

M.R. Ahuja and S.M. Jain (eds.), *Biodiversity and Conservation of Woody Plants*, Sustainable Development and Biodiversity 17, DOI 10.1007/978-3-319-66426-2_8

8.1 Introduction

The genus *Pistacia* L. is a member of the family Anacardiaceae (cashew family) of order Sapindales (Zohary 1952; Pell 2004). The genus encompassing nine species and five subspecies according to the most recent botanical study is one of the most valuable plant genera in the angiosperms (Al Saghir and Porter 2012). Pistachio (*Pistacia vera* L.) and lentisk (mastic tree, *Pistacia lentiscus*) are the members of the genus that are cultivated and grown commercially for the production of pistachio nuts and mastic gum, respectively. The other species are generally used as rootstocks of *P. vera*, and they can be easily distinguished from *P. vera* with their much smaller seeds. The genus *Pistacia* is indigenous to the eastern Mediterranean (Cyprus and Turkey to Israel and Syria), Iran, Central Asia, and Xinjiang, but trees of the genus can be also found naturally in several regions of the world such as from North Africa to the Philippines and in Honduras and Mexico to Texas (Al Saghir and Porter 2012). Trees of the genus are dioecious, deciduous, and xerophytic and grow up to 10 m in height. Moreover, the fruits and resins of several species of the genus *Pistacia* which have potential usage to cure numerous human ailments have been used since ancient times because there are several products obtained from the fruit and different parts of the trees of the genus. Therefore, an extensive cultivation of especially *P. vera* L. has been made in the different regions of the world. Major centers of commercial pistachio production are Iran, Turkey, and California (USA.). World production of pistachios in 2015/16 has been estimated almost 524,000 metric tons (MT) (in shell basis), that is, 8% down compared with the previous season and 24% up over 2005/06. Furthermore, in 2015, Iran's production as a main producer was 210,000 MTs (40% share) and Turkey as a second and USA as a third produced 145,000 MTs (28% share) and 127,000 MTs (24% share), respectively. These countries remain the major producers, accounting for 92% of pistachios world production (Anonymous 2015a). USA and Iran led the exports of pistachios with 106,375 MT and 101,121 MT, respectively, accounting for the 68% of total exports. Top five pistachio producing countries in the world were shown in Fig. 10.1. The second cultivated species of the genus is *P. lentiscus* var. Chia, and it is considered native to the Canary Islands, most Mediterranean coastal countries such as Israel, Tunisia, Morocco, Albania, Greece, and the coastal areas of the Aegean region of Turkey. Lentisk produces an aromatic resin known as mastic, which has a long history of usage; for example, it is noted to have been used in ancient Egypt as part of the embalming process (Hanelt et al. 2001). Although there is no endangered (EN) *Pistacia* species, it has been noted that there are five *Pistacia* species under conservation status according to the IUCN Red List of Threatened Species (Anonymous 2015b). These are *P. aethiopica* (NT, near threatened) *P. mexicana* (VU, vulnerable), *P. malayana* (LC, least concern), *P. cucphuongensis* (VU), and *P.vera* (NT). On the basis of the IUCN conservation criteria, extinction of and declines in *Pistacia* diversity are increasing due to several factors such as increased population growth, high rates of habitat loss and degradation, unplanned urbanization, deforestation, overexploitation of natural resources, pollution,

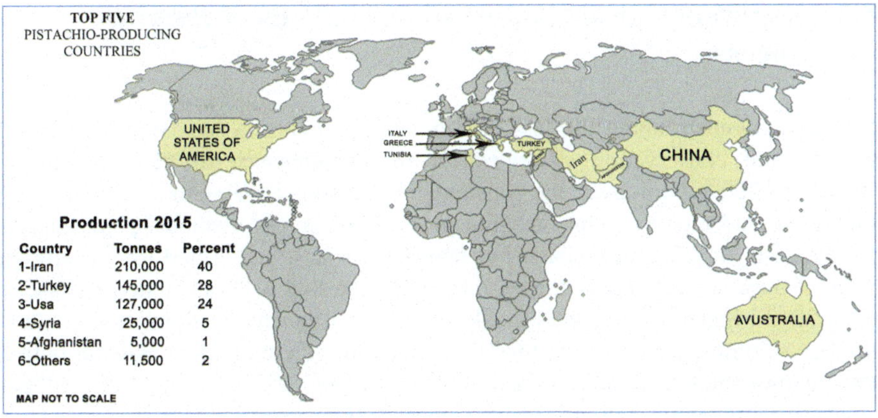

Fig. 8.1 Main pistachio producing countries on the world (The countries were shown in yellow)

diseases, climate change (Khan and Hussain 2013). All these negative factors lead to an international initiation as IPGRI Project in 1994 for the characterization, collection, and conservation of *Pistacia* genetic resources to maintain its biodiversity (Padulosi et al. 1996; Özden-Tokatli et al. 2010). In the frame of the Project, several in situ, ex situ, and private collections were constituted for maintaining the conservation strategies especially in valuable cultivars (Barazani et al. 2003; Ghorbel et al. 2001). Complementary successful unconventional in vitro methods such as organogenesis, somatic embryogenesis and micrografting (Koç et al. 2014a; Onay et al. 2016), and slow growth storage conditions for medium-term preservation and long-term preservation of *P. vera* L. (Akdemir et al. 2013) were previously reported. However, the majority of studies carried out on *Pistacia* genus were related to unconventional biotechnological methods such as micropropagation, organogenesis, somatic embryogenesis, protoplast culture, micrografting, in vitro conservation, and cryopreservation. There are also other biotechnological approaches for the conservation of *Pistacia* germplasm such as molecular marker technology, additional markers, plant DNA bank, and genetic manipulation approaches for improvement of yield and quality traits.

As above mentioned, plant germplasm can be conserved either in situ or ex situ, however, the few species survive successfully in ex situ (off-site) conservation and in situ (in natural habitat) conservation is difficult for most of the plant species due to changes in climate, habitat and/or diseases (Engelmann 2011). Since environmental and anthropogenic factors are responsible for genetic erosion of *Pistacia* species in a natural habitat, the best and safe strategy for *Pistacia* germplasm conservation is using both in situ and ex situ techniques to complement conventional and unconventional methods. Here, we describe biotechnological conservation methods reported on *Pistacia* species as well as in vitro culture techniques which are convenient for maintaining ex situ germplasm collections using *P. vera* and *P. lentiscus* as case studies.

8.2 Biotechnological Conservation Methods of *Pistacia* Species

Increasing concern about the loss of valuable plant species has stimulated in situ and ex situ conservation strategies for the maintenance of plant germplasm (Paunescu 2009). Moreover, the conservation methods are essential for both of the traditional and modern plant breeding projects. In in situ methods, plant material conserves in their natural habitat but both habitat destruction and the transformations of these natural environments may cause the loss of biodiversity (Cruz-Cruz et al. 2013). Since in situ strategies alone are not sufficient, ex situ methods can be used as a complementary way, thereby maintaining the biological material outside their natural habitats in the any form of plant or plant parts including seeds, pollens, vegetative propagules, tissue, or cell cultures (Paunescu 2009; Cruz-Cruz et al. 2013). Among the various ex situ methods, in vitro conservation of plant germplasm especially for vegetatively propagated and for non-orthodox seed plants allows protection of plant genetic diversity for different periods of time (Engelmann 2011). Usage of in vitro culture techniques offers many advantages such as easy and safe exchange of plant materials, maintaining culture collections in minimum space, providing necessary plant materials for molecular investigations and ecological studies (Tandon and Kumaria 2005; Cruz-Cruz et al. 2013). It is clear that the in vitro conservation methods for short term and medium term and cryopreservation for long term should be developed in *Pistacia* species in order to safeguard their germplasm.

8.2.1 In Vitro *Culture Techniques*

8.2.1.1 Micropropagation

Micropropagation is a tissue culture technique used not only for the production of a large number of plants in an artificial nutrient medium under aseptic conditions but also in vitro conservation method for economically important plant species having special phenotypical characters. Several micropropagation protocols were reported for the genus *Pistacia*, and these protocols include axillary shoot multiplication, cotyledonary node culture, indirect organogenesis, direct organogenesis, and somatic embryogenesis (Ozden-Tokatli et al. 2010 and references therein). Main method for the establishment of in vitro culture was apical or nodal bud culture, and it was based on the formation of rooted plantlets from in vitro regenerated shoot tip explants which is called organogenesis. Production of embryos or embryoids by somatic embryogenesis is the second method of plant regeneration in micropropagation (Onay et al. 1995; Onay and Jeffree 2000a). The other methods used for the in vitro regeneration of plantlets include meristem culture (Çetiner et al. 1996), callus culture (Alderson and Barghchi 1982), suspension culture (Ahmad 1993), embryo culture (Abousalim et al. 1992; Benmahioul et al. 2009; Onay et al. 2007a), and protoplast culture (Chan et al. 2002).

All these methods allow the production of numerous plant species in a limited space and short period of time, and keep alive the regenerated plant materials by regular subculturing. Micropropagation of the different species of the genus *Pistacia* by tissue culture has been reported more than three decades ago (Alderson and Barghchi 1982). Since then, several papers reporting successful propagation of cultivars and rootstocks of *Pistacia* by various medium modifications and culture conditions and the most recent being from Blackwell (2016) were published. Some of these important in vitro studies carried out on mature and juvenile explants of *Pistacia* rootstock and cultivar species are detailed in Tables 8.1 and 8.2.

Micropropagation of *Pistacia* Rootstocks

Since the first in vitro culture study of *Pistacia* species (Barghchi 1982), large numbers of micropropagation studies on different *Pistacia* rootstock species were carried out (Table 8.1). Micropropagation protocols from juvenile *P. vera* (Barghchi 1982; 1986a; Barghchi and Alderson 1983a, b; Barghchi and Alderson 1985; Guerrero et al. 2002; Benmahioul et al. 2009; Chelli and Drira 2002; Garcia et al. 2010), *P. terebinthus* (Pontikis 1984; Gannoun et al. 1995; Guerrero et al. 2002; Benmahioul et al. 2009; Chelli and Drira 2002; Garcia et al. 2010), *P. integerrima*, *P. intergerrima* × *atlantica* (Blackwell 2016), *P. palaestina* (Barghchi 1982; 1986a), *P. atlantica* (Barghchi 1982; 1986a; Picchioni and Davies, 1990; Mederos Molina and López Carreño 1991), *P. khinjuk* Stocks (Barghchi 1982;1986a; Tilkat et al. 2005), *P. mutica* (Barghchi 1982; 1986a), *P. lentiscus* (Yıldırım 2012; Koç et al. 2014a, b; Kılınç et al. 2014, 2015), *Pistacia chinensis* Bunge (Parfitt et al. 2007; Tang et al. 2012), *P. atlantica* × *P. integerrima* (UCB-1) (Almehdi et al. 2002), and adult explants of *P. vera* and *P. terebinthus* (Gannoun et al. 1995; Onay 2000a; Tilkat et al. 2009a; García et al. 2010) have been reported, and all these researchers mentioned about in vitro adaptation problems such as the abundant presence of exudates in the culture medium (Tabiyeh et al. 2006; Marín et al. 2016). Therefore, the developed micropropagation protocols seem that they can be inapplicable for most of the nurseries which have no sufficient facilities to establish micropropagation systems for most species. In the recent years, especially private companies in the USA are offering growers micropropagated *Pistacia* rootstocks from juvenile seedlings which parents are proved their superior characteristics. Although it was reported that micropropagation of pistachio rootstock species and hybrids has been difficult due to lower proliferation rates and difficulties on rooting step (Barghchi and Alderson 1983a), very recently a new and distinct *Pistacia* cultivar of *P. intergerrima* × *P. atlantica* named ‘P.N.B.1’ is disclosed and characterized by increased crop yield when micropropagated plants used as a rootstock (Blackwell 2016).

Table 8.1 Micropropagation studies carried out on different rootstock species of the genus *Pistacia*

Species/Explant type	Age	Results	References
P. vera/AB/ND	J M	Single shoot growth, multiple shoot proliferation, adventive shoots, rooting, plantlet production	Barghchi 1982; 1986a; Barghchi and Alderson 1983a, b, 1985; Gannoun et al. 1995; Onay 1996; Garcia et al. 2012; Guerrero et al. 2002; Benmahioul et al. 2009; Chelli and Drira, 2002; Garcia et al. 2010; Gannoun et al. 1995; Onay 2000; Tilkat et al. 2009a; Parfitt and Almedhi 1994
P. khinjuk Stocks/AB, NB	J	Multiple shoot growth Rooting, Plantlet production	Barghchi 1982, 1986a; Barghchi and Alderson 1983b, Barghchi 1986a; Tilkat et al. 2005
P. chinensis Bunge/ST	J	Plantlet production	Jing-Fang et al. 2011; Parfit et al. 2007
P. terebinthus/AB/SN	J M	Multiple shoot growth Rooting, plantlet	Pontikis 1984; Gannoun et al. 1995; Garcia et al. 2012; Guerrero et al. 2002; Benmahioul et al. 2009; Chelli and Drira 2002; Garcia et al. 2010
P.mutica/AB, NB	J	Multiple shoot growth Few plantlets	Barghchi 1982, 1986a
P. atlantica/AB, NB	J	Shoot growth	Barghchi 1982, 1986a; Mederos Molina and López Carreño 1991; Picchioni and Davies 1990.
P. atlantica × *P. integerrima*	J	Clonal production of rootstocks	Parfitt and Almedhi 1994; Almehdi et al. 2002
P. palaestina/AB, NB	J	Shoot growth	Barghchi 1982, 1986a
P. vera/IK	J	Somatic plantlet	Onay and Jeffree 2000
P. vera/IK	J	Somatic plantlet	Onay et al. 2000
P. vera/IK	J	Embryogenic tissue, synthetic seeds	Onay et al. 1995
P. vera/IK	J	Embryogenic tissue, somatic seedlings	Onay et al. 1996
P. vera/IK	J	Embryogenic tissue, somatic seedlings	Onay et al. 2007a
P. vera/L, leaf	J	Embryogenic tissue, somatic seedlings	Onay and Namlı 1998

Abbreviations: AP, apical bud; NB, nodal bud; ST, stem segments; M, mature; J, juvenile; SN, single nod; IK, immature kernels; L, leaf

Table 8.2 Important micropropagation studies carried out on different cultivars of *P. vera* L

Cultivar/explant type	Age of Explant	Results	References
Kalleghochi/AT, NB	J	Multiple shoot proliferation, rooting, plantlet production	Barghchi 1982; Barghchi and Alderson 1983b, 1985
Ohadi/AT,NB, Callus, IVP Callus,	J	Multiple shoot proliferation, rooting, plantlet production, embryoids/meristemoids	Barghchi 1982; Barghchi and Alderson 1983b, 1985
Kerman/AT,NB	M	Shoot growth	Barghchi 1986b; Barghchi and Martinelli 1984
Peter's/AT,NB	M	Shoot growth	Barghchi 1986b; Barghchi and Martinelli 1984
Lambertin/AT, NB	M	Shoot growth	Barghchi 1986b; Barghchi and Martinelli 1984
Lambertin/INF	M	Callus; some differentiation	Barghchi 1986b; Barghchi and Martinelli 1984
Mesocarps	J	Callus formation	Zaheer et al. 1989
ST,NB	J	Plantlet regeneration	Yang and Lüdders 1993
Kerman/ST	J	Multiple shoot formation	Parfitt and Almehdi 1994
ST	J	Plantlet regeneration	Dolcet Sanjuan and Claveria 1995
Antep/ST,NS	M	Plantlet regeneration	Onay 2000a
Kırmızı/ST,NS, CN,	J	Multiple shoot proliferation and plantlet regeneration	Özden-Tokatli et al. 2003
Kırmızı/NS	J	Multiple shoot proliferation and plantlet regeneration	Özden Tokatli et al. 2004
Kırmızı/NS	J	Multiple shoot proliferation and plantlet regeneration	Özden-Tokatli et al. 2005
Kırmızı/CN	J	Multiple shoot proliferation and plantlet regeneration	Özden-Tokatli et al. 2006
Atlı/AB	M	Multiple shoot proliferation and plantlet regeneration	Tilkat et al. 2008b
Atlı/AT	M	Plantlet regeneration	Tilkat et al. 2009b
Siirt/AL, Axenic leaves	M	Multiple shoot proliferation and plantlet regeneration	Tilkat and Onay 2009
Atlı/AT	M	Plantlet regeneration	Tilkat et al. 2009a
Uzun/IK	J	Embryogenic tissue, somatic embryo, somatic plantlet	Onay et al. 1995
Uzun/IK	J	Encapsulated embryogenic tissue, synthetic seeds, plantlet	Onay et al. 1996

(continued)

Table 8.2 (continued)

Cultivar/explant type	Age of Explant	Results	References
Uzun/L	J	Callus, embryogenic tissue	Onay and Namli 1998
Uzun/IK	J	Callus, embryogenic tissue, somatic embryo, somatic plantlets	Onay and Jeffree 2000
Kırmız/IK/MK	J	Somatic embryogenesis, plantlet regeneration	Onay et al. 2007a, b
Kırmızı/AL	J	Development of somatic embryos from single epidermal cell	Onay 2000b
P. vera/JL	J	Protoplast culture	Chan et al. 2002
P. vera/Seeds	J	Embryo culture	Benmahioul et al. 2016

Abbreviations: AT, apical tip; ND, nodal bud; INF, inflorescence; IVP, in vitro produced; Cot, cotyledon, Ad, adventitious shoots; CN, cotyledonary nodes; AL, axenic leaves; IK, immature kernels; L, leaves, IK, immature kernel; MK, mature kernels; JL, juvenile leaf

Micropropagation of *Pistacia* Cultivars

The cultivars of *P. vera* L. are used for the production of edible pistachio nuts in most of the reported in vitro studies. Some of these important in vitro studies carried out on juvenile or mature explants of *Pistacia* spp. are detailed in Table 8.2. It is clear that the most of the propagation studies were carried out using juvenile explants of *P. vera* cultivars (Barghchi and Alderson 1983a, b; Barghchi and Alderson 1985; Barghchi 1986a; Yang and Lüdders 1993; Parfitt and Almehdi 1994; Ozden-Tokatli et al. 2003, 2004, 2005, 2006, Table 8.2). Even the micropropagation of mature *P. vera* L. cultivars has been more difficult compared to using juvenile explants due to same difficulties observed in rootstock propagation, a few studies have been already reported for some cultivars (Onay 1996; Onay 2000b; Barghchi and Alderson 1983a). Micropropagation protocols for *Pistacia* species used as rootstocks and cultivars have also been reported for several species of the genus *Pistacia* (Tables 8.1 and 8.2). The general approach for the main method (organogenesis) involves four stages, which are presented in Fig. 8.2.

(1) **Surface sterilization and culture initiation**: The surface sterilization or disinfection is the primary stage of micropropagation, and it is critically affecting the success of the protocol. While insufficient sterilization results in contaminated shoot tips, oversterilization causes the death of the explants. At the beginning of sterilization, since tissues collected from fields are more contaminated than the explants grown in greenhouse, the explants are usually pre-washed in running tap water for several hours. Then, the surface sterilization is accomplished with a commercial sterilization agent like sodium/calcium hypochlorite solution followed by repeated washes with sterile distilled water. Concentrations and duration in the solution depend on plant species, type and age of explant, and the place and date where and when the explant was collected. After sterilization step, the explants are placed in an

Fig. 8.2 Micropropagation of *Pistacia*: A. Shoot formation from the forced apical shoot tips of *P. vera* L. cv. Atlı cultured on the culture initiation medium after 4 weeks of culture (Tilkat et al. 2013); **B.** Shoot proliferation from in vitro-derived leaf explants of pistachio, P. vera L. cv. Siirt (Tilkat and Onay 2009); **C.** Rooting microshoots of *Pistacia vera* L. 'Atli' on modified MS Medium supplemented with 2 mg l^{-1} IBA (Tilkat et al. 2008a); **D.** Plantlets acclimatized from microshoots following establishment in a mixture of sterile compost (soil:peat, 3:1)

artificial sterile growth medium. In the sterilization protocol, endogenic bacterial contamination causes serious problems especially when the explants are used from mature trees. The problem can be partially avoided by using the forced flushing buds grown under greenhouse conditions (Onay 2000a) and the explants harvested from current year shoots in the spring. Moreover, it was reported that usage of small meristematic tips (approximately 1 mm) from actively growing shoots of mature *P. vera* plants can totally eliminate the contamination problems in tissue culture conditions (Tilkat et al. 2013). The forced shoot tips expanded at the base and extended in length 10–19 mm within 4 weeks of culture (Fig. 8.2a) when the forced shoot tips were used for the initiation of axenic cultures. Four weeks after culture, the newly formed shoots were cut into segments containing a shoot tip, and these microshoots were used for further proliferation or transferred to fresh MS medium.

(2) **Shoot proliferation and maintenance**: Following to surface sterilization, the explants (0.1–1-cm-long shoot tips or nodal bud segments containing one shoot) are placed on a solidified Murashige & Skoog medium (MS, Murashige

and Skoog 1962) containing agar or gelrite, sucrose, and plant growth regulator (s) (PGRs) such as combination of auxin and cytokinin. The type and level of cytokinins are important for optimization of the protocol. The levels from 1 to 4 mg/L of cytokinins (such as benzyl-adenin, BA) will favor shoot formation and multiplication while an addition of a low-level auxin to multiplication medium will favor proliferation of most *Pistacia* species (Barghchi and Alderson 1989; Tilkat and Onay 2009; Onay 2003; Tilkat et al. 2012). Additionally, vitamins and amino acids are sometimes added to specific media. Most media used for the in vitro culture of *Pistacia* species have been modified from MS or DKW (Driver Kuniyuki Walnut medium, Driver and Kuniyuki 1984) medium (Onay 1996). Manipulation of micronutrient levels in MS medium leads to superior shoot multiplication in mature *P. vera* (Onay 2000a). Tilkat and Onay (2009) later developed the method by enhancing release and proliferation of axillary buds cultured in MS medium. The maximum number of shoots (5.20) was obtained from explants on a MS medium containing 1.0 mg/L BA, followed by 4.96 shoots from the medium with 2.0 mg/L BA (Fig. 8.2b).

(3) **Rooting**: Although there have been a few successful reports on the production of *Pistacia* species from stem cuttings (Dunn et al. 1996a, b) and *P. vera* (Al Barazi and Schwabe 1982) under in vivo conditions, there is no applicable method described for mass propagation. Because cuttings are not useful for pistachio propagation, elite clones grafted onto heterozygous rootstocks and these grafted buds are used for clonal propagation (Onay et al. 1995). Therefore, presently, all pistachio rootstocks in all producing countries are produced from seeds. However, individual seedlings developed from heterozygotic seeds are genetically variable since pistachio is a natural out-breeder. The development of protocols for clonally propagation of *Pistacia* rootstocks is highly fascinating to obtain genetically elite genotypes. The reports on in vitro rooting of juvenile materials of several *Pistacia* species (Pontikis 1984; Barghchi and Alderson 1989; Picchioni and Davis 1990; Parfitt and Almehdi 1994; Onay 1996; Onay 2003; Tokatli et al. 2004; Tilkat et al. 2005; Tokatli et al. 2009; Onay 2005) and mature materials of *P. vera* (Tilkat et al. 2008a, 2009a; Tilkat and Onay 2009) can be summarized below: After a multiplication phase, the shoot tips (generally longer than 2 cm) were separated and moved to a solidified rooting medium containing generally sucrose as source of carbon and an auxin as a PGR, followed by acclimation in a peat-based solid axenic substrate. During rooting period, the type and level of auxin directly affect the initiation of roots of regenerated pistachio shoots (Barghchi and Alderson 1989; Parfitt and Almehdi 1994; Onay 2000a; Ozden' Tokatli et al. 2005; Tilkat et al. 2008a). All those previous studies showed that auxins of intermediate stability (i.e., indole-3-butyric acid IBA) were found to be more effective for the induction of rooting in pistachio than auxins of lower (i.e., indole-3-acetic acid, IAA) or higher stability (i.e., 1-naphthaleneacetic acid, NAA). In accordance with the reported results, the regenerated material of most *Pistacia* species may be readily rooted in vitro in the presence of IBA.

The highest rooting frequencies (92% and 94%) of axenic microshoots of *Pistacia vera* L. 'Atlı' (Fig. 8.2c) and *P. lentiscus* L. were recorded at microshoots among all those reported studies, respectively (Tilkat et al. 2008a; Kılınç et al. 2015). Generally, root formation capacity by means of number and length may directly be affected by applied medium strength, type of auxin, and explant size.

(4) **Acclimatization**:
The final stage of micropropagation is the acclimation of rooted explants. Generally, a mixture of solid compost containing soil and peat or soil, peat, and sand medium, watered with a balanced nutrient solution supplemented with half strength MS solution, was used for the acclimatization of *Pistacia* species. In these conditions, the regenerated plants did not show visible variation in morphological or growth characteristics compared with the donor plants (Fig. 8.2d, Tilkat et al. 2009a). The details of the successfully developed acclimatization methods are emphasized or described in the literature (Onay 2000a; Onay 2005; Tilkat and Onay 2009; Tilkat et al. 2009a; Tilkat et al. 2008a, b).

Somatic Embryogenesis

Somatic embryogenesis is another asexual form of plant propagation reported for the *Pistacia* rootstocks. A few studies have claimed somatic embryogenesis in *Pistacia* (Onay et al, 1995, 2004a, 2007a, b; Onay 2000a; Onay and Jefree 2000a; Onay 2000b, 2005; Onay and Namlı 1998). The effects of in vitro culture conditions and nutrient composition in medium on induction of somatic embryogenesis are detailed in the following publications (Onay and Jeffree 2000a; Onay 2005; Tilkat et al. 2012). These reports mainly for *P. vera* L. have been supported by convincing proofs such as histologically (Onay 2000c) and cytologically (Onay and Jeffree 2000a), but not with molecular markers. Spontaneous formation of somatic embryos that resemble zygotic embryos may be obtained by direct embryogenesis which were developed without intervening callus phase, or by indirect embryogenesis including an intermediary callus phase (Onay and Jeffree 2000a; Onay et al. 2007a). It is obtained that the type and concentrations of plant growth regulators (PGRs) are critically important as well as the type of basal medium and other additives, light, and other abiotic conditions. The cytokinin BA and TDZ (thidiazuron) currently proved the most potent and useful PGRs for the regeneration of embryogenic tissues from mature fruits (Onay et al. 2007b), immature fruits (Onay et al. 1995), and juvenile leaf explants (Onay and Namlı 1998) of *Pistacia vera* L. As a plant regeneration method, somatic embryogenesis may be utilized in large-scale propagation, synthetic seed production, and as a source of protoplasts (Onay et al. 1996). Moreover, this protocol has various applications on low-temperature storage and cryoconservation for maintaining plant germplasm.

Micrografting

Several horticultural taxa are traditionally propagated by grafting (Hartmann et al. 1997) to overcome rooting problems and speed up growth of mature fruit trees. Micrografting can be defined as the rejuvenation of the mature miniaturized scion shoot tip onto the in vitro or in vivo grown juvenile rootstock. This method was developed in the 1980s (Jonard 1986) and studied in vivo as well as in vitro. Micrografting may be used for the rejuvenation and/or the invigoration of mature shoot materials, the production of disease-free scions, the study of graft unions and the enhanced potential for true to type cloning mature plants (Onay et al. 2003). Barghchi (1986b) first attempted to rejuvenate mature materials of *Pistacia vera* through micrografting, but slow growth of the scion was observed and no elongation was obtained. Micrografting of *P. vera* was also investigated in vivo as well as in vitro by Abousalim and Mantell (1992), but no rejuvenation was reported when mature *P. vera* trees were used as scions. Pistachio micrografting success depends on numerous factors: humidity, temperature, oxygenation, rootstock diameter, or rootstock activity. Onay et al., (2004b) tested the success of various in vitro micrografting methods of shoot tips of pistachio (*P. vera* L. var. Siirt) by examining the variables including grafting method, size of microscion, effects of time of the year at which shoot tips effects of culture medium and gave all details of the developed protocol in Onay et al. (2007c). Several papers reported the successful micrografting of *P. vera* using various medium modifications and culture conditions and the most recent being from Onay et al. (2016). This most recent study reported the regeneration of mature lentisk genotypes. These studies described on the micrografting of *Pistacia* taxa, especially *P. vera* indicate that the production of micrografted plants may also contribute to ex situ conservation of the possible threatened species of *Pistacia*.

8.2.2 *Conservation of* Pistacia *Species*

Biotechnological advances provide several advantages for plants including selection genotypes with desirable traits, identification, and transfer of genes to improve yield and quality, multiplication, and conservation genetic biodiversity (Gonzalez-Arnao et al. 2014). However, establishment and management of large in vitro collections have important drawbacks such as loosing of germplasm due to contamination or loosing of plant genetic stability through occurrence of somaclonal variations (Scowcroft 1984; Gonzalez-Arnao et al. 2014). In this case, in vitro conservation techniques including cryopreservation offer safest alternative to maintain plant germplasm (Gonzalez-Arnao et al. 2014).

8.2.2.1 Medium-Term Conservation

Plant germplasm can be stored in a period—a few months up to a few years using synthetic seed production (Fig. 8.3a) or slow growth storage conditions such as incubation on lower temperature and light intensity (sometimes in dark) and culture in reduced nutrients for medium-term conservation (Ashmore 1997; Cruz-Cruz et al. 2013). In vitro preservation of pistachio was firstly reported by Barghchi (1986c). In this study, pistachio plantlets stored in the medium supplemented with abscisic acid and mannitol for up to 18 months at 4 °C but recovery rates of the conservation period were not addressed. Similarly, pistachio plantlets were stored in standard culture medium (MS) containing 30 g/L sucrose as the sole carbon source at 4 °C in the dark and finally 91.3% of the in vitro-conserved microshoots of male 'Atlı' and 100% of female 'Siirt' cultivar remained green even after 12 months (Akdemir et al. 2013). Since 97% of proliferation was obtained following to transfer of shoot apices excised from the 12 months stored microshoots to standard proliferation conditions, the results indicated that in vitro-grown pistachio microshoots can be stored for longer periods at low temperature for medium-term conservation (Akdemir et al. 2013). As above mentioned, encapsulation of plant explants (i.e., somatic embryos, shoot tips, nodal segments) is another option (Standardi and Piccioni 1998) for the medium-term conservation of plant species at low temperatures. Somatic embryos and embryogenic masses (Onay et al. 1996) and axillary buds of pistachio (Ozden-Tokatli et al. 2009) have been encapsulated with using Na-alginate. 14% of the synseeds containing somatic embryos were germinated, while more than 75% of the synseeds containing axillary buds were able to convert to shoots after 4 weeks of culture. Encapsulated somatic embryos are conserved up to 2 months at 4 °C with 14% recovery rate (Onay et al. 1996). In the recent report, shoot apices of mature pistachio 'Atlı' and 'Siirt' cultivars were encapsulated in 3% Na-Alg (w/v) and 0.4 M sucrose-containing MS liquid salt solution(Ca^{2+-}free) and then conserved in the cold room at 4°C in dark for different durations up to 12 months (Akdemir et al. 2013). Since high percentages of recovery rates (more than 88%) were obtained for both cultivars even after

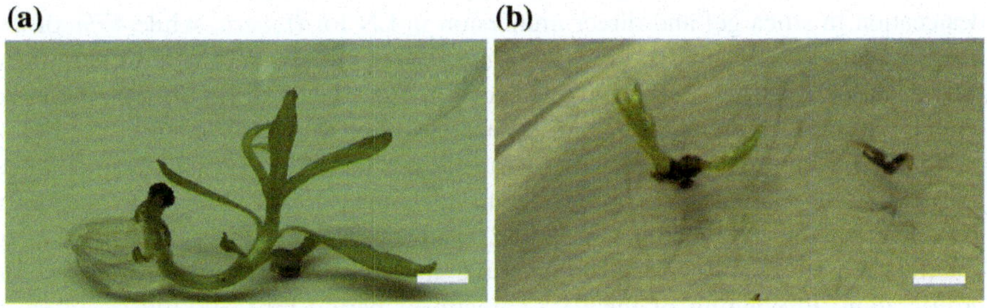

Fig. 8.3 **A.** Pistachio plantlet recovered from medium-term conserved (bar: 0.25 cm) **B.** Synseeds regrowth of cryopreserved and thawed shoot tips of pistachio using droplet-vitrification method (bar: 0.5 cm). (unpublished data)

12 months' storage, it seems that encapsulation procedure can be effectively used for medium-term conservation in pistachio (Akdemir et al. 2013). The same encapsulation protocol has been recently used for medium-term conservation of lentisk shoot tips, and it was possible to preserve the shoot tips at 4 °C without illumination for up to 6 months with a frequency of 87.5% plant regrowth (Koç et al. 2014b). Both slow growth storage and synthetic seed technology seem to be a promising approach for short- and medium-term conservation of some *Pistacia* germplasm.

8.2.2.2 Long-Term Conservation

Plant germplasm can be cryopreserved by using vegetative tissues, cell cultures, and seeds at ultra-low temperatures, normally using liquid nitrogen (LN, −196 °C) for unlimited periods theoretically (Sarasan et al. 2006).

A cryopreservation protocol includes several steps:

(1) Cold hardening at generally 4 °C,
(2) Preculture generally in sucrose-containing medium,
(3) Reduction of the intracellular water physically by dehydration over silica gel or chemically through the use of highly concentrated plant vitrification solutions (i.e., PVS2),
(4) Rapid immersion in LN,
(5) Thawing, and
(6) Conversion to plantlets (Akdemir et al. 2013).

A few studies have reported on the cryopreservation of *Pistacia* species (Table 8.3). There is only one cryopreservation protocol available for seeds of different species of *Pistacia* by using dehydration-based one-step freezing technique (Ozden-Tokatli et al. 2007). In this protocol, the seeds were surface-sterilized and desiccated to reduce MC of the seeds using activated silica gel, then directly immersed into LN and incubated for at least 1 h at this temperature. After thawing, the seeds were cultured in MS medium for germination. Finally, 90% germination was achieved in this following the reduction of MC to 11.7% from 20% with 8-h desiccation in silica gel and direct immersion in LN for *P. vera*, while 47% of the Spanish *P. lentiscus* and 16% of the Italian *P. terebinthus* were able to germinate after relatively shorter periods of desiccation (15 min and 1 h, respectively, to reduce their MC to about 20%) and cryopreservation (Ozden-Tokatli et al. 2007). Apart from cryopreservation of seeds, pistachio axillary buds were firstly cryostored by using vitrification and one-step freezing technique (Ozden-Tokatli et al. 2009). Treatment of the axillary buds in PVS2 for 60 min at 0 °C resulted in 11% of viability and only callus formation from the all survived propagules without shoot proliferation. Prolonging of the PVS2 treatment period up to 90 min resulted in relatively less viability (6.7%), but higher shoot proliferation (3.3%). After that, both encapsulation-dehydration and droplet-vitrification techniques were applied to

Table 8.3 Cryopreserved *Pistacia* species

Pistacia species	Explant type	Conservation method	Proliferation medium	Germination*/ Proliferation** (%)	References
P.vera 'Bianca'	Seeds	Dehydration/ one-step freezing	MS (hormone free)	90.0*	Özden-Tokatli et al. 2007
P. lentiscus	Seeds	Dehydration/ one-step freezing	MS (hormone free)	47.0*	Özden-Tokatli et al. 2007
P. terebinthus	Seeds	Dehydration/ one-step freezing	MS (hormone free)	16.0*	Özden-Tokatli et al. 2007
P. vera 'Siirt'	Axillary buds	Vitrification/one-step freezing	MS containing 2 mg/L BA	3.3**	Özden-Tokatli et al. 2009
P. vera 'Atlı'	Shoot apices	Encapsulation-dehydration/one-step freezing	MS containing 1 mg/L BA and 0.5 mg/L GA3	5.0**	Akdemir et al. 2013
P. vera 'Atlı'	Shoot apices	Droplet-vitrification/one-step freezing	MS containing 1 mg/L BA and 0.5 mg/L GA3	13.6**	Akdemir et al. 2013
P. lentiscus	Basal buds	Droplet-vitrification/one-step freezing	MS containing 1 mg/L BA	17.6**	Koc et al. 2013

*Germination percentage of seeds following the cryopreservation procedure
**Proliferation percentage of seeds following the cryopreservation procedure

shoot tips of mature pistachio 'Atlı' cultivar (Akdemir et al. 2013). Cryopreserved encapsulated shoot apices of 'Atlı' showed only 5% viability when MC of beads dropped to 18% prior to plunge to LN. As regards droplet-vitrification, 13.6% recovery was obtained with incubation of explants in the droplets of vitrification solution for 150 min at 0 °C followed by direct immersion in liquid nitrogen (LN), rapidly thawed, and then cultured on MS medium containing 1 mg/L BA and 0.5 mg/L GA_3 (gibberellic acid) (Akdemir et al. 2013). The same droplet-vitrification and one-step freezing technique were also assessed for long-term conservation of lentisk (Koç et al. 2013), and 17.6% of regeneration was achieved with incubation of basal shoots in the droplets of PVS2 (plant vitrification solution) for 45 min at 0 °C followed by direct immersion in liquid nitrogen. Figure 8.3b represents regrowth of cryopreserved and thawed shoot tips of pistachio using droplet-vitrification method (unpublished data). Even the droplet-vitrification technique appeared as a promising procedure for long-term preservations of pistachio and lentisk, the most recent developed cryoplate procedure (Yamamoto et al. 2011) which combines encapsulation-dehydration and droplet-vitrification method should be used for all *Pistacia* species.

8.2.3 Assessment of Genetic Integrity in Micropropagated and Conserved Pistacia *Species*

The assessment of genetic integrity is critically important to obtain genetically stable plant material following to in vitro culture and conservation procedures. The experimental steps applied in tissue culture such as explant type (meristems, leaves, roots, etc.) and preparation, propagation method, type and concentration of plant growth regulators, duration in subculture as well as plant species and genotype may cause tissue culture-induced somaclonal variations (Pierik 1987; Akdemir et al. 2013). In case of cryopreservation, applied methods may also lead to occurrence of genetically instable regenerants. Similar to tissue culture conditions, plantlets obtained from cryopreserved germplasm have subjected to several treatments including in vitro culture, pre-growth, desiccation, or vitrification depending on cryopreservation method, freezing/thawing, regrowth, and finally tissue culture conditions to enhance regeneration, and it is clear that all these factors have also potential on occurrence of somaclonal variation in conserved plant materials (Harding 1997).

DNA-based molecular marker technology was utilized in assessment of genetic variations at molecular level in different plant species. Inter-simple sequence repeat (ISSR, Zietkiewicz et al. 1994) markers were used, and low rates of polymorphism were obtained between regenerants of long-term propagated London plane tree and donor plant (Huang et al. 2009). Usage of two marker systems [ISSR and randomly amplified polymorphic DNA (RAPD, Williams et al. 1990)] revealed genetic instability between tissue culture-propagated bonnet bellflower plantlets and donor plant. In case of *Pistacia* species, occurrence of somaclonal variation in micropropagated juvenile pistachio shoots was firstly detected by RAPD primers (Ozden-Tokatli et al. 2006). Recently, reasonable polymorphism was found in long-term (up to 7 years) micropropagated mature pistachio via RAPD, ISSR, and amplified fragment length polymorphism (AFLP, Vos et al. 1995) markers (Akdemir et al. 2016). Figure 8.4 shows representative gel profile of AFLP primer pairs (E-AAG and M-CAT) from long-term micropropagated mature male pistachio plantlets (Akdemir 2013). Similarly, genetic instability was detected among in vitro cultured (up to 12 months) four genotypes of lentisk plantlets by using inter-retrotransposon amplified polymorphism (IRAP) markers which revealed the variations in inter-retrotransposon regions (Kalendar and Schulman 2006, Kılınç et al. 2015). Regarding conserved *Pistacia* species, RAPD analysis revealed that cryopreserved mature pistachio plantlets by droplet-vitrification method had high levels of genetic stability even after they were subcultured for at least 3 months following cryopreservation (Akdemir et al. 2013). In another study, evaluation of genetic stability of micropropagated lentisk shoot cultures which have been cold stored for 6 months was established by inter-retrotransposon amplified polymorphism (IRAP) and AFLP marker systems (Koç et al. 2014b).

Fig. 8.4 Example of AFLP patterns assessed with E-AAG and M-CAT primer combinations from long-term micropropagated mature male pistachio plantlets; lanes D: donor plant, G: greenhouse material, 1-10: five-year micropropagated plants, 20-26: seven-year micropropagated plants, M: marker (some polymorphic bands were shown with arrowheads) (from Akdemir 2013)

The study showed that similarity value of the non-stored plant and cold stored clones ranged from 0.66 to 0.84 and reasonable genetic instability after 6 months of storage (Koç et al. 2014b). Since each molecular marker screens particular portion of genome, genetic stability assessments should be performed using more than one molecular marker. But as a quality measure and risk management practice, the assessment of genetic integrity should also be accomplished by biochemical, cytological, and morphological markers.

8.2.4 Potential Problems of Conservation Strategies

Unconventional conservation methods of plant germplasm are considered as a suitable method to maintain plant germplasm. Several techniques such as micropropagation, organogenesis, somatic embryogenesis, micrografting, in vitro conservation, and cryopreservation can be applied for the conservation plant species of economically important crops. Despite the various attempts to conserve *Pistacia* germplasm by in vitro conservation or cryopreservation, most of studies focused on specific species of the genus. The main obstacle on application of cryopreservation protocols is the lack of availability of in vitro systems for mass proliferation (Kartha 1985) since in vitro conservation and cryopreservation are generally proceeded by tissue culture techniques (Niino and Arizaga 2015) and sometimes necessity of development of species-specific conservation techniques. One of the main constraints to the application of micropropagation technology to the pistachio industry is in vitro recalcitrance due to the expression of the adult phase. In some cases, it has been shown that it is possible to rejuvenate adult apical tips utilizing serial in vitro micrografting techniques. Therefore, the production of micrografted plants may contribute also to ex situ conservation of *Pistacia* species. As stressed in Sect. 2.3., somaclonal variation may also be induced in in vitro culture of *Pistacia* taxa. Somaclonal variation produced during in vitro culture reduces the economic value of any regenerated plants. Therefore, the regenerated plantlets should have minimum somatic variations through micropropagation method by reducing the number of subculturing and axillary shoot tip or bud culture. Although several attempts have been performed in *Pistacia* conservation lately, the studies still need to optimize the protocols in order for cryobanking.

8.2.5 Conclusions and Future Perspectives

Climate change with the combination of increased population growth, habitat loss and degradation, unplanned urbanization, overexploitation of valuable species, pollution and diseases are the major causes of the loss of plant biodiversity. Moreover, since environment and environmental factors will change faster than the most of the plant's adaptation, conservation strategies become critically essential for preservation of plant species (Hawkins et al. 2008). We described the current status of biotechnological conservation approaches for *Pistacia* species. Since in vitro conservation and most of cryopreservation methods rely on tissue culture-based methods, we described here micropropagation, micrografting, and somatic embryogenesis procedures in addition to in vitro conservation and cryopreservation in *Pistacia* species. Because there is no one 'universal protocol,' biotechnological conservation studies for *Pistacia* are still limited with a few species of the genus as described in this chapter. One of the major obstacles in *Pistacia* conservation is to develop successful regeneration protocol, and it is not available for some *Pistacia* species. The investigations should conduct optimization and standardization of regeneration and conservation procedures together in order to apply the technology for all species of the genus. Development and optimization of appropriate techniques especially for long-term preservation of different *Pistacia* species can lead to establish cryobanks to conserve germplasm. Furthermore, the developed technologies will provide to prevent possible genetic erosion of *Pistacia* species.

Conflict of interest The authors declare no conflicts of interest.

References

Abousalim A, Mantell SH (1992) Micrografting of pistachio (*Pistacia vera* L. cv. Mateur). Plant Cell Tiss Org Cult 29:231–234

Abousalim A, El Mahboul B, Walali LD (1992) Germination in vitro de graines et croissance de plantules de pistachier (*Pistacia vera* L.). Rev Res Amelior Prod Agr Milieu Aride 4:17–23

Ahmad Z (1993) A study of the chemical manipulations involved in the clonal propagation of Pistacia vera. Doctoral dissertation, University of the Punjab, Lahore

Akdemir H, Süzerer V, Onay A, Tilkat E, Ersali Y, Özden Çiftçi Y (2016) Detection of variation in long-term micropropagated mature pistachio via DNA-based molecular markers. Appl Biochem Biotechnol. doi:10.1007/s12010-016-2168-7

Akdemir H, Süzerer V, Tilkat E, Yildirim H, Onay A, Özden Çiftçi Y (2013) In vitro conservation and cryopreservation of mature pistachio (*Pistacia vera* L.) germplasm. J Plant Biochem Biotechnol 22(1):43–51

Akdemir H (2013) Development of biotechnological methods for micropropagation, in vitro conservation and cryopreservation of mature pistachio (*Pistacia vera* L.). Ph.D. Thesis, Institute of Science, Gebze Technical University (formerly Gebze Institute of Technology), Kocaeli, Turkey, 173 p. (in Turkish)

Al Barazi Z, Schwabe WW (1982) Rooting softwood cuttings of adult *Pistacia vera*. J Hort Sci 57:247–252

Alderson PG, Barghchi M (1982) Potential for *Pistacia* propagation through shoot, leaf and callus culture. In plant tissue culture 1982: proceedings, 5th international congress of plant tissue and cell culture held at Tokyo and Lake Yamanake, Japan, July 11–16, 1982/ed Akio Fujiwara. Tokyo: Japanese Association for Plant Tissue Culture

Almehdi AA, Parfitt DE, Chan H (2002) Clonal propagation of pistachio rootstock by rooted stem cuttings. Sci Hortic 96:359–363

Al-Saghir MG, Porter DM (2012) Taxonomic revision of the genus Pistacia L. (Anacardiaceae). American. Journal of Plant Sciences 3(1):12
Anonymous (2015a) INC International Nut & Dried Fruit official website. https://www.nutfruit.org/.../global-statistical-review-2015
Anonymous (2015b) The IUCN Red List of Threatened Species. Version 2015–4. www.iucnredlist.org
Ashmore SE (1997) Status report on the development and application of in vitro techniques for the conservation and use of plant genetic resources. IPGRI, Rome, Italy
Barazani O, Atayev A, Yakubov B, Kostiukovsky V, Popov K, Golan-Goldhirsh A (2003) Genetic variability in Turkmen populations of *Pistacia vera* L. Genet Resour Crop Evol 50:383–389
Barghchi M, Alderson PG (1985) In vitro propagation of *P. vera* L. and commercial varieties of Ohadi and Kalleghochi. J Hort Sci 60:423–440
Barghchi M, Alderson PG (1983a) In vitro propagation of *Pistacia vera* L. from seedling tissues. J Hort Sci 58:435–445
Barghchi M, Alderson PG (1983b) In vitro propagation of *Pistacia* species. Acta Hort 131:49–60
Barghchi M (1982) In vitro propagation of *Pistacia* species. Ph.D. Thesis, Nottingham University, UK
Barghchi M (1986a) In vitro propagation of Pistacia rootstocks. Proc Int Plant Pro Soc 35:336–337
Barghchi M (1986b) In vitro culture of mature commercial varieties of *Pistacia vera* L. Proc Int Plant Prop Soc 35:331–333
Barghchi M (1986c) In vitro storage of plant genetic resources. Plant Physiology Division Biennial Report, Palmerston North, Department of Scientific and Industrial Research. p. 52
Barghchi M, Alderson PG (1989) Pistachio (*Pistacia vera* L.). In: Biotechnology in agriculture and forestry, vol. 5. Trees II, Ed Bajaj Y.P.S., Springer, Berlin, pp: 68–98
Barghchi M, Martinelli A (1984) In vitro propagation of mature *Pistacia vera* varieties of Kerman (female) and Peter's (male) pistachio. In: 41st Easter school symposium on plant tissue culture and its agricultural applications. University of Nottingham, Nottingham, p. 75
Benmahioul B, Kaid-Harche M, Daguin F (2016) In vitro regeneration of *Pistacia vera* L. from nodal explants. J Forest Sci 62(5):198–203
Benmahioul B, Kaid-Harche M, Dorion N, Daguin F (2009) In vitro embryo germination and proliferation of pistachio (*Pistacia vera* L.). Sci Hortic 122:479–483
Blackwell B (2016) U.S. Patent No. 20,160,021,805. Washington, DC: U.S. Patent and Trademark Office
Çetiner S, Yücel H, Aka-Kaçar Y, Yalçin-Mendi Y (1996) Clonal propagation of *Pistacia* rootstocks by meristem and shoot culture. V Temperate Zone Fruit in the Tropics and Subtropics 441:333–336
Chan HM, Almehdi AA, Parfitt DE (2002) Isolation of protoplasts from pistachio (*Pistacia vera* L.) for production of somatic embryos. XXVIth International Horticultural Congress Program and Abstracts. p. 84
Chelli A, Drira N (2002) Culture media and hormone effects on the micropropagation of two *Pistacia* species. Acta Hort 591:395–398
Cruz-Cruz CA, González-Arnao MT, Engelmann F (2013) Biotechnology and conservation of plant biodiversity. Resources 2:73–95
Dolcet-Sanjuan R, Claveria E (1995) Improved shoot tip micropropagation of *Pistacia vera* L. and the beneficial effects of methyl Jasmonate. J Am Soc Hortic Sci 120:938–942
Driver J, Kuniyuki A (1984) In vitro propagation of Paradox walnut rootstock. HortScience 19:507–509
Dunn ED, Cole JC, Smith MW (1996a) Position of cut, bud retention and auxin influence rooting of *Pistacia chinensis*. Sci Hortic 67:105–110
Dunn ED, Cole JC, Smith MW (1996b) Timing of *Pistacia chinensis* Bunge. rooting using morphological markers associated with calendar date an degree days. J Am Soc Hort Sci 12:269–273
Engelmann F (2011) Use of biotechnologies for the conservation of plant biodiversity. In Vitro Cell Dev Biol Plant 47:5–16

Gannoun S, Lionakis N, Gerasopolulos D, Kaska N, Kuden S, Ferguson L, Michailides T (1995) Aspects of in vitro culture of *Pistacia terebinthus* and *P. vera*. Acta Hort 419:201–206

García E, Imbroda I, Lorente P, Marín JA, Arbeloa A, Padilla IMG, Andreu P (2012) Micropropagation and in vitro grafting techniques to assist the selection of a pistachio rootstock from a population of terebinth (*Pistacia terebinthus* L.) collected in the SE of Spain. Acta Hort 961:245–252

García E, Lorente P, Marín JA, Arbeloa A, Andreu P (2010) Micropropagación e injerto in vitro de pistachio. ITEA 106:294–302

Ghorbel A, Ben Salem-Fnayou A, Chatibi A, Twey M (2001) Genetic resources of *Pistacia* in Tunisia. In: Padulosi S,Hadj-Hassan A (eds) Project on underutilized Mediterranean species. *Pistacia*: towards a comprehensive documentation of distribution and use of its genetic diversity in Central and West Asia, North Africa and Mediterranean Europe. Report of the IPGRIWorkshop, Irbid, Jordan, pp. 62–72

Gonzalez-Arnao MT, Martinez-Montero ME, Cruz-Cruz CA, Engelmann F (2014) Advances techniques for the long-term preservation of plant biodiversity. In: Ahuja MR, Ramawat KG (eds) Biotechnology and Biodiversity. Springer, Switzerland, pp 129–170

Guerrero J, Couceiro JF, Moriana A (2002) Selection of Terebinth (*Pistacia terebinthus* L.) trees as seed producers for pistachio (*Pistacia vera* L.) rootstocks in the Castilla La Mancha (Spain). FAO-Nucis-Newsletter 11:25–29

Hanelt P, Büttner P, Mansfeld R, Kilian R (2001) Mansfeld's encyclopedia of agricultural and horticultural crops. Springer, Berlin

Harding K (1997) Stability of ribosomal RNA genes in *Solanum tuberosum* L. plants recovered from cryopreservation. CryoLetters 18:217–230

Hartmann HT, Kester DE, Davis FT (1997) Plant production, principles and practices. 5th Ed. Prentice-Hall International Inc. Englewood Cliffs, NJ. p. 647

Hawkins B, Sharrock S. Havens K (2008) Plants and climate change: which future? Botanic Gardens Conservation International, Richmond, UK

Huang WJ, Ning GG, Liu GF, Bao MZ (2009) Determination of genetic stability of long-term micropropagated plantlets of *Platanus acerifolia* using ISSR markers. Biol Plant 53(1):159–163

Jing-fang HE, Shu-chai SU, Ping-sheng LENG, Qian-long JI, Yong-kun YU, Wen-ke LIU (2011) On rapid propagation for tissue culture seedlings of *Pistacia chinensis*. J Beijing University of Agriculture 3:015

Jonard R (1986) Micrografting and its applications to tree improvement. In: Bajaj YPS (ed) Biotechnology in agriculture and forestry, trees I. Springer, Berlin

Kalendar R, Schulman AH (2006) IRAP and REMAP for retrotransposon-based genotyping and fingerprinting. Nat Protoc 1:2478–2484

Kartha KK (1985) Meristem culture and germplasm preservation. In: Cryopreservation of plant cells and organs, Kartha, KK (ed), CRC Press, ISBN 0-8493-6102-8, Florida, USA, pp: 116–133

Khan M, Hussain F (2013) Conservation status of plant species in Tehsil Takht-e-Nasrati, District Karak, Khyber Pakhtunkhawa, Pakistan. Int J Biodiversity and Conservation 5(1):20–26

Kılınç FM, Süzerer V, Çiftçi YÖ, Onay A, Yıldırım H, Uncuoğlu AA, Koc I, Metin ÖK (2015) Clonal micropropagation of *Pistacia lentiscus* L. and assessment of genetic stability using IRAP markers. Plant Growth Regul 75(1):75–88

Kılınç FM, Süzerer V, Ozden Çiftçi Y, Koç I, Akdemir H, Yildirim H, Tilkat E, Onay A (2014) Improved shoot multiplication of lentisk (*Pistacia lentiscus* L.) using different carbohydrates and media. Acta Hort 1028:153–160

Koç İ, Akdemir H, Onay A, Özden Çiftçi Y (2014a) Cold-induced genetic instability in micropropagated *Pistacia lentiscus* L. plantlets. Acta Physiol Plant 36(9):2373–2384

Koç İ, Onay A, Özden Çiftçi Y (2014b) In vitro regeneration and conservation of the lentisk (*Pistacia lentiscus* L.). Turkish J Biol 38:1–11

Koc I, Özden Çiftçi Y, Onay A (2013) Cryopreservation of *Pistacia lentiscus* L. (Mastic Tree). In: Vitro Cell Dev Biol (5–9 June 2013, Rhode Island, USA) 2013 In Vitro Biology Meeting Abstract Issue S60–S61

Marín JA, García E, Lorente P, Andreu P, Arbeloa A (2016) A novel approach for propagation of recalcitrant pistachio cultivars that sidesteps rooting by ex vitro grafting of tissue cultured shoot tips. Plant Cell Tiss Org Cult 124(1):191–200

Mederos Molina S, López Carreño I (1991) Control of organogenesis "in vitro" of *Pistacia atlantica* Desf. Rootstock. Acta Hort 289:135–136

Murashige T, Skoog F (1962) A revised medium for rapid growth and bioassay with tobacco cultures. Physiol Plant 15:473–497

Niino T, Arizaga MV (2015) Cryopreservation for preservation of potato genetic resources. Breed Sci 65(1):41–52

Onay A, Jeffree CE, Yeoman MM (1996) Plant regeneration from encapsulated embryoids and an embryogenic mass of pistachio *Pistacia vera* L. Plant Cell Rep 15(9):723–726

Onay A (2000b) Somatic embryogenesis in cultured kernels of pistachio, *Pistacia vera* L. cv Siirt. In: Proceedings of the 2nd Balkan Botanical Congress, İstanbul, May 14–16, 2000 Plants of the Balkan Peninsula: into the next Millennium 11:109–115

Onay A (2000a) Histology of somatic embryo initiation and development in pistachio (*Pistacia vera* L.). Turkish J Bot 24:91–95

Onay A (2000d) Somatic embryogenesis in cultured kernels of pistachio, *Pistacia vera* L. cv Siirt. In: Proceedings of the 2nd Balkan Botanical Congress, Istanbul, May 14–16, Plants of the Balkan Peninsula: into the next Millennium, vol II. pp. 109–115

Onay A (2000b) Micropropagation of pistachio from mature trees. Plant Cell Tiss Org Cult 60:159–162

Onay A (2000d) Somatic embryogenesis from mature seed cultures of *Pistacia atlantica*. Turk J Agric For 24:465–473

Onay A, Jeffree CE, Theobald C, Yeoman MM (2000) An analysis of the effects of maturation treatments on the probabilities of somatic embryo germination and embling development in pistachio, *Pistacia vera* L. using a logistic regression method. Plant Cell Tiss Org Cult 60 (2):121–129

Onay A (2005) Pistachio, *Pistacia vera* L. In: Mohan Jain S. and P. Gupta Protocols for somatic embryogenesis in woody plants. Forestry Sciences, vol. 77, Kluwer Academic Publishers, The Netherlands, pp. 289–300

Onay A, Jeffree CE (2000) Somatic Embryogenesis in Pistachio. In: Mohan Jain S, Gupta PK and Newton RJ Somatic embryogenesis in woody plants. (Forestry Sciences), Chapter 10, Section B, vol. 6, Kluwer Academic Publishers, The Netherlands, pp. 361–390

Onay A, Jeffree CE, Yeoman MM (1995) Somatic embriyogenesis in cultured immature kernels of pistachio, *Pistacia vera* L. Plant Cell Rep 15:192–195

Onay A, Pirinç V, Yıldırım Tilkat E, Aktürk Z, Yıldırım H (2004a) Somatic embryogenesis of pistachio from female flowers. J Horticultural Sci Biotechnol 79(6):960–964

Onay A, Jeffree CE (2000) Somatic embryogenesis in pistachio. In: Mohan Jain S, Gupta PK, Newton RJ (eds) Somatic embryogenesis in woody plants (Forestry Sciences), vol. 6, The Netherlands, Kluwer Acad Publ, Chapter 10, Section B, pp. 361–390

Onay A, Namli S (1998) Somatic embryogenesis in pistachio leaf (*Pistacia vera* L.). XIV. National Biology Congress, Samsun, vol. 2. September 7–10, Samsun (in Turkish), pp. 44–53

Onay A, Tilkat E, Yıldırım H, Süzerer V. (2007a) Indirect somatic embryogenesis from mature Embryo cultures of Pistachio, *Pistacia vera* L. Propag Ornamental Plants 7(2): 68–74

Onay A, Tilkat E, Işıkalan Ç, Namlı S (2007c) Micrografting of pistachio (*Pistacia vera* L. cv. Siirt).In: Mohan Jain S, Häggman H, (eds) Protocols for micropropagation of woody trees and fruits. Springer, The Netherlands, pp. 289–298

Onay A (1996) In vitro organogenesis and embryogenesis of pistachio, Pistacia vera L. Ph.D. Thesis, University of Edinburgh, Edinburgh, UK, p. 198

Onay A (2003) Micropropagation of pistachio. In: Mohan Jain S and Ishii K Micropropagation forest trees and fruits, Forestry Sciences, vol.75, Kluwer Academic Publishers, The Netherlands, Section C, pp. 565–588
Onay A, Pirinç V, Işıkalan C, Adıyaman F, Tilkat E, Başaran D (2003) In vivo and in vitro micrografting of pistachio, *Pistacia vera* L. cv. Siirt. Tur J Biol 27:95–100
Onay A, Pirinç V, Yıldırım H, Başaran D (2004b) In vitro micrografting of pistachio (*Pistacia vera* L. var. Siirt. Plant Cell Tiss Org Cult 77:215–219
Onay A, Tilkat E, Süzerer V, Karakaş Metin Ö, Özden Çiftçi Ö, Kılınç FM, Koç I, Şakiroğlu M, Yıldırım H, Altınkut Uncuoğlu A, Çalar N, Akdemir ÖF (2016) Rejuvenation of mature lentisk by micrografting and evaluation of genetic stability. Turkish J Biol 40:781–796
Onay A, Tilkat E, Yildirim H (2007b) Effect of genotype on somatic embryogenesis in pistachio (Pistacia vera L.). Propag Ornamental Plants 7(4):204–209
Onay A, Tilkat E, Yildirim H, Süzerer V (2007d) Indirect somatic embryogenesis from mature embryo cultures of pistachio. *Pistacia vera* L. Propag Ornamental Plants 7(2):68–74
Özden-Tokatli Y, Akdemir H, Tilkat E, Onay A (2010) Current status and conservation of *Pistacia* germplasm. Biotechnol Adv 28:130–141
Özden-Tokatli Y, Özudogru EA, Akdemir H, De Carlo A, Gümüşel F (2009) Application of cryopreservation technology for pistachio germplasm conservation. Acta Hort 839:245–251
Özden-Tokatli Y, Ozudogru EA, Gümüşel F, Lambardi M (2007) Cryopreservation of *Pistacia* spp. seeds by dehydration and one-step freezing. CryoLetters 28(2):83–94
Özden-Tokatli Y, Özudogru EA, Akcin A (2006) Optimization of an efficient micropropagation protocol and assessment of plant genetic fidelity by RAPD markers in pistachio (Pistacia vera L.). Adv Horticultural Sci 20(2):162–169
Özden-Tokatli Y, Özudogru EA, Akcin A (2003) Enhancement of regeneration in pistachio (Pistacia vera L.) with silver nitrate. In: Fifth International Symposium in the Series "Recent Advances in Plant Biotechnology", September 7–13, High Tatras, Slovak Republic. p. 21
Özden-Tokatli Y, Ozudogru EA, Akcin A (2004) In vitro regeneration of pistachio (*Pistacia vera* L.) through organogenesis: effect of silver nitrate, polyvinylpyrrolidone and citric acid. In Vitro Cell Dev Biol 40:46A–46A
Özden-Tokatli Y, Ozudogru EA, Akcin A (2005) In vitro response of pistachio nodal explants tosilver nitrate. Sci Hortic 106:415–426
Padulosi S, Caruso T, Barone E. (1996) Taxonomy, distribution, conservation and uses of Pistacia genetic resources. Report of a workshop, 29–30 June, 1995, Palermo, Italy. International Plant Genetic Resources Institute, Rome, Italy; p. 69
Parfitt E, Helen MC, Almehdi AA (2007) Development of micropropagation and acclimation protocols for the commercialization of a New Bonsai ornamaental tree for the California market. Authors: Dan 1 Pomologist in AES, Dept. of Plant Sciences, UC Davis 2 Lab Asst., Dept. of Plant Sciences, UC Davis 3 SRA Dept. of Plant Sciences, UC Davis This is a final report for work performed from July 1, 2006 to June 30, 2007
Parfitt DE, Almehdi A (1994) Use of high CO_2 atmospheric and medium modifications for the successful micropropagation of pistachio. Sci Hortic 56:321–329
Paunescu A (2009) Biotechnology for endangered plant conservation: a critical overview. Rom Biotech Lett 14:4095–4103
Pell SK (2004) Molecular systematics of the cashew family (Anacardiaceae). Ph. D. Dissertation. Department of Biological Sciences, Louisiana State University, Baton Rouge, LA, USA
Picchioni GA, Davis FT Jr (1990) Micropropagation of *Pistacia atlantica* shoots from axillary buds. Plant Prop Nesl 2:14–15
Pierik RLM (1987) In vitro culture of higher plants. Martinus Nijhoff Publishers, Dordrecht
Pontikis CA (1984) In vitro propagation of *Pistacia terebinthus* L. Plant Propagator 30:14–15
Sarasan V, Cripps R, Ramsay MM, Atherton C, McMichen M, Prendergast G, Rowntree JK (2006) Conservation in vitro of threatened plants—progress in the past decade. In Vitro Cell Dev Biol Plant 42:206–214

Scowcroft WR (1984) Genetic variability in tissue culture: impact on germplasm conservation and utilization. International board for plant genetic resources (IBPGR) technical report AGPGIBPGR/84/152, Rome

Standardi A, Piccioni E (1998) Recent perspectives on synthetic seed technology using nonembryogenic in vitro-derived explants. Int J Plant Sci 159:968–978

Tabiyeh DT, Bernard F, Shacker H (2006) Investigation of glutathione, salicylic acid and GA3 effects on browning in *Pistacia vera* shoot tips culture. Acta Hort 726:201–204

Tandon P, Kumaria S (2005) Prospects of Plant Conservation Biotechnology in India with Special Reference to Northeastern Region. In: Tandon P and Kumaria S (eds) Biodiversity: Status and Prospects, Norasa Publishing House: New Delhi, India, 79–91

Tang M, Zhang P, Zhang L, Li M, Wu L (2012) A potential bioenergy tree: *Pistacia chinensis* Bunge. Energy Procedia 16:737–746

Tilkat E, Ozden-Tokatli Y, Onay A (2009b) In vitro rooting improvement of adult pistachio, *Pistacia vera* L. cv. “Atlı”. Acta Hort 839:245–251

Tilkat E, Onay A, Yıldırım H, Ayaz E (2009a) Direct plant regeneration from mature leaf explants of pistachio. Pistacia vera L. Sci Hortic 121(3):361–365

Tilkat E, Onay A (2009) Direct shoot organogenesis from in vitro derived mature leaf explants of pistachio. In Vitro Cell Dev Biol Plant 45:92–98

Tilkat E, Işikalan Ç, Onay A (2005) In vitro propagation of khinjuk pistachio (*Pistacia khinjuk* stocks) through seedling apical shoot tip culture. Propagation of Ornamental Plants 5(3):124–128

Tilkat E, Onay A, Özden Tokatli Y (2008a) In vitro rooting improvement of adult pistachio, Pistacia vera L’.Atli’. In I International Symposium on Biotechnology of Fruit Species: Biotechfruit 839:215–221

Tilkat E, Süzerer V, Akdemir H, Ayaz Tilkat E, Ozden Ciftci Y, Onay A (2013) A rapid and effective protocol for surface sterilization and in vitro culture initiation of adult male pistachio (*Pistacia vera* L. cv. “Atlı”). Academia. Journal of Scientific Research 1(8):134–141

Tilkat E, Onay A, Yıldırım H, Özen HC (2008b) Micropropagation of mature male pistachio *Pistacia vera* L. J Hortic Sci Biotechnol 83(3):328–333

Tilkat E, Ozden-Tokatli Y, Akdemir H, Onay A, Ayaz E (2012) Novel Methods in Micropropagation of Pistachio, Crop Production for Agricultural Improvement (CPAI), Editors: Ashraf M., Öztürk M., Ahmad M.S.A., Aksoy A., Springer. pp: 379–393

Vos P, Hogers L, Bleeker M, Van De Lee T, Hornes M, Frijters A, Pot J, Peleman J, Kuiper M, Zabeau M (1995) AFLP: a new technique for DNA fingerprinting. Nucleic Acids Res 23:4407–4414

Williams JG, Kubelik AR, Livak KJ, Rafalski JA, Tingey SV (1990) DNA Polymorphisms Amplified by Arbitrary Primers Are Useful Genetic Markers. Nucleic Acids Res 18:6531–6535

Yamamoto S, Rafique T, Priyantha WS, Fukui K, Matsumoto T, Niino T (2011) Development of a cryopreservation procedure using aluminum cryo-plates. CryoLetters 32:256–265

Yang Z, Lüdders P (1993) In vitro propagation of Pistachio (*Pistacia vera* L.). Gartenbauwissenschaft 59(1):30–34

Yıldırım H (2012) Micropropagation of *Pistacia lentiscus* L. from axenic seedling-derived explants. Sci Hortic 137:29–35

Zietkiewicz E, Rafalski A, Labuda D (1994) Genome fingerprinting by simple sequence repeat (SSR)-anchored polymerase chain reaction amplification. Genomics 20:176–183

Zaheer A, Nasreen A, Shah FH (1989) Callus formation from the mesocarp tissues of *Pistacia vera* L. Pak. J. Sci Ind Res 32(8):549–550

Zohary M (1952) A monographical study of the genus *Pistacia*. Pal J Bot 5:187–228

Chapter 9
Biodiversity and Conservation of *Elaeis* Species

Maizura Ithnin, Norhalida Mohamed Serdari, Norziha Abdullah, Ahmad Kushairi and Rajinder Singh

Abstract The current oil palm planting materials yield three times more than the figure recorded six decades ago. This was partly achieved through genetic improvement and selection procedures imposed on the oil palm breeding populations. The repeated cycles of selection applied had removed a lot of 'unwanted' alleles from the breeding populations. However, this might have affected the populations' ability to withstand environmental threat and genetic gains in the future. Realizing the important role of oil palm germplasms in harmonizing these effects, several research organizations had accumulated a large number of *Elaeis guineensis* and *Elaeis oleifera* accessions from their centers of origin. These materials exhibit greater genetic diversity and can offer genes for many economically important traits valuable for future oil palm improvement. These potential germplasm materials should be strategically conserved to ensure effective and continuous accessibility, in line with the long-term breeding cycle of the oil palm. Core collection can effectively preserve the genetic diversity of the oil palm germplasms. Apart from maintaining the genetic resources in the living ex situ field plots, in vitro methods of storing genetic materials within smaller size facilities are also available. These methods offer conservation of high numbers of materials, protection from diseases and low cost of maintenance. Both, the ex situ and in vitro methods provide complementary means for effective long-term conservation of the oil palm genetic materials for posterity.

Keywords Oil palm germplasm · *E. guineensis* · *E. oleifera* · Biodiversity
Ex situ collections · In situ collection · In vitro conservation · Core collection

9.1 Introduction

Oil palm is the major source of vegetable oil in the world. Being the most efficient oil crop, oil palm yields on average 3.61 tonnes of oil per hectare (Oil World 2016a), which is about four to seven times more than other oilseed crops. In terms

M. Ithnin (✉) · N.M. Serdari · N. Abdullah · A. Kushairi · R. Singh
Malaysian Palm Oil Board, P.O. Box 10620, 50720 Kuala Lumpur, Malaysia
e-mail: maizura@mpob.gov.my

M.R. Ahuja and S.M. Jain (eds.), *Biodiversity and Conservation of Woody Plants*,
Sustainable Development and Biodiversity 17, DOI 10.1007/978-3-319-66426-2_9

of production, oil palm contributes to about 40% of the world's vegetable oils, but occupies only 8% of the total area planted with oilseed crops. In 2015, Indonesia recorded 8.63 million ha of land under mature oil palm cultivation, which constitutes approximately 49.8% of the land area worldwide planted with mature oil palm (Fig. 9.1). This is closely followed by Malaysia with 4.79 million ha of mature oil palm, while South and Central America has 1.2 million ha and another 2.67 million ha in other countries. The main producers of palm oil are Indonesia and Malaysia, accounting for 85% (53.3 million tonnes) of the global palm oil production. Approximately 3.8 million tonnes of palm oil was contributed by Central and South America and another 5.3 million tonnes was produced by other countries.

Among the major oilseeds, palm oil alone contributed 62% (47.7 million tonnes) to the world vegetable oil exports in 2015 (Fig. 9.2) (Oil World 2016b). Indonesia and Malaysia are also the main exporters, accounting for about 90% of the total palm oil export. The major importers are India, European Union, China, Pakistan, Egypt, Bangladesh, and USA.

9.2 Taxonomy of *Elaeis* Species

The oil palm is a monocotyledonous perennial tree belonging to the genus *Elaeis* under the Arecaceae family in the order of Arecales. *Elaeis* is placed together with Cocos genus under the Cocoineae tribe (Kahn 1997; Schultes 1990). The word '*Elaeis*' was derived from the Greek word 'elaion,' which means oil. There are two species in the genus *Elaeis,* namely *Elaeis guineensis* Jacq and *Elaeis oleifera* (H.B. K) Cortés. The species name, 'guineensis' is attributed to the Guinea coast

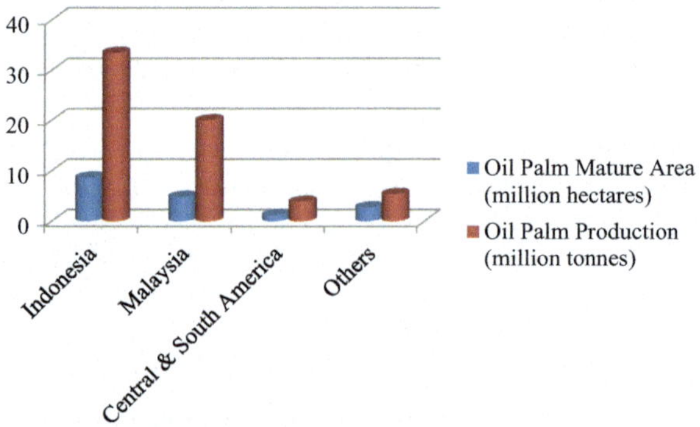

Fig. 9.1 Total area worldwide cultivated with mature oil palm and production for year 2015 (Oil World 2016a)

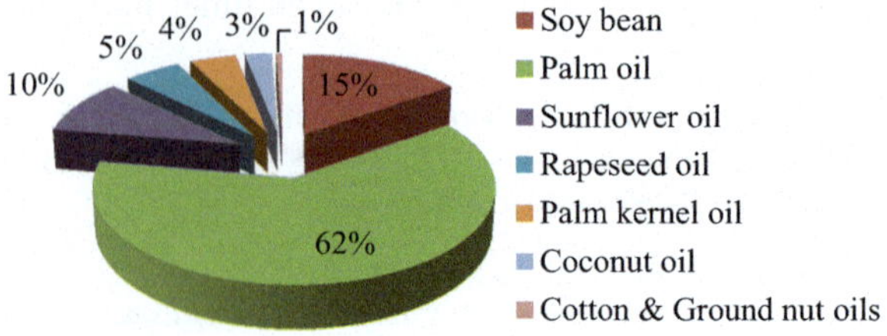

Fig. 9.2 World exports of eight vegetable oils (Oil World 2016b)

(Schultes 1990; Corley and Tinker 2003; Hartley 1988). Occasionally, *Elaeis guineensis* and *Elaeis oleifera* are also addressed according to their origins, African and American oil palm, respectively. The African oil palm is commercialized mainly in Southeast Asia and Africa while the hybrids of *E. oleifera* and *E. guineensis* (OxG) are more commonly planted in the plantations in South America due to the susceptibility of *E. guineensis* to the bud rot disease in the region.

9.3 Natural Distribution and Origin of *Elaeis* Species

The natural distribution of the African oil palm or *E. guineensis* is greatly influenced by rainfall. The rapid decline of rainfall northward has resulted in narrow belt of oil palm in West Africa. In this region, oil palm natural groves can be found in Sierra Leone, Liberia, Ivory Coast, Ghana, Togoland, Nigeria, and Cameroon. In Central Africa, oil palm grows along the Congo region and Angola. They also occupy the forest fringes of lowland regions especially near the rivers, freshwater swamp, generally close to human habitation (Gerritsma and Wessel 1997). Oil palm also grows well in areas with sufficient water supply, but not stagnant as the species is unable to tolerate permanent high water tables in impermeable soils (Hartley 1988). Natural populations of oil palm were not found in eastern Africa due to the dry climate and extreme altitude. Nevertheless, oil palm was introduced to Uganda, Kenya, Tanzania, Rwanda, and Burundi as well as Madagascar Island. The wide distribution of the oil palm in this region was believed to be associated with tribal migration and Arab slave trade (Smith et al. 1992). Owing to these introductions, oil palm groves are found in the region between 16° north in Senegal to 15° south in Angola as well as eastwards to Zanzibar and Madagascar.

The natural populations of the *E. oleifera* or American oil palm are located at the tropical regions of South and Central America, particularly in Brazil, Colombia, Venezuela, Panama, Costa Rica, Nicaragua, Honduras, French Guyana, and Surinam. In Colombia, *E. oleifera* grows under stress condition between rolling areas of fallow land. Pure and dense stand of palms were found in swampy areas near the riverbanks. The palms cover large areas of the San Juan and Escondido riverbanks in Nicaragua. In Costa Rica, the palms can be found on sloping or steep fallow lands but are less dense in tracts of unoccupied riverside and semi-swamp regions. The palms in Brazil were commonly found in the central of Amazonas above Manaus while the palms in Surinam are relatively smaller in size and can be found in dense stand on poor and sandy soil (Corley and Tinker 2003; Ooi et al. 1981).

There have been several postulations on the basis of fossil, historical, and linguistic evidences with regards to the origin of the oil palm. Fossil evidences were reported by Zeven (1964), Elenga et al. (1994), and Reynaud-Farrera et al. (1996). Oil palm pollen was claimed from sediments (Zeven 1964, Elenga et al. 1994), dated 2200 BP and more increasingly observed around 1600 BP in the forest of southwest Cameroon (Reynaud-Farrera et al. 1996). Sowunmi (1999) reported that the oil palm nut shell, dated to originate in the late Holocene periods, was discovered in the rainforest of West Africa. Despite these reports, Corley and Tinker (2003) favored a linguistic evidence of oil palm described in the Ca' da Mosto, a compilation of plants description recorded by Crone (1937).

9.4 Importance of Genetic Diversity in the Oil Palm

The commercially planted oil palm is hybrid seeds, known as *tenera* (thin shell), which are derived from crossing two parental palms, *dura* (thick shell) and *pisifera* (shell-less). In oil palm conventional breeding, two basic plans are used namely, Modified Reciprocal Recurrent Selection and Modified Recurrent Selection (Soh 1999) whereby the parental palms are evaluated and selected from separate populations. The parental palms are subjected to progeny testing to estimate progeny performance and combining abilities. *Duras* and *pisiferas* that show good combining ability with outstanding progenies are used for commercial DxP seed production.

Oil palm breeders worldwide have been depending on the Deli *dura* populations derived from very restricted origin. The Deli *dura* sub-populations include Chemara, Dami, Socfindo, Socfin, Dabou, and Banting. However, the Nigerian Institute for Oil Palm Research (NIFOR) also actively uses non-Deli *duras* in their breeding programme. The Deli *dura* populations are descended from four palms planted at the Bogor Botanical Garden, introduced as ornamental plants in 1848 from West Africa. Some examples of the common *pisiferas* used are AVROS, Ekona, La Me, Yangambi, and NIFOR (Rajanaidu et al. 2007). These populations are relatively more genetically diverse than the *duras,* but still were restricted to only the breeding stocks utilized in the improvement programs carried out in the West African countries such as Zaire, Ivory Coast, and Nigeria. Joint research programs were established since 1950s whereby genetic materials and breeding populations were introduced, exchanged, and sold among the participating agencies from Malaysia, Indonesia, South and Central America as well as West Africa. Through these initiatives, these oil palm breeding populations were widely distributed across the regions and have been applied in the genetic improvement programs for a number of generations. This likely contributed to the global average oil yield of 3.61 tonnes per hectare, (Oil World 2016a) recorded in 2015.

In Malaysia, the average palm oil yields have fluctuated between 3.5 and 4.0 tonnes per hectare for more than two decades (Kushairi et al. 2010). These figures reflect yield improvement of approximately three times (Davidson 1993), attributed to conventional genetic improvement initiated since 1920s as well as better management practices. These factors are equally important in determining productivity in oil palm plantations. Improved planting materials established in plantations that have poor management practices would result in low yield recovery and vice versa. In this regards, several research groups have reported impressive oil yield from the commercial materials planted in experimental plots that applied superior management practices. The best *tenera* progenies developed in Malaysia possess between 5 and 11 tonnes/hectare/year oils, e.g., Isa et al. (2005, 2013), Chin et al. (2005), Donough et al. (2005), and Zuraini et al. (2013). Oil palm seed producers in Indonesia have also reported oil yield ranging from 8.5 to 11.6 tonnes/hectare/year, e.g., Purba et al. (2013), Sharma et al. (2016), and Turnbull et al. (2016). These demonstrate that despite the narrow genetic base of their progenitors, current *tenera* hybrids are capable of yields higher than the global average yield. Therefore, superior planting materials should be rapidly adopted and coupled with appropriate plantation management, significant yield improvement can be realized, nationally and globally.

Although satisfactory yield is attainable from the present oil palm planting materials, the high selection procedure applied in breeding has led to the narrowing of the genetic diversity of the breeding populations. Although palm uniformity is favorable to the breeders and plantations, the disappearance of allelic variants can have harmful effects on the ability of the species to withstand any future environmental changes and challenges. These effects can be exacerbated by the active exploitation of clonal method to propagate elite *tenera* palms and selected parental lines for clonal (mono/semi and bi) production of oil palm commercial hybrid seeds (Wong et al. 2010; Musa et al. 2011; Mohaimi 2011; Zulhermana et al. 2011; Prasetyo et al. 2011). The clonal propagation method generates large number of oil palms that are genetically similar. Establishment of these clones commercially would result in the lower genetic diversity in oil palm plantations.

One of the greatest challenges faced by the oil palm industry worldwide is the economic loss due to disease occurrences. In Malaysia, the economic loss due to basal stem rot (BSR) caused by *Ganoderma* species was estimated at USD 375 million (Idris and Norman 2015). There is an increasing incidence of BSR in Malaysia and Indonesia especially in new replantings. The high incidence of this disease is believed to be due to the high susceptibility of the parental populations used to produce oil palm commercial seeds. Although some breeding populations recorded putative tolerance to diseases (Normahnani and Siang 2016; Rajanaidu et al. 2012; Norziha et al. 2012; Purba et al. 2012; Turnbull et al. 2014), the tolerance has shown to be not stable across fungal isolates or pathotypes and environments. The wild oil palm materials may provide additional genetic source for the development of stable tolerant breeding populations and consequent commercial varieties.

The global drastic changes in weather patterns attributed to climate change have caused extreme fluctuations in temperature and rainfall in various parts of the world. These changes have caused oil palm to become more prone to diseases and possibly induce aggressiveness of several secondary pests and diseases. In fact, climate change has been attributed to the increased outbreak of pest and diseases, reduction of soil fertility, occurrence of bunch rot diseases, and overall reduction in oil palm productivity (Nuzul 2015). The outbreak of several other species of leaf eating caterpillars, replacing the major nettle caterpillar was reported in oil palm plantations in Indonesia (Sudharto et al. 2016). Excessive rainfall had also led to increased damage to oil palm bunches caused by *Marasmius palmivorus* (Nuzul

2015). Purportedly, due to climate change, bud rot disease became epidemic in Tumaco and Puerto Wilches of Colombia (Mauricio and Fontanilla 2015). These incidences point to the need for developing oil palm planting materials tolerant to drought, heat, cold, and other abiotic stresses. The genes of these traits may not be available in the current genetic pools. As such, oil palm breeders have to rely on wild and unimproved populations in seeking for palms that possess tolerance to abiotic stresses to be introgressed into the commercial oil palm.

There appears to be a great necessity to apply new genetic stocks from the oil palm wild populations in dealing with the threat caused by diseases and climate change. The inclusion of these materials will enhance the genetic diversity of both, the breeding as well as the commercial materials. This will overall improve agro-biodiversity and adaptation to climate change in oil palm plantations (Soh et al. 2015; Heslop-Harrison 2012).

9.5 Collection of *Elaeis* Genetic Resources

Many oil palm research groups had initiated collection programs for genetic materials in Africa and South America. Initially, a small collection of *E. guineensis* was obtained from a few sites by some oil palm workers in the Belgian Congo (Vanderweyen 1952; Pichel 1956). In the early 1960s, the Nigerian oil palm breeders collected 72 open-pollinated progenies from the eastern part of Nigeria. These materials were planted and evaluated at the NIFOR main station. Some outstanding palms were selected and introduced into their current breeding programs (Okwuagwu 1986).

In Cameroon, Blaak (1967) sampled oil palm materials in Bamenda Highlands and parts of the collections were planted at Lobe, Cameroon. Between 1974 and 1975, Institute de Researches pour les Huiles et Oleagineux (IRHO) prospected *E. guineensis* materials from the western region of Cameroon. The French oil palm workers systematically evaluated natural stand of oil palms in Ivory Coast, and selected palms were progeny-tested and utilized as new foundation materials in their breeding program (Meunier 1969; Meunier and Baudouin 1986). It was reported that IRHO also selected four palms in this country and another 38 palms in Dahomey and used them as their original *tenera* stock.

A number of systematic prospections to collect oil palm genetic materials were mounted by researchers from the Malaysian Palm Oil Board (MPOB). The objective of the explorations was not only to broaden the genetic base of the current oil palm breeding materials, but also to ensure conservation of oil palm genetic resources for posterity. The first attempt was accomplished in Nigeria (Rajanaidu 1986a) in 1976, followed by collections from some other countries in Africa namely, Cameroon, Zaire, Tanzania, Madagascar (Rajanaidu and Rao 1987), Angola, Senegal, Gambia, Sierra Leone, and Guinea Conakry (Rajanaidu 1995) and Ghana (Rajanaidu et al. 2013). In the expeditions accomplished between 1983 and 1996, a total of 1838 accessions were collected. In 1991, NIFOR initiated another expedition in the peripheral areas of oil palm natural distribution in Nigeria. A total of 80 accessions were collected (Ataga et al. 2000). Pillai et al. (2000) prospected *E. guineensis* genetic materials in Cameroon, Guinea Bissau, Tanzania, and Zambia. Oil palm bunches were taken from 60 sites across the four countries. These materials were believed to be tolerant to cold and drought. Later, Rey et al. (2004) explored five regions in Angola and assembled 44 accessions. A group of researchers from the oil palm research institute in Ghana had also taken the initiative to collect 22 accessions from five locations in the northern region of the country (Sapey et al. 2012). This region is well known for its irregular rainfall and high temperature. The materials assembled may be useful for developing drought planting materials. The Indonesian team had also performed collection of genetic materials in Cameroon and Angola (Purba et al. 2013) and a total of 230 accessions were gathered. In 2010, MPOB together with the governments of Indonesia and Angola also carried out prospection of oil palm materials in Angola. One hundred and twenty-seven accessions were collected (Marhalil and Rajanaidu 2011). Recently, Arias et al. (2011) collected 74 oil palm accessions from 21 sites in Cameroon.

Efforts to assemble the *E. oleifera* genetic materials from the species' natural areas of distribution were initiated as early as in 1967. The collection was mounted in batches by the United Brands Company. *E. oleifera* seeds were gathered from the natural populations of *E. oleifera* in Costa Rica, Panama, and Colombia, followed by other populations in Surinam, Honduras, Nicaragua, and Brazil between 1975 and 1976 (Escobar 1982). The entire collection covered 36 zones and a total of 326 accessions were assembled. An oil palm private company in Malaysia acquired *E. oleifera* genetic materials from Colombia, Costa Rica, Panama, and Brazil (Sharma 2000). The materials were delivered in batches by IRHO between 1975 and 1983. A total of 141 accessions were sampled covering 23 sites. In 1982, Andrade prospected *E. oleifera* materials from 53 populations dispersed across 13 geographic groups in the Amazon basin in Brazil (Andrade 1982). In the same year, a team from MPOB explored Honduras, Costa Rica, Nicaragua, Panama, and Colombia (Rajanaidu 1986b) and collected 167 accessions from 59 sites. The same group once again visited Colombia as well as Peru, Ecuador, and Brazil in 2004 to specifically collect genetic materials with excellent fruit set, high mesocarp to fruit ratio, and reasonable stalk length (Rajanaidu et al. 2013).

9.6 Classical Genetic Characterization of *Elaeis* Species and Harnessing Valuable Traits

In oil palm breeding and selection, data on fresh fruit bunch (FFB) yield, bunch quality, vegetative and physiological characteristics are collected for each experimental individual palm. FFB refers to the weights of bunches produced by a particular palm. FFB and oil to bunch (O/B) ratio are the two main determinants for palm oil yield (OY). O/B is determined based on the values of three bunch quality components, namely, fruit to bunch (F/B), mesocarp to fruit (M/F), and oil to wet mesocarp (OTWM) ratios. Any increase in the values of these components will also increase O/B. Correlation analysis also showed strong relationship for several other pair of traits. For instance, M/F is negatively correlated with shell to fruit (S/F) ratio. S/F however, showed positive correlation with kernel to bunch (K/B) ratios (Kushairi et al. 1999). Hence, higher O/B can be attained from bunches that recorded high M/F, but low shell to fruit (S/F) ratios. Oil palm breeders critically examine these component traits for selecting potential parental palms for commercial seed production.

The value of the oil palm genetic resources relies on the massive work of collecting and measuring the appropriate phenotypic parameters. Despite the high cost and massive labor required for this exercise, the information obtained are highly valuable and could certainly enhance efficient exploitation of the germplasm materials. Numerous efforts in evaluating the genetic diversity of oil palm breeding populations have been reported. Phenotypic characterization of the Nigerian oil palm germplasm collections assembled by the Malaysian team revealed individual palms with high yield (>10 tonnes/hectare/year) and low height increment (<25 cm/year) (Rajanaidu et al. 2000). The incorporation of these materials into breeding programs revealed remarkable oil yields. Selected *tenera* hybrids derived from crosses involving short and high yielding Nigerian palms revealed 7–10 tonnes oil/hectare (Musa and Gurmit 2008; Junaidah et al. 2008). In addition to the Nigerian materials, Ang (2003) also exploited some *pisiferas* derived from Ghana germplasm. Combining these genetic materials with Deli *duras* had resulted in yield potential of more than 9 tonnes oil/hectare. Some Nigerian derived progenies exhibited shorter rachis length (Junaidah et al. 2008; Veriappan et al. 2008) and half of the average height of the standard cross (Isa et al. 2008) which, opened up the possibility of developing compact palms suitable for higher density planting. The selected Nigerian palms were combined with Deli *duras* to obtain reasonable compactness, bunch number, and bunch size. This may overall improve oil palm productivity.

Slow height increment palms, approximately 15 cm/year were also identified among the Nigerian genetic populations collected by NIFOR (Ataga et al. 2000). Genes for slow height increment are also available from *E. oleifera* germplasm materials. As low as 4.6 cm/year of height increment was recorded for individual *E. oleifera* palms from Colombia (Mohd Din et al. 2000). The current maximum economic lifespan of the oil palm is about 25 years as beyond this age, the oil palm trees are too tall, causing difficulties in bunch harvesting. Introgressing these materials into breeding programs will generate planting materials with reduced height increment which facilitate harvesting and extends the economical lifespan of the crop. The efficiency of harvesting can also be enhanced if the oil palm bunches have longer stalk. Such trait is available in the MPOB Angola oil palm germplasm collection. Some palms in this collection posess stalk length of 20–30 cm as compared to 10–15 cm recorded in the present planting materials (Kushairi et al. 2003c). The long stalk trait can be introgressed into *virescens* oil palms. Such palms produce fruit with distinct color change, from the immature green to reddish orange when ripe (Rao 1996) which, facilitates identification of mature bunches timely for harvesting. Low lipase genotypes are also available in MPOB (Kushairi et al. 2011). Low lipase activity among oil palm ripe fruits is desirable to delay free fatty acid (FFA) accumulation which could reduce the palm oil quality (Ebongue et al. 2008). Morcillo et al. (2013) reported that bunches harvested from selected low lipase palms recorded FFA below 5% even when processing is delayed thus, preserving the palm oil quality. Non-abscising genotypes of oil palm (Roongsattham et al. 2012) are interesting as incorporating the trait into the commercial varieties will reduce the need for frequent harvesting rounds and minimize lost due to inefficient loose fruit collection in the palm plantations.

Several other superior yield related traits were discovered in the oil palm germplasm collection. Palms with low S/F ratio, between 2.8 and 7.4% were identified in the MPOB germplasm collection from Tanzania (Kushairi et al. 2003a). Low S/F results in higher M/F if K/B and fruit size are constant. Some palms from Angola showed larger fruit size, between 24 and 34 g (Kushairi et al. 2003b). This trait, measured based on the mean fruit weight (MFW), can also increase the M/F ratio of a bunch. The current planting materials contain 12% and 10 g, S/F and MFW, respectively. Introgressing the selected palms that possess low S/F and high MFW into breeding programs can increase M/F ratios and subsequently increase OY.

Escobar (1982) reported that the *E. oleifera* germplasm from the Central America also showed interesting bunch composition. Superior fresh fruit bunch (FFB) production was observed among palms from Colombia with average value of 123.8 kg/palm/year. This is considered high as compared to the germplasm of other countries that showed mean FFB between 90.8 and 103.2 kg/palm/year. Similar results were also described by Sharma (2000) who attained 138 kg/palm/year of FFB. Higher O/B, at 4.59 and 4.92% were recorded for populations from Costa Rica and Panama, respectively. In addition, palms from Brazil showed low S/F ratio as well as large fruit size (mean = 10.91 g) which contributed to higher O/B and OTDM ratios. These findings signify that valuable genetic resources for oil improvement via the interspecific hybrids (OxG) route, are available from the *E. oleifera* germplasm collections.

Apart from oil yield, the wild genetic materials also offer genes for high iodine value (IV). Some palms from Nigeria and Madagascar exhibited iodine value (IV) of more than 60, which is relatively higher than the current levels (<50) (Kushairi et al. 2003c). Introgressing these materials into the present breeding populations have resulted in increased IV (57–58) in the progenies (Musa and Gurmit 2008). Genes for high IV are also accessible from the *E. oleifera* germplasm collection. At the country level, the *E. oleifera* palms recorded an impressive mean IV of more than 80 (Mohd Din et al. 2002). IV is a measure of the saturation levels of palm oil. Oil with higher iodine value is more unsaturated and remains liquid in temperate countries. This feature provides opportunity to expand palm oil markets in cold climatic countries. In addition, the ability to produce unsaturated palm oil naturally can significantly reduce the cost incurred by industries utilizing palm olein for frying. Currently, palm olein is produced using the expensive multi-stage fractionation method.

Oil palm generates two types of oils, namely, palm and palm kernel oils, respectively, extracted from the mesocarp and kernel of the oil palm fruit. Oil palm kernel is one of the major sources of lauric oils. The physical and chemical properties of oil palm kernel are similar to coconut oil and have become an important source of raw materials for the oleochemical industry. The current oil palm planting materials has kernel to bunch (K/B) ratio of 4–8%. Palms with K/B ratio of more than 10% were identified in the Nigerian germplasm collection. Economic analysis revealed that increasing the K/B ratio from 5 to 10% would generate a 3% increase in the gross income based on the kernel oil recovery, estimated for every 100 tonnes of fresh fruit bunch (Rajanaidu and Jalani 1994). Palm kernel cake (PKC) is a solid leftover from the extraction process of oil from palm kernel. PKC is an important ingredient for feedstock production. The current palm kernel cake contains about 16–18% protein. Thus, any feedstock produced using PKC is classified as energy feed. In Malaysia, the production of protein feed relies on the imported ingredients such as soybean and corn. Increasing the protein content of PKC would provide domestic raw materials for protein feed production at a reduced cost. Genetic materials that posess more than 20% crude protein in the kernel were identified in germplasm collections from Senegal (Noh et al. 2008). The introduction of these materials into commercial populations can improve the protein content in the palm kernel.

9.7 Developing Oil Palm with Novel Traits

The oil palm germplasm collections also offer the genetic foundation for developing planting materials with value added palm oil. Such oil can diversify the use of palm oil and thus, increase revenue for the industry. Selected palms from the germplasm collections produced oils with high vitamin E and carotene content. Vitamin E is measured based on the level of tocopherols and tocotrienols in the palm oil. The current planting materials produce oil with 600–1000 ppm tocopherols and tocotrienols which are lower than that observed in selected palms from the Nigeria and Cameroon germplasm collections. These palms possess more than 1000 ppm of both tocopherol and tocotrienols (Kushairi et al. 2003c). The mean carotene content of the *E. oleifera* palms collected by the researchers from Malaysia ranged between 1285 and 2220 ppm (Mohd Din et al. 2000). These palms can be exploited to improve the carotene contents of the current planting materials which are reported at 500–700 ppm. The improved materials would produce palm oil applicable for the pharmaceutical and nutraceutical industries. Diversification of the palm oil uses would intensify returns and enhance sustainability and competitiveness.

Rachier et al. (2009) reported some wild oil palms grown in the highlands of Lake Tanganyika in Tanzania and high altitude area in Cameroon that are believed to be well adapted to cold temperatures. These palms have been exploited in the oil palm breeding programs in Kenya. The MPOB germplasm materials assembled from Senegal, Nigeria, Cameroon, Tanzania, Madagascar, and Angola had experienced extreme low rainfall (Marhalil et al. 2015). Palms from these collections are potentially useful for development of drought and heat tolerant planting materials. Krishnan (2015) described some Nigerian prospected materials that showed positive response to drought. These materials also possess high bunch number which indicated potential sustainable yields under unfavorable weather conditions. Recently, Alvarado and Escobar (2016) reported several crosses derived from Deli dura, Ekona, and selected palms from Ghana, Nigeria, and Tanzania that showed high tolerance to drought, low temperature, and low solar radiation. Xianhai et al. (2016) identified some high yield and cold tolerant palms in the province Guangdong in Southern China. New genetic foundation is therefore available in the oil palm germplasm materials and could be exploited to create planting materials with better adaption to climate change.

Basically, there are three main diseases which occur in oil palm plantations. Basal stem rot is the major oil palm disease in South East Asia. Vascular wilt and bud rot have impeded the oil palm commercial development in Africa and South America, respectively. Implementing protection measures are not easy due to the requirements for specific devices and trained labor. Therefore, development of new planting materials resistant to oil palm diseases may provide longer term solution. Genetic materials resistant to vascular wilt were identified in oil palm genetic populations originating from Angola, Ivory Coast, Nigeria, Cameroon, and Congo (Durand-Gasselin et al. 2000). Idris et al. (2004) described a partially resistant progeny against basal stem rot disease that was derived from a cross between palms from Congo and Cameroon. Some *E. oleifera* materials collected from Ecuador showed in situ resistance to bud rot (Julian Barba 2012). Alvarado et al. (2012) reported oil palm hybrids derived from mother palm originating from Manaus, Brazil crossed with compact *pisiferas* revealed high tolerant to bud rot. Another set of interspecific hybrid palms from Coari, Brazil also showed partial resistance to bud rot disease (Restrepo et al. 2012). The genes for development of planting materials tolerant to some oil palm diseases are available. This could allow sustainable oil palm cultivation in regions with high occurrence of diseases.

The identification of palms carrying valuable traits described above signifies the importance of incorporating new source of genetic materials in the oil palm breeding programs. Undeniably, the impact from improved planting materials can only be realized after several decades. This is due to the long (25 years) economic cycle of a plantation, and replanting usually only covers 2–5% of existing acreage. Nevertheless, the oil palm industry should rapidly exploit the genetic potential of the germplasm materials to achieve improved productivity and remain competitive and sustainable in the future.

9.8 Molecular Genetic Diversity of *Elaeis* Species

In addition to evaluating phenotypic traits, DNA-based methods are increasingly being used for assessing genetic diversity of oil palm. Some examples include Restriction Fragment Length Polymorphisms (RFLPs), Random Amplified

Polymorphic DNAs (RAPDs), Amplified Fragment Length Polymorphisms (AFLPs), Simple Sequence Repeats (SSRs), and Single Nucleotide Polymorphisms (SNPs).

Despite its laborious protocol, RFLPs have been applied to evaluate the genetic diversity of oil palm in the early days. The genetic diversity and structure of *E. guineensis* and *E. oleifera* accessions were studied using three pairs of AFLP primers and 37 RFLP probes developed from *E. guineensis* (Barcelos et al. 2000, 2002). Allele specific to each of the *Elaeis* species were detected. Both markers revealed higher level of diversity in the *E. oleifera* accessions than in the *E. guineensis*. Among the *E oleifera* accessions analyzed, high polymorphism was attained for the palms from Brazil. The factorial analysis of correspondence (FAC) showed grouping of the accessions based on country. The AFLP dendrogram clearly distinguished the African accessions from the American groups. Using 16 RFLP probe and enzyme combinations, Maizura et al. (2006) screened 55 genetic populations collected from 11 African countries. Highest polymorphism was detected among *E. guineensis* populations from Nigeria populations. These populations also recorded the most number of rare alleles suggesting that Nigeria may likely be the center of diversity for *E. guineensis*. Most of the germplasm populations revealed higher diversity than the *dura* breeding populations indicating that there was much genetic variation in the germplasm that could be exploited for oil palm improvement programs. The dendrogram revealed five broad genetic groups. The genetic differentiation between oil palm populations was 0.143 which also showed that 85.7% of the total variation was due to differences within palms in the populations.

Moretzsohn et al. (2002) analyzed *E. oleifera* germplasm from the Amazon River Basin in Brazil and several *E. guineensis* accessions from Africa using 95 polymorphic RAPD markers. The clustering results clearly showed the dispersal of palms that followed the Amazon River network. Similar to the report by Maizura et al. (2006), most of the variation observed in the samples was attributed to the differences within populations. Kularatne et al. (2001) extended the diversity study of the populations analyzed by Maizura et al. (2006) using eight primer combinations of AFLPs. Similarly, high genetic diversity was attained for samples from Nigeria. The diversity levels gradually decreased toward the west and east of the oil palm natural distribution in Africa. This finding supported the earlier evidence which point to Nigeria as the center of diversity for the African oil palm. The AFLP analysis also revealed higher portion of genetic diversity within than between populations. Lower diversity was also detected in the Deli *dura* populations.

Several research groups have developed microsatellite (SSR) markers using the genomic information of both *E. guineensis* (Billotte et al. 2001; Rajinder et al. 2008; Ting et al. 2010); and *E. oleifera* (Mohd Zaki et al. 2010, 2012). Application of these markers revealed successful cross-species and cross-taxa amplification suggesting the markers' ability for various genetic studies and phylogenetic analysis of *Elaeis* and several other palm species.

SSRs analysis of the oil palm germplasm collections assembled from the African continent generally revealed a separation of populations collected from the west from those assembled from the central-east regions (Cochard et al. 2009; Zulkifli et al. 2012; Bakoume et al. 2015). Using 14 SSRs, Cochard et al. (2009) attained a third cluster that consisted of 99 samples of the advanced breeding materials. Bakoume et al. (2015), however, reported the grouping of the advanced materials into the central cluster. Both Bakoume et al. (2015) and Zulkifli et al. (2012) described the separation of Madagascar samples from the other germplasm collections analyzed in their studies. Both also proposed Nigeria as the center of diversity for *E. guineensis* based on the high number of rare alleles and polymorphism attained in their analyses. SSR diversity analysis of the oil palm germplasm also revealed greater variation within than between populations (Cochard et al. 2009). Greater variability within population was also attained from the genetic diversity studies involving oil palm germplasm populations collected from Cameroon (Arias et al. 2011) and Angola (Rey et al. 2004) using SSR markers.

The increased improvement achieved in sequencing technologies and bioinformatics analysis had provided researchers with the ability to apply SNPs for genotyping the oil palm populations. Maizura et al. (2013) analyzed selected oil palm populations from Nigeria, Tanzania, Angola, and Madagascar using 1164 SNPs. Population structure analysis revealed two sub-groups with populations from the central Africa forming the biggest cluster, separated from those coming from Madagascar. Using 815 SNP markers, Alves et al. (2015) reported higher variability within than between populations. Moderate level of heterozygosity (0.185) was recorded indicating that the populations analysed were derived from unrelated individuals. The higher genetic variability within populations has implication on strategies for future collections and conservation (as discussed later) while the ability to separate populations (Wong et al. 2015) facilitates the development of heterotic populations for hybrid breeding and production.

Reports on the genetic diversity analysis among the advanced oil palm breeding populations mainly focused on applying SSR markers. Analysis of several DxP hybrid populations revealed clustering according to the parental lines used to create them (Norziha et al. 2008; Arias et al. 2012). Based on 21 SSRs, oil palm populations from China showed lower diversity than that from Malaysia (Zhou et al. 2015). The oil palm was introduced to China in 1926. The batch of seeds that were deposited in the country at that time may represent only a part of the entire diversity that exist in the breeding populations in Malaysia. This explained the lower diversity of the populations in China. The principle component analysis indicated one big cluster signifying close genetic relationship among the populations of the two countries. Population structure analysis of hybrid populations derived from Deli *dura* and AVROS genetic backgrounds exhibited four groups, corresponding to the different seed producers in Thailand (Taeprayoon et al. 2015). This implied the ability of the SSRs in distinguishing the genetic background of the populations. This was also demonstrated by Wong et al. (2015) using 11 SSR markers. The genetic diversity estimation of oil palm parental palms was also reported (Wening et al. 2013). Using eight SSR markers, the *pisifera* parents were found to be more variable than the *duras*. The markers are applicable for determining parental palms with highest homozygosity, which are more favorable to the breeders.

9.9 Conservation Initiatives for *Elaeis* Species

Basically, there are two approaches in conserving plant genetic resources, ex situ and in situ conservations. In situ conservation refers to preserving germplasm in their natural habitats. The ex situ conservation, on the other hand, denotes to the preservation of genetic materials outside their natural habitats (Engelman 1991). Some approaches of ex situ conservation include field genebanks, seed storage, pollen storage, in vitro storage, and botanical gardens, while the in situ approach encompasses of genetic reserves, on-farm and home garden conservation (Maxted et al. 1997).

9.9.1 *Oil Palm* Ex Situ *Living Collection*

As far as the *Elaeis* species is concerned, ex situ conservation was established by several oil palm research centers. Among others are MPOB in Malaysia, French Agricultural Research Centre for International Development (CIRAD) in France, Indonesian Oil Palm Research Institute (IOPRI) in Indonesia, and the Empresa Brasileira de Pesquisa Agropecuaria (EMBRAPA) in Brazil. At a planting density between 136 and 148 palms per hectare, these conservation efforts inhabit several hundred or even thousand of hectares of land. The MPOB and IOPRI oil palm germplasm collections (Fig. 9.3) occupy approximately 600 and 150 ha of land, respectively. Conservation cum evaluation of these large germplasm resources is a daunting but nevertheless important task.

Apart from the expensive cost involved in maintaining these living collections, the genetic materials are also exposed to diseases and extreme weather conditions. Excellent and valuable materials may be lost at any time. The current germplasm collections may also contain duplications (materials that are genetically similar), populations with low diversity, and individual palms that do not possess economically valuable traits, which reduce land utilization efficiency. Further, at 25th year, the oil palms also become very tall (>10 m) and are not easily accessible for crossing and/or introgression programs. These issues pose a great challenge to the genebank managers, oil palm breeders, and researchers who have interest in the species. As these technical groups hold diverse responsibilities toward the germplasms, appropriate consultation among them is crucial in drawing up conservation strategy to achieve effective management of land use and access of the genetic materials in the future.

One of the ways to achieve efficient management and access of germplasm collections is through the establishment of core collection. The term core collection was first coined by Frankel (1984) which was defined as a limited set of accessions representing the genetic diversity of a crop species and/or its wild relatives with minimum repetitiveness. The basic target of core collection is to capture the maximum genetic diversity of the whole collection while minimizing accession

Fig. 9.3 Oil palm ex situ living collection maintained at Malaysian Palm Oil Board

number and redundancy to a manageable size of living collections. More importantly, the goal is to ensure effective function of the collection in the future. For the oil palm germplasm collection, phenotypic evaluation is a mandatory activity as this information can directly point to the individuals that carry valuable traits for use in breeding and improvement. In addition, some of the germplasm materials have also been assessed using molecular marker systems. Combining both, the phenotypic as well as molecular diversity evidences will greatly enhance the efficiency in selecting individuals and/or populations for core collection. Nowadays, powerful software such as M-Strat (Gouesnard et al. 2001), genetic distance sampling (Jansen and van Hintum 2007), PowerCore (Kim et al. 2007), and CoreHunter (Thachuk et al. 2009) are available that could be applied in analyzing the phenotypic and marker data and subsequently shortlist samples for core collection. The resulting individuals from this analysis will fulfill the aspirations of genebank managers, i.e., to conserve the diversity of the collections within a reduced population size.

The selection of core samples using computer softwares certainly has merits. Nonetheless, the genetic variability parameters computed from molecular marker analysis applied on the germplasm collections will also help in the identification of diverse populations. Previous reports showed that oil palm natural populations from Nigeria and Brazil revealed higher diversity (Maizura et al. 2006; Moretzsohn et al. 2002). It may be worthwhile to include some populations from these countries into the core collection. The molecular marker analysis also indicated that greater variability in oil palm populations was attributed to within rather than between populations. For future conservation of oil palm, sampling more individuals within a few populations is therefore recommended. Previous work indicated that there was no significant difference in terms of the genetic diversity values estimated for oil palm with population size 20 against 30 (Hayati et al. 2004). At this juncture, preserving 20 palms would sufficiently cover the genetic diversity that is present in the population. Alternatively, conserving 16 palms per population can also be considered, as this is the typical number of palms per cross that are field planted in one replicate in an oil palm breeding plot.

Genetic similarity values among the populations were also estimated from the molecular marker data analysis. These values are often illustrated in the form of phylogenetic tree or principle component analysis. Populations that are genetically similar were grouped together or distributed closely to each other. Depending on the size of the group or distribution area, sampling individuals from one or several populations from each group would cover a wider range of variability than selecting at random.

Apart from the software-based and marker-based selections, germplasm palms that exhibit economically important traits are precious resources for oil palm improvement. From the oil palm breeders' point of view, these palms deserve special attention for core collection and are justified for systematic long-term conservation. In order to ensure the survival of these gene pools, selected palms can be selfed or intercrossed with palms from the same family. The procedure has to be initiated once every 20 or 25 years for continuous accessibility of the important genes, in line with the development of new and improved planting materials that require approximately 30–40 years.

9.9.2 In Vitro *Conservation Method of* Elaeis *Species*

9.9.2.1 Cryopreservation Method

Conservation of large size species like oil palm in the form of living collections provides direct access of the genetic materials to the oil palm breeders. Nevertheless, this method requires large amount of land, man power, and high cost of maintenance. Cryopreservation is an alternative method that serves as a desirable storage protocol of biological materials under extremely low temperature (−196 °C). This method has the advantage of limited space requirement, protection of plant materials from diseases, cost efficiency, and storage for an unlimited period of time. Cryopreservation has become an important method for maintaining plant genetic resources that complements the field preservation techniques (Engelmann 2000). Plant cryopreservation was pioneered by Sakai (1965) who froze mulberry twigs in liquid nitrogen. Since then, a huge variety of plants have been successfully cryopreserved for long-term conservation (Bajaj 1984; Makeen et al. 2005; Normah et al. 1994; Radhamani and Chandel 1992). This method has been used for the storage of many different types of plant material such as cells, protoplasts, shoot apices, somatic embryos, seed, or excised zygotic embryos (Engelmann 2004). Cryopreservation appears to be the most practicable method for storing recalcitrant seeds and species that are vegetatively propagated.

Oil palm seeds are neither recalcitrant nor orthodox. This intermediate type of seed can only withstand partial dehydration and is also cold sensitive (Ellis et al. 1990, 1991). Previous reports indicated that such seeds have embryos that are relatively more tolerant to desiccation and cryoexposure than whole seeds (Bajaj 1984; Radhamani and Chandel 1992; Normah et al. 1994, Makeen et al. 2005). The in vitro technology for the conservation of oil palm germplasm did not only focus on zygotic embryos (Grout et al. 1983), but also on somatic embryos (Engelmann et al. 1985) and pollen (Tandon et al. 2007). These tissues are relatively smaller in size as compared to seeds and kernels thus, less sensitive to desiccation and freezing. The cryopreservation of oil palm zygotic embryos was first reported by Grout et al. (1983). Using the silica gel method, Norziha et al. (2011) successfully reduced the moisture content of zygotic embryos to 10–20% and subsequently attained 70% survival rate of the cryopreserved materials. The method is currently being applied to cryopreserve more than 68,000 zygotic embryos excised from selected individual palms originating from Tanzania, Guinea, Gambia, Madagascar, Cameroon, Angola, Sierra Leone, Senegal, Nigeria, Ghana, and Zaire (Fig. 9.4).

In 1993, Dumet et al. (1993) reported that dehydration could be performed either by placing oil palm somatic embryos for 0–10 h under the laminar flow or in air-tight boxes containing silica gel for 0–18 h. Polyembryoids of oil palm were cryopreserved with a successful revival rate of 68% using the droplet vitrification technique (Gantait 2015). Duval et al. (2000) described the cryopreservation of 80 different genotypes of oil palm polyembryogenic cultures in the cryopreservation facilities in CIRAD. A total of 150 clones had been tested for cryopreservation in

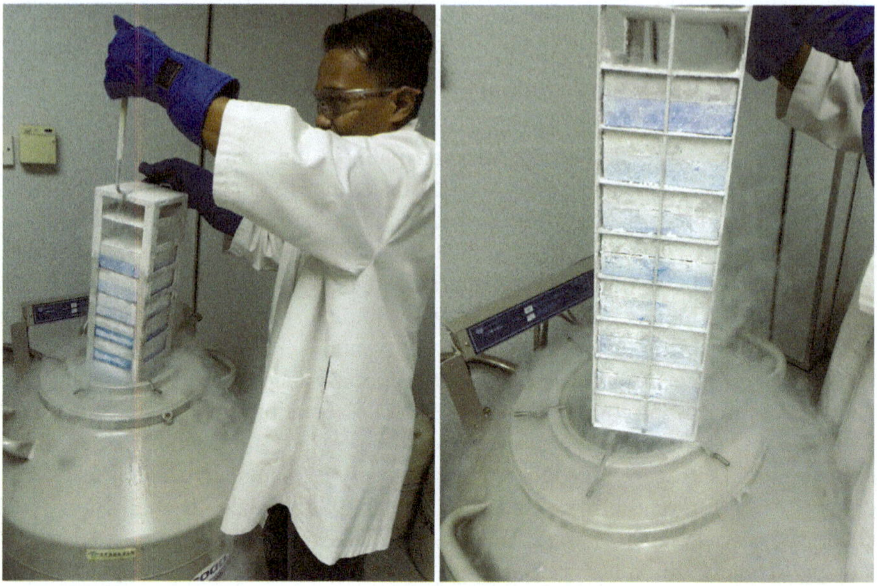

Fig. 9.4 Cryopreservation facilities at Malaysian Palm Oil Board

Indonesia, Malaysia, Ivory Coast, and France (Engelmann 1991). The materials were stored at ultra-low temperature for 52 months without any effect on the recovery rate (Engelmann 1992).

9.9.3 Pollen Conservation

Oil palm pollen was shown to be tolerant to desiccation and low temperature. After drying at 37 °C for 2–8 h, the pollen could be kept for 12 months at −15 °C (Ekaratne and Senathirajah 1983). Tandon et al. (2007) described the feasibility of storing oil palm pollen for long periods in liquid nitrogen without any significant loss in viability and germination ability. Cryopreserved pollen grains recorded as high as 54% viability (compared with 62% before storage) and 49% in vitro germination ability (compared with 52% before storage). In general, an ideal survival rate is achieved when samples are frozen with water content of between 10 and 20% on fresh weight basis (Engelmann 1992). Depending on the species and tissues applied for cryopreservation, survival rates between 20% (Golmirzaie and Panta 2000) and 70% (Malik et al. 2012) could be attained. More importantly, since the method developed is genotype specific, researchers need to carry out modification when dealing with new genotypes or genetic materials.

9.9.4 Maintenance Under Growth Limitation

In vitro maintenance of whole plants is another method for conserving plant materials. This method applies minimal growth conditions to slow down metabolism which allow the plants to be kept for short and medium term. Using genetic materials obtained from the oil palm genebank established at EMBRAPA, Camillo and Scherwinski-Pereira (2015) had attempted minimal growth conditions on 10 cm size plants regenerated from oil palm zygotic embryos. The effect of low temperature and various types and concentrations of carbohydrates on the survival rate of the materials were tested. It was found that keeping the plants at 20 °C in media containing 3% sucrose significantly reduced the growth of the plants. Applying these parameters had successfully generated normal looking plants with a survival rate of 95%.

9.9.5 Artificial Seeds

Artificial seed is a technology that involves encapsulating somatic embryos, shoot tips, axillary buds, stem, and root segments using coating materials. Some examples of coatings used are polyoxyethylene, sodium alginate complexed with calcium chloride or calcium nitrate, sodium alginate mixed with gelatin complexed in calcium chloride, carrageenan mixed with locus bean gum and complexed with potassium or ammonium chloride and Gelrite gel (Ravi and Anand 2012). This technology offers a solution for long-term and high-scale storage of genetic materials at low cost (Ghosh and Sen 1994). A method to encapsulate oil palm secondary somatic embryos (SSEs) was developed by Inpuay and Te-chato (2012). The SSEs were attained from green mature somatic embryos known as haustorium embryos (Te-chato and Hilae 2007; Chehmalee and Te-chato 2008). It was reported that the most suitable protocol for encapsulation of these materials was 2.5% sodium alginate polymerized in 100 mM calcium nitrate for 15 min. Encapsulated colyoptile SSEs gave a high germination rate of 73% after cultured on germination medium for 1 week.

As described above, methods have been developed for preserving various oil palms tissues including zygotic embryos, pollen, shoot tips, segments of stems and roots, small oil palm plants as well as tissue-culture derived materials such as somatic embryos, polyembroids, and secondary somatic embryos. These reports provided useful references to genebank managers or researchers who wish to apply low temperature, minimal growth, and cryopreservation methods to preserve oil palm genetic materials.

9.9.6 DNA Banking

Plant DNA banks have emerged as new resource with great potential for characterizing and utilizing biodiversity. MPOB has also taken the initiative to preserve

DNA of important palms used in crossing programs as well as the germplasm materials maintained in the ex situ living collections. These DNA collections will become reference samples for analyses of parentage or verification of specific genotypes or crosses. Leaf samples of the palms selected for cryopreservation are also targeted for DNA banking. The extracted DNA is stored in a 1.5 ml tubes and kept at 4 °C. The leaf and DNA samples are stored at two different locations. To date, MPOB had collected a total of 3312 leaf samples from its germplasm collections and extracted DNA from 1854 samples.

A database that contained large number of dura, pisifera, and tenera palms from different sources such as Deli, Yangambi, La Me, AVROS, and Ekona selected over many generations was established (Seng et al. 2009). The DNA from these palms was also extracted. Coupled with molecular marker analysis, the database is useful for routine illegitimacy tests in oil palm breeding crosses, seed production, and fingerprinting of tissue-cultured derived clones.

9.10 Conclusion

In the course of selection, oil breeding populations have lost numerous alleles and this was clearly demonstrated through molecular marker analysis. This may one way or another affect the ability of the crops to adapt to the changing environments that may be stressful to the plant. This will also cause the crop to be more prone to diseases, especially since climate change is predicted to occur worldwide. Therefore, efforts to improve the diversity of the present breeding populations by introgressing the underutilized germplasm materials into the breeding schemes should be initiated actively.

Phenotypic characterization revealed enormous number of valuable genes present in the oil palm germplasm collections. Among others are high oil yield, slow height increment, longer stalk, tolerance to diseases, high iodine value, high carotene, high kernel, and high vitamin E contents. Palms with ability to efficiently utilize resources and adapt effectively to climate change are likely present in the germplasm. Preliminary efforts to incorporate some of these materials with interesting phenotypes into breeding schemes revealed encouraging results. Adopting such planting materials in commercial plantations would take decades to achieve due to the perennial nature of oil palm. Nonetheless, introgressing these genes into the current breeding programs will overall enhance diversity, productivity, environmental adaptation and add value to palm oil and oil palm products.

There is also increasing concerns over the loss of genetic resources of crop plants. The progress of crop improvement depends on conservation efforts of these resources and their effective utilization by plant breeders. Several oil palm research organizations had attempted to accumulate *E. guineensis* and *E. oleifera* genetic resources from their centers of natural distributions. As far as ex situ conservation is concerned, computer software that make use of phenotypic and molecular marker data are available to help genebank managers and oil palm researchers select palms for establishment of core collection. This collection represents the genetic diversity of the main collection in a minimum population size. In addition, selected palms that carry genes of economic values should also be continuously maintained in the form of selfed or introgressed progenies to ensure long-term and effective accessibility. The in vitro conservation methods have been developed for oil palm tissues. These are alternatives which offer low cost for maintaining large number of samples for long periods at smaller facilities. Despite the specific expertise required for the management of such facility, methods have been established for various oil palm tissues. Maintaining palms or tissues in such a way ensures the genetic materials are free from external threats such as diseases and extreme weather condition.

Such conservation methods will ensure that oil palm breeders will always have the necessary genetic resources to breed for the sustainability of the crop and industry in a climate change world.

References

Alvarado A, Escobar R (2016) Seed production and oil palm breeding in ASD Costa Rica. Proceedings of the international seminar on oil palm breeding and seed production and field visits, Kisaran, Indonesia, pp 21–38, 29–30 September 2016

Alvarado A, Escobar R, Henry J (2012) The OxG hybrid: a new alternative for areas affected by bud rot disorder. Proceedings of the international seminar on breeding for oil plam disease resistance and field visits, Bogota, Colombia, 21–24 September 2012

Alves AA, Pereira VM, Filho JAF, Leao AP, Formighieri EF, Souza MT Jr, Rios SA (2015) Genetic diversity patterns and genetic structure among subpopulations of *Elaeis oleifera* based on genome-wide SNP markers. Proceedings of the 8th Brazilian congress of plant breeding, Goiania, Brazil, 3–6 August 2015

Andrade EB (1982) Relatorio da Expedicao para Coleta de Germoplasma de Caiaue *Elaeis oleifera* (H.B.K.) Cortes, na Amazonia brasileira. EMBRAPA-CNPSD. Manaus, p 31

Ang BB (2003) Early performance of PT Asian Agri DxP test-cross progeny trials on peat and alluvial soils. Paper presented at seminar on progress of oil palm breeding and selection, Medan, Indonesia, 6–9 October 2003

Arias D, Montoya C, Rey L, Romero H (2012) Genetic similarity among commercial oil palm materials based on microsatellite markers. Agronomia Colombiana 30(2):188–195

Arias DM, Daza ES, Montoya C, Romero HM (2011) Genetic collection of oil palm materials from Cameroon. PALMAS 32:27–37

Ataga CD, Okwuagwu CO, Okolo EC (2000) Oil palm genetic resources in Nigeria—a review. Proceedings of the international symposium on oil palm genetic resources and their utilization, Kuala Lumpur, Malaysia, pp 142–156, 8–10 June 2000

Bajaj YPS (1984) Introduction of growth in frozen embryos of coconut and ovules of citrus. Cryoletters 23:1215–1216

Bakoumé C, Wickneswari R, Siju S, Rajanaidu N, Kushairi A, Billote N (2015) Genetic diversity of the world's largest oil palm (*Elaeis guineensis* Jacq.) field genebank accessions using microsatellite markers. Genet Resour Crop Evol 62:349–360

Barcelos E, Amblard P, Berthaud J, Seguim M (2000) The genetic diversity of the American oil palm *Elaeis oleifera* (Kunth) cortes revealed by nuclear RFLP markers. Proceedings of the international symposium on oil palm genetic resources and their utilization, Kuala Lumpur, Malaysia, pp 544–555

Barcelos E, Amblard P, Berthaud J, Seguim M (2002) Genetic diversity and relationship in American and African oil palm as revealed by RFLP and AFLP molecular markers. Pesq Agropec bras Brasilia 37(8):1105–1114

Billotte N, Risterucci AM, Barcelos E, Noyer JL, Amblard P, Baurens FC (2001) Development, characterization and across-taxa utility of oil palm (*Elaeis guineensis* Jacq.) microsatellite markers. Genome 44:413–425

Blaak G (1967) Oil palm prospection tour in the Bamenda highlands of West Cameroon. Internal Report, Unilever, London

Camillo J, Scherwinski-Pereira JE (2015) In vitro maintenance, under slow-growth conditions of oil palm germplasm obtained by embryo rescue Pesquisa Agropecuária Brasileira. 50(5), 426–429

Chehmalee S, Te-chato S (2008) Induction of somatic embryogenesis and plantlet regeneration from cultured zygotic embryo of oil palm. J Agric Technol 4:137–146

Chin CW, NG WJ, Junaidah J, Shuhaimi S, Mohd Nasruddin M (2005) Developing high oil yield DxP: the Felda experience. Proceedings of the 2005 national seminar on advances in breeding and clonal technologies for super yielding planting materials, Seri Kembangan, Selangor, Malaysia, pp 152–174

Cochard B, Adon B, Rekima S, Billotte N, Desmier R, Koutou A et al (2009) Geographic and genetic structure of African oil palm diversity suggests new approaches to breeding. Tree Genet Genomes 5:493–504

Corley R, Tinker P (2003) The oil palm, 4th edn. Blackwell Publishing, New Jersey

Crone GR (1937) The voyages of Cadamosto and other documents on Western Africa in the second half of the fifteenth century. Hakluyt Soc Series II, 80 [1.1.2]

Davidson L (1993) Management for efficient cost—effective and productive oil palm plantations. Proceedings of the PORIM international palm oil conference, Kuala Lumpur, Malaysia, pp 153–170

Donough CR, Chia CC (2005) Breeding oil palms for high oil yield DxP in IOI group: first cycle development of OPGL-derived materials. Proceedings of the 2005 national seminar on advances in breeding and clonal technologies for super yielding planting materials, Seri Kembangan, Selangor, Malaysia, pp 175–216

Dumet D, Engelmann F, Chabrillange N, Duval Y (1993) Cryopreservation of standard oil palm (*Elaeis guineensis* Jacq.) somatic embryos involving a desiccation step. Plant Cell Rep 12:352–355

Durand-Gasselin T, Diabate S, de Franqueville H, Cochard B, Adon B (2000) Assessing and utilizing sources of resistance to Fusarium wilt in oil palm (*Elaeis guineensis* Jacq.) genetic resources. Proceedings of the International Symposium on oil palm genetic resources and their utilization, Kuala Lumpur, Malaysia, pp 446–470

Duval Y, Chabrillange N, Dumet D, Engelmann F (2000) Ex situ conservation of oil palm (*Elaeis guineensis* Jacq.) genetic resources using biotechnology. Proceedings of the International Symposium on oil palm genetic resources and their utilization, Kuala Lumpur, Malaysia, pp 471–479

Ebongue FN, Koona P, Nouy B, Zok S, Carriere F, Zollo PH et al (2008) Identification of oil palm breeding lines producing oils with low fatty acids. Eur J Lipid Sci Technol 110:505–509

Ekaratne SNR, Senathirajah S (1983) Viability and storage of pollen of the oil palm. *Elaeis guineensis* Jacq. Ann Bot 51(5):661–668

Elenga H, Schwartz D, Vincens A (1994) Pollen evidence of late quarternary vegetation and inferred climate changes in Congo. Palaeogeogr Palaeoclimatol Palaeoecol, pp 345–356

Ellis RH, Hong TD, Roberts EH (1990) An intermediate category of seed storage behaviour? I. Coffee. J Exp Bot 41:1167–1174. doi:10.1093/jxb/41.9.1167

Ellis RH, Hong TD, Roberts EH, Soetisna U (1991) Seed storage behaviour in *Elaeis guineensis*. Seed Sci Res 1:99–104. doi:10.1017/S0960258500000726

Engelmann F (1991) In vitro conservation of tropical plant germplasm—a review. Euphytica 57:227–243

Engelmann F (1992) Cryopreservation of embryos, reproductive biology and plant breeding. In: Dattee C, Dumas C, Gallais A (eds), Springer, Berlin, pp 281–290

Engelmann F (2000) Importance of cryopreservation of plant genetic resources. In: Engelmann F, Takagi, H (eds) Cryopreservation of tropical plant germplasm. Current research progress and applications, Rome, JIRCAS, Tsukuba/IPGRI, pp 8–20

Engelmann F (2004) Plant cryopreservation: progress and prospects. In Vitro Cell Dev Biol-Plant 40:427–433

Engelmann F, Duval Y, Dereuddre J (1985) C.R. Acad. Sci. Paris 301, Sér. DI, 111–116. Survieet proliferation d'embryons somatiques de palmier huile (*Elaeis guineensis* Jacq.) après congélationdansl'azoteliquide. Comptes Rendus de I'Acadéniie des Sciences Paris, Série 111 301, 111–116

Escobar R (1982) Preliminary results of the collection and evaluation of the American oil palm *Elaeis oleifera* (HBK, Cortes) in Costa Rica. In: Pushparajah E, Chew PS (eds) The oil palm in agriculture in the eighties, vol 1. Incorporated Society of Planters, Kuala Lumpur, pp 79–93

Frankel OH (1984) Genetic perspectives of germplasm conservation. In: Arber W, Llimensee K, Peacock WJ, Starlinger P (eds) Genetic manipulation: impact on man and society. Cambridge University Press, Cambridge, pp 161–170

Gantait S, Sinniah UR, Suranthran P et al (2015) Improved cryopreservation of oil palm (*Elaeis guineensis* Jacq.) polyembryoids using droplet vitrification approach and assessment of genetic fidelity. Protoplasma 252:89

Gerritsma W, Wessel M (1997) Oil palm: domestication achieved? Neth J Agric Sci 45:463–475

Ghosh B, Sen S (1994) Plant regeneration from alginate encapsulated somatic embryos of Asparagus cooperi baker. Plant Cell Rep 9:189–194

Golmirzaei AM, Panta A (2000) Advances in potato cryopreservation at CID. In: Engelmann F, Takagi, H (eds) Cryopreservation of tropical plant germplasm. Current Research Progress and Application, Japan International Research Center for Agricultural Sciences, Tsukuba, Japan, pp 250–254

Gouesnard B, Bataillon TM, Decoux G, Rozale C, Schoen DJ, David JL (2001) MSTRAT: An algorithm for building germplasm core collections by maximizing allelic or phenotypic richness. J Hered 92:93–94. doi:10.1093/jhered/92.1.93

Grout BWW, Sheltan K, Prichard HW (1983) Orthodox behaviour of oil palm seed and cryopreservation of the excised embryos for genetic conservation. Ann Bot 52:381–384

Hartley C (1988) The oil palm, 3rd edn. Longman Scientific & Technical, London

Hayati A, Wickneswari R, Maizura I, Rajanaidu N (2004) Genetic diversity of oil palm (*Elaeis guineensis* Jacq.) germplasm collections from Africa: implications from improvement and conservation of genetic resources. Theor Appl Genet 108:1274–1284

Heslop-Harrison JSP (2012) Traits with ecological functions. Ann Bot 110:139–140

Idris AS, Kushairi A, Ismail S, Ariffin D (2004) Selection for partial resistance in oil palm progenies to Ganoderma basal stem rot. J Oil Palm Res 16(2):12–18

Idris AS, Norman K (2015) Upstream management practices to reduce Ganoderma disease outbreak—the industry's approach. Book of abstract of the PIPOC 2015: Agriculture, Biotechnology & Sustainability (ABS) Conference, Kuala Lumpur, Malaysia. pp 9–10

Inpuay K, Te-chato S (2012) Primary and secondary somatic embryos as tool for the propagation and artificial seed production of oil palm. J Agric Technol 8(2):597–609

Isa ZA, Kushairi A, Chin, CW, Sharma M, Chia CC, Rajanaidu N, Mohd Din A (2005) Performance of Malaysian DxP—third round evaluation. Proceedings of the 2005 national seminar on advances in breeding and clonal technologies for super yielding planting materials, Seri Kembangan, Selangor, Malaysia, pp 95–137

Isa ZA, Ong KP, Norasyikin I, Suboh O (2008) Performance of MPOB-Nigerian population 12—an update. Proceedings of the 3rd seminar on performance of MPOB PS1 and PS2 materials and elite germplasm, Bangi, Selangor, Malaysia, pp 139–149

Isa ZA, Suboh O, Azhar D, Norasykin I (2013) Performance of Dami DxP planted in inland soil in Tereh Utara, Kulim Plantations. Proceedings of the international seminar on oil palm breeding—yesterday, today and tomorrow, Kuala Lumpur, Malaysia, pp 40–63

Jansen J, van Hintum T (2007) Genetic distance sampling: a novel sampling method for obtaining core collections using genetic distances with an application to cultivated lettuce. Theor Appl Genet 114:421–428

Julian Barba R (2012) Ecuadorian oleiferas minimize losses caused by bud rot in Latin America. Proceedings of the international seminar on breeding for oil plam disease resistance and field visits, Bogota, Colombia, 21–24 September 2012

Junaidah J, Chin CW, Rafii MY, Syuhada WH (2008) Performance and utilization of MPOB-Nigerian oil palm materials in FELDA. Proceedings of the 3rd seminar on performance of MPOB PS1 and PS2 materials and elite germplasm, Bangi, Selangor, Malaysia, pp 73–90

Kahn F (1997). The palms of eldorado (Edition Champlour ed) Orstom

Kim KW, Chung HK, Cho GT, Ma KH, Chandrabalan D, Gwag JG, Kim TS, Cho EG, Park YJ (2007) PowerCore: a program applying the advanced M strategy with a heuristic search for establishing core sets. Bioinformatics 23:2155–2162. doi:10.1093/bioinformatics/btm313

Krishnan S (2015) Fine tuning the Nigerian germpalsm breeding programme at FGV for climate change in Malaysia. Proceedings of the international seminar on gearing oil palm breeding and agronomy for climate change, Kuala Lumpur, Malaysia, p 8

Kularatne RS, Shah F, Rajanaidu N (2001) The evaluation of genetic diversity of Deli dura and African oil palm genrmplasm collection by AFLP markers. Trop Agric Res 13:1–12

Kushairi A, Rajanaidu N, Jalani S, Mohd Rafii Y, Mohd Din A (1999) PORIM oil palm planting materials. PORIM Bull 38:1–13

Kushairi A, Rajanaidu N, Mohd Din A, Isa ZA, Noh A, Junaidah J (2003a) Performance of Tanzania germplasm. Proceedings of the seminar on progress of PS1 and PS2 planting materials and release of elite germplasm to the industry, Bangi, Selangor, Malaysia, pp 91–112

Kushairi A, Rajanaidu N, Mohd Din A, Isa ZA, Noh A, Junaidah J (2003b) Performance of Angola genetic materials. Proceedings of the seminar on progress of PS1 and PS2 planting materials and release of elite germplasm to the industry, Bangi, Selangor, Malaysia, pp 75–90

Kushairi A, Rajanaidu V, Mohd Din A (2003c) Mining the germplasm. Paper presented at ISOPB seminar on the progress of oil palm breeding and selection, Medan, Sumatra, Indonesia, 6–9 October 2003

Kushairi A, Tarmizi A H, Zamzuri A, Ong-Abdullah M, Rohani O, Samsul Kamal R, Ooi S E, Ravigadevi S, Mohd Basri W (2010) Current status of oil palm tissue culture in Malaysia. Proceedings of the clonal and quality replanting material workshop: towards increasing the national productivity by one tonnes FFB/ha/year, Kuala Lumpur, Malaysian, pp 3–14, 10 August 2010

Kushairi A, Mohd Din A, Ranaidu N (2011) Oil palm breeding and seed production. In: Mohd Basri W, Choo YM, Chan KW (eds) Further advances in oil palm research (2000-2010), vol 1. Malaysian Palm Oil Board, Kuala Lumpur, pp 47–93

Maizura I, Rajanaidu N, Zakri A, Cheah S (2006) Assessment of genetic diversity in oil palm (*Elaeis guineensis*) using Restriction Length Polymorphism (RFLP). Genet Resour Crop Evol 56:187–195

Maizura I, Suzana M, Ting NC, Leslie Ooi CL, Norziha A, Rajinder S. (2013) Population structure of selected oil palm germplasm as revealed by SNP markers, a pre-requisite for association mapping. Proceedings of MPOB International Palm Oil Conference (PIPOC), Kuala Lumpur, Malaysia

Makeen AM, Normah MN, Dussert S, Clyde MM (2005) Cryopreservation of whole seeds and excised embryogenic axes of Citrus suhuiensis cv. Limau langkat in accordance to their desiccation sensitivity. Cryoletters 26:259–268

Malik SK, Chaudhury R, Pritchard HW (2012) Long-term, large-scale banking of citrus species embryos: comparisons between cryopreservation and other seed banking temperatures. Cryoletters 33(6):453–464

Marhalil M, Mohd Din A, Rajanaidu N, Kushairi A (2015) Utilization of MPOB oil palm germplasm for climate change. Proceedings of the international seminar on gearing oil palm breeding and agronomy for climate change, Kuala Lumpur, Malaysia, p 7

Marhalil M, Rajanaidu N (2011) Oil palm (*Elaeis guineensis*) germplasm in Angola: second prospection. Proceedings of the international seminar on breeding for sustainability in oil palm, Kuala Lumpur, Malaysia, pp 133–154

Mauricio M, Fontanilla C (2015) Bud rot technology validation in oil palm plantations from the Colombian Central Zone. Book of abstract of the PIPOC 2015: Agriculture, Biotechnology & Sustainability (ABS) Conference, Kuala Lumpur, Malaysia, pp 24–25

Maxted N, FordLlyod BV, Hawkes JG (1997) Plant genetic conservation: the *in situ* approach. Chapman and Hall, London, UK

Meunier J (1969) Etude des populations naturelles d'*Elaeis guineensis* en Cote d'Ivoire. Oleagineux 24(2):195

Meunier J, Baudouin L (1986) Evaluation and utilization of Yocoboue population of *Elaeis guineensis*. Proceedings of the international workshop on oil palm germplasm and utilization, Bangi, Selangor, Malaysia, pp 144–152

Morcillo F, Cros D, Billotte N, Ngando-Ebongue GF, Domonhédo H, Pizot M et al (2013) Improving palm oil quality through identification and mapping of the lipase gene causing oil deterioration. Nat Commun 4:2160. doi:10.1038/ncomms3160

Mohaimi M (2011) Updates of the evaluation of semi-clonal DxP progenies in Sime Darby. Proceedings of international seminar on breeding for sustainability in oil palm, Kuala Lumpur, Malaysia, pp 101–111

Mohd Din A, Rajanaidu N, Jalani S (2000) Performance of *Elaeis oleifera* from Panama, Costa Rica, Colombia and Honduras. J Oil Palm Res 12(1):71–80

Mohd Din A, Rajanaidu N, Kushairi A, Rafii M, Isa ZA, Noh A (2002) PS4- High Carotene *E. oleifera* planting Materials. MPOB Information Series No. 154. MPOB TT No. 137. MPOB, Bangi, Selangor

Mohd Zaki N, Ismail I, Rosli R, Chin TN, Singh R (2010) Development and characterization of *Elaeis oleifera* microsatellite markers. Sains Malaysiana 39(6):909–912

Mohd Zaki N, Singh R, Rosli R, Ismail I (2012) *Elaeis oleifera* genomic-SSR markers: exploitation in oil palm germplasm diversity and cross-amplification in arecaceae. Int J Mol Sci 13:4069–4088. Available online: doi:10.3390/ijms13044069

Moretzshon MC, Ferreira MA, Amaral ZPS, Coelho PJA, Grattapaglia D, Ferreira ME (2002) Genetic diversity of Brazillian oil palm (*Elaeis oleifera* H.B.K.) germplasm collected in the Amazon forest. Euphytica 124:35–45

Musa B, Gurmit S (2008) Utilization of MPOB germplasm at UPB. Proceedings of the 3rd seminar on performance of MPOB PS1 and PS2 materials and elite germplasm, Bangi, Selangor, Malaysia, pp 43–72

Musa B, Sritharan K, Xaviar A (2011) Performance of bi-clonal DxP planting material at United Plantations Berhad. Proceedings of international seminar on breeding for sustainability in oil palm, Kuala Lumpur, Malaysia, pp 78–100

Noh A, Kushairi A, Mohd Din A, Maizura I, Marhalil M, Osman A, Rajanaidu N (2008) Genetic variation for long stalk and high protein kernel in oil palm germplasm. Proceedings of the 3rd seminar on performance of MPOB PS1 and PS2 materials and elite germplasm, Bangi, Selangor, Malaysia, pp 150–167

Normah MN, Reed BM, Yu YL (1994) Seed storage and cryoexposure behaviour in hazelnut (Corylusavellana L. cv. Barcellona). Cryoletters 15:315–322

Normahnani MN, Siang SC (2016) Screening for Ganoderma disease tolerant oil palm progenies in Sime Darby. Program book of the 6th IOPRI-MPOB International seminar on current research and management of pests, Ganoderma and pollination in oil palm for higher productivity, Medan, Indonesia, pp 32, 27–29 September 2016

Norziha A, Marhalil M, Maizura I (2011) Dessication and cryopreservation of oil palm (*Elaeis guineensis* Jacq.) embryos. Proceedings of agriculture, biotechnology and sustainability conference PIPOC 2011, Kuala Lumpur, Malaysia, pp 300–303

Norziha A, Noh A, Fadila AM, Zulkifly Y, Mohd Din A (2012) Ganoderma infection studies in selfed and intercrossed Deli dura progenies. Proceedings of the international seminar on breeding for oil palm disease resistance and field visits, Bogota, Colombia, 21–24 September 2012

Norziha A, Rafii MY, Maizura I, Ghizan S (2008) Genetic variation among oil palm parent genotypes and their progenies based on microsatellite markers. J Oil Palm Res 20:533–541

Nuzul HD (2015) Agricultural practices strategies in oil palm plantations to adapt climate change. Proceedings of the international seminar on gearing oil palm breeding and agronomy for climate change, Kuala Lumpur, Malaysia, pp 9–16

Oil World (2016a) Fading growth in palm oil raises the dependence on seed oil 59(4):37–39 (29 Jan 2016)

Oil world (2016b) Soya oil temporarily gaining market share from palm oil 59(8):89–92 (26 Feb 2016)

Okwuagwu CO (1986) The genetic base of the NIFOR oil palm programme. Proceedings of international workshop on oil palm germplasm, Bangi, Selangor, Malaysia, pp 228–237

Ooi S, Da Silva E, Muller A, Nascimento J (1981) Oil palm genetic resources—native *E. oleifera* populations in Brazil offer promising sources. Pesq agropec bras 16(3):385–395

Pichel R (1956) L'Ameriolation du Palmier a huile du Congo Belge Conf. Franco-Brittanique sur le palmier a huile. Bull Agron Min Fr d'outre Mer 14:59–66

Pillai RSN, Blaak G, Paul Closen H (2000) Collection of oil palm (*Elaeis guineensis* Jacq.) germplasm from Africa. Int Jour of Oil Palm 1(1&2):23–37

Prasetyo JHH, Sitepu B, Iswandar HE, Djhujana J, Nelson SPC (2011) Performance of Sumbio semi-clonal progenies. Proceedings of the agriculture, biotechnology and sustainability conference PIPOC 2011, Kuala Lumpur, Malaysia, pp 212–216

Purba AR, Arif M, Yenni Y (2013) Current oil palm breeding programs and commercial planting materials. Proceedings of the International Seminar on oil palm breeding—yesterday, today and tomorrow, Kuala Lumpur, Malaysia, pp 30–42

Purba AR, Setiawati U, Susanto A, Rahmaningsih M, Yenni Y, Tahmadi HY, Nelson SPC (2012) Indonesia's experience of developing Ganoderma tolerant/resistant oil palm planting materials. Proceedings of the international seminar on breeding for oil palm disease resistance and field visits, Bogota, Colombia, 21–24 September 2012

Rachier GO, Wabule MN, Wasilwa L, Odenya JO, Orondo SG, Mambiri G, Wanjalla AW, Cheboi P, Makheti PT (2009) Introduction and promotion of cold tolerant oil palm (*Elaeis guineensis*) in Western Kenya. Proceedings of agriculture, biotechnology and sustainability conference PIPOC 2009, Kuala Lumpur, Malaysia, pp 1035–1044

Radhamani J, Chandel KPS (1992) Cryopreservation of embryonic axes of trifoliate orange (Poncirus trifoliate (L) Raf.). Plant Cell Rep 11:204–206

Rajanaidu N (1986a) The oil palm (*Elaeis guineensis*) collections in Africa. Proceedings of international workshop on oil palm germplasm and utilization, Bangi, Selangor, Malaysia, pp 59–83

Rajanaidu N (1986b) *Elaeis oleifera* collection in Central and South America. Proceedings of international workshop on oil palm and utilization, Bangi, Selangor, Malaysia, pp 84–94

Rajanaidu N (1995) PORIM oil palm gene bank. Palm Oil Tech Bull 1(4):7

Rajanaidu N, Ainul MM, Kushairi A, Mohd Din A (2013) Historical review of oil palm breeding for the past 50 years—Malaysian journey. Proceedings of the international seminar on oil palm breeding—yesterday, today and tomorrow, Kuala Lumpur, Malaysia, pp 11–28

Rajanaidu N, Jalani BS (1994) Prospects for breeding for kernels in oil palm (*Elaeis guineensis*). The Planter 70(820):307–318

Rajanaidu N, Kushairi A, Chan KW, Mohd Din A (2007) Current status of oil palm material in the world and future challenges. Proceedings of agriculture, biotechnology and sustainability conference, vol 1, Kuala Lumpur, Malaysia, pp 503–520

Rajanaidu N, Kushairi A, Mohd Din A, Norziha A, Ainul M (2012) Breeding for disease resistance in oil palm—current status. Proceedings of the international seminar on breeding for oil palm disease resistance and field visits, Bogota, Colombia, 21–24 September 2012

Rajanaidu N, Kushairi A, Rafii M, Mohd Din A, Maizura I, Jalani BS (2000) Oil palm breeding and genetic resources. In: Basiron Y, Jalani BS, Chan KW (eds), Advances in oil palm research. Malaysian Palm Oil Board, Kuala Lumpur, pp 171–237

Rajanaidu N, Rao V (1987) Oil palm genetic collections: their performance and use to the industry. Proceedings of international oil palm/palm oil conference in agriculture, Kuala Lumpur, Malaysia, pp 59–85

Rajinder S, Noorhariza MZ, Ting NC, Rozana R, Tan SG, Low ETL, Maizura I, Cheah SC (2008) Exploiting an oil palm EST database for the development of gene-derived SSR markers and their exploitation for assessment of genetic diversity. Biologia 63(2):227–235

Ravi D, Anand P (2012) Production and applications of artificial seeds: a review. Int Res J Biol Sci 1(5):74–78

Rao V (1996) Ripening in the virescens oil palm. Proceedings of the international conference on oil and kernel production in oil palm: a global perspective. Bangi, Selangor, p 226

Restrepo EF, Navia EA, Avila RA, Daza EE (2012) Use of epidemiological tools to differentiate the ability of various oil palm materials to respond to bud rot disease under field conditions. Proceedings of the international seminar on breeding for oil palm disease resistance and field visits, Bogota, Colombia, 21–24 September 2012

Rey L, Gómez PL, Ayala I, Delgado W, Rocha P (2004) Colecciones genéticas de palma de aceite *Elaeis guineensis* (Jacq.) y *Elaeis oleifera* (H.B.K.) de Cenipalma: Características de importancia en el sector palmicultor. PALMAS 25(2):39–48

Reynaud-Farrera I, Maley J, Wirrmann D (1996) Vegetation and climate in the forest of South-West Cameroon since 4770 years BP: pollen analysis of sediments from Lake Ossa. C. R. Acad. Sci. Ser. IIA. Sciences de le terre et des plantes 322:749

Roongsattham et al (2012) Temporal and spatial expression of galacturonase members reveals divergent regulation during fleshy fruit ripening and abscission in the monocot species oil palm. BMC Plant Biol 12:150

Sakai A (1965) Determining the degree of frost-hardiness in highly hardy plants. Nature 185:393–394

Sapey E, Adusei-Fosu K, Agyei-Dwarko D, Okyere-Baoteng G (2012) Collection of oil palm (*Elaeis guineensis* Jacq.) germplasm from the northern regions of Ghana. Asian J Agric Sci 4 (5):325–328

Schultes RE (1990) Taxonomic, nomenclatural and ethnobotanic notes on *Elaeis*. Elaeis 2(1):172–187

Seng TY, Siti Hawa MS, Junaidah J, Chin CW, Sharifah SRSA (2009) Developing a DNA database of Felda's critical breeding palms. Proceedings of the international seminar on oil palm genomics and its application to oil palm breeding, Kuala Lumpur, Malaysia

Sharma M (2000) Exploitation of (*Elaeis oleifera*) germplasm in improving the quality of palm oil. Proceedings of the international symposium on oil palm genetic resources and their utilization, Kuala Lumpur, Malaysia, pp 322–398

Sharma M, Beng AB, Harkingto, Sidhu M (2016) Performance of Asian Agri Group's 2nd generation DxP planting materials—moving to the next level of improved DxP. Proceedings of the international seminar on oil palm breeding and seed production and field visits, Kisaran, Indonesia, p 7

Smith NJH, Williams JT, Plucknett DL Talbot JP (1992) Rubber, oils and resins. In: Tropical forests and their crops. Cornell University Press, New York, pp 207–263

Soh AC (1999) Breeding plans and selection method in oil palm. Proceedings of the symposium on the science of oil palm breeding. Montpellier, France, pp 65–95

Soh AC, Walker S, Ong CK (2015) Breeding for sustainability. Book of abstracts PIPOC 2015 Kuala Lumpur, Malaysia, p 5

Sowunmi M (1999) The significance of the oil palm (*Elaeis guineensis* jacq.) in the late Holocene environments of west and central Africa: a further consideration. Veg Hist Archaeobotany 8:199–210

Sudharto P, Giono, Alam S, Taofik M, Salman M, Pujangga, Advento AD, Caliman MN (2016) Research and management of major oil palm pests: strategy for control of bagworm in oil palm plantation. Program book of the 6th IOPRI-MPOB international seminar on current research and management of pests, Ganoderma and pollination in oil palm for higher productivity, Medan, Indonesia, p 16

Taeprayoon P, Tanya P, Lee SH, Srinives P (2015) Genetic background of three commercial oil palm breeding in Thailand revealed by SSR markers. Aust J Crop Sci 9(4):281–288

Tandon R, Chaudhury R, Shivana KR (2007) Cryopreservation of oil palm pollen. Curr Sci 92(2), pp 182–183 (Ref. 15)

Te-chato S, Hilae A (2007) High-frequency plant regeneration through secondary somatic embryogenesis in oil palm (*Elaeis guineensis* Jacq. var. tenera). J Agric Technol 3:345–357

Thachuk C, Crossa J, Franco J, Dreisigacker S, Warburton M, Davenport GF (2009) Core hunter: an algorithm for sampling genetic resources based on multiple genetic measures. BMC Bioinform 10:243. doi:10.1186/1471-2105-10-243

Ting NC, Zaki NM, Rosli R, Low ETL, Ithnin M, Cheah SC, Tan SG, Singh R (2010) SSR mining in oil palm EST database: application in oil palm germplasm diversity studies. J Genet 89 (2):135–145

Turnbull N, Cazemajor M, Guerin C, Louise C, Amblard P, Cochard B, Durand-Gasselin, T (2016) Oil palm breeding and seed production in PalmElit. Proceedings of the international seminar on oil palm breeding and seed production and field visits, Kisaran, Indonesia, pp 8–20

Turnbull N, de Franqueville H, Breton F, Jeyen S, Syahoutra I, Cochard B, Poulin V, Durand-Gassellin T (2014) Breeding methodology to select oil palm planting material partially resistant to *Ganoderma boninense*. Paper presented at the 5th quadrennial international oil palm conference, Bali Nusa Dua Convention Center, Indonesia, 17–19 June 2014

Vanderweyen R (1952) La prospection des palmeraies congolaises et ses premier results. Bull Inf INEAC 1:357–382

Veriappan A, Rajanaidu N, Johari O, Ahmad SF (2008) Performance of MPOB population 12 in SPAD. Proceedings of the 3rd seminar on performance of MPOB PS1 and PS2 materials and elite germplasm, Bangi, Selangor, Malaysia, pp 91–116

Wening S, Wilkinson MJ, Filianti H, Djuhjana J, Forster BP, Nelson SPC, Caligari PDS (2013) Genetic diversity of commercial oil palm seed production parents. Oil Palm Bulletin 67:1–4

Wong WC, Jit TC, Kien WC, Mayes S, Singh R, Soh AC (2015) Development of an effective SSR-based fingerprint system for commercial planting materials and breeding applications in oil palm. J Oil Palm Res 27(2):113–127

Wong CK, Choo CN, Liew YR, Ng WJ, Krishnan K, and Tan CC (2010) Benchmarking best performing semi-clonal seeds against tenera clones: avenue for superior clonal ortet identification. Proceeding of the advances in oil palm tissue culture, Yogyakarta, Indonesia, 29 May 2010

Xianhai Z, Genglang P, Zhao L, Junming C, Weifu L (2016) Evaluation of cold tolerant high yielding oil palm germplasm in Guangdong province of South China, a northern tropical region. J Oil Palm Res 28(3):266–280

Zeven A (1964) On the origin of the oil palm (*Elaeis guineensis* Jacq.). Grana Palynologica, pp 121–123

Zhou LX, Xiao Y, Xia W, Yang YD (2015) Analysis of genetic diversity and population structure of oil palm (*Elaeis guineensis*) from China and Malaysia based on species-specific simple sequence repeat markers. Genet Mol Res 14(4):16247–16254

Zulhermana, Puspitanigrum Y, Sirait MR, Raganata AP, Asmady, Asmono D (2011) Current progress on Sampoerna Agro DxP semi-clones. Proceedings of the agriculture, biotechnology and sustainability conference PIPOC 2011, Kuala Lumpur, Malaysia, pp 266–270

Zulkifli Y, Maizura I, Rajinder S (2012) Evaluation of MPOB oil palm (*Elaeis guineensis* Jacq.) germplasm populations using EST-SSR. J Oil Palm Res 24:1368–1377

Zuraini S, Jayne JJ, Hoong HW (2013) Exploitation of sabah breeding programme in enhancing the performance of Sawit Kinabalu DxP planting materials. Proceedings of the international seminar on oil palm breeding—yesterday, today and tomorrow, Kuala Lumpur, Malaysia, pp 40–63

Chapter 10
Conservation of Woody Plant Diversity in Kenya

Kimberly E. Medley, John K. Maingi and Christine M. Mutiti

Abstract Environmental conditions along humid to arid gradients in Kenya contribute to woody plant diversity patterns and complicate their conservation as an important resource for human livelihoods. Applied conservation research links floristic and ethnobotanical inventories of woody plant species, measures landscape ecological complexity and the value of woody vegetation types as a cultural resource, and effectively integrates research with application in the context of global environmental policies. For Kenya, and through a case study of an Afromontane/bushland corridor between Tsavo East and West national parks, the review highlights the importance of "water tower" forests but also recognizes a diversity of other woody vegetation types that show unique floristic affinities, high measures of cultural plant resources derived from significant local knowledge, and complex ecological patterns across landscapes in relation to physical–environmental and human resource relationships. Much woody vegetation occurs outside the area measured as forest or woodland cover by international and national authorities, substantiating a need to spatially broaden policies for woody plant conservation and better engage local stakeholders. Global to national policies and practices for carbon accrual, especially with financial inflows from the UN-REDD + program, strengthen the need to conserve woody plants in Kenya and promote adaptive co-management practices that contribute to the country's sustainable development.

Keywords Adaptive co-management · Biogeography · Ecological diversity Ethnobotany · Landscape analyses · REDD+ · Sustainable development

K.E. Medley (✉) · J.K. Maingi
Department of Geography, Miami University, Oxford, OH 45056, USA
e-mail: medleyke@miamioh.edu

J.K. Maingi
e-mail: maingijk@miamioh.edu

C.M. Mutiti
Department of Biological and Environmental Sciences, Georgia College, Milledgeville, GA 31061, USA
e-mail: christine.mutiti@gcsu.edu

M.R. Ahuja and S.M. Jain (eds.), *Biodiversity and Conservation of Woody Plants*, Sustainable Development and Biodiversity 17, DOI 10.1007/978-3-319-66426-2_10

10.1 Introduction

Kenya represents an interesting case example to examine the conservation of woody plant diversity. Straddling the equator in Tropical East Africa, much of the country represents an "impressive climatic anomaly" (Trewartha 1981, p. 134). Rather than falling under the influence of the equatorial low and the consistent convective showers that should maintain tropical wet conditions, over 80% of the country is classified as arid to semiarid with rainfall much below 1000 mm and distinct dry seasons that dominate from 5 to 8 months of each year. Indigenous high stature "humid" forests are estimated at just over 2% of the land area and conserve about 50% of Kenya's woody plant diversity (Wass 1995). These forests are restricted to a very narrow belt along the Indian Ocean coast, isolated mountains that can capture and destabilize the moisture that flows westward as monsoonal winds off the Indian Ocean, and an isolated extension of the central African rainforest to lowlands bordering Lake Victoria. Humid to arid latitudinal, longitudinal, and elevational gradients across the country support the occurrences of woody plants in evergreen to deciduous forest, woodland, bushland, thorn scrub, and wooded grassland settings (Lind and Morrison 1974; Beentje 1994). Environmental heterogeneity over short distances in Kenya contributes significantly to geographic patterns of diversity among woody plants and complicates their conservation.

Under global conservation priority plans, Kenya captures parts of three "biodiversity hotspots" out of eight mapped for the African continent (Mittermeier et al. 2011) and six "global 200 ecoregions" (Olson and Dinerstein 2002). Following a "politics of efficiency," these classification schemes recognize regions of rich species densities, endemic species occurrences, and high human threat (Youatt 2015, p. 31). Their overlap in Kenya highlights the fragmented occurrences of different vegetation types and an equally important reliance on these ecosystems by human populations. Conservation as sustainable development in the country needs to be sensitive to the indigenous rights and knowledge of local populations (Laird 2002; Adams and Hutton 2007), better understand how woody plants contribute as an important entitlement to livelihood conditions (Cunningham 2001) and recognize "[t]he poor conservation outcomes that followed decades of intrusive resource management strategies" (Agrawal and Gibson 2001, p. 1). Narratives and counter-narratives that debate fortress and community-based management provide context for the evaluation of conservation strategies for woody plants in Kenya (Western et al. 1994; Adams and Hutton 2007). They also elevate the importance of conservation research that includes local, culturally sensitive interpretations of woody plant diversity, and its value as a livelihood resource (see Medley and Kalibo 2007 for "Global Localism;" Nabhan 2016 on "Ethnobiology for the Future"). The conservation agenda would be to move from the view of humans as a threat to their role as partners in sustainable development strategies (Brown 2002) and aim to achieve a better focus on "long-term communal interests" (Bell 1987, p. 79).

We structure our review to show how physical–environmental conditions and human resource relationships define diversity patterns for woody plants in Kenya and guide a critique of opportunities and constraints for conservation. First, the chapter reviews how woody plant botanical and ethnobotanical inventories establish *species* distribution patterns across the country, showing Kenya's unique floristic biogeography and highlighting important plant resources for human livelihoods especially across the more expansive arid and semiarid lands. Second, an analysis of ecological research focused on the composition, and structure of woody plant *vegetation types* across complex landscapes jointly considers the role of physical–environmental conditions and human resource practices as influences on diversity patterns. For example, we collaboratively review research findings for an Afromontane/bushland setting in southeast Kenya. The chapter concludes with an analysis of how global initiatives for sustainable development can promote more *adaptive* policies, planning, and practice that engage countries, like Kenya, in the conservation of woody plants.

10.2 Floristic Biogeography and Cultural Plant Resources

The *Flora of Tropical East Africa* (FTEA 1948–2012) for the countries of Kenya, Uganda, and Tanzania represents globally one of the largest regional–tropical floras, providing records for 12,500 plant species, 12,000 collection localities (Polhill 1988) and 2700 plant collectors in the region since 1793 (Polhill and Polhill 2015). The work globally positioned botanical research in Kenya at the East African Herbarium (EA), National Museums of Kenya (NMK), which now holds 700,000 plant specimens and significant expertise to support field inventories. Beentje's (1994) *Kenya trees, shrubs, and lianas* completed at EA and published by NMK provides the most definitive account of woody plants in the country, building off the 1000 trees and shrubs reported by Dale and Greenway (1961) to include descriptions and distribution records for "some 1800 species" of woody plants (p. 3), and accordingly a foundation for the record of new woody plant occurrences. Moreover, the *Journal of East African Natural History* published by the East Africa Natural History Society and NMK provide a "natural home and repository for biodiversity information of the East African region" (Bytebier 2005, p. 1) that greatly supports any floristic review. Following the guidelines under the convention on biological diversity signed by 150 countries at the 1992 Rio Earth Summit, the United Nations Environmental Program (UNEP) helped to establish the Center for Biodiversity at NMK to serve as the coordinator of biodiversity research, encouraging effective integration between the recognition of biological and cultural plant resources.

For woody plants in Kenya, highest concentrations of species are reported for high stature indigenous forest (Wass 1995). Between 1991 and 1994, the Kenya Indigenous Forest Conservation Program (KIFCON) compiled comparative biodiversity records for trees and shrubs from sample plots in western rainforest,

Afromontane, and coastal forests. Their sites, following the vegetation map prepared by White (1983), show distinct taxa from Guineo-Congolian, Afromontane, and Zanzibar-Inhambane floristic regions, but each forest differs uniquely in its biogeographical mix. Luke (2005) compiled an annotated checklist for Shimba Hills in southeast Kenya that includes 21% of Kenya's total and 44% of the coastal flora (1396 species), of which 67% are documented as woody plants (152 tree species), and none shows affinities to the Guineo-Congolian or Afromontane region. Fischer et al. (2010), working under Biodiversity Monitoring Transect Analysis in Africa (BIOTA) framework (www.biota-africa.org), report an annotated checklist of 986 species (15.2% of Kenya's flora) for Kakamega Forest in western Kenya that is "floristically strongly related to the Central and West African lowland rainforest, but is also further enriched by Afromontane forest elements from the Rift Valley escarpment" (p. 133). For Kenya's Eastern Arc Mountains, Medley and Maingi (2014) report nearly equal proportions of tree species with Afromontane and coastal affinities at Mt. Kasigau (c.f. Rodgers and Homewood 1982 for the East Usambaras in Tanzania) in contrast to the absence of trees with coastal affinities reported for the Taita Hills, Kenya located 50 km to the west (Aerts et al. 2013; Mbuthia 2003; c.f. Faden et al. 1988). Botanical surveys in other isolated "hill" forests also document unique species occurrences (e.g., Bagine 1998 for Ramogi Hill; Bytebier and Bussmann 2000 for Mt. Nyiru; Githae et al. 2008 for Mt. Marsabit), further substantiating a complex biogeography that contributes to regional and local patterns of species endemism across the country.

Less is known about diversity patterns across the more broadly distributed "other" woody vegetation types located mostly outside of protected areas and under the direct influence of local human resource relations. Collaborative mixed method approaches that link "local" vernacular naming with "scientific" plant names show significant contributions (c.f. Gachathi 2007; Kokwaro and Johns 2013). For example, "A checklist of indigenous trees and shrubs at Bura, Tana River District, Kenya, with Malakote, Orma, and Somali names" includes a record for 228 woody plant species in semiarid eastern Kenya (Gachathi et al. 1994), and the "Vernacular names and uses of plants in northern Kenya" documents names and uses for 92% of 888 species from the Pokot, Turkana, and Marakwet (Timberlake 1994). Working under the 1992–2004 people and plants initiative (see http://peopleandplants.org/), Maunder et al. (2001) conducted an extensive ethnobotanical survey of Loita Maasai that explores the names and uses of over 250 plant species from the Loita Hills, gains measures of plant values to resident populations, and applies these qualitative and quantitative data toward community management of natural resources in the region. Ethnobotanical surveys help to locally guide conservation strategies particularly through their identification of "cultural keystone" plants (Garibaldi and Turner 2004; see Kokwaro 1995 on "Ethnobotany in Africa") and places (Cuerrier et al. 2015; e.g. Sheridan and Nyamweru 2008 on *African Sacred Groves*), and the inherent value that plants have to human livelihoods (e.g., Maunder et al. 1999 on *Traditional food plants of Kenya* compiled under the indigenous food plants program, 1989–1992). Recognizing the role of indigenous knowledge to biodiversity conservation, the Kenya Resource Center for Indigenous

Knowledge (KENRIK) was established in 1995 under the Center for Biodiversity at NMK "to document and preserve the endangered/threatened indigenous knowledge held by different communities in Kenya, which has traditionally served an important role in environmental conservation, natural resource management, food security, and traditional healthcare systems" (http://www.museums.or.ke/content/view/117/83/1/1/).

10.3 Landscape Ecological Complexity among Woody Vegetation Types

Beentje's (1994) *Kenya trees, shrubs, and lianas* provide distribution records for woody plant species by 55 km^2 (0.5°) grid cells that can be used to map out spatial patterns of diversity for the country (Kissling et al. 2008). A modeling approach proposed by Field et al. (2005), and the physical–environmental hypotheses tested by Kissling et al. (2008) show the importance of environmental, particularly topographic, heterogeneity for predicting spatial patterns of species richness (γ-diversity, regional) across the country. Prinzing et al. (2003), employing a "higher taxon approach," found that plant family richness may be used to predict species richness for a grid cell (α-diversity, habitat) but higher taxon predictions of species turnover by grid (β-diversity, landscape) were low for the country and even within defined biogeographic regions. These studies support hierarchical analyses of species diversity (c.f. Whittaker et al. 2001 for regional, landscape, and habitat distinctions) and substantiate ecological research that compares diversity patterns among woody vegetation types across heterogeneous landscapes. For example, ecological studies in Kenya show the important role of floodplain hydrology in determining riverine forest communities and landscape heterogeneity along the Tana (Medley and Hughes 1996; Maingi and Marsh 2006) and Turkwel (Stave et al. 2003) rivers, the role of landscape and climatic factors as they influence woody vegetation across the drylands of Turkana district (Coughenour and Ellis 1993), and the role of elevation and aspect in a syntaxonomic classification of forest types on Mt. Kenya (Bussmann and Beck 1995).

10.3.1 Ecological Complexity in the Kasigau Corridor

We review landscape ecological research conducted since 2002 for Mt. Kasigau and surrounding semiarid *Acacia–Commiphora* bushland that exemplifies how complex physical–environmental conditions and human resource activities contribute to heterogeneity among woody plant vegetation types and their value as a resource for human livelihoods. Biogeographically, tropical seasonal forests–woodlands should comprise a major vegetation type (biome) across seasonally wet–dry lands in Kenya (Lind and Morrison 1974). Woody plants are dominant (>50%) across the bushland landscape (Lind and Morrison 1974), but the ecological transitions among vegetation types and their seasonal dynamics are complex in response to total precipitation, the length of the dry season(s) and the amount of rain received during the wet season(s) (Walter 1971). On isolated hills, elevational gradients that capture structural and phenological differences are compressed over short distances in response to temperature changes and consequent moisture availability (Lomolino 2001).

Located in southeastern Kenya, Mt. Kasigau is the most north-eastern mountain in the ancient crystalline Eastern Arc Mountains, a global 200 ecoregion (Olson and Dinerstein 2002) that is also part of the Eastern Afromontane biodiversity hotspot (Mittermeier et al. 2011). The Eastern Arc Mountains are among the oldest mountains on the African continent and noted for high species densities and a high concentration of endemic species in isolated montane forest fragments (Küper et al. 2004; Newmark 2002). The study area includes montane forests along a steep elevational gradient on Mt. Kasigau to its summit at 1640 m, five village settlements and farms at the mountain's base, and lowland bushland thicket vegetation in the Kasigau Trust lands, Rukinga Wildlife Sanctuary, privately owned and managed by Wildlife Works (http://www.wildlifeworks.com/), and Kasigau ranch, a cooperative (by payment) grazing area (Fig. 10.1). Lowland bushland thicket vegetation forms a wildlife corridor between Tsavo East and Tsavo West national parks as part of the northern *Acacia–Commiphora* ecoregion (Burgess et al. 2004).

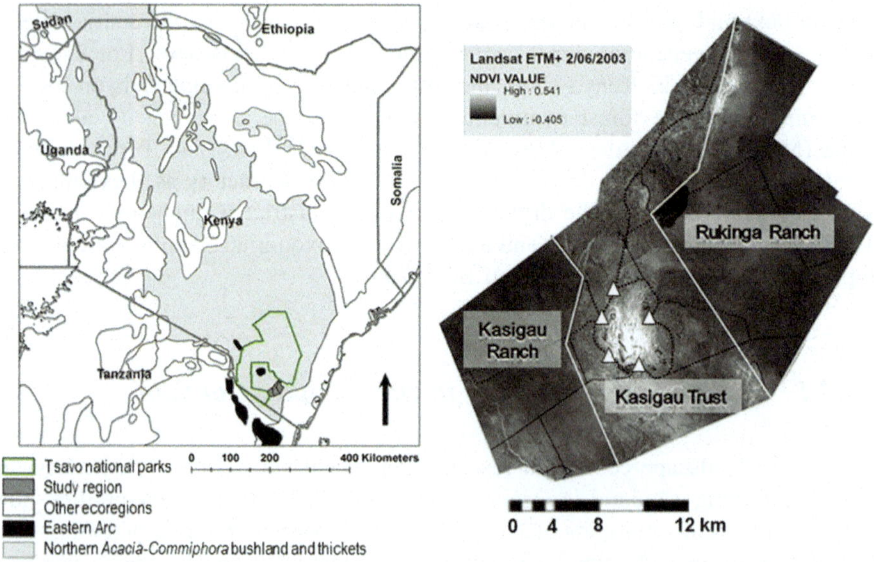

Fig. 10.1 Kasigau corridor between Tsavo East and West national parks in the Northern *Acacia–Commiphora* and Eastern Arc ecoregions, showing a study region that includes the five villages around Mt. Kasigau in the trust lands, Rukinga ranch, and Kasigau ranch

A nearby weather station at Voi, Kenya reports an annual average of 549 mm of precipitation, which ranged from 200 to 800 mm between 2002 and 2015.

Field research at Mt. Kasigau investigated landscape patterns of diversity in woody plants by integrating a floristic and ethnobotanical survey of all woody plants (Medley et al. 2007) with ecological measures on woody plant composition and structure as they varied among 55 plot samples for trees >10 cm diameter at breast height (dbh; 0.1 ha plots) and all woody plants >1 m height and <10 cm dbh (0.02 ha plots) (Medley et al. 2010; Medley and Maingi 2014), participatory maps (Kalibo and Medley 2007), and the collection of landscape ethnoecological narratives about places (Medley and Kalibo 2007). Working with Kasigau Taita local experts and residents, we employed cooperative exercises and collaborative learning to share information about woody plants and interpret landscape changes along an elevational gradient that extended from the lowest farm and bushland sites (~480 m) to cloud forest at the mountain's summit (1640 m; Medley and Kalibo 2005; Medley et al. 2010). Medley et al. (2007) report 338 woody plants in 74 families and 208 genera provide Kasigau Taita names for 74% (252) of those plants and describe 758 material uses and over 200 ecosystem services (see also Medley et al. 2013). Most woody plant uses were reported for the bushland plots, and the greatest number of woody plant species with uses occurs in lower- to mid-montane plots, but the research identifies locations, all below evergreen forest, that are particularly species rich and/or valued for certain resources (Medley et al. 2010). Mt. Kasigau is unique among Afromontane locations because of its near contiguous forest cover on the mountain, an absence of plantation tree crops (c.f. Omoro et al. 2010 for Taita Hills), and cumulatively high diversity of plant species and cultural plant resources over such a small area and elevational range. Local residents helped to map 139 named places of historical–cultural significance from the five villages that added to our collaborative interpretations of diversity patterns.

From a record of 140 trees species >10 cm dbh, 56% occurred in one or two plots, suggesting much heterogeneity at the landscape scale (Medley and Maingi 2014). Cluster and indicator species analyses identified woody vegetation types characterized by the relative importances of these trees (Medley and Maingi 2014). Elevation partly explained differences among the community types, where we measured steep lapse rates (annual = 8.1 °C/1000 m in 2011–2012 and 8.6 °C/1000 m in 2012–2013), among the highest in the documented literature, and corresponding increases in relative humidity in just over 1000 m of elevation gain to the summit (Fig. 10.2; see Henkin et al. 2015 for 2011–2012 data). Cloud forest shows a narrow and biophysically discrete occurrence near the summit of this relatively low peak in the Eastern Arc Mountains, and the elevational change from deciduous to evergreen vegetation types is distinctive on the mountain. Among the montane woodland vegetation types, however, we found continua of change in relation to local topography and a complex history of human activities. Lower montane and semi-evergreen woodland plots showed high β-diversity. Highest γ-diversity, cumulative species richness for all plots, was calculated for the lower montane woodlands (Medley and Maingi 2014).

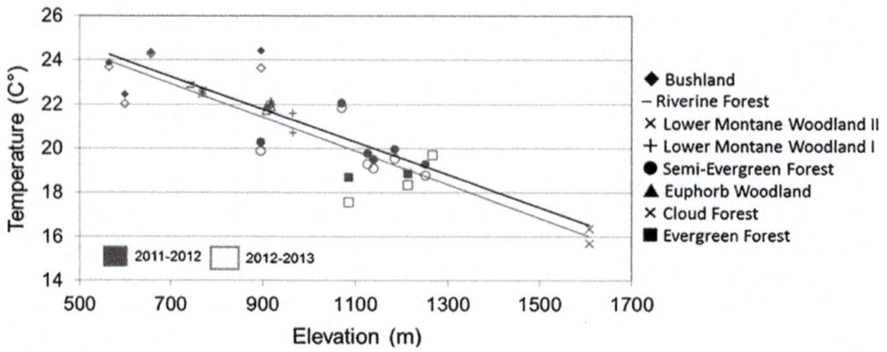

Fig. 10.2 Elevational gradients in temperature measured from HoboPro v2 data loggers placed in the different montane vegetation types

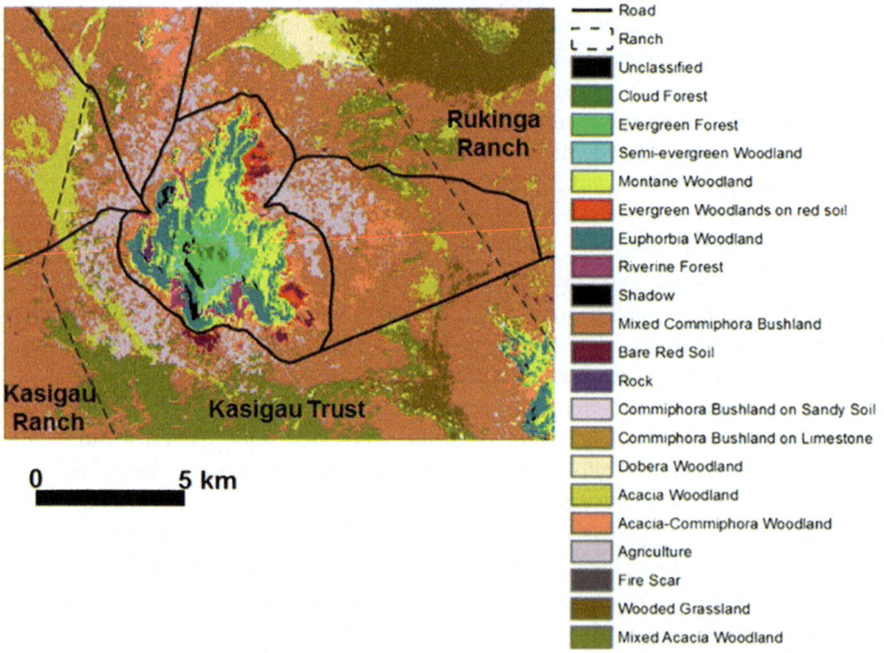

Fig. 10.3 Land cover map for Mt. Kasigau, showing the heterogeneity of vegetation types on the mountain and in the surrounding bushland

At Mt. Kasigau, we applied ecological data toward the creation of a fine-resolution land cover map based on the classification of a layer stack of wet (12/31/12) and dry (10/07/10) Landsat images (Fig. 10.3). Landsat TM/ETM + spectral bands were radiometrically corrected to top-of-atmosphere reflectances (Moran et al. 1992) and also for atmospheric effects, using the dark object subtraction method (Chavez 1988). In order to reduce topographic shadowing found in the rugged Mt. Kasigau landscape, we applied the c-factor topographic normalization technique (Teillet et al. 1982). The classification procedure also used several image enhancements that included tasselled cap brightness, greenness, and wetness images generated by applying Kauth–Thomas transform techniques (Kauth and Thomas 1976). Ancillary data on elevation from an ASTER-derived DEM and slope, aspect, and topographic position computed from the DEM were included in the layer stack for classification. Land cover classification relied on Garmin GPS training data obtained from the field in June–July 2012, December 2012, and June–July 2013. Additional training data were obtained from a high spatial resolution WorldView-2 image acquired in July 2010. A supervised classification procedure, employing a neural network classifier with adaptive boosting, was found to be the best classifier. Field- and image-based data, not used in training the classifier, were used for an assessment of the land cover map, producing an overall accuracy of 92.7% and a kappa coefficient of 0.919.

The land cover classes matched the forest community types empirically derived from the measured ecological plots except that the Lower montane woodland II and riverine forest classes were merged into a single class (Fig. 10.2), representing mostly concave intermittent to permanent stream drainages on the mountain (Fig. 10.3). Moreover, spectral signatures identified in the land cover classification identified heterogeneity in the "bushland" lowlands, where further ecological samples confirmed differences in the relative importances of trees that occur on black cotton, limestone, and lateritic soils. Also "Evergreen Woodland on red soils" at the base of the mountain, locally called *magagarenyi,* occurs on heavily weathered and eroded basement rock of volcanic origin where we found open occurrences of stunted trees and shrubs. These findings further substantiate a diversity of woody plant vegetation types in this montane setting and our capacity to effectively integrate ecological sampling, field survey, and remote sensing technology for fine-scale mapping of diversity patterns.

Working within a corridor between Tsavo East and Tsavo West national parks, Mutiti (2010) employed remote sensing and landscape ecological techniques in GIScience to explore spatial patterns of heterogeneity among bushland vegetation types at a more regional scale. Photos and vegetation descriptions compiled during a field survey were used to guide an unsupervised classification of a 30 m resolution Landsat ETM + image acquired for February 6, 2003 (see Fig. 10.1). The acquisition date fell within the dry season, which made it possible to distinguish between areas dominated by deciduous species such as *Commiphora* species and those that mostly remain evergreen, including some *Acacia* species. Landsat ETM + data were converted to reflectance values for bands 1–5 and 7 (Moran et al. 1992) and then put in a layer stack with a mapped soil and atmospherically resistant vegetation index (SARVI-2, Huete et al. 1997). Distinctive reflectance signatures were used to classify and map eight vegetation types (Fig. 10.4): (1) ***Commiphora*** stands (29% of the land area) where *Commiphora* species were dominant often with

Lannea and *Boswellia* species, but with no *Acacia*; (2) ***Acacia*** stands (18%) where *Acacia* species occurred alone or were notably dominant with *Commiphora, Cordia* and *Boscia* species; (3) **Mixed** stands (10%) characterized by both *Acacia* and *Commiphora* species; (4) **Open bushland** (9%) stands with open occurrences of *Acacia* and *Commiphora* species; (5) **Thickets** (7%) with a dense cover of deciduous (*Terminalia* species) or evergreen (*Manilkara mochisia, Diospyros consolatae*) trees at the base of the mountain or in lowland clusters; (6) **Forests** (3%) mapped on Mt. Kasigau; (7) **Grasslands** where grasses dominated with a low occurrence of trees; and (8) **Shrublands** with *Premna, Bauhinia,* and *Sericocomopsis* species that are either erect or spreading with occasional trees such as *Delonix elata, Melia volkensii,* and *Acacia* species. Most of the lowlands are mapped as closed canopy "bushland" but landscape statistics confirm a lack of dominance by any particular vegetation type and high patchiness among bushland types (47 patches/km^2). Different spatial patterns of diversity among the three land units relate partly to the underlying lithology and consequent soil conditions, but also suggest consequent effects from different land utilization practices. For example, agricultural lands included on the map as "other" predominate from the study villages at the base of Mt. Kasigau and along the main road to the northwest, and local utilization practices on the Kasigau Trust lands may partly explain the greater abundances of *Acacia*-dominated stands near their farms (Fig. 10.4). In Tsavo East national park, deliberate actions to increase wildlife populations through prescribed burning, and the construction of temporary catchment basins expanded open grasslands (Belsky 1989) in contrast to its much lower occurrence in the wildlife sanctuary recently designated for Rukinga ranch (Mutiti 2010). Landscape ecological research focused at different scales of analyses in "bushland" shows much potential to better understand spatial patterns of diversity in this valuable resource for woody plant conservation in Kenya.

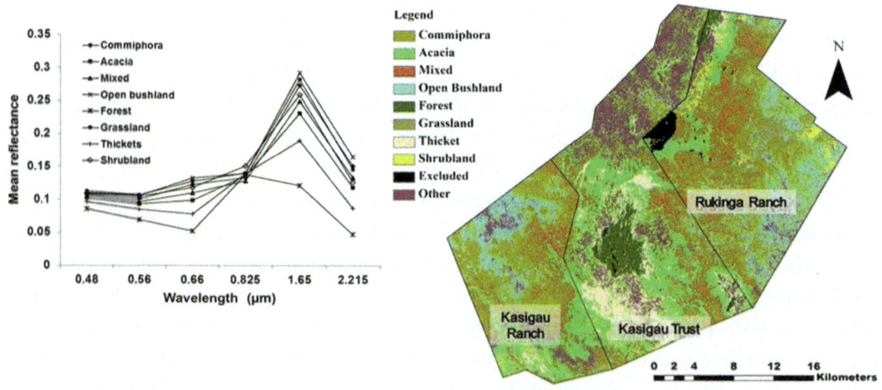

Fig. 10.4 Spectral signatures and classified vegetation map for Rukinga Ranch, Kasigau Trust lands, and Kasigau Ranch

10.4 Adaptive Policies, Planning, and Practice for Woody Plant Conservation

The UN conferences on the environment and development, especially since the 1992 Earth Summit (UNCED), influence policy and planning in tropical countries like Kenya as they move from "legitimacy" to "effectiveness concerns" in the North–South debates for sustainable development (Najam 2005). Under UNCED's Agenda 21, Kenya's National Environmental Action Plan for 2009–2013 (National Environmental Management Authority 2009) shows country support as a signatory under the "global" Convention for Biological Diversity (CBD) and the UN Framework Convention for Climate Change (UNFCCC), but also elaborates on how the plan supports Kenya's Vision 2030 (http://www.vision2030.go.ke/) to be a "globally competitive and prosperous nation." The UN FAO International Panel on Forests (1995–1997) guided the preparation of Kenya's Ministry of Environment and Natural Resources (MENR 2016) National Forest Program (NFP) as a collaborative document that includes representation from the Kenya Forest Service (KFS), Kenya Forest Research Institute (KEFRI), and other public–private stakeholders. The NFP frames an inter-sectoral "tree" of support for forest conservation and management that is rooted in national policy and responsive to the UN 2016–2030 Sustainable Development Goals (SDG; https://sustainabledevelopment.un.org/).

First, we highlight a shift in global forest monitoring from measures of deforestation to a focus on forest growth achieved through protection, conservation, and restoration (c.f., SDG 15.1 and 15.2; Chazdon 2008). A strategic goal under Kenya's NFP (MENR 2016) and for the Kenya Forest Service (KFS) is to increase forest cover in the country to 10%. Forest cover, however, relates to how forests are measured as a woody vegetation type. Earlier estimates of forest cover from Wass (1995) at 2.1% measure the distribution of indigenous closed canopy high stature (>10 m) stands of trees. Kenya's NFP adopts a broader interpretation from the FAO Forest Resource Assessments (FRA 2000) to include a:

> *"land area of more than 0.5 ha, crown cover of 10%, trees of at least 2.5 m height, which is not primarily under agricultural or other specific non-forest land use. In line with the Constitution, the NFP also operates with 'tree cover,' e.g. trees on farms."* (MENR 2016, p. xi).

The Kenya Forest Service (KFS) presents a narrower definition (i.e., 15% cover with trees at least 5 m height; see "history of forestry in Kenya" at http://www.kenyaforestservice.org/) but the amount of land reported by MENR (2016), and KFS is the same at 6.99%. The protection of high-canopy forest vegetation on the "water towers" that includes Mt Kenya, Mt Elgon, Aberdare Ranges, Mau Complex, and Cherangani Hills remains a top national priority (MENR 2016; Pearce 2015) but "forest" monitoring by KFS also includes tree growth on farms and land areas covered by seasonal woodlands along rivers and drier hills (see Gachathi 1996; Wass 1995).

At our study location in the corridor between Tsavo East and West national parks, gazetted evergreen forest above 1000 m and all montane woodland types on Mt. Kasigau would be included as "forest" in this assessment, but the diversity of plant communities that comprise the lowland bushland are not included under the National Forest Program (MENR 2016). Interestingly, the 2010 FAO Forest Resource Assessment for Kenya reports that 6% of the land is in forest, comparable to Kenya's report, but that 50% of the country's lands are in "other wood land." The estimation of forest and/or other woody vegetation types is certainly complicated by the country's heterogeneous landscape. For woody plants, all land types show unique and important contributions to floristic–ecological diversity protection and for the effective use and management of culturally valued plant resources.

Second, Kenya's Draft National Forest Policy, 2015 (ROK 2015, p. 5) supports the UN's definition of sustainable forest management that values "equitable sharing of accrued benefits for the present and future generations." Kenya's Forest Act 2005, which established the Kenya Forest Service as a corporate (parastatal) agency under the Ministry of the Environment and Natural Resources, also provided a mechanism to decentralize forest management toward participatory "bottom-up" decision making under community forest associations. The forestry sector's contribution to the national economy is largely undocumented, supporting a need for more "local" insight on how resources are acquired to meet subsistence needs or traded in an informal economy. In practice, participatory forest management activities at coastal Arabuke Sokoke (Matiku et al. 2013), north-central Ngare Ndare (Chomba et al. 2015); central Mau (Mutune and Lund 2016) and western Kakamega (Guthiga et al. 2015) forests question intended support for local livelihoods, reduced vulnerability, and poverty alleviation.

Community forest associations, estimated at 325 by Pearce (2015), gain responsibility for "diverse management activities" but not the "decisions-making, revenue streams, and overall resource control rights" that guide those activities (Mogoi et al. 2012, p. 182). Problems arise from the "elite capture" of benefits that weaken the devolution process and contribute to the marginalization of certain groups (Chomba et al. 2015; Guthiga et al. 2015). Moreover, research in political ecology substantiates a need to deepen inquiry on the "chains of explanation" and "webs of relation" as they are influenced by extra-local forces on local adaptive practices (Rocheleau 2008). In order to avoid critiques of "participation as tyranny" (Cooke and Kothari 2001), all stakeholders need to be engaged early in the process of problem recognition and maintain better control over management outcomes. We recommend methodologies that promote participation as a research approach that is jointly focused on the process and outcome of new learning (e.g., Nemarundwe and Richards 2002), better validates local "community" knowledge (Medley and Kalibo 2007) and effectively applies that knowledge for the adaptive comanagement of resources (Armitage et al. 2009).

Third, we emphasize global policy and practices as they financially value woody plants as an ecosystem/environmental service. Payment for ecosystem services (PES), while under some critique as a "win–win" option (Wunder 2013), does present opportunities in Kenya to meet livelihood needs for income in support of broader conservation goals (Kairo 2008). Namirembe et al. (2014), in a comparative review of 50 African projects, distinguish three types of tree-based services as commoditization for products, compensation for non-use, and co-investment through joint management. In Kenya, the World Bank Global Environmental Facility funded a commodity-focused "Integrated Ecosystem Management Project" in 2005 for watershed protection in the Lake Victoria basin and in 2006 provided BioCarbon funds to the Green Belt Movement (2005), mobilizing community groups to *co-invest* in closing 2000 ha of degraded land with indigenous trees in the Afromontane Aberdare forests (Namirembe et al. 2014; Jindal et al. 2008).

Conference of Parties (COP) agreements under the UN Framework Convention for Climate Change (UNFCCC) formulated and approved UN-REDD+ as a market-based mechanism for Reducing Emissions from Deforestation and forest Degradation, plus forest conservation, sustainable management of forests and enhancement of forest carbon stocks. REDD+ offers "performance based payments," which can challenge the "international *willingness to pay* and national *willingness to play*" (Angelson 2009, pp. xii, xvii). Jindal et al. (2008, p. 116) identifies East Africa as a "preferred destination for carbon investors" and highlights potential financial inflows for the region. In response, Kenya Forest Service's (KFS 2010) "REDD Readiness Preparation Proposal" emphasizes carbon gains by increasing forest land area (up to 10%), emphasizes the protection/restoration of forest cover on the water towers, and shows interest in promoting and better monitoring tree growth promoted through agroforestry practices on farms (c.f. Minang et al. 2014).

In 2006, the private-sector Kasigau Corridor REDD+ project under Wildlife Works Carbon began working in the same semiarid corridor between Tsavo East and West that we investigated for its diversity of woody plant resources, heterogeneous bushland landscape, and significant human-resource relations as a "first project ever to issue carbon credits under an internationally accepted carbon standard" (Veronesi et al. 2015). Recent assessments of the Kasigau Corridor [demonstration] project hypothesize inequities that might arise when land tenure arrangements are unclear or where structural hierarchies promote marginalization and the 'elite capture' of project benefits (Chomba et al. 2016). Moreover, top-down PES (payment for ecosystem services) or integrated conservation and development (ICDP) activities may lose local participation with consequent landscape effects (Atela et al. 2015). A focus on the positive, however, would certainly show how financial REDD+ inflows, when continued over a long time period, *could* and *should* have the potential to alleviate poverty and promote the conservation of woody plants in the country. The conservation challenge would be to apply lessons learned toward adaptive co-management practices that contribute to sustainable development.

Acknowledgements Field research at Mt. Kasigau was supported by grants from the National Geographic Society (2003–2004), the Critical Ecosystem Partnership Fund (Conservation International, 2005–2007), and the National Science Foundation (NSF GSS 1061407, 2011–2015). We recognize affiliation with the East African Herbarium at the National Museums of Kenya and the Department of Geography and Environmental Studies at the University of Nairobi, under most recent approval by the Kenya Ministry of Education, Science, and Technology (NCST/RR1/12/1/ES-011/11). Ph.D. dissertation research by co-author C. Mutiti in the Kasigau corridor received financial support from the Department of Botany, Turrell Herbarium (MU) and the Graduate School at Miami University, and the International Foundation for Science (IFS). We especially thank the Kasigau Taita residents and staff at the former Taita Discovery Center on Rukinga Ranch for their expert knowledge and logistical support.

References

Adams WM, Hutton J (2007) People, parks and poverty: political ecology and biodiversity conservation. Conservation Soc 5:147–183

Aerts R, Thijs KW, Lehouck V, Beentje H, Bytebier B, Matthysen E, Gulinck H, Lens L, Muys B (2013) Woody plant communities of isolated Afromontane cloud forests in Taita Hills, Kenya. Plant Ecol 212:639–649

Agrawal A, Gibson CC (2001) The role of community in natural resource conservation. In: Agarwal A and Gibson C (eds) Communities and the environment. Ethnicity, Gender, and the State in Community-Based Conservation. Rutgers University Press, New Brunswick, NJ, p 1–31

Angelsen A, with Brockhaus M, Kanninen M, Sills E, Sunderlin WD, Wertz-Kanounnikoff S (eds) (2009) Realising REDD+ : National strategy and policy options. CIFOR, Bogor, Indonesia

Armitage DR, Plummer R, Berkes F, Arthur RI, Charles AT, Davidson-Hunt IJ, Diduck AP, Doubleday NC, Johnson DS, Marschke M, McConney P, Pinkerton EW, Wollenberg EK (2009) Adaptive co-management for social-ecological complexity. Front Ecol Environ 7:95–102

Atela JO, Quinn CH, Minang PA, Duguma LA (2015) Implementing REDD+ in view of integrated conservation and development projects: leveraging empirical lessons. Land Use Policy 48:329–340

Bagine R (1998) Biodiversity in Ramogi Hill, Kenya, and its evolutionary significance. Afr J Ecol 36:251–263

Beentje HJ (1994) Kenya Trees. Shrubs and Lianas, National Museums of Kenya, Nairobi

Bell R (1987) Conservation with a human face: conflict and reconciliation in African land use planning. In: Anderson D, Grove R (eds) Conservation in Africa, people, policies, and practice. Cambridge University Press, Cambridge, pp 79–101

Belsky AJ (1989) Landscape patterns in a semi-arid ecosystem in East Africa. J. Arid Environments 17:265–270

Brown K (2002) Innovations for conservation and development. Geographical J 168:6–17

Burgess N, Hales JD, Underwood E, Dinerstein E, Olson D, Itoua I, Schipper J, Ricketts T, Newman K (2004) Terrestrial Ecoregions of Africa and Madagascar. Island Press, Washington, A Conservation Assessment

Bussmann RW, Beck E (1995) The forests of Mt. Kenya (Kenya), a phytosociological synopsis. Phytocoenologia 25:467–560

Bytebier B (2005) Editorial. J East Afr Nat Hist 94:1–2

Bytebier B, Bussmann RW (2000) The vegetation of Mount Nyiru (Samburu District, Kenya): a checklist and syntaxonomical survey. J East Afr Nat Soc 89:45–71

Chavez P (1988) An improved dark-object subtraction technique for atmospheric scattering correction of multispectral data. Remote Sens Environ 24:459–479

Chazdon RL (2008) Beyond deforestation: restoring forests and ecosystem services on degraded lands. Science 320:1458–1459

Chomba S, Treue T, Sinclair F (2015) The political economy of forest entitlements: can community based forest management reduce vulnerability at the forest margin. Forest Policy Econom 58:37–46

Chomba S, Kariuki J, Lund JF, Sinclair F (2016) Roots of inequity: how the implementation of REDD+ reinforces past injustices. Land Use Policy 50:202–213

Cooke B, Kothari U (2001) The case for participation as tyranny. In: Cooke B, Kothari U (eds) Participation: the new Tyranny?. Zed Books, London, UK, pp 1–15

Coughenour MB, Ellis J (1993) Landscape and climatic control of woody vegetation in a dry tropical ecosystem: Turkana district, Kenya. J Biogeogr 20:383–398

Cuerrier A, Turner NJ, Gomes TC, Garibaldi A, Downing A (2015) Cultural keystone places: conservation and restoration in cultural landscapes. J Ethnobiology 35:427–448

Cunningham AB (2001) Applied Ethnobotany. Earthscan Publications Ltd., London, U.K.

Dale IR, Greenway PJ (1961) Kenya Trees and Shrubs. Buchanan's Kenya Estates Limited, Nairobi

Faden RB, Beentje HJ, Nyakundi DO (1988) Checklist of the forest plant species. In: Beentje HJ (ed) An ecological and floristic study of the forests of the Taita Hills, Kenya. Utafiti Occasional Papers of the National Museums of Kenya 1:43–66

Field R, O'Brien EM, Whittaker RJ (2005) Global models for predicting woody plant richness from climate: development and evaluation. Ecology 86:2263–2277

Fischer E, Rembold K, Althof A, Obholzer J, Malombe I, Mwachala G, Onyango JC, Dumbo B, Tehisen I (2010) Annotated checklist of the vascular plants of Kakamega Forest, Western Province, Kenya. J East African Nat Hist Soc 99:129–226

FRA (Forest Resource Assessment Program) (2000) On definitions of forest and forest change. Working Paper 33. FAO, Rome. ftp://ftp.fao.org/docrep/fao/006/ad665e/ad665e00.pdf

Gachathi FNM (1996) Conservation priorities in the arid and semi-arid lands: the case of the hilltop forests of Kenya. In: van der Maesen, LJG, van der Burgt XM, van Medenbach de Rooy LJG. (eds) The biodiversity of African plants. Proceedings XIVth AETFAT Congress. Kluwer Academic Publishers, London, p. 313–316

Gachathi FNM (2007) Kikuyu Plant Dictionary, 2nd edn. Tropical Botany, Gituamba, Kenya

Gachathi FN, Johansson SG, Alakoski-Johansson GM (1994) A check-list of indigenous trees and shrubs of Bura, Tana river district, Kenya with Malakote, Orma and Somali names. J East Afr Nat Soc 83:117–141

Garibaldi A, Turner N (2004) Cultural keystone species: implications for ecological conservation and restoration. Ecology Soc **9**(3):1 [online] http://www.ecologyandsociety.org/vol9/iss3/art1/

Githae EW, Chuah-Tetiot M, Mworia JK, Odee DW (2008) A botanical inventory and diversity assessment of Mt. Marsabit forest, a sub-humid montane forest in the arid lands of northern Kenya. Afr J Ecol 46:39–45

Guthiga P, Nyangena W, Juma OM, Sikei G (2015) Local community participation under reformed forest management in Kenya. Lessons and policy implications. In: Bluffstone R, Robinson EJZ (eds) Forest Tenure and reform in Asia and Africa: local control for improved livelihoods, forest management, and carbon sequestration. RFF Press Routledge, New York, pp 142–150

Henkin MA, Medley KE, Maingi JK (2015) Biophysical analysis of afromontane forest community types at Mt. Kasigau. Afr J Ecology 53:454–464

Huete AR, Liu HQ, Batchily K, van Leeuwen W (1997) A comparison of vegetation indices global set of TM images for EOS-MODIS. Remote Sens Environ 59:440–451

Jindal R, Swallow B, Kerr J (2008) Forestry-based carbon sequestration projects in Africa: potential benefits and challenges. Nat Resour Forum 32(2):116–130

Kairo JG (2008) Experiences with PES in Kenya. ETFRN News (September 2008):107–111

Kalibo HW, Medley KE (2007) Participatory resource mapping for adaptive collaborative management at Mt. Kasigau, Kenya. Landscape Urban Plann 82:145–158

Kauth RJ, Thomas GS (1976) The Tasseled Cap—a graphic description of the spectral-temporal development of agricultural crops as seen by Landsat, LARS. In: Proceedings of the symposium on machine processing of remotely sensed data, Purdue University, West Lafayette, IN, p 4B-41–4B-51

Kissling WD, Field R, Böhning-Gaese K (2008) Spatial patterns of woody plant and bird diversity: functional relationships or environmental effects? Glob Ecol Biogeogr 17:327–339

Kokwaro JO (1995) Ethnobotany in Africa. In: Schultes RE, von Reis S (eds) Ethnobotany. Timber Press, Portland, OR, Evolution of a Discipline, pp 216–225

Kokwaro JO, Johns T (2013) Luo-English Botanical Dictionary. East African Educational Publisher, Nairobi

Küper W, Sommer JH, Lovett JC, Mutke J, Linder H, Beentje HJ, Van Rompaey RSAR, Chatelain C, Sosef M, Barthlott W (2004) Africa's hotspots of biodiversity redefined. Ann Mo Bot Gard 91:525–535

Laird S (ed) (2002) Biodiversity and Traditional Knowledge. Equitable Partnerships in Practice. Earthscan Publications Ltd., London

Lind EM, Morrison MES (1974) East African vegetation. Longman Group Limited, London

Lomolino MV (2001) Elevation gradients of species-density: historical and prospective views. Glob Ecol Biogeogr 10:3–13

Luke Q (2005) Annotated checklist of the plants of the Shimba hills, Kwale district, Kenya. J East Afr Nat Hist Soc 94:5–120

Maingi JK, Marsh SE (2006) Composition, structure and regeneration patterns in a gallery forest along the Tana River near Bura, Kenya. For Ecol Manage 236:211–228

Matiku P, Caleb M, Collistus O (2013) The impact of participatory forest management on local community livelihoods in the Arabuke-Sokoke Forest, Kenya. Conservation Soc 11:112–129

Maunder PM, Ngugi GW, Kabuye CHS (1999) Traditional food plants of Kenya. Kenya, Resource Centre for Indigenous Knowledge (KENRICK), National Museums of Kenya, Nairobi

Maunder PM, Berger D, Saitabau C, Nasieku J, Kipelian M, Mathenge S, Morimoto Y, Hoft R (2001) Ethnobotany of the Loita Maasai. UNESCO, Paris

Mbuthia KW (2003) Ecological and ethnobotanical analyses for forest restoration in the Taita Hills, Kenya. PhD Dissertation, Miami University, Oxford, Ohio, USA

Medley KE, Hughes FMR (1996) Riverine forests. In: McClanahan TR, Young TP (eds) East African ecosystems and their conservation. Oxford University Press, New York, pp 361–383

Medley KE, Kalibo HW (2005) An ecological framework for participatory ethnobotanical research at Mt. Kasigau, Kenya. Field Methods 17:302–314

Medley KE, Kalibo HW (2007) Global localism: recentering the research agenda for biodiversity conservation. Nat Resour Forum 31:151–161

Medley KE, with village residents at Mt. Kasigau. (2013). Some Kasigau woody plants and their uses. Printed with support from the National Science Foundation, critical ecosystem partnership fund, National geographic society, wildlife works, and the Turrell Herbarium fund, Miami University. Miami University, Oxford, OH, USA. http://www.users.muohio.edu/medleyke/kasiplantbook_may2013_website.pdf)

Medley KE, Maingi JK (2014) Biogeographic patterns of forest diversity at Mt. Kasigau, Kenya. J East Afr Nat Hist 103:1–24

Medley KE with Mwandoe Z, Mwamodo M, Zungi J, Mwatate D, Njege N (2010) Interpreting resource gradients and patches for the conservation of woody plant diversity at Mt. Kasigau, Kenya. Ethnobotany Res Appl 8:49–60

Medley KE, Kalibo HW, with Mwandoe Z, Mwamodo M, Zungi J, Mwatate D, Njege N, Muasya J, Gachathi FN, Mungai G, Mwachala G (2007) Ethnobotanical survey of 'wild' woody plant resources at Mt. Kasigau, Kenya. J East Afr Nat Hist 96:149–186

MENR [Ministry of Environment and Natural Resources] (2016) National forest programme of Kenya. MENR, Nairobi, Kenya. http://www.environment.go.ke/wp-content/uploads/2016/07/Kenya-NFP-draft-doc-2016-07-12-small-v2-1.pdf

Minang PA, Duguma LA, Bernard F, Mertz O, van Noordwijk O (2014) Prospects for agroforestry in REDD+ landscapes in Africa. Curr Opin Env Sust 6:78–82

Mittermeier RA, Turner WR, Larsen FW, Brooks TM, Gascon C (2011) Global biodiversity conservation: the critical role of hotspots. In: Zachos FE, Habel JC (eds) Biodiversity Hotspots. Distribution and Protection of Conservation Priority Areas, Springer, NY, pp 3–22

Mogoi J, Obonyo E, Ongugo P, Oeba V, Mwangi E (2012) Communities, property rights and forest decentralization in Kenya. Early lessons from participatory forest manage 10:182–194

Moran MS, Jackson RD, Slater PN, Teillet PM (1992) Evaluation of simplified procedures for retrieval of land surface reflectance factors from satellite sensor output. Remote Sens Environ 41:169–184

Mutiti CM (2010) Landscape structure and dynamics in *Acacia Commiphora* bushland, southeastern Kenya. PhD Dissertation, Miami University, Oxford, Ohio USA

Mutune JM, Lund JF (2016) Unpacking the impacts of 'participatory' forestry policies: evidence from Kenya. Forest Policy Econom 69:45–52

Nabhan GP (ed) (2016) Ethnobiology for the Future. The University of Arizona Press, Tucson, AZ

Najam A (2005) Developing countries and global environmental governance. From contestation to participation to engagement. Int Environ Agreements 5:303–321

Namirembe S, Leimona B, van Noordwijk M, Bernard F, Bacwayo KE (2014) Co-investment paradigms as alternatives to payments for tree-based ecosystem services in Africa. Curr Opinions Environ Sustainability 8:89–97

National Environmental Management Authority (2009) National Environment action plan framework, 2009–2013. Government of Kenya, Nairobi. http://theredddesk.org/countries/plans/national-environment-action-plan-kenya-2009-2013

Nemarundwe N, Richards M (2002) Participatory methods for exploring livelihood values derived from forests: potential and limitations. In: Campbell BM, Luckert MK (eds) Uncovering the hidden harvest. Valuation methods for woodland and forest resources. Earthscan Publications Ltd, London, p 168–197

Newmark WD (2002) Conserving biodiversity in East African forests. A study of the eastern arc mountains. Springer, Berlin Heidelberg

Olson DM, Dinerstein E (2002) The global 200: priority ecoregions for global conservation. Ann Mo Bot Gard 89:199–224

Omoro LM, Pellikka PK, Rogers PC (2010) Tree species diversity, richness, and similarity between exotic and indigenous forests in the cloud forests of Eastern Arc Mountains, Taita Hills, Kenya. J Forestry Res 21:255–264

Pearce F (2015) In Kenya's mountain forests, a new path to conservation. Yale Environment 360 (February 26). http://e360.yale.edu/feature/in_kenyas_mountain_forests_a_new_path_to_conservation/2850/

Polhill D (1988) Flora of tropical East Africa index of collecting localities. Royal Botanic Gardens, Kew, UK

Polhill D, Polhill R (2015) East African Plant Collectors. Royal Botanic Gardens, Kew, UK

Prinzing A, Klatz S, Stadler J, Brandl R (2003) Woody plants in Kenya: expanding the Higher Taxon Approach. Biol Cons 110:307–314

Rocheleau DE (2008) Political ecology in the key of policy: from chains of explanation to webs of relation. Geoforum 39:716–727

Rodgers WA, Homewood KM (1982) Species richness and endemism in the Usambara mountain forests, Tanzania. Biol J Lin Soc 18:197–242

ROK [Republic of Kenya] (2015) Draft National Forest Policy, 2015. Ministry of the environment, water, and natural resources, Nairobi. http://www.environment.go.ke/wp-content/uploads/2015/04/Draft-National-Forest-Policy-2015-23-03-2015.pdf

Sheridan MJ, Nyamweru C (2008) African sacred groves. Ohio University Press, Athens

Stave J, Oba G, Bjorå CS, Mengistu Z, Nordal I, Stenseth N (2003) Spatial and temporal woodland patterns along the lower Turkwel River, Kenya. Afr J Ecol 41(3):224–236

Teillet PM, Guindon B, Goodenough DG (1982) On the slope-aspect correction of multispectral scanner data. Can J Remote Sens 8:84–106

Timberlake J (1994) Vernacular names and uses of plants in Northern Kenya. J East Afr Nat Hist 83:31–69

Trewartha GT (1981) The Earth's problem climates, 2nd edn. The University of Wisconsin Press, Madison

Veronesi M, Reutemann T, Zabel A, Engel S (2015) Designing REDD+schemes when forest users are not forest landowners: evidence from a survey-based experiment in Kenya. Ecol Econ 116:46–57

Walter H (1971) Ecology of tropical and subtropical vegetation. Oliver & Boyd, Edinburgh
Wass P (ed) (1995) Kenya's indigenous forests. IUCN, Gland, Switzerland, Status, Management and Conservation
Western D, Wright RM, Strum SC (eds) (1994) Natural connections: perspectives in community-based conservation. Island Press, Washington DC
White F (1983) The vegetation of Africa. Unesco, Paris
Whittaker RJ, Willis KJ, Field R (2001) Scale and species richness: towards a general, hierarchical theory of species diversity. J Biogeogr 28:453–470
Wunder S (2013) When payments for environmental services will work for conservation. Conservation Lett 6:230–237
Youatt R (2015) Counting Species: Biodiversity in Global Environmental Politics. University of Minnesota Press, St. Paul, MN

Chapter 11
Woody Plant Diversity and Density in Selected Eucalyptus and Other Plantation Forest Species in Ethiopia

Shiferaw Alem and Muhammad Nakhooda

Abstract *Eucalyptus* spp. forests play a vital role in meeting the global demands for wood, paper, pulp and timber products. They are established in many countries worldwide in varying climates and are significant contributors to the regional and global economies due to their relatively diverse gene pool. Despite this, *Eucalyptus* spp. are often considered to be detrimental to the diversity and density of indigenous woody plants than other exotic plantations since they may inhibit the natural regeneration of local flora. Through the consolidation of various data, this review highlights the impact of *Eucalyptus* spp. forests on woody plant diversity and density. Data from three sites in Ethiopia were investigated: plantations of *E. camaldulensis* in the Bedele area and Abelti Gibie Valley, and *E. grandis* in Belete Gera Forest. The results indicated that far from being inhibitory, *Eucalyptus* spp. plantations sustain a greater number of woody plants in terms of density and diversity compared with other exotic plantation forests. In all cases, the Shannon Diversity Index (*H'*) within an investigated *Eucalyptus* spp. plantation was above 1.57. Furthermore, *Eucalyptus* spp. forests were shown to contain densities of at least 2293 woody plant species/ha, which compares favourably against other plantation forests containing species such as *Pinus* or *Acacia*. It is suggested that when managed well, with appropriate species-to-site establishment, *Eucalyptus* spp. forests may favour the selection of woody plants for sustainable agroforestry practices and should be considered in conjunction with conservation initiatives.

S. Alem (✉)
Department of Forest Botany, Dendrology and Geo-Biocenology,
Mendel University in Brno, Zemedelska, 3, 61300 Brno, Czech Republic
e-mail: shife19@gmail.com

M. Nakhooda
Department of Biotechnology and Consumer Sciences, Faculty of Applied Sciences,
Cape Peninsula University of Technology, Cape Town, South Africa

M. Nakhooda
School of Life Sciences, College of Agriculture, Engineering and Science,
University of KwaZulu-Natal, Durban, South Africa

M.R. Ahuja and S.M. Jain (eds.), *Biodiversity and Conservation of Woody Plants*,
Sustainable Development and Biodiversity 17, DOI 10.1007/978-3-319-66426-2_11

Keywords Conservation · Density · Diversity · *Eucalyptus camaldulensis* *Eucalyptus grandis* · Woody plants

11.1 Introduction

Plantation forests are cultivated forest ecosystems established by planting and/or seeding, which aids in the process of afforestation and reforestation for the purposes of wood biomass production, soil and water conservation and/or wind protection (FAO 2012). Plantations comprise either introduced species (all planted stands) or intensively managed stands of indigenous species, which meet the following criteria: one or two species of uniform age that are regularly spaced and are characterised by relatively short rotations when compared with natural forests (FAO 2012). At present, the area used for plantation forests globally totals 277.9 million ha, which represents only 6.95% of total forest area (Payn et al. 2015). The majority of planted forests comprise native species, with only 18–19% of the total area comprising introduced species (Payn et al. 2015). The vast majority of the world's plantation forests are monocultures, with just a small number of widely used tree genera such as *Eucalyptus*, *Pinus* and *Tectona* spp. (FAO 2001; Evans and Turnbull 2004).

In addition to their direct benefits for timber production, plantation forests have also been integrated into biodiversity management strategies through the reduction of logging pressure on natural forests and their role towards the restoration of degraded lands (Evans and Turnbull 2004). Plantation forests, adjacent to exposed remnants of indigenous forests, can provide shelter, reduce edge effects, increase connectivity between forest fragments and accommodate edge-specialist and generalist forest species (Davis et al. 2000; Norton 1998; Christian et al. 1998; Telila et al. 2015). In Ethiopia, as in many other parts of Africa where such plantations are established, forests are also an important source of income for local communities, providing wood for fuel and construction (Davidson 1995). The sale of *Eucalyptus* poles and products has substantial potential to raise farm incomes, reduce poverty, increase food security and diversify smallholder farming systems (Jagger and Pender 2000). In some areas of central Ethiopia, forestry species such as *Eucalyptus globulus* provide high yields, i.e. 1.3–1.7 times that of land area and 1.2–1.5 times that of labour input, as compared to sole wheat plantations (Selamyihun 2004). Furthermore, by providing material for firewood, plantation forests significantly reduce the pressure on natural and indigenous forests for such products (Evans 1989).

Despite the many uses of plantation forests, intensive monocultures of exotic species are widely viewed as unfavourable to biological diversity conservation (Carnus et al. 2003). Among the different plantation forest species established worldwide *Eucalyptus* has faced the most criticism (FAO 2001; Evans and Turnbull 2004), primarily due to the claim that such forests suppress ground vegetation and thus impede diversity (Jagger and Pender 2000). The present review therefore seeks to consolidate various studies with a view to: (1) assessing the woody plant diversity and density under different *Eucalyptus* spp. plantation forests established in various localities in Ethiopia, (2) comparing these *Eucalyptus* spp. plantations with that of other exotic and indigenous plantation forests in Ethiopia and (3) evaluating the role of *Eucalyptus* plantations with respect to indigenous woody plant conservation. The density and diversity of woody plants under *E. grandis* and *E. camaldulensis* plantations, and relative to exotic and indigenous plantations, was also investigated.

11.2 Diversity of Woody Plants

The Shannon Diversity Index (*H′*) was used for comparison of woody plants under the canopies of different *Eucalyptus* spp. plantations in different localities in Ethiopia (summarised and presented in Tables 11.1 and 11.2). The diversity of woody plants in *E. camaldulensis* plantations was previously reported to be 1.57 for a 26-year-old plantation in the Bedele area, Southwestern Ethiopia (Alem et al. 2015), 1.64 for a 27-year-old plantation in the Abelti Gibie Valley (Alem and Pavlis 2012) and 2.02 for a 33-year-old plantation in the Belete Gera forest (Hundera 2010). This indicated that among the investigated *E. camaldulensis* plantations, woody plant diversity was greatest in Belete Gera forest, followed by Abelti Gibie valley and then Bedele. A tentative indication is that the diversity of woody plants in *E. camaldulensis* plantations appears to increase with plantation age. However, this does not seem to be consistent with *E. saligna* plantation diversity data. Senbeta et al. (2002a) showed that the diversity of woody plants in 11-, 22- and 27-year-old *E. saligna* plantations established in southern Ethiopia was 1.98, 1.89 and 1.79, respectively (Table 11.2). This indicates a decrease in diversity with increasing age of the *E. saligna* plantation. Nevertheless, Hundera (2010) reported a woody plant diversity index of 2.13 in a 33-year-old *E. saligna* plantation, located in southwestern Ethiopia. This suggests that the diversity of woody plants in *E. saligna* plantations in Ethiopia varies with plantation locality, most likely associated with factors such as differences in the amount of rainfall, distance from natural forest and also with the previous land use history.

For *E. globulus* plantations in southern Ethiopia, Senbeta et al. (2002a) showed that the diversity index of woody plants was 1.86, 1.72 and 1.03 for 13-, 16- and 22-year-old plantations, respectively. In the central parts of Ethiopia, however, a 17-year-old *E. globulus* plantation had a woody plant diversity index of 2.3 (Senbeta and Teketay 2001), whilst in the northern parts of Ethiopia woody plant diversity was recorded as 1.8 for an *E. globulus* plantation that is more than 20 years old. Second and fourth coppices of *E. globulus* plantations yielded 1.99 and 1.98 woody plant diversity indices, respectively (Yirdaw and Luukkanen 2003). A variety of factors, such as age of the forest, location and geographical parameters, distance from the natural forest, and previous land use history could be

Table 11.1 List of woody plant species and their respective densities (per ha) in *Eucalyptus* spp. plantations from various localities in Ethiopia [From: Alem and Woldemariam (2009), Alem and Pavlis (2012) and Alem et al. (2015)]

No.	Species	Family	Tree density/hectare		
			E. camaldulensis (Bedele)	*E. camaldulensis* (Gibie Valley)	*E. grandis* (Belete Forest)
1	*Acacia abyssinica* Hochst. Ex. Benth	Mimosaceae		3	
2	*Acacia nilotica* (L.) Del.var. adansonii O. Ktze	Mimosaceae		18	
3	*Acacia tortilis* (Forsk.) Hay	Mimosaceae		3	
4	*Acokanthera schimperii* (DC.) Olive.	Apocynaceae		3	
5	*Albizia grandibracteata* Taub.	Fabaceae	3		
6	*Albizia gummifera* (J. F.Gmel.) C.A. Sm.	Mimosaceae		3	53
7	*Allophylus abyssinicus* (Hochst.) Radlk.	Sapindaceae			25
8	*Apodytes dimidiata* E. Mey. Ex Benth.	Icacinaceae			4
9	*Bersama abyssinica* Fres.	Melianthaceae	256		24
10	*Brucea antidysenterica* J.f. Mill	Simaroubaceae	11		
11	*Calpurnia aurea* (Lam.) Benth.	Fabaceae	22	3	35
12	*Carissa edulis* (Forsk.) vahl	Apocynaceae	78		
13	*Cassipourea malosana* (Bak.) Alston	Rhizophoraceae			4
14	*Catha edulis* (Vahl) Forsk., ex Endl.	Celastraceae			4
15	*Celtis africana* Burm. f.	Ulmaceae			14
16	*Clausena anisata* (Willd.) Hook. F. ex Benth	Rutaceae	122		74
17	*Coffea arabica* L.	Rubiaceae	22		1035

(continued)

Table 11.1 (continued)

No.	Species	Family	Tree density/hectare		
			E. camaldulensis (Bedele)	*E. camaldulensis* (Gibie Valley)	*E. grandis* (Belete Forest)
18	*Combretum ghasalense* Engl.& Diels	Combretaceae		9	
19	*Cordia africana* Lam.	Boraginaceae			4
20	*Croton macrostachyus* Hochst.ex A. Rich.	Euphorbiaceae	111	9	11
21	*Deinbollia kilimandscharica* Taub.	Sapindaceae		15	
22	*Dichrostachys cinerea* (L.) Wight & Arn.	Mimosaceae		212	
23	*Diospyros abyssinica* (Hiern.) White	Ebenaceae			3
24	*Diospyros mespiliformis* Hochst. Ex Dc	Ebenaceae		3	
25	*Dodonaea angustifolia*	Sapindaceae		4268	
26	*Dracaena afromontana* Mildbr.	Agavaceae		6	3
27	*Dracaena steudneri* Schweinf. Ex Engl.	Agavaceae			8
28	*Ehretia cymosa* Thonn.	Boraginaceae	22		32
29	*Ekebergia capensis* sparrman.	Meliaceae	22	3	3
30	*Entada abyssinica* Steud.ex A.rich	Mimosaceae		12	
31	*Euclea schimperii ex.* Robyns	Ebenaceae		729	
32	*Ficus sur* Forsk	Moraceae		26	19
33	*Ficus vasta* Forsk.	Moraceae			2
34	*Flacourtia indica* (Burm.f.) Merr	Flacourtiaceae	56	6	
35	*Flueggea virosa* (Roxb. Ex Willd.) voigt	Phyllanthaceae		9	
36	*Galiniera coffeoides* Del.	Rubiaceae			20
37	*Galiniera saxifraga* (Hochst.) Bridson	Rubiaceae	67		249
38	*Glinus lotoides* L.	Aizoaceae			2

(continued)

Table 11.1 (continued)

No.	Species	Family	Tree density/hectare		
			E. camaldulensis (Bedele)	*E. camaldulensis* (Gibie Valley)	*E. grandis* (Belete Forest)
39	*Grewia bicolor* Juss.	Tiliaceae		38	
40	*Grewia ferruginea* Hochst	Tiliaceae		97	
41	*Jasminum abyssinicum* Hochst. Ex Dc.	Oleaceae			34
42	*Maesa lanceolata* Forsk	Myrsinaceae	11		17
43	*Maytenus gracilipes* (loes.) Sebsebe	Celastraceae		24	190
44	*Maytenus ovatus* var. argutus (Loes.) Blakelock	Celastraceae	2022		
45	*Maytenus senegalensis* (Lam.) Excell	Celastraceae		24	
46	*Millettia ferruginea* (Hochst.) Bak	Papilionaceae	89	12	164
47	*Nuxia congesta* R. Br. Ex Fres	Loganiaceae		74	
48	*Olea capensis* L.	Oleaceae		50	14
49	*Oncoba spinosa* Forsk.	Flacourtiaceae		3	
50	*Paveta abyssinica* Fres.	Rubiaceae			43
51	*Phytolacca dodecandra* L' Herit	Phytolaccaceae			2
52	*Pittosporum viridiflorum* Sims.	Pittosporaceae			7
53	*Polyscias fulva* (Hiern.) Harms	Araliaceae			7
54	*Premna schimperia* Engl.	Verbenaceae		41	
55	*Prunus africana* (Hook.f.) Kalkm.	Rosaceae	44		
56	*Pterolobium stellatum* (Forsk.) Chiov.	Fabaceae			2
57	*Rhamnus prinoides* L'Herit	Rhamnaceae			3
58	*Rhus natalensis* Benth. Ex Krauss	Anacardiaceae		447	

(continued)

Table 11.1 (continued)

No.	Species	Family	Tree density/hectare		
			E. camaldulensis (Bedele)	*E. camaldulensis* (Gibie Valley)	*E. grandis* (Belete Forest)
59	*Rubus apetalus* poir	Rosaceae			3
60	*Schefflera abyssinica* (Hochest.) ex A.Rich) Harms	Araliaceae		3	3
61	Solanum species	Solanaceae	11		
62	*Stereospermum kunthianum* Cham.	Bignoniaceae		6	
63	*Syzygium guineense* (Willd.) DC.	Myrtaceae			2
64	*Terminalia brownii* Pers.	Combretaceae		421	
65	*Vepris dainellii* (Pic. Serm) kokwaro	Rutaceae	11		3
66	*Vernonia amygdalina* Del.	Asteraceae			26
67	*Vernonia thomsoniana* oliv. & Hiern	Asteraceae	1222		145
68	*Woodfordia uniflora* (A. Rich.) Koehne	Lythraceae	11		
69	*Ximenia americana* L.	Oleaceae		9	
70	*Ziziphus spina-christi* (L.) Willd.	Rhamnaceae		15	
71	Aba bira©		11		
72	Cheyi©		189		
	Total		4413	6604	2293

© Species with common names only

responsible for the wide range of woody plant diversity indices observed among the different *Eucalyptus* plantations.

The mean diversity of woody plants (arranged by species) under different plantation forests in various regions of Ethiopia is presented in Fig. 11.1. Collectively, the mean diversity of woody plants under *E. saligna, E. camaldulensis* and *E. globulus* plantations in Ethiopia was 1.95, 1.74 and 1.5, respectively. This clearly shows a decrease in woody plant diversity, which suggests that *E. saligna* plantations are able to support the highest mean diversity of woody plants in Ethiopia as compared with the other *Eucalyptus* species. Mean results also indicate that the diversity of woody plants in the *E. saligna* (1.95) plantation was higher than

Table 11.2 Diversity (H'), density and number of woody plant species in forest plantations in different localities in Ethiopia

Plantation forest	Number of woody plant species	Diversity (H')	Density of woody plants/ha	Number of plantation trees/ha	Location (latitude, longitude, locality)	Rainfall (mm/year)	References
Eucalyptus saligna (11 years)	18	1.98	3575	450	78° 13′N, 38° 37′E Munessa-Shashemene Forest, Southern Ethiopia	1250	Senbeta et al. (2002a)
Eucalyptus saligna (22 years)	23	1.89	10,100	900			
Eucalyptus saligna (27 years)	25	1.79	18,650	100			
Eucalyptus globulus (13 years)	16	1.86	6550	275			
Eucalyptus globulus (16 years)	13	1.72	2300	900			
Eucalyptus globulus (22 years)	17	1.03	13,400	625			
Pinus patula (10 years)	18	2.3	2325	750			
Pinus patula (21 years)	16	1.84	3750	500			
Pinus patula (28 years)	15	1.89	2525	475			
Cupressus lusitanica (9 years)	30	2.65	7325	900			
Cupressus lusitanica (17 years)	22	2.08	7375	575			
Cupressus lusitanica (25 years)	16	1.72	5950	475			
Cupressus lusitanica (33 years)	40	2.5	7200	1000	7° 30′N to 7° 45′N and 36° 15′E to 36° 45′E, in Belete Gera Forest, South west Ethiopia	1800–2300	Hundera (2010)
Eucalyptus saligna (33 years)	24	2.13	3274	675			
Eucalyptus camaldulensis (33 years)	36	2.02	6113	456			
Pinus patula (33 years)	40	1.99	8564	190			
Cupressus lusitanica (14 years)	18	1.67	5770	980	9° 00′N and 38° 35′E, in Menagesha Suba Forest, Central Ethiopia	1225	Senbeta and Teketay (2001)
Eucalyptus globulus (17 years)	27	2.3	7730	790			
Pinus radiata (24 years)	15	2.25	3130	860			
Pinus patula (24 years)	17	1.97	3940	790			
Juniperus procera (42 years)	27	2.24	18,270	710			
Cupressus lusitanica (24 years)	17	1.97	3940	820			

(continued)

Table 11.2 (continued)

Plantation forest	Number of woody plant species	Diversity (*H* ′)	Density of woody plants/ha	Number of plantation trees/ha	Location (latitude, longitude, locality)	Rainfall (mm/year)	References
Cupressus lusitanica	2	0.39	–	–	11° 12′N and 39° 40′E, Tehuledere district, South Wello, North Ethiopia	1030	Abiyu et al. (2011)
Eucalyptus globulus	4	0.22	–	–			
Eucalyptus globulus	41		932	–	9° 04′ to 9° 05′N and 38° 43′ to 38° 47′E, Entoto mountain, Central Ethiopia	1233.1	Debush et al. (2015)
Eucalyptus grandis	39	2.19	2293	535	7° 31′N, 36° 33′E Belete Forest	1547	Alem and Woldemariam (2009)
Eucalyptus camaldulensis	34	1.57	6604	524	8° 10′N, 37° 34′E Abelti Gibie Valley	1287	Alem and Pavlis (2012)
Eucalyptus camaldulensis	22	1.64	4413	822	8° 27′N and 36° 21′E, Bedele, South west Ethiopia	1974	Alem et al. (2015)
Cupressus lusitanica (>20 years)	19	1.96	5790	–	10° 18′N and 37° 45′E, Yirba Forest, North Ethiopia	1200	Mulugeta and Alemayehu (2014)
Acacia melanoxylon (>20 years)	15	1.81	1510	–			
Acacia decurrens (>20 years)	15	1.53	2590	–			
Eucalyptus globulus (>20 years)	11	1.80	1090	–			
Eucalyptus globulus (11 years)	20	1.1	–	1706	9° 11′N and 36° 46′E, Chancho, Central Ethiopia	1270	Yirdaw and Luukkanen (2003)
Eucalyptus globulus (2nd coppice crop)	22	1.99	–	4472	98° 00′N and 38° 35′E, Menagesha, Central Ethiopia	1225	
Eucalyptus globulus (2nd coppice crop)		1.93	–	4297			
Eucalyptus globulus (4th coppice crop)		1.98	–	4207			

(continued)

Table 11.2 (continued)

Plantation forest	Number of woody plant species	Diversity (H')	Density of woody plants/ha	Number of plantation trees/ha	Location (latitude, longitude, locality)	Rainfall (mm/year)	References
Pinus patula (16 years)	30	2.41	4495	1125	7° 6′N and 38° 37′, Wondo Genet, South Ethiopia	1200	Yirdaw (2001)
Cupressus lusitanica (16 years)	34	1.07	8218	1150			
Cupressus lusitanica (14 years)				1500			
Cupressus lusitanica (15 years)				1258			
Juniperus procera (14 years)	32	2.16	5470	1650			
Juniperus procera (15 years)				1978			
Grevillea robusta (16 years)	29	2.00	5498	1718			
Cupressus lusitanica (27 years)	20	1.52	4645	444	8° 27′N and 36° 21′E, Bedele, South west Ethiopia	1974	Alem et al. (2015)
Cordia africana (28 years)	19		5320	198	7° 25′N and 38°55′E, Degaga Forest, South Ethiopia	990	Lemenih et al. (2004)
Eucalyptus saligna (31 years)	17		6280	210			

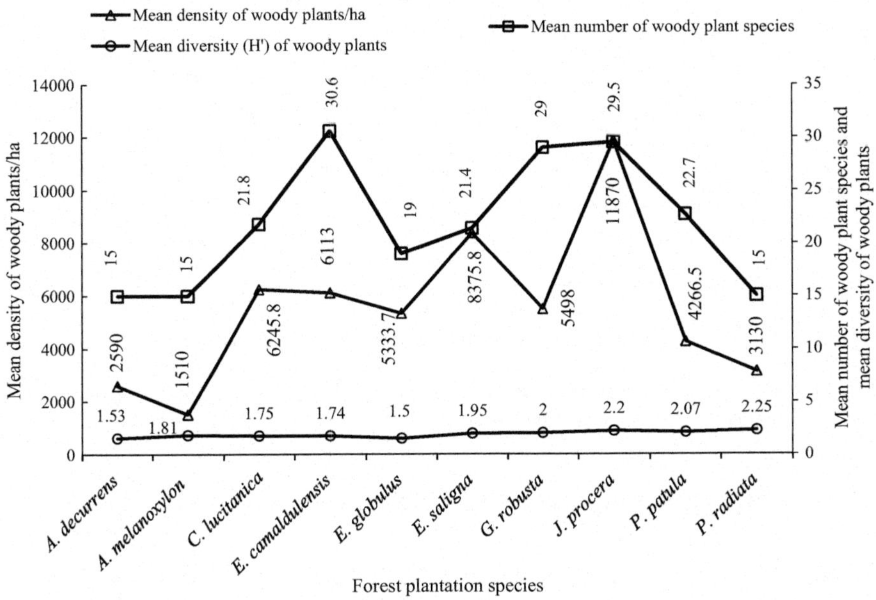

Fig. 11.1 Mean diversity (*H′*), mean density (per ha) and mean number of woody plant species under different plantation forests in Ethiopia

that of *Cupressus lusitanica* (1.75) and *Acacia decurrens* (1.53) plantations, suggesting that some *Eucalyptus* plantations in Ethiopia may be more diverse than certain coniferous and nitrogen-fixing plantation forest species (Fig. 11.1).

According to Kent and Coker (1994), the optimum value of the Shannon Diversity Index (*H*') lies between 1.5 and 3.5. The results from Fig. 11.1 indicate that most *Eucalyptus* spp. plantations in Ethiopia display a diversity value greater than 1.5. This supports the claim that *Eucalyptus* plantations in Ethiopia may not be poor with respect to woody plant diversity, as is often the criticism. The lack of uniformity in woody plant diversity among the various *Eucalyptus* plantations, between and within different locations, is indicative of external factors influencing this parameter, such as availability of seed sources, distance from the natural forest, previous land use history, disturbances and local climatic conditions. A regression analysis result (Fig. 11.2) indicates a poor relationship ($R^2 = 0.011$) between the density of *Eucalyptus* plantation forest trees/ha with the diversity of woody plants recorded in the plantations. Generally, the diversity of woody plants in *Eucalyptus* plantations in Ethiopia is comparable with several other studies within Ethiopia (Yirdaw 2001; Senbeta and Teketay 2001; Senbeta et al. 2002a, b; Yirdaw and Luukkanen 2003) and in other countries such as Congo (Loumeto and Huttel 1997), Hawaii (Harrington and Ewel 1997), China (Chen et al. 2003) and in certain tropical countries (Lugo 1992; Parrotta 1995; Parrotta et al. 1997).

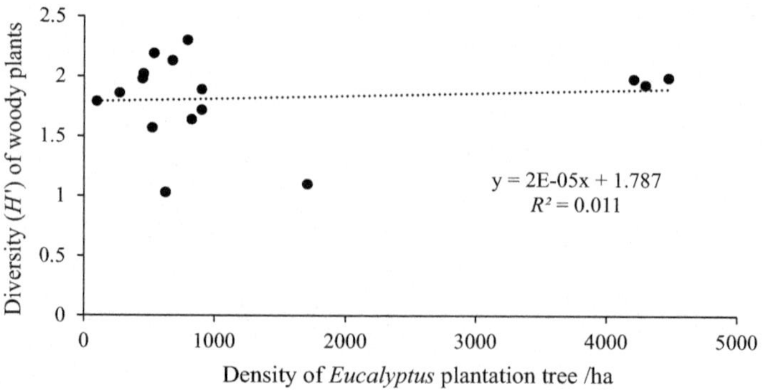

Fig. 11.2 Regression analysis showing the relationship between density of *Eucalyptus* trees/ha and the diversity of woody plants under their canopies in Ethiopia

11.3 Floristic Composition of Undergrowth Woody Plants

In terms of the distribution, composition and relationship of plant species among the investigated sites, 34 woody plant species, representing 20 families, were recorded in the *E. camaldulensis* plantation in Gibie valley, 22 woody plant species (19 families) in the *E. camaldulensis* plantation in the Bedele area and 39 woody plant species (26 families) in the *E. grandis* plantation in Belete forest (Table 11.1). Four species were common among the investigated plantations: *Calpurnia aurea*, *Ekebergia capensis*, *Croton macrostachyus* and *Millettia ferruginea*. A number of woody plant species were found exclusively within each of the three sites investigated, i.e. 9 species in the *E. camaldulensis* plantation (Bedele), 23 in the *E. camaldulensis* plantation (Gibie Valley) and 21 in the *E. grandis* plantation (Belete forest) (Table 11.1).

Certain *Eucalyptus* plantations were found to sustain a greater number of woody plant species compared with non-eucalypt forests (Table 11.2). For example, 19 woody plant species were recorded in a 24-year-old *Cordia africana* plantation in southern Ethiopia (Lemenih et al. 2004), 20 in a 27-year-old *Cupressus lusitanica* (exotic) plantation in southwestern Ethiopia (Alem et al. 2015), 15–18 in a *Pinus patula* plantation ranging between 10 and 28 years old in central Ethiopia (Senbeta et al. 2002a), 15 and 17 in 24-year-old *Pinus radiata* and *P. patula* plantations in central Ethiopia, respectively (Senbeta and Teketay 2001) and 17 in a 24-year-old *Cupressus lusitanica* plantation in central Ethiopia (Senbeta and Teketay 2001) (Table 11.2). This shows that a higher number of woody plant species were recorded for *E. grandis* plantations in Belete forest (Alem and Woldemariam 2009) and *E. camaldulensis* plantations in Gibie Valley (Alem and Pavlis 2012) than in certain indigenous and exotic non-eucalypt plantation forests across Ethiopia. This could indicate that some *Eucalyptus* plantations may have the potential to harbour higher numbers of woody plant flora than previously thought.

The mean number of woody plant species recorded in different *Eucalyptus* and non-eucalyptus plantation forest species in Ethiopia is presented in Fig. 11.1. The average number of woody plant species recorded in 25–33-year-old *E. camaldulensis* plantations, 11–22-year-old *E. globulus* plantations and 11–33-year-old *E. saligna* plantations was 30.6, 19 and 21, respectively. This could show that among these plantations in Ethiopia, those of *E. camaldulensis* appear to sustain the highest mean number of woody plant species, followed by *E. saligna* and finally *E. globulus*. Furthermore, these eucalypt plantations possess even greater numbers of woody plants than plantations containing nitrogen-fixing species, such as *Acacia decurrens* and *A. melanoxylon* (Table 11.2). Overall, the number of identified woody plant species under the canopies of different *Eucalyptus* plantation forests in Ethiopia was comparable to the total number of woody plants recorded under the canopies of *Eucalyptus* species from countries such as Zimbabwe (Tyynela 2001), Vietnam (Van et al. 2005), China (Daun et al 2010) and South Africa (Geldenhuys 1997).

11.4 Vegetation Structure and Density of Woody Plants

The relative density of woody plants/ha in different diameter classes in *E. camaldulensis* (Bedele and Gibie Valley) and *E. grandis* (Belete forest) plantations, presented in Fig. 11.3, is taken from several sources (Alem and Woldemariam 2009; Alem and Pavlis 2012; Alem et al. 2015). Results in Fig. 11.3 indicate that the relative density of woody plants in the stem diameter classes of 0.1–1 cm was

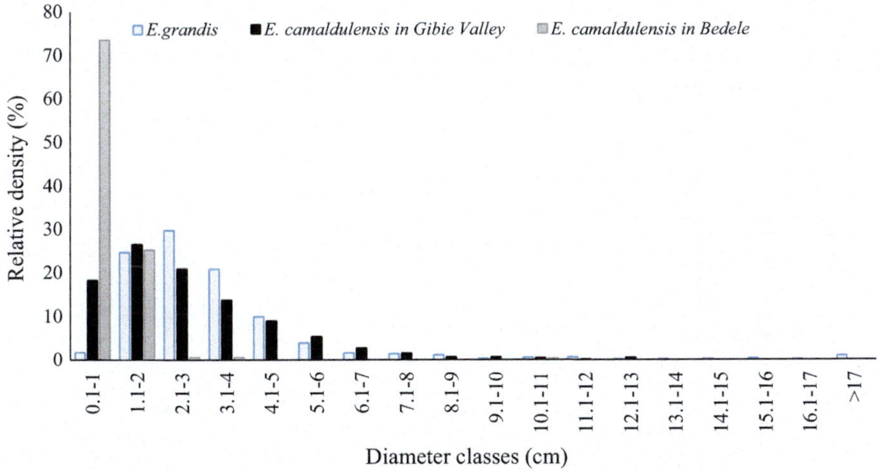

Fig. 11.3 Relative density (%) of woody plants in different diameter classes (cm), under the *E. grandis* plantation in Belete forest, and in the *E. camaldulensis* plantations in Gibie Valley and Bedele, Ethiopia

1.9% for *E. grandis* in Belete forest, 18.3% for *E. camaldulensis* in Gibie Valley and 73.5% for *E. camaldulensis* in Bedele. The relative density of woody plants (stem diameter class: 10.1–11 cm) in the *E. grandis* plantation (Belete forest) and *E. camaldulensis* plantations (Gibie valley and Bedele) was 0.6, 0.4 and 0.3%, respectively. This revealed that for *Eucalyptus* plantations in Ethiopia, the relative density of woody plants decreases as the stem diameter classes increase (Fig. 11.3).

The total density of woody plants/ha in the *E. camaldulensis* plantation in Gibie Valley, *E. camaldulensis* plantation in Bedele and *E. grandis* plantation in Belete forest was 6604/ha, 4413/ha and 2293/ha, respectively (Table 11.1). These findings could indicate that, from the investigated sites, the highest relative density of native woody plant species was found in the *E. camaldulensis* plantation in Gibie Valley, followed by the *E. camaldulensis* plantation in Bedele, and lastly the *E. grandis* plantation in Belete forest. These differences in the densities of regenerated woody plants/ha could potentially result from factors such as differences in the density of *Eucalyptus* trees/ha (Table 11.2). Varying *Eucalyptus* densities could create differences in the light penetrating the canopies, thereby influencing the recruitment of native woody plant species. Lemenih et al. (2004) found that differences in canopy density coupled with canopy-influenced variations in understory environmental factors in the plantation stands were responsible for many of the discrepancies in the regeneration parameters of woody plants in various plantation forests. This further indicates that those plantations with lighter canopies allow for greater diversity and density of naturally regenerated native woody species, larger breast-height trunk diameter and increased total height, when compared with species with denser/heavier canopies. Despite this, a regression analysis result of the presently investigated *Eucalyptus* plantations in Ethiopia revealed no significant relationship between density of *Eucalyptus* trees/ha and density of regenerated woody plants/ha (Fig. 11.4). This implies that besides the density of plantation trees other factors such as availability of seed sources, age of the plantation trees, cutting cycles, etc., could have influences on the regeneration of woody plants which in turn affects the density.

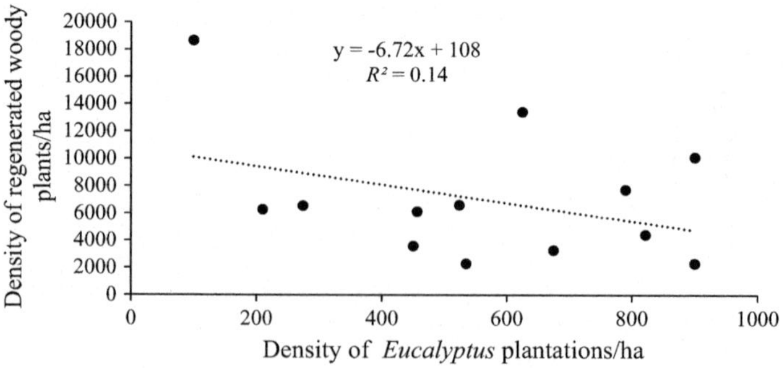

Fig. 11.4 Regression analysis graph showing the relationship between density of *Eucalyptus* plantation trees/ha and density of regenerated native woody plants/ha in Ethiopia

The results from Table 11.1 further indicate that *Maytenus ovatus* in the *E. camaldulensis* plantation (Bedele), *Dodonaea angustifolia* in the *E. camaldulensis* plantation (Gibie Valley) and *Coffea arabica* in the *E. grandis* plantation (Belete forest) had the highest density of woody stems/ha when compared with other regenerated woody plants found in those plantations. This suggests that *E. grandis* plantations have the potential to accommodate coffee, one of the most important economic crops in Ethiopia. An evaluation of coffee grown in *E. grandis* plantations versus that grown in a natural forest showed no significant difference in the bean size or the cup quality of the coffee (Alem and Woldemariam 2010); hence, some *Eucalyptus* species plantations can serve more than a single agroforestry purpose, which goes a long way towards sustainable land use practices. Other studies have also shown that the consumable products of some crops, grown alongside and/or within eucalypt plantations, are not affected by the dual-purpose plantation environment. Stocking (1993) showed that in Sri Lanka, *E. grandis* forms a successful agroforestry combination with cardamom. Zeng (1992) reported a highly favourable economic yield when sweet potato was inter-cropped in 208.7 ha of *Eucalyptus* plantations in China. In some parts of Costa Rica, *Eucalyptus deglupta* is commonly used as coffee shade, and no evidence of diminished growth, yield and mineral nutrition has been recorded (Schaller et al. 2003). The use of some *Eucalyptus* trees in agroforestry (silvo-pastoral systems) is also fairly common in the savannah region of Nigeria (Igboanugo et al. 1990).

In the *E. camaldulensis* plantation in the Bedele area, tree densities of selected species were recorded as follows: *Vernonia thomsoniana*: 1222/ha; *Bersama abyssinica*: 256/ha; *Clausena anisata*: 122/ha; *Croton macrostachyus*: 111/ha; *Millettia ferruginea*: 89/ha; *Carissa edulis*: 78/ha; *Galinaria saxifrage*: 67/ha; and *Flacourtia indica*: 56/ha (Table 11.1). Similarly, in the *E. camaldulensis* plantation in Gibie valley, the stem densities for selected species were recorded as follows: *Euclea schimperii*: 729/ha; *Rhus natalensis*: 447/ha; *Terminalia brownii*: 421/ha, *Dichrostachys cinerea*: 212/ha, *Grewia ferruginea*: 97/ha; *Nuxia congesta*: 74/ha; and *Olea capensis*: 50/ha (Table 11.1). Finally, in the *E. grandis* plantation located in Belete forest, selected woody plant species stem densities were: *Galiniera saxifrage*: 249/ha; *Maytenus gracilipes*: 190/ha; *Millettia ferruginea*: 164/ha; *Vernonia thomsoniana*: 145/ha; *Clausena anisata*: 74/ha; and *Albizia gummifera*: 53/ha (Table 11.1). These results suggest that different *Eucalyptus* plantation species may favour the recruitment of different woody plant species under their canopies. It may also indicate that, as an over story tree, the different *Eucalyptus* plantations foster the natural regeneration and growth of these selected species. The enhanced recruitment of the selected woody plant species could be associated with the phytochemicals (allelochemicals) produced by the different *Eucalyptus* trees, since these may sometimes be stimulatory to the growth of other plants (Rizvi et al. 1992). Different studies have indicated that allelochemicals, which inhibit the growth of some species at certain concentrations, may promote the growth of other species at lower concentrations (Narwal 1994). The stimulatory allopathic effects of any plant on another plant can be used to develop eco-friendly, cheap and effective 'Green Growth Promoters' (Oudhia et al. 1998). There is evidence to indicate that some *Eucalyptus* plantations have the potential to enhance the recruitment, establishment and success of certain native woody species, such as *Dodonaea angustifolia*, a species commonly used for soil conservation (Loumeto and Huttel 1997; Yirdaw 2001; Senbeta et al. 2002a; Mulugeta and Alemayehu 2014). On the other hand, some species, such as *Ehretia cymosa* and *Dracaena afromontana* were relatively sparse in the various *Eucalyptus* plantations (Table 11.1). This could be attributed to the shade intolerance of the regenerated woody plant species, poor soil seed bank of the species, etc. These results tentatively suggest that, as a management option, it is possible to assess and identify the species which display favourable densities for a given *Eucalyptus* plantation and establish these within those particular *Eucalyptus* plantations to increase overall woody plant density and diversity.

The density of regenerated woody plants/ha in different plantation forests in Ethiopia is presented in Table 11.2. For certain sites in southern Ethiopia, the number of woody plants per hectare was recorded as 3575 in an 11-year-old *E. saligna* plantation, 2300 on a 13-year-old *E. globulus* plantation, 2325 on a 10-year-old *Pinus patula* plantation, 3750 on a 21-year-old *Pinus patula* plantation and 2525 on a 28-year-old *Pinus patula* plantation (Senbeta et al. 2002a). Similarly, the same authors recorded a total number of woody stems/ha of 10,100 in a 22-year-old *E. saligna* plantation, 18,650 in a 27-year-old *E. saligna* plantation and 13,400 in a 22-year-old *E. globulus* plantation in southern Ethiopia, whilst Hundera (2010) noted a total of 8564 woody stems/ha in a 33-year-old *P. patula* plantation. Senbeta and Teketay (2001) recorded 18,270 woody stems/ha in a 42-year-old indigenous *J. procera* plantation, located in Central Ethiopia. The fact that the highest number of woody stems/ha was recorded in a *E. saligna* plantation (Table 11.2) shows that in some cases *Eucalyptus* spp. plantations show potential for sustaining greater numbers of woody stems per ha when compared with other indigenous and exotic plantation forests.

The mean density of woody plants/ha in the different plantation forests in Ethiopia was calculated as 2590 stems/ha, 6113 stems/ha, 8375.8 stems/ha in the *Acacia decurrens*, *E. camaldulensis* and *E. saligna* plantations in Ethiopia, respectively (Fig. 11.1). However, the mean density of woody plants/ha in the *Juniperus procera*, *Pinus patula* and *P. radiata* plantations of different ages in Ethiopia was 11,870 stems/ha, 4266.5 stems/ha and 3130 stems/ha, respectively (Fig. 11.1). This could indicate that some *Eucalyptus* plantation forests have the potential to facilitate the recruitment of indigenous woody plants under their canopies more effectively than some non-eucalyptus and exotic plantation forest species. In general, these results indicate that *Eucalyptus* plantations in Ethiopia are not 'green deserts' as some authors have expressed but should be better considered as plantation forests which can harbour high numbers of woody plant stems/ha. Figure 11.5 also indicated a photo that shows the regeneration of different woody plant species under the canopies of different *Eucalyptus* plantation forests in different localities of Ethiopia

Fig. 11.5 Photo that shows the regeneration of woody plants under different *Eucalyptus* plantation forests in Ethiopia. **A** = *E. grandis* in Wondo genet forest; **B** = *E. globulus* in Shashemene; **C** = *E. camaldulensis* in Wondo genet; **D** = *E. camaldulensis* in Bedele

11.5 The Role of Plantation Forest in Indigenous Woody Plant Conservation

In Ethiopia, considerable debate has focussed on the effects of *Eucalyptus* plantations on conservation initiatives. These concerns have resulted in government nurseries abandoning *Eucalyptus* production from the early 2000s, despite it being the main tree seedling produced for over a decade (Mekonnen 2010). The major worry was the allelopathic effects that *Eucalyptus* plantations are often criticised for, such as high moisture and nutrient requirements which result in drier and nutrient-depleted soils. The assumption would then be that these plantations would inhibit the establishment of native woody species; however, the present analysis indicates the other way. Although forestry plantations are not expected to compete with the fecundity of natural native forests, certain indigenous woody plant species such as *Catha edulis*, *Phytolacca dodecandra*, *Rubus apetalus*, recorded within *Eucalyptus* plantations, were not recorded in natural forests (Alem and Woldemariam 2009; Alem and Pavlis 2012), thus the potential of *Eucalyptus* plantation forests to complement conservation initiatives cannot be ignored. Yirdaw

(2002) and Mekonnen (2010) have demonstrated that, with appropriate management practices, *Eucalyptus* plantations may even help restore degraded lands to accommodate native woody plants. Bremer and Farley (2010) have stated that when *Eucalyptus* plantations are established on degraded lands, as opposed to replacing natural forest ecosystems, they becomes key contributors to land restoration due to their resilient nature.

In order to realise the potential of *Eucalyptus* plantations in indigenous woody plant conservation, the following evidence-based management options are recommended:

11.5.1 Density of the Plantation Forest

Dessie and Erkossa (2011) stated that planting less density of *Eucalyptus* trees per ha is important to allow more light to reach in the understory canopy which will support the establishment of understory vegetation. An increase in the diversity of indigenous species has been recorded in Ethiopia in a *E. globulus* plantation where the density of the plantation was decreased (Debush et al. 2015). As the density of *Eucalyptus* trees/ha increases (Fig. 11.6) the potential of the plantations in recruiting indigenous woody plants and to serve for conservation purposes may decrease.

11.5.2 Age of Plantation and Duration of Cutting Cycles

Since nutrient and water consumption increases with cutting cycles in *Eucalyptus* plantations, it is suggested that extending the time between cutting cycles, which also reduces recurrent litter raking, may reduce any negative impact on native woody plants. Selwyn and Ganesan (2009) indicated that as the age of the

Fig. 11.6 High density of *Eucalyptus*/ha in Kajimma Umbullo kebele, Sidama Zone, Southern Ethiopia

Eucalyptus plantations increases, the number of regenerated indigenous woody plants also increases. The cutting cycle of *Eucalyptus* plantations also affects indigenous woody plant conservation, as demonstrated by Zewdie (2008). In this study of an Ethiopian *E. globulus* plantation, vegetation cover, species diversity and richness decreased with increased cutting cycles.

11.5.3 Availability of Seed Sources

The presence of native soil seed banks is important for the regeneration of indigenous woody plants under *Eucalyptus* plantation canopies. Senbeta and Teketay (2001) found that *Eucalyptus* plantations can foster the regeneration of native woody species and increase biodiversity in the stands when seed sources are available in the vicinity of the plantations. Similarly, Mulugeta and Alemayehu (2014) indicated that *Eucalyptus* plantation forests can support natural regeneration and succession provided that there is an adjacent seed source and dispersal agents. Importantly, Yirdaw (2002) demonstrated the crucial role of the small remnant patches of natural forests as a source of diaspores for the restoration of the woody species diversity in different *Eucalyptus* plantations that where established in degraded areas of the Ethiopian highlands.

11.6 Conclusions and Future Prospects

The synthesised results of the different published works indicate that the diversity and density of indigenous woody plants in different *Eucalyptus* plantations in Ethiopia is comparable, and in certain cases, greater than other indigenous and exotic plantation forests. If the plantation forest species is well managed, it can serve not only for timber production, but also complement conservation efforts, which is particularly true if *Eucalyptus* plantations are established on degraded lands. In order to realise the complete value of *Eucalyptus* plantations and improve the diversity of woody plants under the canopies of different *Eucalyptus* plantations in Ethiopia, the following points are recommended: (1) establish *Eucalyptus* at low densities of trees/ha; (2) establish *Eucalyptus* plantations close to a natural forests where the natural forest can serve as a seed source for the recruitment of woody plants under the canopy; (3) thinning of *Eucalyptus* plantations to allow more light to filter through the canopy, and thus permit the growth of native vegetation; (4) identify eco-friendly woody plants and shade-tolerant species, and establish these alongside *Eucalyptus* plantations to create mixed plantation forests; and (5) establish soil and water conservation structures in *Eucalyptus* plantation forests to increase the availability of soil moisture, which could in turn enhance the regeneration and growth of woody plants underneath the canopies.

Acknowledgements The Authors would like to thank Dr. Taryn Ralph of the National Research Foundation (NRF), South Africa, for proofreading this manuscript and for improving upon the language used, as well as all other researchers who graciously shared their knowledge with us, in putting this publication together.

References

Abiyu A, Lemenih M, Gratzer G, Aerts R, Teketay D, Glatzel G (2011) Status of native woody species diversity and soil characteristics in an exclosure and in plantations of *Eucalyptus globulus* and *Cupressus* lusitanica in Northern Ethiopia. Mt Res Dev 31(2):144–152

Alem S, Woldemariam T (2009) A comparative assessment on regeneration status of indigenous woody plants in *Eucalyptus grandis* plantation and adjacent natural forest. J For Res 20(1):31–36

Alem S, Woldemariam T (2010) Regeneration of *Coffea arabica* and quality of coffee found in a *Eucalyptus grandis* plantation. In: Gil L, Tadesse W, Tolosana E, López R, (eds.). Eucalyptus Species Management, History, Status and Trends in Ethiopia. Proceedings from the congress held in Addis Ababa. September 15th –17th, Addis Ababa

Alem S, Pavlis J (2012) Native woody plants diversity and density under *Eucalyptus camaldulensis* plantation, in Gibie Valley. South Western Ethiopia. Open J For 2(4):232–239

Alem S, Pavlis J, Urban J, Kucera J (2015) Pure and mixed plantations of *Eucalyptus camaldulensis* and *Cupressus lusitanica*: their growth interactions and effect on diversity and density of undergrowth woody plants in relation to light. Open J For 5(4):375–386

Bremer LL, Farley KA (2010) Does plantation forestry restore biodiversity or create green deserts? A synthesis of the effects of land-use transitions on plant species richness. Biodivers Conserv 19:3893–3915

Carnus JM, Parrotta J, Brockerhoff EG, Arbez M, Jactel H, Kremer A, Lamb D, Hara KO, Walters B (2003) Planted Forests and Biodiversity: UNFF inter-sessional experts meeting on the role of planted forests in sustainable forest management. March 25–27, Wellington, New Zealand

Chen X, Li BL, Lin ZS (2003) The acceleration of succession for the restoration of the mixed-broadleaved Korean pine forests in Northeast China. For Ecol Manage 177:503–514

Christian DP, Hoffman W, Hanowski JM, Niemi GJ, Beyea J (1998) Bird and mammal diversity on woody biomass plantations in North America. Biomass Bioenerg 14(4):395–402

Daun W, Ren H, Shenglei F, Wang J, Zhang J, Yang Y, Haung C (2010) Community comparison and determinant analysis of understory vegetation in six plantations in South China. Restor Ecol 18(2):206–214

Davidson J (1995) Eucalyptus Tree Improvement and Breeding. Ministry of Natural Resources Development and Environmental Protection, Forestry Research Centre, Addis Ababa, Ethiopia

Davis AJ, Huijbregts H, Krikken J (2000) The role of local and regional processes in shaping dung beetle communities in tropical forest plantations in Borneo. Glob Ecol Biogeogr 9(4):281–292

Debush F, Soromessa T, Argaw M (2015) Structure, species composition and status of naturally regenerated woody species in *Eucalyptus globulus* Labill. (Myrtaceae) plantation at Entoto Mountain, Ethiopia. J Envir Earth Sci 5(15):41–52

Dessie G, Erkossa T (2011) Planted Forests and Trees Working Papers: Eucalyptus in East Africa Socio-economic and environmental issues. Food and Agricultural Organisation of the United Nations, Rome

Evans J (1989) Community forestry in Ethiopia; the Bilate project. Rural Development in Practice 1(4):7–8

Evans J, Turnbull JW (2004) Plantation Forestry in the Tropics. Oxford University Press, Oxford, UK

FAO (2001) State of the World's Forests. Food and Agricultural Organisation of the United Nations, Rome

FAO (2012) Forest Resources Assessment Working Paper 180, FRA 2015: Terms and Definitions, Food and Agricultural Organisation of the United Nations, Rome

Geldenhuys CJ (1997) Native forest regeneration in pine and eucalypt plantations in Northern Province, South Africa. For Ecol Manag 99:101–115

Harrington RA, Ewel JJ (1997) Invasibility of tree plantations by native and nonindigenous plant species in Hawaii. For Ecol Manag 99:153–162

Hundera K (2010) Status of indigenous tree species regeneration under exotic plantations in Belete forest, south west Ethiopia. Ethiop J Edu Sci 5(2):19–28

Ighoanugo AB, Omijeh JE, Adegebehin JO (1990) Pasture floristic composition in different eucalyptus species plantations in some parts of the northern Guinea savanna zone of Nigeria. Agro For Sys 12(3):257–268

Jagger P, Pender J (2000) The Role of Trees for Sustainable Management of Less Favoured Lands: the Case of eucalypts in Ethiopia. International Food Research Institute, Washington, DC

Kent M, Coker P (1994) Vegetation Description and Analysis: A Practical Approach. Wiley, England

Lemenih M, Gidyelew T, Teketay D (2004) Effects of canopy cover and understory environment of tree plantations on richness, density and size of colonizing woody species in southern Ethiopia. For Ecol Manag 194:1–10

Loumeto J, Huttel C (1997) Understory vegetation in fast growing tree plantations on savanna soils in Congo. For Ecol Manag 99:65–81

Lugo AE (1992) Tree Plantation for Rehabilitating Damaged Forest Lands in the Tropics. In: Wali MK (ed) Ecosystem rehabilitation: ecosystem analysis and synthesis. SPB Academic Publishing, Hague, pp 247–255

Mekonnen Z (2010) Community opinion, marketing and current debates on Eucalyptus in Huruta District, Arsi Zone of Oromia Region, Ethiopia. In; Gil L, Tadesse W, Tolosana E, López R, (eds). Eucalyptus Species Management, History, Status and Trends in Ethiopia. Proceedings from the Congress held in Addis Ababa. September 15th–17th, 2010, pp. 131–145

Mulugeta G, Alemayehu B (2014) Status of native woody species regeneration in the plantation stands of Yeraba Priority State Forest, Amhara Region, Ethiopia. J Nat Sci Res 4(16):2224–3186

Narwal SS (1994) Allelopathy in Crop Production. Scientific Publishers, India

Norton DA (1998) Indigenous biodiversity conservation and plantation forestry: options for the future. N Z J For 43:34–39

Oudhia P, Kolhe SS, Tripathi RS (1998) Allelopathic effect of *Blumea lacera* L. on Rice and common Kharif weeds. Oryza 32(2):175–177

Parrotta JA (1995) Influence of overstory composition on understory colonization by native species implantations on a degraded tropical site. J Veg Sci 6:627–636

Parrotta JA, Turnbull WJ, Jones N (1997) Catalysing native forest regeneration on degraded tropical lands. For Ecol Manag 99:1–7

Payn T, Carnus JM, Peter FS, Kimberley M, Kollert W, Liu S, Orazio C, Rodriguez L, Silva LN, Wingfield MJ (2015) Changes in planted forests and future global implications. For Ecol Manag 352:57–67

Rizvi SJH, Haque H, Singh VK, Rizvi V (1992) A Discipline Called Allelopathy: Basic and Applied Aspects. Chapman and Hall, London

Schaller M, Schroth G, Beer J, Jimenez F (2003) Species and site characteristics that permit the association of fast growing trees with crops: the case of *Eucalyptus deglupta* as coffee shade in Costa Rica. For Ecol Manag 175:205–215

Selamyihun K (2004) Using Eucalyptus for Soil and Water Conservation in the Highland Vertisols of Ethiopia: Academic Dissertation. Wageningen, The Netherlands

Selwyn MA, Ganesan R (2009) Evaluating the potential role of *Eucalyptus* plantations in the regeneration of native trees in southern Western Ghats. India. Trop Ecol 50(1):173–189

Senbeta F, Teketay D (2001) Regeneration of indigenous woody species under the canopies of tree plantations in Central Ethiopia. Trop Ecol 42(2):175–185

Senbeta F, Teketay D, Naslund BA (2002a) Native woody species regeneration in exotic tree plantations at Munessa-Shashemene Forest, southern Ethiopia. New For 24:131–145

Senbeta F, Beck E, Lüttge U (2002b) Exotic trees as nurse-trees for the regeneration of natural tropical forests. Trees 16(4):245–249

Stocking M (1993) Eucalyptus-Cardamom system in Sri Lanka. J Agrofor 7:10–19

Telila H, Hylander K, Nemomissa S (2015) The potential of small Eucalyptus plantations in farm scapes to foster native woody plant diversity: local and landscape constraints. Restor Ecol 23 (6):918–926

Tyynela TM (2001) Species diversity in *Eucalyptus camaldulensis* woodlots and miombo woodland in Northern Zimbabwe. New For 22(3):239–257

Van DT, Lee DK, Van TH (2005) Rehabilitation of the native tree species in the forest plantations and Denuded Hills of Namlau Commune in Onla Province, Vietnam. For Sci Tech 1(1):51–58

Yirdaw E (2001) Diversity of naturally-regenerated native woody species in forest plantations in the Ethiopian highlands. New For 22(3):159–177

Yirdaw E (2002) Restoration of the Native Woody-species Diversity, Using Plantation Species as Foster Trees, in the Degraded Highlands of Ethiopia: Academic Dissertation. University of Helsinki, Helsinki, Finland

Yirdaw E, Luukkanen O (2003) Indigenous woody species diversity in *Eucalyptus globulus* Labill. ssp. Globulus plantations in the Ethiopian highlands. Biodivers Cons 12(3): 567–582

Zeng Q (1992) An effective measure for controlling soil degradation in eucalyptus plantation. In: research on site degradation of timber plantation: Edited by the specialized committee of forest ecology, Chinese Forestry Association, Chinese scientific and technical publishing house, Beijing

Zewdie M (2008) Temporal Changes of Biomass Production, Soil Properties and Ground Flora in Eucalyptus globulus Plantations in the Central Highlands of Ethiopia: Academic Dissertation. Swedish University of Agricultural Sciences, Uppsala, Sweden

Chapter 12
Biodiversity and Conservation of Date Palm

Adel A. Abul-Soad, S. Mohan Jain and Mushtaque A. Jatoi

Abstract Date palm is one of the main horticultural crops of the ancient and recent world and an important source of food. Geographical distribution between 24° and 34°N latitude is considered as the main biodiversity centers of date palm distribution. There are about 5000 date palm cultivars besides countless varietal strains throughout the world, but every date-producing country has only few top commercial cultivars being cultivated and exported in all over the world. Commercial laboratories of tissue culture-derived date palm have contributed in spreading the cultivation and exchange safely cultivars of date palm worldwide. However, the date palm diversity is nowadays facing crucial problems such as Bayoud disease and red palm weevil (RPW) besides water shortage, urban settlement and disruption, rapid soil and genetic erosion and uncontrolled commercial exploitations of cultivating only elite cultivars. Numerous efforts have been made for in situ/on-farm conservation methods though effective but quite expensive and hence found only on small scale. The ex situ conservation methods of seed and DNA storage are not applicable due to their recalcitrant seeds and heterozygosity nature. Nevertheless, the in vitro conservation of date palm in form of shoot tips, axillary buds, embryos, and callus for longer periods by using slow growth or cold storage in vitro cultures and cryopreservation techniques is promising but might undergo somaclonal variation and less viability after thawing. The maintenance of field gene banks is found to be highly effective for conservation and providing the long-term preservation but relatively expensive and requiring huge space.

Keywords Aswan · Biodiversity · Conservation · Genetic diversity Oasis · *Phoenix dactylifera* L.

A.A. Abul-Soad (✉)
Horticulture Research Institute, Agricultural Research Center, Giza, Egypt
e-mail: adelaboelsoaud@gmail.com

S.M. Jain
Department of Agricultural Sciences, Helsinki University, Helsinki, Finland

M.A. Jatoi
Date Palm Research Institute, Shah Abdul Latif University, Khairpur, Sindh, Pakistan

M.R. Ahuja and S.M. Jain (eds.), *Biodiversity and Conservation of Woody Plants*,
Sustainable Development and Biodiversity 17, DOI 10.1007/978-3-319-66426-2_12

12.1 Introduction

The date palm (*Phoenix dactylifera* L.) is an important horticultural crop as well as heart of oasis cultivation systems in the date palm-growing countries (Fig. 12.1). It is widely cultivated for edible sweet fruit mainly in the arid or semiarid zones of the world as the date fruit ripening requires long and hot summer with very low humidity. The 200–250-mm rainfall is ideal for date palm cultivation. However, rainfall during flowering and ripening stages may lead to the loss of fruit production and quality; e.g., in Pakistan (mainly Sindh Province), the date production is severely affected by monsoon rains (Abul-Soad et al. 2010). It can grow on diverse range of soils, but sandy loam and clayey loam soils are considered highly suitable though it can even grow on highly saline soils. With life span of 100–150 years, it may reach to 30 m in height (Gepts 2002). However, the commercial production age of trees is less (mostly from 15 to 30 years) and the height above 15–20 m increases the risk and efforts to reach the crown for pollination and harvesting. Nevertheless, there were date palm plants seen commercially productive for 50 years. Date palm productivity is degrading due to gradual decline in the number of emerged spathes every year. Date palm is heterozygous in nature and hence required cross-pollination either naturally (wind pollination) as in the case of traditional oasis or artificially by hand. Spathes from a single palm can be utilized to pollinate 40–50 female trees (Jatoi et al. 2009).

12.1.1 Problems Facing Date Palm Biodiversity Maintenance and Utilization

Despite of rich cultural and natural diversity, the date palms are facing some maintenance problems and its utilization is sometimes hampered by different biotic

Fig. 12.1 Date palm at 'Ghardaia'oasis in Algeria

and abiotic stresses globally, which varies in level and nature of the problems as per cultivar, location, climate conditions, and cultural practices (Zaid and de Wet 2002). Moreover, date palm presence has recently endangered by modernization and industrial activities. In Egypt, date palm is widely cultivated on both banks of river Nile (Fig. 12.2), and hundreds of trees have been destroyed due to the construction of new buildings even after the restrictions imposed by the government. Date palm tree is considered as a wealth that must be conserved on-farm in all date palm-producing countries. Although the state legislations and laws prohibiting export of adult trees of commercial cultivars, Egypt is on the top of the countries getting off the non-commercial adult for export. Before the current new legislation of 6-month incubation at isolated place before export, amount of about 50 thousands trees/year were been exported to other countries mainly for landscaping purposes (Fig. 12.3). Another case study in the deserts of Baluchistan Province, where date palm is the main crop specie in the dispersed Bedouins communities are

Fig. 12.2 Seeded date palm trees growing on both banks of river Nile at Aswan governorate in Egypt

Fig. 12.3 Depletion of productive date palm trees of 3–6 m height through exportation from Egypt for landscaping purposes

threatened for the survival. This is due to fall in water table and rainfall fluctuations and may be the mining activities in the copper belt (Abul-Soad et al. 2009).

Pests and diseases invade date palm plantations, especially in date palm-growing countries. Bayoud disease, caused by *Fusarium oxysporum* f.sp. albedinis, has destroyed millions of palm trees in Morocco and Algeria during the last few decades (Sedra 2011). The spreading like a fire disease has declined thousands of trees in Sindh Province, Pakistan, which is caused by *Fusarium solani*. Symptoms start with yellowing the midrib of outer fronds followed by drying out the leaflets toward the central fronds. This disease can infect the date palm at any age and kill the plant within short period of few months (Abul-Soad et al. 2011c), while Al-Wijam disease is caused by phytoplasma mainly reported in Saudi Arabia. The common symptoms are leaf stunting, yellow streaking, and a marked reduction in fruit and stalk size, which leads to failure in fruit production at harvest (Alhudaib et al. 2007). The red palm weevil (RPW) spread rapidly from India to Arabian Gulf countries, and then Egypt in the North Africa in early nineteenth century, and southern Europe through shipment of adult seedling palm for landscaping purposes in the last decade (Fig. 12.3). Although a number of methods have been used to control this RPW, but keep watching the trees by the growers to detect any early infestation (oozing brown sap with bad smell from the trunk) and control it immediately remained the only acceptable application. By destroying the date palm seedlings at the date palm biodiversity centers—Elche Valley of Spain and Aswan Governorate in Upper Egypt—may forever lose naturally created elite palms. Micropropagation technique could help to conserve rapidly such nature-created models. This can be done through using the floral buds as in vitro explants without destroying the parent unique plant (Abul-Soad 2011a). Nevertheless, the demand of well-known date palm cultivars propagated by tissue culture has been increasing, which creates a fear of extinction of rich biodiversity of date palm worldwide. In addition to developing the supercultivars of date palm through genetic engineering with resistance to Bayoud and RPW, drought, high salinity, aged trees, and genetic erosion are the major threats to combat the loss of date palm biodiversity in global scenario nowadays.

12.2 Date Palm Domestication and Diversification in Ancient and New World

Date palm is the oldest companion of the mankind and believed to be cultivated since the dawn of human civilization. Due to its complicated historical background, extensive cultivation trend, distribution, and exchange of date cultivars make it difficult to know the exact origin of date palm (Popenoe 1924; Lee 1963; Chao and Krueger 2007; Jatoi et al. 2009; Zabar and Borowy 2012; Jatoi 2013). Nonetheless the historical evidences depicted in ancient sculptures and date seeds explorations in some ancient archaeological sites of different ancient civilizations, the researchers assumed Mesopotamia (ancient southern Iraq) (Nixon 1959), Nile valley (Riad 1993; Zohary and Hopf 2000), Arabian Peninsula (Zohary and Hopf 2000), or Indus civilization (currently Pakistan) (Marshal 1931; Wrigley 1995; Jatoi et al. 2009) the center of origin of date palm. The Mesopotamian region is the most likely date palm center, and several mud tablets depicted with date palm and cultural trend are found in ancient Assyrian, Sumerian, and Babylonian eras as well as the well-known Code of Hammurabi, which delimited laws relating to date palm trend and sales (Popenoe 1973). However, the date palm and its cultivation have been known from the earliest records of predynastic Egypt (before about 3000 BC). Mahmud (1958) reported the date palm presence in Indo-Pak subcontinent may be the consequence of dispersal from other centers of origin. However, the explorations of date seeds from ancient archaeological sites of Mohenjo Daro (2500 BC) and Mehrgarh (3000 BC) in Pakistan advocating its presence at that time and claiming Pakistan as one of the possible center of origin of date palm as well (Marshal 1931; Costantini 1985; Beech 2003).

The Arabs and Nomads used to move long distances in large caravan and always carried the date fruits as the source of their daily complete nutritious diet. Dates were easy to carry long distances without proper storage facilities due to its long-lasting storage features. They threw out the date seeds extensively during travel from the ancient business routes, and it seems to be the cause of its extensive cultivation throughout the known world at that time. Later on, military expeditions from Muslim world (from the seventh to the twelfth century A.D) to some European, Asian, and African regions further extended its cultivation to those regions. As a result, the world largest seeded palm grove still present in Elche Valley of Spain containing 180,000 adult trees (Rivera et al. 2015). The Spanish voyages to the new world further distributed this specie into some warmer parts of North and South American continents. The small patches of its cultivation still found near the old missionaries' buildings and in old botanical gardens or home gardens of old buildings (Nixon 1959). In early twentieth century, a large number of suckers were imported from main date-producing countries and planted in California and Arizona states of USA. The dates that showed better yield and quality were further commercially adopted and distributed in other dispersed warmer areas of these two states. Still Coachella Valley of California produces a large number of quality dates, and they are exported all over the world (Fig. 12.4). In 1957, the Coachella Valley accounted for 85% of the date acreage in the USA (Nixon 1959). A tree sample at least from each variety was conserved in a collection farm of the date palm and citrus repository in Coachella Valley (Fig. 12.5). The fruit quality of many cultivars indicates the suitability of the climate to grow many commercial varieties (Fig. 12.6a). But date palm in this area is mainly grown for landscaping purpose (tennis courts) which made particular cultivars dominate others such as 'Deglet Nour,' 'Medjool,' and 'Barhee' (Fig. 12.6b).

Fig. 12.4 Date palm groves of 'Deglet Nour' in Coachella Valley, California, USA

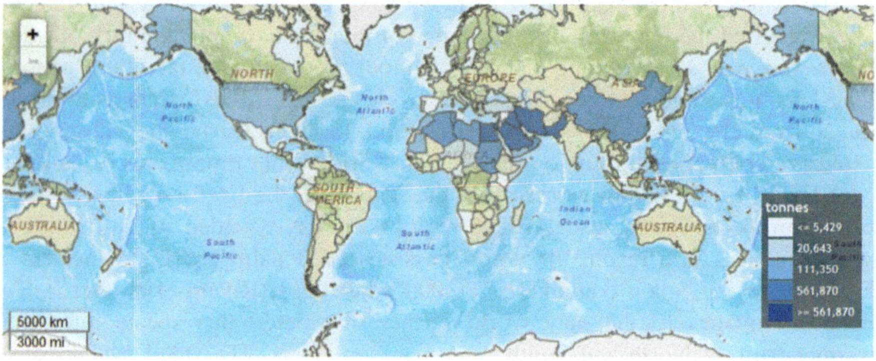

Fig. 12.5 Date palm geographical distribution in the world (FAO 2014)

The recent commercial importance of date fruits and advancement of date palm tissue culture has attracted some non-date-producing countries to start date palm cultivation in the new environmental conditions such as Afghanistan (Osmanzai et al. 2010), Kenya (Gammell 1989; Wasilwa et al. 2007), Namibia (Zaid and de Wet 2002), North of Nigeria (Aisueni et al. 2009; Sani et al. 2010), South Africa (McCubbin 2007), Cameroon (Bourou et al. 2006; Elhoumaizi 2007; Bourou et al. 2015), Chile and Peru (Escobar and Valdivia 2015), Bangladesh (Das 2009), Thailand (koratdatepalm.com 2016). Mostly, the cultivation in these countries has started by seed; then, recently, systematic plantation has been developed by importing healthy elite cultivars, derived from tissue culture. It is worth to mention that superior date palm genotypes grown from seeds are mostly adapted with the local climate and can be conserved by tissue culture technique using the inflorescence explants (Abul-Soad and Mahdi 2010, 2012; Abul-Soad 2011a).

(a) Different cultivars in Date Palm Repository at Coachella Valley, USA.

(b) 'Medjool' cv. native to Morocco.

(c) 'Saidi' or 'Siwy' cv. at New Valley in Egypt.

(d) Dry dates of (1) 'Pertamoda', (2) 'Malakaby', (3) 'Sakoty' cvs. at South of Egypt.

(e) 'Samany' cv. at Date Palm Repository, Coachella Valley which is native to Egypt.

(f) 'Aseel' cv. at 'Khairpur' in Pakistan.

Fig. 12.6 Fruit quality of date palm for top world cultivars, from **a** to **f**. **a** Different cultivars in date palm repository at Coachella Valley, USA. **b** 'Medjool' cv. native to Morocco. **c** 'Saidi' or 'Siwy' cv. at New Valley in Egypt. **d** Dry dates of *1* 'Pertamoda,' *2* 'Malakaby,' *3* 'Sakoty' cvs at South of Egypt. **e** 'Samany' cv. at date palm repository, Coachella Valley which is native to Egypt. **f** 'Aseel' cv. at 'Khairpur' in Pakistan

12.3 Date Palm Biodiversity

Generally, date palm can be found in the warmer areas of all continents (excluding Antarctica) but varying in area and production. Sawaya (2000) suggested the importance of latitude and altitude to explain its geographical distribution. The date palm distribution for both Northern and Southern Hemispheres are between 10°N (Somalia) and 39°N (Elche/Spain or Turkmenistan) rendering to the latitude, whereas the promising regions are found between 24° and 34°N in Morocco, Algeria, Tunisia, Libya, Egypt, Iraq, Iran, and Pakistan (Zohary and Hopf 2000). The sea level from 392 m below to 1500 m above with an altitude range of 1892 m is suitable for date palm cultivation. The major date production areas are therefore from the Middle East, Northern Africa, and Pakistan, while some small production comes from dispersed parts of North America and South Europe (Fig. 12.5).

12.3.1 The Genus **Phoenix** *or Species Diversity*

It is a monocot angiosperm woody diploid ($2n = 2x = 36$) dioecious species of Arecaceae (palm family) plant family. There is a confusion regarding the exact number of species in genus *Phoenix*. In this genus, 19 species have been known so far but most often mentioning 12 valid species (Miller et al. 1930; Chevalier 1952; Moore 1963; Munier 1973; Dowson 1982; Jaradat 2011; Zabar and Borowy 2012) with little variations in the names of species. Among them, widely accepted species are *P. acaulis*, *P. canariensis*, *P. dactylifera*, *P. paludosa*, *P. reclinata*, *P. rupicola,* and *P. sylvestris,* while there is a confusion over the other spp. However, Chevalier (1952) and Munier (1973) accepted *P. atalantica* as a valid species, though it is perhaps a hybrid of *P. dactylifera* that does not deserve species status (Krueger 2001). According to Dowson (1982), the family contains 200 genera, among which *Phoenix* contains 12 of the total 1500 species, while The Plant List Web site mentions 185 genera among which *Phoenix* contains 14 species of the total accepted 2522 species. Uhl and Dransfield (1987) reported the genus comprises approximately 17 species. There is also a bit variation in names of species of genus *Phoenix* given by several researchers due to the synonyms of the species as listed in Table 12.1. The main specimen type of this genus is *P. dactylifera* L. which is commercially cultivated worldwide. It is characterized and distinguished by pinnated leaves and conduplicate leaflets with acute tips (Uhl and Dransfield 1987) as compared to two other common palm species, *P. canariensis* and *P. sylvestris*. *P. theophrasti*, an endemic palm specie native to southern Greece (Crete islands) and southwestern Turkey (Datça Peninsula) before its acceptance status as a separate specie was usually mixed with *P. dactylifera* as almost similar in overall appearance but usually shorter, 15 m tall, with small fruit with only a thin, sour-tasting flesh layer around the seed. The leaves of *P. theophrasti* are also used in Palm Sunday celebrations like date palm (Barrow 1998). *P. atlantica* or Cape

Table 12.1 Accepted or recognized species of genus *Phoenix* with their synonyms and distribution by several prominent resources or investigators

Species no.	Species name	Mentioned/accepted	Synonyms declared by other sources/authors	Common name	Distribution
1	*P. abyssinica Drude*	Krueger (2001)	Synonym of *P. reclinata* (The Plant List 2012)	–	Poorly known specie Ethiopia (Krueger 2001)
2	*P. acaulis* Roxb	Martius (1836–1853), Beccari (1890), Miller et al. (1930), Chevalier (1952), Mowry (1952), Moore (1963), Munier (1973, 1974), Bailey (1976), (Dowson 1982), Barrow (1998), Krueger (2001), The Plant List (2012)	–	Dwarf date palm, stemless/trunk less date palm	India (Assam and Uttar Pradesh), Nepal, Bangladesh, Myanmar
3	*P. atalantica* A. Chev	Dowson (1982), Krueger (2001), The Plant List (2012)	–	Cape Verde palm	Cape Verde Islands (Africa)
4	*P. canariensis* Chabaud or *P. canariensis* Hort ex Chab	Beccari (1890), Miller et al. (1930), Chevalier (1952), Mowry (1952), Moore (1963), Munier (1973, 1974), Bailey (1976), Barrow (1998), Dowson (1982), Krueger (2001), The Plant List (2012)	–	Canary date palm, Canary Island palm, palmera Canaria	Native to Spain (Canary Islands-Africa) but also found in Italy, Australia, Bermuda
5	*P. dactylifera* L.	Martius (1836–1853), Beccari (1890), Miller et al. (1930), Chevalier (1952), Mowry (1952), Moore (1963), Munier (1973, 1974), Bailey (1976), Dowson (1982), Barrow (1998), Krueger (2001), The Plant List (2012)	–	Date palm, Nakhl (Arabic), khajoor (Urdu)	Widely cultivated species found in Arabian Peninsula, Middle East, Northern and Central African countries, Pakistan. Small populations found in southern Europe (Spain), USA (California, Arizona), South America (Peru, Chile), Australia
6	*P. farinifera* Roxb	Martius (1836–1853), Beccari (1890), Chevalier (1952), Moore (1963), Munier (1973, 1974), Dowson (1982)	Synonym of *P. pusilla* (The Plant List 2012)	–	Poorly known specie India (Krueger 2001)
7	*P. hanceana Naudin*	Dowson (1982), Krueger (2001)		–	–

(continued)

Table 12.1 (continued)

Species no.	Species name	Mentioned/accepted	Synonyms declared by other sources/authors	Common name	Distribution
			Synonym of *P. loureiroi var. loureiroi* (The Plant List 2012)		
8	*P. humilis* Royle	Miller et al. (1930), Chevalier (1952), Munier (1973, 1974), Dowson (1982), Krueger (2001)	Synonym of *P. loureiroi var. pedunculata* (The Plant List 2012) *P. loureii var. humilis* (Barrow 1998)	–	–
9	*P. loureiroi* Kunth	Martius (1836–1853), Mowry (1952), Moore (1963), Bailey (1976), Krueger (2001), The Plant List (2012)	*P. loureiri var. loureiri* (Barrow 1998), *P. formosana*, *P. Hanceana*, *P. humilis*, *P. ouseleyana* (Krueger 2001)	Loureiro's palm	China (Guangdong, Guangxi, Yunnan, Hong Kong), Taiwan, Philippines, Bangladesh, India, Nepal, Pakistan, Cambodia, Myanmar, Thailand, Vietnam
10	*P. ouseleyana* Griff	Martius (1836–1853), Krueger (2001)	Synonym of *P. loureiroi var. pedunculata* (The Plant List 2012)	–	–
11	*P. paludosa* Roxb.	Martius (1836–1853), Beccari (1890), Miller et al. (1930), Chevalier (1952), Mowry (1952), Moore (1963), Munier (1973, 1974), Bailey (1976), Dowson (1982), Barrow (1998), Krueger (2001), The Plant List (2012)	–	Mangrove date palm	Bangladesh, India (Orissa, West Bengal, Andaman and Nicobar Islands), Myanmar, Thailand, Vietnam, Indonesia (Sumatra), Malaysia
12	*P. pumila* Hort	Miller et al. (1930), Krueger (2001)	Synonym of *P. reclinata* (The Plant List 2012)	–	–

(continued)

Table 12.1 (continued)

Species no.	Species name	Mentioned/accepted	Synonyms declared by other sources/authors	Common name	Distribution
13	*P. pusilla* Gaertn	Martius (1836–1853), Beccari (1890), Miller et al. (1930), Mowry (1952), Moore (1963), Bailey (1976), Barrow (1998), Krueger (2001), The Plant List (2012)	*P. zeylanica* (Krueger 2001)	Ceylon date palm, flour palm	India (Kerala, Tamil Nadu), Sri Lanka
14	*P. reclinata* Jacq.	Martius (1836–1853), Miller et al. (1930), Dowson (1982), Barrow (1998), Beccari (1890), Chevalier (1952), Mowry (1952), Moore (1963), Munier (1973, 1974), Bailey (1976), Krueger (2001), The Plant List (2012)	*P. senegalensis*, *P. spinosa*, *P. zanzibarensis*, *P. madagascariensis* (Krueger 2001)	Senegal date palm, wild date palm	Africa: Kenya, Tanzania, Uganda, Eritrea, Ethiopia, Somalia, Angola, Malawi, Mozambique, Zambia, Zimbabwe, Botswana, South Africa (Cape Province, KwaZulu-Natal, Transvaal), Benin, Cote D'Ivoire, Gambia, Ghana, Guinea, Guinea-Bissau, Nigeria, Senegal, Sierra Leone, Burundi, Cameroon, Central African Republic, Gabon, Rwanda, Zaire, Comoros, Madagascar Asia: Arabian Peninsula: Saudi Arabia, Yemen
15	*P. roebelenii* O'Brein	Miller et al. (1930), Chevalier (1952), Moore (1963), Munier (1973, 1974), Bailey (1976), Dowson (1982), Barrow (1998), Krueger (2001), The Plant List (2012)	–	Miniature date palm, pygmy date palm, roebelenii palm	China (Yunnan), Laos, Myanmar, Thailand, Vietnam
16	*P. rupicola* T. Anderson	Miller et al. (1930), Beccari (1890), Chevalier (1952), Mowry (1952), Moore (1963), Munier (1973, 1974), Bailey (1976), Dowson (1982), Barrow (1998), Krueger (2001), The Plant List (2012)	–	Cliff date, East Indian wine palm, Indian date palm, wild date palm	Bhutan, India (West Bengal)

(continued)

Table 12.1 (continued)

Species no.	Species name	Mentioned/accepted	Synonyms declared by other sources/authors	Common name	Distribution
17	*P. spinosa FC schum*	Martius (1836–1853), Krueger (2001)	Synonym of *P. reclinata* (The Plant List 2012)	–	–
18	*P. sylvestris* (L) Roxb	Martius (1836–1853), Beccari (1890), Miller et al. (1930), Chevalier (1952), Mowry (1952), Moore (1963), Munier (1973, 1974), Bailey (1976), Dowson (1982), Barrow (1998), Krueger (2001), The Plant List (2012)	–	Date sugar palm, Indian date, silver date palm, wild date palm	India (Bihar, Punjab, Rajasthan, Tamil Nadu, Uttar Pradesh, West Bengal), Nepal
19	*P. zeylanica* Trimen	Miller et al. (1930), Mowry (1952), Bailey (1976), Krueger (2001)	Synonym of *P. pusilla* (The Plant List 2012)		
20	*P. andamanensis* S. Barrow	Barrow (1998), The Plant List (2012)	–		India (Andaman and Nicobar Islands)
21	*P. theophrasti* Greuter	Barrow (1998), The Plant List (2012)	–	Cretan date palm	Turkey, Greece (Crete)
22	*P. caespitosa* Chiov.	Barrow (1998), The Plant List (2012)	–		Djibouti, Somalia, Saudi Arabia, Yemen, Oman

Verde palm is a close relative of *P. dactylifera* that merely represent the feral date populations of date palm (Jaradat 2011) or might be a product of spontaneous hybridization between *P. dactylifera* and other *phoenix* spp. Henderson et al. (2006) revealed the genetic discontinuity with microsatellite markers that supported the status of *P. atlantica* as a separate species, which suggests its isolation long ago without any gene flow even though having very similar characteristics like its mainland relative, *P. dactylifera.*

12.3.2 Genetic Diversity

Genetic diversity is defined as the genetic variations between species, subspecies, cultivars, populations, or individual clones that can be measured at the morphological, physiological, biochemical, and molecular levels (Jaradat 2015). The present-day date palm cultivars are the outcome of thousands of years of selection practice of seedlings holding desired features. As per statement of Wrigley (1995), each cultivar is derived from a unique single seed, cloned and multiplied vegetatively. There are about 3000 (Zaid and de Wet 2002) or 5000 (Bashah 1996) date palm cultivars exist globally but sometimes might be synonyms of one cultivar found in different countries under a different name, but about 10% of them of a commercial importance (Johnson 2011). However, each country got its own top elite cultivars of commercial value. The situation gets critical when the same cultivars being cultivated in other countries by replacing or altering its Arabic names into their local languages which makes the task difficult to know the exact numbers of cultivars found in the Arabian Peninsula, Northern Africa, and Middle East (Chao and Krueger 2007). Like 'Abattamoda' in Sudan, 'Bertamoda' or 'Pertamoda' in Egypt is the altered name for cv. 'Barakawi,' which is the most famous cultivar of Sudan (Khairi 2015). Another example from Egypt is the most famous cultivar 'Siwy' which is widely cultivated in Siwa oasis, Al Wahat (el-Bahariya oasis), and Al-Fayum and Giza governorates and was originally named 'Saidi' in the main export area with low humid air 'New Valley' governorate (Fig. 12.6c). Fruit of cv. 'Saidi' is characterized with a brown ring that made the fruit having two colors. 'Saidi' cultivar is present in Libya and Date Palm Repository in USA too.

The date cultivars around the famous date-producing countries evaluated so far estimated about 450 in Saudi Arabia (Bashah 1996), 400 in Iran (Hajian and Hamidi-Esfahani 2015), 400 in Iraq (Zabar and Borowy 2012), 1000 in Algeria (Bouguedoura et al. 2015), 250 in Tunisia, 244 (Zaid and de wet 2002) or 453 (Sedra 2015) in Morocco, 95 in Libya (Battaglia et al. 2015), 400 in Sudan (Osman 1984; Elshibli 2009), 250 in Oman (Al-Yahyai and Al-Khanjari 2008; Al-Yahyai and Khan 2015), 321 in Yemen (Ba-Angood 2015), 52 cultivars in Egypt (Rabei et al. 2012) 300 in Khairpur, Pakistan (Mahar 2007; Markhand et al. 2010; Abul-Soad et al. 2015), besides numerous cultivars in other date-producing countries. Each date-growing country has its own won cultivars in addition to various other cultivars (Abul-Soad et al. 2015; Al-Khayri et al. 2015a, b), and their distribution is restricted to these regions due to numerous reasons (Al-Khalifa 2006).

Dates are categorized into soft, semidry, and dry cultivars as per their moisture content, texture, fruit appearance, and sugar content. For instance, the dry substances in soft cultivars are nearly 80% of invert sugars (mixture of equivalent extent of fructose and glucose) with high moisture (>30%) and soft flesh, e.g., cvs. 'Barhi,' 'Khadrawy,' and 'Hayany.' The second group is the semidry cultivars that maintain about 40% invert sugars and 40% sucrose with firm flesh and fairly low moisture (20–30%), e.g., cvs. 'Medjool,' 'Siwy,' 'Aseel,' 'Khalas.' Fruits of this group are in the top of export cultivars due to the taste and reasonable good shelf life. The dry cultivars are distinguished by having around 20–40% invert sugars and 40–60% of sucrose with hard or dry flesh and low moisture (<20%), e.g., cvs. 'Barakawi,' (Bender and Bender 2005; Zabar and Borowy 2012), 'Sakoty,' 'Malakaby,' and 'Pertamoda' in Upper Egypt (Fig. 12.6d). The cultivars of later group are limited to very hot areas such as South Egypt (in Aswan), North Sudan, and West Pakistan. Cultivars are also considered as early, mid-season, and late cultivars on the basis of duration of time required to produce mature fruit (Jatoi 2013).

According to Panga (2014), the date palm has normally 36 chromosomes ($n = 18$; $2n = 36$), but polyploidy cases are also reported with some date palm cultivars of Iraq ($n = 64$). In addition, differences among cultivars have also been found like cv. 'Sayer' ($2n = 32$) which is an early cultivar as compared to a late cultivar 'Khasab' ($2n = 36$). In addition, aneuploidy and euploidy in both cultivars were also reported, e.g., cv. 'Sayer' (32, 34, 36, and 64) and 'Khasab' (32 and 36).

The climatic conditions (air temperature and humidity) and soil type play significant role in the fruit properties of any cultivar during months of ripening. The cv. 'Deglet Noor' develops black nose (blackening and shriveling of the tip) and fruit checks (small, linear scars near the apex) when grown in humid conditions. 'Deglet Noor' is a native cultivar of Algeria and Tunisia where it performs as a dry cultivar, while it is generally a semidry cultivar under USA environmental conditions (Krueger 2015). It is noticed that the soft cultivar of Egypt 'Samany' showed semidry fruit quality when cultivated in the date palm repository in USA (Fig. 12.6e). Moreover, under non-appropriate environmental conditions, cv. 'Deglet Noor' yields and quality were often unsatisfactory as in case when cultivated in 'Punjab' Province in Pakistan, 'Wadi an Natrun' in Egypt. Sometimes these differences happened even in small intra-zones like, Khairpur Sindh, Pakistan, where 'Aseel' is a predominant cultivar (Fig. 12.6f). Cv. 'Aseel' is mainly confined to Khairpur district of Sindh; surprisingly, the 'Aseel' fruits from the eastern part of Khairpur district (near to hills or mountain range) contain low moisture and subsequently having the longer shelf life in comparison with same cultivar in the rest of the district and province (Markhand et al. 2010). Brac de la Perrière and Benkhalifa (1989) also found some intra-cultivar variability due to environmental effects on fruits of cv. 'Deglet Noor' in Algeria and observed that the fruits from Tolga or Biskra oasis are excellent in quality, i.e., semi dry with 20–30% moisture content, buttery, and shiny as compared to the same cultivar grown in M'zab region which were largely drier and smaller and thus of less quality (Fig. 12.7).

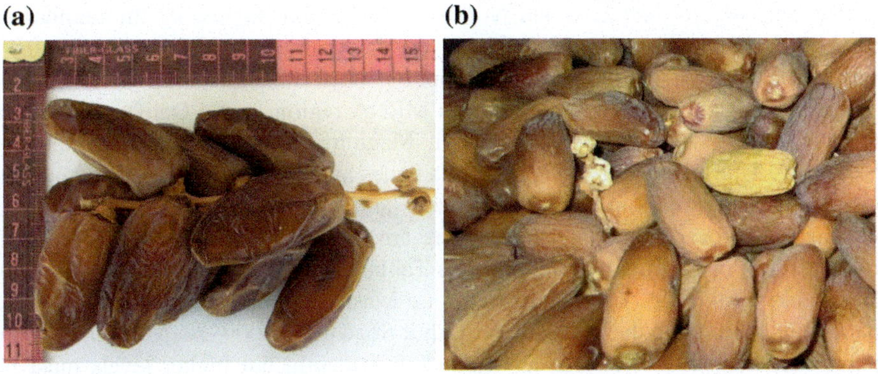

Fig. 12.7 Intra-cultivar variability due to environmental conditions on fruits of cv. 'Deglet Noor' in Algeria. **a** At Biskra oasis with buttery and semidry properties, and **b** at 'Ghardaia' oasis with drier and less fruit quality

12.3.3 Natural, Oasis Eco-diversity

Date palm being adopted to long dry summers and mild winters thrives well in desert and oasis environment (Al-Bakr 1972). These groves or oasis are vital ecological niches for indigenous wildlife and play a critical role in the desert ecosystems (Chao and Krueger 2007). Date palm trees create their own protection from the sun, accumulate water moisture, attract insects, produce humus which they often feed on, protect the ground from atmospheric agents, i.e., soil fixation, and allow cultivation under their foliage, which further increases the positive dynamics triggered by the palms. This is the oasis effect, capable of creating self-reproducing and self-sustaining life cycles in conditions of scarce resources (Sirjacobs 1988). An oasis should be understood as an artificial creation that exists because of the sophisticated environmental knowledge of the humans who live there (Mancini 2010). The natural oasis system usually comprises date palm originated from seeds and hence presents very important diversity as the agronomic traits differ greatly normally with inferior date quality. Like the series of oasis (Siwa, Baharia, Farafra, Dakla, Kharga, and Fayoum) in Egypt where total number of trees are approx. 700,000 and about 50% of them originated from seeds (Riad 1993). However, the this number has been increased now into about 4 million date palm trees. Al-Abbad et al. (2011) mentioned approximately three million date palm trees in the Al-Hassa oasis (eastern Saudi Arabia) where the cv. 'Khalas' represents more than 50% of the area. Depending on only one cultivar or monoculture farming system might increase the probability of pests and diseases attack in the future. However, Mikki (1998) suggested that the date cultivars, grown in the Middle East oasis, are well adapted to the local climate, edaphic, and socioeconomic conditions and quality of their fruit. In Libya, each date palm grove is represented by a distinct cultivar composition resulted from local selection within the oases (Battaglia et al. 2015).

The MENA (Middle East and North Africa) region possessing the number of distinctive oasis agro-ecosystems having the date palm as the main crop covers 1 million ha area which is small but supports the livelihood of about 10 million inhabitants. Some of the famous oases in MENA region are Al-Qatif and Al-Ahsa (Saudi Arabia), Al-Ain (UAE), Buraimi, Maghta and Bahla (Oman), Bahraiya, Siwa (Fig. 12.8) and Kharga (Egypt), Ghadames and Kufra (Libya), Ouargla, Taut, Ghardaia and Timimoun (Algeria), Tozeur and Tamerza (Tunisia), Tafilalt and Ourzazat (Morocco) (Jaradat 2011). There are some scattered patches of oases in Balochistan Province of Pakistan such as Tahlab, Washab, Gualishtop (Fig. 12.9), Brouhook, Esa Tahir, Yak-Mach, Hamun-e-Mashkhel, and Nokkundi where date palm relies mainly on the groundwater (Abul-Soad et al. 2009, 2015). Roots of date palm trees can reach water table within 7–8 m depth, but higher levels must be pumped to provide irrigated plantations with their water requirements. Consequently, date palm productivity will be dropped according to water availability (Abul-Soad et al. 2009).

Fig. 12.8 Siwa oasis19 m below sea level in Egypt in the middle of desert

Fig. 12.9 Scattered date palm oases at Balochistan, Pakistan. In photograph 'Gualishtop oasis,' Balochistan, Pakistan

The date palm oases are threatened by number of factors: frequent and lethal infection with Bayoud in the Northern African countries particularly Morocco and Algeria, declining date palm populations and genetic diversity and hence destabilizing the integrity of the oasis ecosystem. Recently, RPW infestation extended to reach Siwa oasis in Egypt and Mauritania. The depletion of groundwater level, loss of vegetation cover, soil erosion due to overgrazing, urban settlement and disruption, salinity and environmental pollution add additional problems in declining oasis ecosystem.

12.3.4 Induced Diversity of Date Palm

The historic Elche palm grove is an example of introduced oasis system in Europe (Johnson et al. 2013) containing about 180,000 adult date palm trees (Rivera et al. 2015). The palm grove of Elche (Southern Spain) is comprised of completely different genotypes, which are originated from seeds and hence exhibited very high phenotypic diversity. Unlike commercial orchards to get just fruit production, the Elche grove is consisting of old mature palms due to no economic interest mainly for the production of white leaves or lulavs (palma blanca in Spanish) used in Palm Sunday for Christians and Jewish Sokkot religious festivals as there is no or less market value of dates produced. Recently, efforts have been taken by the local Phoenix station and administrations to replace the old palms by the selective genotypes produced from tissue culture like cv. 'Medjool' (Ferry et al. 2002; Rivera et al. 2015). However, seeds were used for the creation of new elite genotypes. In Sindh, Pakistan, one progressive grower sown about 18,000 seeds of different Arabian cultivars in his private orchard. They were sorted out at the time of fruiting and reduced to about 20 superior new lines. In Egypt and Sudan especially in the oasis, and non-date-producing countries, still seeds are used for propagation purpose and emerged trees of inferior quality for animal feeding.

12.3.5 Maintenance of Date Palm Diversification

Studies on the manifestation of the date palm diversification and their maintenance in natural ecosystems and community uses of uncultivated varieties in agro-diversity are limited. However, there are some cultivars which constitute a fairly large component of several natural and induced oasis systems and agricultural areas worldwide. Further, date palm is serving as a source of shelter for both plants and animals and playing an important role in the interdependency among pollinating, seed dispersing, and other ecological functions. Breeding and selection for improving oasis varieties need basic knowledge, and there are several desired characteristics of those cultivars that could be selected for domestication and commercial purposes. To improve the diversity needs, these populations should be maintained under natural or oasis conditions so that they can continue to cross with each other and in the long term continue to evolve new gene pools. Oasis and agro-biodiversity status of date palm on global level are being exposed by rapid genetic erosion because of natural habitat devastation and further financial and cultural pressures, such as habitat destruction, extension of agriculture, conversion of biodiversity rich sites for human settlement and industrial development, and uncontrolled commercial exploitation. Eco-lodge is a recent concept of housing created to overcome the negative impact of urbanization on oasis ecosystem. All materials in the eco-lodge were made from natural substances to keep oasis nature and at same time encourage the tourism at these places (Fig. 12.10).

Commercialization has increased the cultivated area under monoculture condition and reduced species diversity in recent years. Those modern commercial cultivars tend to replace the traditional farmers' cultivars, reducing available diversity and many associated values. An extreme example comes from cv. 'Aseel' (Sindh, Pakistan) replaced numerous other cultivars being utilized before its huge commercialization for 'Chuhhara' purpose in Khairpur district. Establishment of orchards comprised of only single cultivar has increased uniformity and simultaneously susceptibility to pests and diseases. Likewise, cv. 'Aseel' found to be more susceptible to sudden decline in disease of date palm than cv. 'Karblian' and 'Fasli' in Sindh Province, Pakistan (Maitlo 2015). Nevertheless, the famous cultivars such as 'Medjool' and 'Barhee' can replace the seed-originated palms with inferior qualities which usually been used for grazing animals. This is in addition to the good selections from the locally evaluated seed palms with competitive fruit quality. Needless to say seed is not preferred as propagation part for date palm as it is heterozygous except for breeding purpose which is difficult due to long generation of a date palm tree to mature (about 7 years).

Fig. 12.10 The eco-lodge maintains oasis system and encourage tourism

The effective conservation of date palm cultivars presents particular problems, and satisfactory methods have not as yet been developed. In situ/*On-farm* conservation is extremely costly and requires field gene banks because its seeds being heterozygous nature cannot be stored. Furthermore, these gene banks suffer severe losses due to disease and other causes and need constant replacement. The development of in vitro methods for date palm has been reported on large scale, but still there is an urgent need to develop and upscale successful ex situ and in vitro strategies to conserve both natural and agro-biodiversity of date palm.

Most of the date palm institutes and centers are investing in collection and maintenance of date palm diversity in field gene banks as it holds the key for characterization and evaluation and for their utilization in breeding improvement programs. Date Palm Research Institute (DPRI) in Pakistan has set up a collection farm containing large number of Pakistani cultivars. However, they are the commercial and well-known cultivars only. Therefore, trees at this farm were under the continuous attack of RPW which inflamed the cost to maintain such ex situ farm.

Using the nature-gifted trait of seedling-germinated date palms that sometimes possessing fabulous fruit quality or natural resistance to a threat would maintain the date palm biodiversity and can lead to develop new commercial cultivars too. When Bayoud disease has destroyed millions of date palm trees in Morocco and Algeria, some cultivars such as cvs. BSTN and IKL are claimed to be resistant (Saaidi 1992). In fact, these valuable individual date palms can be considered at risk of extinction and should be protected through rapid propagation. In a research project funded by Science, Technology, and Development Fund (STDF), Egypt, a number of superior date palm genotypes from Siwa and Aswan have been selected and evaluated and after then subjected to micropropagation using inflorescence explants (Abul-Soad et al 2016). Production of date palm derived from tissue culture maintains these valuable domestic genotypes from disappearance, as in Fig. 12.11 (Abul-Soad 2011b).

Fig. 12.11 Tissue-cultured date palm plants produced using inflorescence explants ready to be planted in open field

12.3.6 Characterization of Date Palm Cultivars

The characterization of date palm cultivars is of paramount importance toward the conservation of diversity of date palm. There are thousands of cultivars existing, but most of them are either synonyms of one cultivar being cultivated in inter- or intra-zones or varietal lines. The characterization of these cultivars can be done by evaluating the fruit morphological traits, viz. shape, size, weight, color, and texture (El-Houmaizi et al. 2002; Markhand et al. 2010), and by using some biochemical markers such as isozymes and proteins (Baaziz and Saaidi 1988). But the accuracy of such markers is not reliable as sometimes same cultivar grown under two different ecological conditions might exhibit some variations in general appearance. The use of molecular markers for studying genetic diversity of date palm cultivars is a recent trend started few decades ago. It helps to construct genetic maps, which in turn is an important selection tool for breeding programs as the genetic improvement of any crop depends on the ability to select promising plant material (Eissa et al. 2009). Various molecular techniques have been used for date palm cultivar identification from several countries (Bodian et al. 2012a, b) such as random amplified polymorphic DNA (RAPD) markers for Pakistani and Moroccan cultivars (Mirbahar et al. 2014; Sedra et al. 1998), amplified fragment length polymorphism (AFLP) for Egyptian cultivars (El-Assar et al. 2005), inter-simple sequence repeats (ISSR) for Iraqi cultivars (Khierallah et al. 2014), random amplified microsatellite polymorphism (RAMPO) for Tunisian cultivars (Sakka et al. 2004), and microsatellites for Moroccan cultivars (Bodian et al. 2012a).

However, there are huge technical differences in terms of cost, speed, amount of DNA needed, technical labor, degree of polymorphism, accuracy of genetic linkage estimates, and the statistical power of tests (Garcia et al. 2004).

12.3.6.1 Phenotypic

The phenotypic or morphological identification is a traditional but effective and economic way of characterization of date palm germplasm. But these morphological characteristics are usually influenced by the ecological factors (Sedra et al. 1993, 1996) with the limitation of observation only in adult palms (Hammadi et al. 2009). There are numerous reports available on documentation of date palm cultivars using vegetative, flowering, and fruit characteristics. Al-Doss et al. (2001) studied 21 vegetative growth parameters, 13 flowering and yield characteristics, and 11 fruit properties to evaluate the variations among 17 different date palm cultivars from Saudi Arabia. Similarly, Sulieman et al. (2012) studied the physicochemical differences of five Sudanese date palm cultivars, viz. 'Gondeila', 'Barakawi', 'White Gau', 'Red Gau', and 'Black Gau', and recorded significant differences among them in terms of fruit and seed weight, flesh thickness, and fruit length, but the chemical constituents were more or less same. Markhand et al. (2010) conducted an extensive study using several prominent fruit characteristic to evaluate 85 date palm cultivars of Khairpur (Pakistan) on the basis of fruit color, fruit shape and size, fruit type (soft, semidry, and dry), and micropyle position as well as categorized them into early, mid- and late cultivars. Most of the Pakistani dates belong to semidry type and are mid-season cultivars and that is why most of the cultivars face monsoon rains during ripening period and need bunch covering (Abul-Soad et al. 2010). Therefore, early fruit-bearing/fruit-ripening cultivars, escape the monsoon season, may be highly suitable for the cultivation during this season. Ahmed et al. (2010) found significant morphological variations among 21 date palm cultivars from Mauritania and revealed that leaflets length and spine, fruit, and seed sizes accounted for a large proportion of perceived variability. Ismail et al. (2013) performed physicochemical and sensory evaluation of six Tunisian date cultivars and found that each cultivar has its own distinctive characteristics.

Despite of number of studies on phenotypic characterization, it is still very hard to recognize date cultivars, particularly without observing fruiting stage. Therefore, true to type of the planting materials, i.e., offshoots, is necessary in commercial plantation. So far, there is no viable identification method to determine the sex of date palm at the seedling stage before commence of flowering after 4–7 years of planting. In fact, due to the great adaptive flexibility of date palm, many growers and researchers cannot distinguish cultivars outside their oasis and orchards (Munier 1973). Nonetheless, Haider et al. (2012a) proposed an innovative system for grouping of dates on the basis of images by computer vision and pattern recognition in UAE date-processing factories to make ease in sorting of dates during processing.

12.3.6.2 Genotypic or Molecular Marker

Recent advancement in molecular biology and development of different molecular markers enable scientists to accurately identify dates cultivar with their genetic diversity and phylogenetic relationships. Genetic conservation is reliant on understanding the extent, and distribution of the genetic diversity exists in the prevailing germplasm (Jubrael et al. 2005). In addition, due to the accessibility of nuclear (Al-Dous et al. 2011) and chloroplast (Yang et al. 2010; Khan et al. 2011; Soumaya et al. 2014) genome sequences of date palm, a number of molecular marker systems can be applicable for an effective assessment of the genetic diversity for germplasm conservation and exploitation of true-to-type elite material (Khierallah et al. 2011).

RAPD is a PCR-based marker technique, requires only small amount of genomic DNA (usually 10 bp), and is a quite inexpensive way as it does not require blotting and radioactive material like other molecular techniques (Mirbahar et al. 2014; Emoghene et al. 2015), (Fig. 12.12). Several studies conducted using RAPD markers to detect genetic diversity and polymorphism in date cultivars of Saudi Arabia (Al-Khalifah and Askari 2003; Al-Moshileh et al 2004; Abdulla and Gamal 2010; Munshi and Osman 2010), Algeria (Benkhalifa 1999), Iraq (Jubrael 2001; Al-Khateeb and Jubrael 2006; Khierallah et al 2014), Tunisia (Trifi et al. 2000), Egypt (El-Tarras et al. 2002; Soliman et al. 2006; Younis et al. 2008; Moghaieb et al. 2010), Bahrain (Pathak and Hamzah 2008), Syria (Haider et al 2012b),

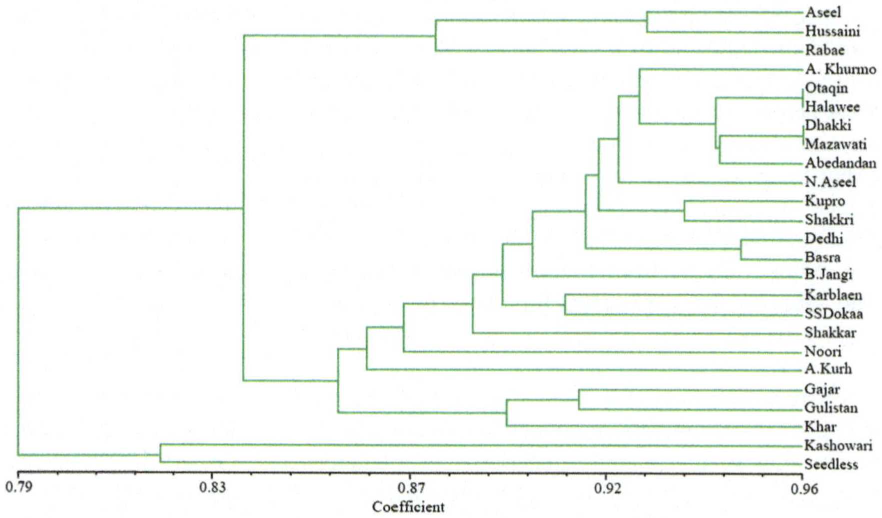

Fig. 12.12 Phylogenetic tree of 25 Pakistani date palm cultivars derived from UPGMA cluster analysis of RAPD markers (Mirbahar et al. 2014)

Morocco (Sedra et al. 1998; Sedra 2013), Nigeria (Emoghene et al. 2015). The main drawback of this marker is dominance, as it hardly discovers heterozygosity and genetic diversity and moderately reliable as the results usually differ from other molecular marker techniques. Soliman et al. (2006) identified the genetic polymorphism for semidry date cultivars in Egypt using RAPD and ISSR markers. Al-Moshileh et al (2004) found RAPD markers reliable to detect polymorphism for identification of data palm cultivars in Saudi Arabia. Furthermore, they found that cv. 'Rothana' was the closest to cv. 'Nabtet Ali,' while cv. 'Sokkari' cultivar exhibited the greatest genetic difference from cvs. 'Barhi' and 'Ajwa.'

The AFLP markers have been successfully employed in many crops to produce high-density linkage maps with ease in detecting even lower levels of variations, especially when large inter-varietal polymorphism among date palm cultivars is screening (Mueller and Wolfenbarger 1999; Diaz et al. 2003; Khierallah et al. 2011). Several reports are available on detecting polymorphism in date palm cultivars using AFLP markers (El-Kharbotly et al. 1998; Cao and Chao 2002; El-Khishin et al. 2003; Diaz et al. 2003; Jubrael et al. 2005; El-Assar et al. 2005; Saker et al. 2006; Khierallah 2007; Khierallah et al. 2011; Sabir et al. 2014a). Elhoumaizi et al. (2006) confirmed that 'Medjool' is a landrace cultivar in Morocco by AFLP analyses of 'Medjool' accessions from Morocco, Egypt, and USA. They found that 'Medjool' is not genetically uniform and most probably evolved by natural selection from a mixture of genotypes under the grown environmental condition. AFLP analysis is more laborious and time–consuming, but having the advantage over RAPD is their high reproducibility and powerful approach to detect polymorphism.

The RFLP markers are co-dominant, reproducible, and highly locus-specific and able to show strong molecular differentiation and phylogenetic analysis between closely related species. The requirement of large amount of DNA (generally 5–10 μg per digest), time–consuming, and high cost are some of the limitations, and hence, very little work has been done on its usage on date palm. Sakka et al. (2003, 2004) used PCR-based RFLP analysis to identify genotype and genetic polymorphisms in plastid DNA from 38 Tunisian female date palm accessions including 5 male ecotypes. Al-Mahmoud et al. (2012) developed early sex detection of date palm using PCR-based RFLP. Corniquel and Mercier (1997) used RFLP for generation of cultivar-specific hybridization of four date palm cultivars.

Microsatellites or simple sequence repeats (SSR) are usually comprised of 1–6 nucleotide repeats. Their abundance and dispersion throughout the diverse genomes, highly polymorphic, species-specific, and co-dominance in inheritance make them ideal DNA markers for cultivar identification and detection of germplasm diversity and genetic mapping. Such potential of SSR in addition to the published linkage maps of SSR markers recently diverted the attention of date palm researchers from other markers for identification and constructed gene-based maps of date palm (Ahmed and Al-Qaradawi 2009; Pintaud et al. 2010; Bodian et al. 2012b; Zhao et al. 2012, 2013; Elmeer and Mattat 2015; Maryam et al. 2016). SSRs are easier to use as compared to RFLPs due to its smaller amount of DNA requirement and ability to automate assay, but the major disadvantage is their high cost of development.

ISSR markers are powerful tool to study the inter- and intra-specific genetic variations in date palm and an easy approach with highly reproducible and multiple genomic loci target ability. Usually, they are dominant in nature involving 16–25 bp as primers. Several works have been done so far on assessing genetic diversity using ISSR markers in date palm cultivars of Saudi Arabia (Kumar et al. 2010; Munshi and Osman 2010), Egypt (Younis et al. 2008), Kutch, India (Srivastav et al. 2013), Pakistan (Mirbahar et al. 2013) Tunisia (Zehdi et al. 2002; Karim et al. 2010; Zehdi-Azouzi et al. 2011; Hamza et al. 2012), Iraq (Khierallah et al. 2014), Morocco (Bodian et al. 2012a), etc.

Single nucleotide polymorphisms (SNPs) are the most potential markers to create a very high-density genetic map of plants. The application of SNPs in studying the date palm genome sequences has been recently employed by group of date palm researchers. Al-Dous et al. (2011) assembled first time the draft genome for cv. 'Khalas' which is also the first resource among the member of order Arecales. Sabir et al. (2014b) revealed heteroplasmy and close phylogenetic relationship among nine Saudi Arabian cultivars by sequencing the mitochondrial and chloroplast genomes using SNP analysis. Mathew et al. (2015) conducted genotyping-by-sequencing from fruit and leaf samples of 70 female cultivars of date palm from different countries and generated genome-wide genotypic data for 13,000–65,000 SNPs. They found that date palm cultivars segregate into two main regions of shared genetic background from North Africa and Arabian Gulf. Recently, Hazzouri et al. (2015) achieved a complete catalog of about seven million SNPs from 62 cultivars of date palm based on whole genome resequencing. The findings indicated the major genetic division between North Africa and the Middle East/South Asian date palms, with evidence of admixture in cultivars from Egypt and Sudan.

12.3.7 Utilization of Genetic Diversity

Date palm sometimes referred as the 'tree of life' due to its ability to thrive in desert environment and its multipurpose utilization by nomadic tribes and the local inhabitants. Dates, apart from its pleasant taste, are very high-energy food containing all important nutritious constituents and still an important diet of people living in desert. Besides nutritional values, almost each and every part of the date palm tree has its utilization. The trunk sections are used to build supporting beams and roofs, tables, and leaves for fences (Fig. 12.13). Trunk beams were noticed inserted between stones to build the Citadel of Qaitbay which was established in 1477 AD as a defensive fortress located on the Mediterranean Sea coast, in Alexandria, Egypt. The leaflets are used for making hand fans, hats, mats, ropes, baskets, and many artifacts. Fruit stalks and leaves basis are used as fuel. Dates yield value-added food products such as date chutney or sweet pickle, date vinegar, date paste for bakery products, dates syrup or liquid sugar, dates honey and used in many confectionery items in date-producing countries. The date seeds pressed or mixed with other fodder are also used for animal feed. Date palm is becoming increasingly a popular tree among landscaping trees as date palm now surrounding new shopping malls, hotels, and housing developments in many countries (Jatoi 2013). Moreover, date palms have been proven as one of high salt tolerant plant and, thus, recommended to tackle desertification phenomena (Bouchireb and Clark 1997). Cultivating date palm trees can change the microclimate of the area which

(a) (b)

Fig. 12.13 Date palm uses. **a** In Oman, using trunk for decoration to make tables and frond midrib for chairs, and **b** trunk beams used to build the Citadel of Qaitbay located on the Mediterranean Sea coast, in Alexandria, Egypt

allow intercropping of other species without interference in fertilizer or irrigation requirements. The rational and sustainable utilization of date palm genetic resources requires extensive characterization, evaluation, and documentation of the germplasm available at a local or regional level (Jaradat 2011). Some seed-derived plants could show characteristics of a commercial value and prove more adaptable with the local climate. These need to be discovered and utilized at same area or similar climates. According to Sudhersan et al. (2015), the diversity is the basis of adaptation that needed to meet the unpredictable environmental changes (climate change) and several pest and disease problems.

In Oasis, the rich diversity of date palm plays a key role as an important source of food and shelter to other plant species and animals, providing stability in complex natural ecosystems (Jaradat 2015; Bouguedoura et al. 2015). In agricultural areas, date palms are vital constituents of multi-crop systems and are rich in diversity having a number of economic, social, cultural, aesthetic, and ecological functions important to livelihoods. Increasing date palm diversity in both oasis and agricultural areas is a sustainable way to conserve date palm biodiversity. From the conservation standpoint, Oasis can serve as a buffer zones to protect loss of biodiversity of date palm. Date palm being components of natural oasis ecosystem diversity is also contributing some ecosystem services, for instance, carbon reclamation, protection of soil erosion, soil formation, nutrient cycling, hydrologic systems, and biotic regulation in the form of pollination ecology and food web. The maintenance of date palm genetic resources therefore becomes a major issue for the development of date palm production and food security in arid and semiarid regions.

Oasis in Arabian Peninsula and North Africa are the centers of origin, and diversity for several leading cultivars possesses very large amounts of genetic diversity. Fruits are important for both local communities and other products such as for specific medicinal uses and livestock feed. Date palm can be grown in marginal conditions and are therefore important to improve agricultural production under low input conditions, especially inability of farmers to afford fertilizers and irrigation in Balochistan (Pakistan) (Abul-Soad et al. 2015), and Jigawa state, Sudan (Adesiji et al. 2013). The livelihood activities and preference in cultivating and maintaining genetic resources often have impact on biodiversity of any crop (Asrat et al. 2010) and thus implicating the cooperation of farmers as a necessary step or strategy toward date palm germplasm conservation (Ezebilo et al. 2013).

12.4 Conservation of Genetic Diversity

The richest diversity of date palm having thousands of cultivars and countless varietal strains is present and actively managed through generations of date palm growers using traditional agricultural knowledge. But, now, this diversity is under tremendous pressure due to several biotic and abiotic factors and extensive commercialization of date cultivation from traditional farming to commercial orchards that depends on a relatively few commercial cultivars. This decline has accelerated throughout the twentieth century. Replacing the traditional varieties by commercial or exotic cultivars is one of the major causes of genetic erosion. Moreover, the popularization of tissue culture technology that only focusing in propagation of very limited number of date palm cultivars is also an alarming situation in near future for great loss of date palm diversity. The implementation of conservation strategies of the date palm diversity is vital for the welfare of the peoples living in both oasis and agricultural areas of dates producing regions. A global assessment of genetic diversity of current date palm accessions therefore has become necessary together with the global strategies for the date palm germplasm conservation (Mancini 2010). Hence, an integrated system based on dynamic ex situ and in situ conservation is the need of the day for maintaining, enhancing the date palm diversity, and utilizing to meet the challenges of climate change, biotic and abiotic challenges.

12.4.1 Methods of Conservation

There are two approaches of conservation of plant genetic resources termed as in situ and ex situ conservation strategies.

12.4.1.1 In Situ Conservation or on-Farm Conservation

In situ as the term indicates literally mean 'in place,' involving conservation of plants in its natural habitat to which it is adapted and maintained by farmers within the traditional agricultural systems and allows the recovery of germplasm in their natural surroundings (Singh 2009; Rao and Sthapit 2012). Date palm growers and lovers are playing an important role in preserving the biodiversity of traditional date palm grove and gardens by continuous use of century-old practices in maintaining the traditional date cultivars and propagation of the newly developed races with distinctive properties. Since the conservation of such agro-biodiversity is carrying out as on-farm, therefore such type of conservation is also termed as on-farm conservation. In such a way, the genetic diversity of target species is managed and wild plants have been maintained within the traditional agricultural or horticultural systems (Maxted et al. 1997). It has helped the species to adapt gradually with new variations in the gene pool caused by environmental conditions such as global warming, changed rainfall patterns (Heywood and Dulloo 2005). Sustainable on-farm and in situ conservation of date palm diversity is only promising when farmers, academia, and government organizations show interest in recognizing the benefits in terms of genetic, economic, social, and environmental point of view and by implementing the private utility benefits to the individual grower or user.

On-farm conservation is encouraging in several Middle East and North African countries as a potential method of date palm conservation strategy. When farmers get motivation from the state, they also show their interest in participating this global cause (Jaradat 2015). Nowadays, in date-producing countries, the major crops growing concern is to exchange the information to develop the date palm sector. The great advance and wide usage of social media programs made groups with direct contact able to exchange the photographs and movies instantly either on national or on international levels. This is expected to not only encourage the on-farm conservation but also support all other activities in date palm. For example, the active group on Facebook which named 'Khobraa Wa Montgi Nakhil Al Balah Fi Misr' which include more than 5000 person working on different aspects of date palm in the Middle East (https://www.facebook.com/groups/1601120226777268/). Sometimes, the problem of language gap is there as the language of this group is Arabic. Another group with English language from Pakistan on WhatsApp is 'SAGP Dates Value Chain' which involve the most active group of Sindh Agricultural Growth project funded by the World Bank to develop the dates sector in Sindh, Pakistan. The recent third group from Saudi Arabia in Arabic language is 'Nadi Sosit Al Nakhil' or in English 'The club of RPW' on WhatsApp. This group includes eminent scientist from 57 Universities and Institutes of 16 countries. Every night during five days from the week, a single voice lecture is delivering by a member and debate to exchange the information about all activities of date palm. It is a corner stone to create a Web site including more than 6000 full articles on date palm till the moment.

The status of on-farm conservation of date palm is still limited. However, there are numerous small-scale conservations or rather germplasm collection stations or farms maintaining the local cultivars in more or less all date-producing countries. This could keep the progeny of elite landraces and commercial cultivars of a limited population at same place and prevent losing such valuable genetic resources. It is a practice for the date palm growers in the non-systematic farms to regularly clean their orchards by detaching the offshoots from parent female productive trees and plant them once again in between the adult trees or establish a new orchard. Same practice was performed at the Date Palm Repository of Agricultural Research Station, Coachella Valley, University of California, USA (Fig. 12.14).

Fig. 12.14 On-farm conservation of the offshoots of different cultivars at the Agricultural Research Station, California, USA

12.4.1.2 Ex Situ Conservation

When conservation of plant genetic resources attempted to perform outside or away from their natural habitat, it is termed as ex situ conservation. It can be done by seed and DNA storage, gene banks, collection farms, in vitro preservation or cryopreservation, and botanical gardens (Bekheet and Taha 2013). There are some efforts have been done in date palm ex situ conservation that can lead to preserve date palm germplasm for the purposes of successful propagation and improvement programs.

Seed Bank

The seed bank conservation is one of the most widespread and valuable ex situ conservation approach maintaining seed viability at low temperatures and by desiccation. As compared to the 'orthodox' seeds which can be stored for longer durations at subzero temperatures (Roberts 1975), the date palm seeds being 'Recalcitrant' and heterozygous nature cannot be stored for the purpose of conserving genetic resources (Bekheet 2011).

In Vitro Conservation/Repository

In vitro conservation or in vitro gene bank of the plant genetic resources, various tissues are used such as shoot tips, axillary buds, embryos, callus, and cell suspension cultures (Singh 2009). They are easy to maintain, less expensive, and effective way of storing the plant genetic resources particularly the dioecious nature plants like date palm. In vitro conservation basically involves two stages: at first the in vitro culture establishment and secondly in vitro storage. The in vitro cultures can be conserved for short time (less than one year) or for years to loss some of the viability after sawing once again because of freezing damage.

(a) **Slow growth or Cold storage-in vitro cultures**

Slow growth methods are used to conserve plant cultures for relatively longtime storage (few months) by reducing the growth parameters either the temperature and light intensity, adding growth inhibitors, reducing O_2 concentration, modifying the nutrient medium which includes dilution of mineral elements, reducing sugar concentrations, and by changing the use of growth regulators, choosing small explants, adding chemicals with osmotic properties (Engelmann 1991, 2011; Wang et al. 1993; Lédo et al. 2014). Depending on the plant species, slow growth technique allows cultures to be held for 1–15 years under tissue culture regimes with periodic subculturing (Jain 2011a, b). However, the high costs of labor and the potential risks of somaclonal variation for some species are the major problems (Cruz-Cruz et al. 2013). There are certain limiting factors of this technique, i.e., reducing temperature cannot be handled effectively when tropical plant species are concerned due to their higher temperature growth habit (Withers 1991). Not all types of explants were tested in date palm. Shoot tip explants and callus cultures were successfully employed through slow growth conservation of cv. 'Zaghloul' for 12 months at 5 °C in the darkness (Bekheet et al. 2001). Callus explants of cv. 'Gundila' were also successfully applied for slow growth conservation for the period of 6 and 12 months. The modified medium contained 0.3 M of different sugars with the recovery of 90.73% after four weeks of thawing in normal conditions (Zaid et al. 2011). Incubation temperature during slow growth conservation is reduced from 27 to 15 °C for callus cultures of cv. 'Sakkoty' (El-Dawayati et al. 2012). It is necessary to get high survival rate after thawing which reached 88. 8% in cv. 'Sakkoty' and 87.7% in cv. 'Bartamoda' after 12 months at 15 °C, but the number of germinated embryos/culture decreased from the conserved embryogenic callus (El-Ashry et al. 2013). However, slowing the growth of callus culture under appropriate conditions could allow enough time for maturation. Subsequently, it will increase the developed somatic embryos. In addition to that, stored callus culture could serve as stock for micropropagation as per need. Also, in vitro cultures can be transferred safely through cargo while they are under slow growth medium. It is worth to mention that the in vitro cultures of date palm are very sensitive for endogenous bacterial contamination. At any time, it may appear specially if the growth condition was inappropriate. Unfortunately, there is no published work on the bacterial contamination or somaclonal variation after thawing in date palm.

(b) **Cryopreservation**

Cryopreservation involves maintaining of living cells and tissue organs at ultralow temperature or in liquid nitrogen (between −79 and −196 °C) for longer periods by halting all the metabolic activities and cell division. Thus, cells will not undergo genetic changes or somaclonal variations during storage as compared to serial sub-culturing where the cultures are exposed to the risks of contamination and handling errors (Engelmann 2004; Cruz-Cruz et al. 2013). There are two modes of cryopreservation protocols based on their physical mechanisms. The classical cryopreservation technique is performed in the presence of ice or ice formation, while the vitrification usually does not involve the ice formation (Gonzalez-Arnao et al. 2008).

Since date palm is a dioecious plant, conservation of its genetic resources, using cryopreservation is the best solution being cost-effective and requires small space with the capacity to store large genetic resources without the fear of natural disasters, disease outbreaks, etc. Depending on the plant species and type of cultivars, the cryopreservation technique involves several steps such as selection of the plant material preferably young rapidly growing material, which show resistance against freezing due to smaller size, fewer or small number of vacuoles, and dense cytoplasm; pretreatment of explants in a medium containing osmotically active compounds for dehydration of the tissues and protection of cell membranes; freezing requires to avoid injuries through ice crystal formation); storing at a freezing point where the metabolism activity is suppresses; thawing is done to prevent damage of the cells from the intracellular ice crystals; and post-cryo-treatment minimizes the toxic effect of cryoprotectants and reduce the osmotic shock.

Several explants have been tested in date palm cryopreservation technique using caulogenic meristem (Fki et al. 2013), friable callus (Bekheet et al. 2007; Subaih et al. 2007; Al-Bahrany and Al-Khayri 2012), pro-embryogenic masses (Fki et al. 2011), somatic embryos (Bekheet et al. 2005), shoot apices (Bagniol et al. 1992), and pollen (Mortazavi et al. 2010).

Cryopreservation of nodular callus of cv. Zaghloul was initially achieved at 0 °C for 2 h and then transferred into liquid nitrogen (−196 °C) for 48 h. The recovery percentage after thawing was 80% on 1 M sucrose-pre-culture medium (Bekheet et al. 2007). Three different cryopreservation protocols, i.e., standard (tube) vitrification, droplet-vitrification, and encapsulation-vitrification, were tested on cv. 'Khenizi.' The standard vitrification gave the maximum recovery rates using small explants (2 mm), while the larger explants (>3 mm) died after thawing stage (Fki et al. 2013). Cryopreservation of polyembryonic masses (PEMs) of cvs. 'Sokary' and 'Sultany' using droplet-vitrification and dehydration cryo-plate techniques has been carried out. There recovery percentages of pro-embryos or PEM of cvs. 'Sokary' were 90.9–98.6% and for 'Sultany' 85.6–88.0% after transfer to the standard culture medium containing 3.3 M glycerol + 2.4 M ethylene glycol + 0.4 M sucrose + 1.9 M dimethyl sulfoxide (Salma et al. 2014). Recently, the effect of salt mixture (NaCl, MgCl, and $CaCl_2$) along with other osmotic stimulators such as mannitol and polyethylene glycol (PEG) for cryopreservation of embryogenic cultures of cv. 'Sakkoty' of date palm was investigated. The highest values of fresh mass were at salt tolerance ratio of 1500 ppm (Bekheet 2015).

Field Genebanks or Field Repository

The date palm genetic resources are maintained by vegetative propagation for maintaining their genetic makeup true to type in the field genebanks, which is also known as germplasm collections, varietal collections, or clonal repositories providing the long-term preservation of the genetic or inter-specific variability. In this conservation strategy, there is always a risk of damage by natural disasters, pest and pathogen problems, etc. (Singh 2009). The field genebanks of date palm is though relatively expensive to maintain and requires huge space but is providing easy and ready access to conserve palms for research as well as for their use.

There are number of field genebanks in almost all date-producing countries such as: the date palm field collection of 49 local date palm cultivars located at stations of the Technical Institute of Saharan Agricultural Development (ITDAS) Ain Ben Noui and Feliache in Biskra and El Arfi ane in Oued Righ (Bouguedoura et al. 2015); the field collection of 25 local date palm cultivars at the nursery of Date Palm Research Institute, Shah Abdul Latif University, Khairpur, Pakistan (Markhand et al. 2010; Abul-Soad et al. 2015); the research stations, located in Degache, Ibn Chabbat I, Ibn Chabbat II, and Atilet, Tunisia, maintaining 100 endangered cultivars (Hamza et al. 2015); the date palm germplasm bank at King Faisal University, in southeastern Al-Hassa (Saudi Arabia), comprising 31 Saudi Arabian cultivars collected from 7 major growing regions (Al-Hassa, Qatif, Madinah, Besha, Najran, Jouf, and Qassim) and 26 exotic cultivars (Al-Ghamdi 2001; Aleid et al. 2015). A project started by General Board of Date Palm (GBDP) with the help of the ministry of Agriculture in Iraq has collected 497 cultivars in various regions of Iraq (Khierallah et al. 2015). The Kuwait Institute for Scientific Research (KISR) at Kuwait university main campus maintains 34 female and 6 male cultivars (Sudhersan et al. 2015). Sixty genotypes/cultivars at CIAH, Bikaner (Rajasthan, India), have been conserved in a national field genebank (Pareek 2015). The establishment of such collection farms can help in evaluating and comparing the fruit quality of alien cultivars at the experimental stations. This will definitely help growers to take the right decisions and introduce valuable alien cultivars that suit the local environment.

12.5 Conclusions and Prospects

Date palm is playing an integral role in the livelihood of the people living in the hot and barren desserts since ancient times, and was exclusively mentioned and depicted in old sculptures in many ancient civilizations. Therefore, many countries claim themselves its center of origin. The Arab nomads and later the military expeditions expanded date palm distribution to the Europe from where it reached to American continent. Recently, the tissue culture technology motivated and enabled its cultivation to other non-date-producing regions of the world.

Date palm belongs to the genus *Phoenix,* family *Arecaceae,* and has more or less 5000 cultivars and countless varietal strains worldwide. They are sometimes similar with the same cultivars being cultivated in different regions and countries, but with new names. In addition, sometimes same cultivars exhibited different fruit types like in case of cv. 'Deglet Noor' which appeared dry type in its place of origin (Morocco) and semi-dry in USA and Pakistan. In such case, characterization of the date cultivars by using the reliable molecular markers is of preference in comparison with the typical phenotypic identification methods and biochemical markers.

The natural oasis consists mainly of seedling palms and usually produces inferior date fruits but exhibits rich diversity which needs to be maintained and conserved in situ or ex situ. They are frequently threatened by several pest and pathogen problems besides numerous abiotic factors which are disrupting and declining the oasis ecosystems. Recently the trend of establishing the date palm orchards with only single or few top commercial cultivars by replacing the poor yielder cultivars has become the cause of rapid loss of numerous cultivars and erosion of natural diversity. For example, in Sindh Province (Pakistan), cv. 'Aseel' has replaced hundreds of other cultivars in 100 years for the sake of commercial purposes. And hence, depending on a single cultivar has increased the risk of sudden decline in disease of date palm since 2 or 3 decades which has destroyed many commercial orchards in the area.

The date palm diversity, maintained by generations of growers in the form of in situ or on-farm conservation using traditional knowledge, is of no use nowadays without the proper interest of growers and encouragement of government organizations and hence urging the importance of adoption of modern way of conservation. The communication revolution could help to gather the efforts through the new applications of social media. The in vitro germplasm conservation with slow growth of cultures and cryopreservation techniques is promising and requires less space and easy maintenance, however may face genetic or somaclonal variation among the regenerated plants. The initiatives of establishing date palm collection farms taken by many dates producing countries are good but not enough, and hence, further efforts needed to maintain the genetic diversity of date palm worldwide.

References

Abdulla M, Gamal O (2010) Investigation on molecular phylogeny of some date palm (*Phoenix dactylifera* L.) cultivars by protein, RAPD and ISSR markers in Saudi Arabia. Aust J Crop Sci 4:23–28

Abul-Soad AA (2011a) Micropropagation of date palm using inflorescence explants. In: Jain SM, Al-Khayri JM, Johnson DV (eds) Date palm biotechnology. Springer, Dordrecht, pp 91–118

Abul-Soad AA (2011b) Date palm in Pakistan, current status and prospective. Report USAID Pakistan

Abul-Soad AA, Mahdi SM (2010) Commercial production of tissue culture date palm (*Phoenix dactylifera* L.) by inflorescence technique. J Genet Eng Biotech 8:39–44

Abul-Soad AA, Mahdi SM (2012) Micropropagation of date palm males and females. Bless Tree 4:80–86

Abul-Soad AA, Markhand GS, Shah SA (2009) Study on potential impact of lowering of groundwater on date palm plantations. Final plantations. Final report, Hagler Bailly Pakistan (Pvt.) Ltd. Islamabad, Pakistan

Abul-Soad AA, Markhand GS, Memon S (2010) DuPont Tyvek® bags impact on dates. Final report. DuPont Pakistan Operations (Pvt.) Ltd. Lahore, Pakistan

Abul-Soad AA, Maitlo WA, Markhand GS, Mahdi SM (2011c) Date palm wilt disease (sudden decline syndrome) in Pakistan, symptoms and remedy. Blessed Tree 3:38–43

Abul-Soad AA, Mahdi SM, Markhand GS (2015) Date palm status and perspective in Pakistan. In: Al-Khayri JM, Jain SM, Johnson DV (eds) Date palm genetic resources and utilization, vol 2. Asia and Europe. Springer, Netherlands, pp 153–205

Abul-Soad AA, Emara KS, Abdallah AS, Mahdi SM (2016) Conservation of superior Egyptian date palm genotypes through *in vitro* culture of inflorescence explants. In: Proceedings of the 2nd international conference for date palm (ICDP), Qassim University, Qassim, Kingdom of Saudi Arabia, 10–12 October 2016 (Under publication)

Adesiji G, Olujide M, Bolarin O, Sanusi M, Komalafe S (2013) An assessment of date palm technologies utilized by the farmers in Dutse local government area of Jigawa state, Nigeria. J Sustain Dev Africa 15:11–24

Ahmed TA, Al-Qaradawi AYK (2009) Molecular phylogeny of Qatari date palm genotypes using simple sequence repeats markers. Biotechnol 8:126–131

Ahmed MVO, Bounaa ZO, Fouteye M, LemineaM DjehaTKO, Mokhtarb T, Salema AOM (2010) Use of multivariate analysis to assess phenotypic diversity of date palm (*Phoenix dactylifera* L.) cultivars. Sci Hort 127:367–371

Aisueni NO, Ikuenobe CE, Okolo EC, Ekhator F (2009) Response of date palm (*Phoenix dactylifera*) seedlings to organic manure, N and K fertilizers in polybag nursery. Afr J Agric Res 4:162–165

Al-Abbad A, Al-Jamal M, Al-Elaiw Z, Al-ShreedF Belaifa H (2011) A study on the economic feasibility of date palm cultivation in the Al-Hassa Oasis of Saudi Arabia. J Dev Agric Econ 3:463–468

Al-Bahrany AM, Al-Khayri JM (2012) Optimizing in vitro cryopreservation of date palm (*Phoenix dactylifera* L.). Biotechnol 11:59–66

Al-Bakr A (1972) The date palm. Its past and present status. Alani Press, Baghdad

Al-Doss AA, Aly MA, Bacha MA (2001) Morphological and agronomical variations among some date palm cultivars grown in Saudi Arabia using principal component and cluster analysis. King Saud Univ Agr J 13:3–18 (In Arabic)

Al-Dous EK, George B, Al-Mahmoud ME, Al-jaber MY, Wang H, Salameh YM, Al-azwani EK, Chaluvadi S, Pontaroli AC, Debarry J, Arondel V, Ohlrogge J, Saie IJ, Suliman-Elmeer KM, Bennetzen JL, Kruegger RR, Malek JA (2011) De novo genome sequencing and comparative genomics of date palm (*Phoenix dactylifera*). Nat Biotechnol 29:521–527

Aleid SM, Al-Khayri JM, Al-Bahrany AM (2015) Date palm status and perspective in Saudi Arabia. In: Al-Khayri JM, Jain SM, Johnson DV (eds) Date palm genetic resources and utilization, vol 2. Asia and Europe. Springer, Netherlands, pp 125–168

Al-Ghamdi AS (2001) Date palm (*Phoenix dactylifera* L.) germplasm bank in King Faisal University, Saudi Arabia. Survival and adaptability of tissue cultured plantlets. Acta Hort 450:241–244

Alhudaib K, Arocha Y, Wilson M, Jones P (2007) Al-Wijam, a new phytoplasma disease of date palm in Saudi Arabia. Bull Insectology 60:285–286

Al-Khalifah NS (2006) Metaxenia: influence of pollen on the maternal tissue of fruits of two cultivars of date palm (*Phoenix dactylifera* L.). Bangladesh J Bot 35:151–161

Al-Khalifah NS, Askari E (2003) Molecular phylogeny of date palm (*Phoenix dactylifera* L.) cultivars from Saudi Arabia by DNA fingerprinting. TheorAppl Genet 107:1266–1270

Al-Khateeb TA, Jubrael JM (2006) The use of RAPD markers for sex and male cultivars identification in (*Phoenix dactylifera* L.) in Iraq. Acta Hort 736:162–170

Al-Khayri JM, Jain SM, Johnson DV (2015a) Date Palm date palm genetic resources and utilization, vol. 2-Asia and Europe. Springer, Netherlands

Al-Khayri JM, Jain SM, Johnson DV (2015b) Date Palm genetic resources and utilization, vol. 1-Africa and Americas. Springer, Netherlands

Al-Mahmoud ME, Al-Dous EK, Al-Azwani EK, Malek JA (2012) DNA-based assays to distinguish date palm (*Arecaceae*) gender. Amer J Bot 99:7–10

Al-Moshileh AM, Motawei MI, Al-Wasel A, Abdel-Latif T (2004) Identification of some date palm (*Phoenix dactylifera* L.) cultivars in Saudi Arabia using RAPD fingerprints. Agr Marine Sci 9:1–3

Al-Yahyai R, Al-Khanjari S (2008) Biodiversity of date palm in the sultanate of Oman. Afr J Agric Res 3:389–395

Al-Yahyai R, Khan MM (2015) Date palm status and perspective in Oman. In: Al-Khayri JM, Jain SM, Johnson DV (eds) Date palm genetic resources and utilization, vol 2. Asia and Europe. Springer, Netherlands, pp 207–240

Asrat S, Yesuf M, Carlsson F, Wale E (2010) Farmers' preferences for crop variety traits: lessons for on-farm conservation and technology adoption. Ecol Econ 69:2394–2401

Ba-Angood S (2015) Date palm status and perspective in Yemen. In: Al-Khayri JM, Jain SM, Johnson DV (eds) Date palm genetic resources and utilization, vol 2: Asia and Europe. Springer, Netherlands, pp 241–263

Baaziz M, Saaidi M (1988) Preliminary identification of date palm cultivars by esterase isoenzymes and peroxidase activities. Can J Bot 66:89–93

Bagnioll S, Engelmann F, Michaux-Ferriè N (1992) Histo-cytological study of apices from in vitro plantlets of date palm during a cryopreservation process. Cryo-Lett 13:405–412

Bailey HLH (1976) Hortus Third. Macmillan, New York

Barrow SA (1998) Monograph of *Phoenix* L. (*Palmae*: *Coryphoideae*). Kew Bull 53:513–575

Bashah MA (1996) Date variety in the Kingdom of Saudi Arabia. Guide booklet for palms and dates. King Abdul aziz Univ Press, Jeddah, pp 1225–1319

Battaglia M, Ghsera B, Mancini M, Bergesio C, Camussi A, Racchi ML (2015) Date palm status and perspective in Libya. In: Al-Khayri JM, Jain SM, Johnson DV (eds) Date palm genetic resources and utilization, vol 1. Africa and the Americas. Springer, Netherlands, pp 223–255

Beccari O (1890) Revista monografica delle species del genera *Phoenix L.* Malesia 3:345–416

Beech M (2003) Archaeological evidence for early date consumption in the Arabian Gulf. The date palm: from traditional resource to green wealth. Emirates Centre for Strategic Studies and Research, Abu Dhabi, pp 11–33

Bekheet SA (2011) In vitro conservation of date palm germplasm. In: Jain SM, Al- Khayri JM, Johnson DV (eds) Date palm biotechnology. Springer, Netherlands, pp 337–360

Bekheet SA (2015) Effect of cryopreservation on salt and drought tolerance of date palm cultured in vitro. Sci Agr 9:142–149

Bekheet SA, Taha HS (2013) Complementary strategy for conservation of date palm germplasm. Glob J Biodivers Sci Manag 3:96–107

Bekheet SA, Taha HS, Saker MM (2001) In vitro long-term storage of date palm. Biol Plant 45:121–124

Bekheet SA, Taha HS, Elbahr MK (2005) Preservation of date palm cultures using encapsulated somatic embryos. Arab J Biotechnol 8:319–328

Bekheet SA, Taha HS, Saker MM, Solliman ME (2007) Application of cryopreservation technique for in vitro grown date palm (*Phoenix dactylifera* L.) cultures. J Appl Sci Res 3:859–866

Bender DA, Bender AE (2005) A dictionary of food and nutrition. Oxford University Press, New York

Benkhalifa A (1999) Gestion de la diversitégénétique du palmier dattierenAlgérie. Paper presented at the workshop constitution etorganisationd'équipes de recherchescientifi que dans les domaines de foresterie et des arbresfruitiers, Marrakech, 13–15 April

Bodian A, Elhoumaizi MA, Ndoye Ndir K, Hasnaoui A, Nachtigall M, Wehling P (2012a) Genetic diversity analysis of date palm cultivars from Figuig oasis (Morocco) using SSR markers. Int J Sci Adv Technol 2:96–104

Bodian A, NdoyeNdir K, Diedhiou PM, Borgel A, Sane D, Sagna M (2012b) Analysis of genetic diversity of date palm (*Phoenix dactylifera* L.) cultivars from Mauritania using microsatellite markers. Int J Sci Adv Technol 2:36–43

Bouchireb N, Clark MS (1997) The application of biotechnology to date palm culture. In: Watanabe KN, Pehu E (eds) Plant biotechnology and plant genetic resources for sustainability and productivity. Academic Press, San Diego, USA, pp 183–195

Bouguedoura N, Bennaceur M, Babahani S, Benziouche SE (2015) Date palm status and perspective in Algeria. In: Al-Khayri JM, Jain SM, Johnson DV (eds) Date palm genetic resources and utilization, vol 1. Africa and the Americas. Springer, Netherlands, pp 125–168

Bourou S, Woin N, Elhoumaizi AM (2006) Synthèse des rapports de projet: Projet FAO/CMR/3003, Promotion de production du palmier dattier dans la zone du Nord-Cameroun

Bourou S, Woin N, Elhoumaizi MA (2015) Date palm status and perspective in Cameroon. In: Al-Khayri JM, Jain SM, Johnson DV (eds) Date palm genetic resources and utilization, vol 1. Africa and the Americas. Springer, Netherlands, pp 257–323

Brac de la Perrière RA, Benkhalifa A (1989) Identification de cultivars de dattiers (*Phoenix dactylifera* L.) du sud-ouestalgérien. Plant Gen Res Newsletter, FAO/IBPGR.78/79:13–20

Cao BR, Chao C-CT (2002) Identification of date cultivars in California using AFLP markers. HortSci 37:966–968

Chao CCT, Krueger RR (2007) The date palm (*Phoenix dactylifera* L.): overview of biology, uses, and cultivation. HortSci 42:1077–1082

Chevalier A (1952) Recherches sur les *Phoenix* africains. Rev Intl Bot Appl 32:205–236

Cornicquel B, Mercier L (1997) Identification of date palm (*Phoenix dactylifera* L.) cultivars by RFLP: partial characterization of a cDNA probe that contains a sequence encoding a zinc finger motif. Int J Plant Sci 158:152–156

Costantini L (1985) Considerazionisualcunireperti di palma da dattero e sulcentro di origine e l'area di coltivazionedella *Phoenix dactylifera* L. In: Gnoli G, Lanciotti L (eds) Orientalia Josephi Tucci Memoriae Dicata. Serie Orientale 56(1). IstitutoItaliano per il Medio ed Estremo Oriente, Rome, pp 209–218

Cruz-Cruz CA, González-Arnao MT, Engelmann F (2013) Biotechnology and conservation of plant biodiversity. Resources 2:73–95

Das S (2009) Past and present status of date palm cultivation in Jessore District. Bangladesh Agricultural University, Mymensingh, pp 1–52

Diaz S, Pire C, Ferrer J, Bonete MJ (2003) Identification of *Phoenix dactylifera L.* varieties based on amplified fragment length polymorphism (AFLP) markers. Cell Mol Biol Lett 8:891–899

Dowson VHW (1982) Date production and protection. Technical paper no. 35, FAO publication, Rome, Italy

Eissa EA, Abdelrazak AB, Elsharabishi SF, Rizk RM (2009) Morphological and molecular genetic characterization of soft date palm (*Phoenix dactylifera* L.) cultivars in Egypt. Egypt J Genet Cytol 38:269–284

El-Ashry AA, Shaltout AD, El-Bahr MK, Abd EL-Hamid A, Bekheet SA (2013) In vitro preservation of embryogenic callus cultures of two Egyptian dry date palm cultivars at darkness and low temperature conditions. J Hort Sci Ornament Plants 5:118–126

El-Assar AM, Krueger RR, Devanand PS, Chao CT (2005) Genetic analysis of Egyptian date (*Phoenix dactylifera* L.) accessions using AFLP markers. Genet Res Crop Evol 52:601–607

El-Dawayati MM, Zaid ZE, Elsharabasy SF (2012) The effect of conservation on steroids contents of callus explants of date palm cv. Sakkoti. Aust J Basic App Sci 6:305–310

Elhoumaizi MA (2007) Promotion de la production du palmier dattier dans la zone Nord du Cameroun. Rapport de projet TCP/CMR/3003

Elhoumaizi MA, Panchanoor SD, Fang J, Chao CCT (2006) Confirmation of ' Medjool ' date as a landrace variety through genetic analysis of Medjool accessions in Morocco. J Am Soc Hortic Sci 131:403–407

El-Houmaizi MA, Saadi M, Oihabi A, Cilas C (2002) Phenotypic diversity of date-palm cultivars (*Phoenix dactylifera* L.) from Morocco. Genet Res Crop Evol 49:483–490

El-Kharbotly A, El-Mardi O, Al-Saidi N, Al-Mahruki Y (1998) Towards the construction of genetic map of date palms using AFLP technique. In: Proceedings of the first international conference on date palm, United Arab Emirates University, Al-Ain, 8–10 March 1998, pp 194–207

El-khishin DA, Adawy SS, Hussein EHA, Elitriby HA (2003) AFLP fingerprinting of some Egyptian date palm (*Phoenix dactylifera L.*) cultivars. Arab J Biotechnol 6:223–234

Elmeer K, Mattat I (2015) Genetic diversity of Qatari date palm using SSR markers. Genet Mol Res 14:1624–1635

Elshibli S (2009) Genetic diversity and adaptation of date palm (*Phoenix dactylifera* L.). Phd thesis, University of Helsinki, Helsinki, Finland, p 177

El-Tarras A, Hossni Y, lbanna EA, SefEldin A (2002) Genetic diversity within Egyptian date palm cultivars based on RAPD analysis. Egypt J Hort 29:205–216

Emoghene B, Asemota O, Idu M, Eke C (2015) Molecular characterization of some date palms in Nigeria using RAPD markers. J App Biol Biotech 3:26–30

Engelmann F (1991) In vitro conservation of tropical plant germplasm: a review. Euphytica 57:227–243

Engelmann F (2004) Plant cryopreservation: progress and prospects. Vitro Cell Dev Biol Plant 40:427–433

Engelmann F (2011) Use of biotechnologies for the conservation of plant biodiversity. vitro Cell Dev Bio 47:5–16

Escobar HA, Valdivia RGJ (2015) Date palm status and perspective in South American Countries: Chile and Peru. In: Al-Khayri JM, Jain SM, Johnson DV (eds) Date palm genetic resources and utilization, vol 1. Africa and the Americas. Springer, Netherlands, pp 257–323

Ezebilo E, Elsafi M, Garkava-Gustavsson L (2013) On-Farm diversity of date palm (*Phoenix dactylifera L*) in Sudan: a potential genetic resources conservation strategy. Sustainability 5:338–356

Ferry M, Gomez S, Jimenez E, Navarro J, Ruiperez E, Vilella J (2002) The date palm grove of Elche, Spain: Research for the sustainable preservation of world heritage site. Palms 46:139–148

Fki L, Bouaziz N, Sahnoun N, Swennen R, Drira N, Panis B (2011) Palm cryobanking. CryoLetters 32:451–462

Fki L, Bouaziz N, Chkir O, Benjemaa-Masmoudi R, Rival A, Swennen R, Drira N, Panis B (2013) Cold hardening and sucrose treatment improve cryopreservation of date palm meristems. Biolog Plantarum 57:375–379

Gammell T (1989) Date palms in Kenya: an economic resource for arid and semi- arid areas. Acts Press, African Technology Studies, Nairobi

Garcia AAF, Benchimol LL, Barbosa AMM, Geraldi IO, Souza CL, de Souza AP (2004) Comparison of RAPD, RFLP, AFLP and SSR markers for diversity studies in tropical maize inbred lines. Genet Mol Biol 27:579–588

Gepts P (2002) The crop of the day: the date, *Phoenix dactylifera*. UC Davis Department of plant sciences, Davis, CA

González-Arnao MT, Panta A, Roca WM, Escobar RH, Engelmann F (2008) Development and large scale application of cryopreservation techniques for shoot and somatic embryo cultures of tropical crops. Plant Cell Tiss Org Cult 92:1–13

Haidar A, Dong H, Mavridis N (2012a) Image-based date fruit classification. In: Proceedings of 4th international congress on ultra-modern telecommunications and control systems and workshops (ICUMT), St. Petersburg, Russia, 3–5 Oct 2012, pp 369–375

Haider N, Nabulsi I, MirAli N (2012b) Phylogenetic relationships among date palm (*Phoenix dactylifera* L.) cultivars in Syria using RAPD and ISSR markers. J Plant Biol Res 1:12–24

Hajian S, Hamidi-Esfahani Z (2015) Date palm status and perspective in Iran. In: Al-Khayri JM, Jain SM, Johnson DV (eds) Date palm genetic resources and utilization, vol 2: Asia and Europe. Springer, Netherlands, pp 19–47

Hammadi H, Mokhtar R, Mokhtar E, Ali AF (2009) New approach for morphological identification of date palm (*Phoenix dactylifera* L.) cultivars from Tunisia. Pak J Bot 41:267–2681

Hamza H, Jemni M, Benabderrahim MA, Mrabet A, Touil S, Othmani A, Salah MB (2015) Date palm status and perspective in Tunisia BT—date palm genetic resources and utilization: volume 1: Africa and the Americas. In: Al-Khayri JM, Jain SM, Johnson DV (eds) Springer, Netherlands, Dordrecht, pp 193–221

Hamza H, Benabderrahim MA, Elbekkay M, Ferdaous G, Triki T, Ferchichi A (2012) Investigation of genetic variation in Tunisian date palm (*Phoenix dactylifera* L.) cultivars using ISSR marker systems and their relation with fruit characteristics. Turk J Biol 36:449–458

Hazzouri KM, Flowers JM, Visser HJ, Khierallah HSM, Rosas U, Pham GM, Meyer RS, Johansen CK, Fresquez ZA, Masmoudi K, Haider N, El Kadri N, Idaghdour Y, Malek JA, Thirkhill D, Markhand GS, Krueger RR, Zaid A, Purugganan MD (2015) Whole genome re-sequencing of date palms yields insights into diversification of a fruit tree crop. Nat Commun 6:11

Henderson SA, Billotte N, Pintaud J-C (2006) Genetic isolation of Cape Verde Island *Phoenix atlantica* (*Arecaceae*) revealed by microsatellite markers. Conserv Genet 7:213–223

Heywood VH, Dulloo ME (2005) In situ conservation of wild plant species a critical global review of good practices. International Plant Genetic Resources Institute (IPGRI) Technical Bulletin 11, Rome, Italy

Ismail HB, Djendoubi N, Kodia A, Hassine DBe, Slama MB (2013) Physicochemical characterization and sensory profile of 7 principal Tunisian date cultivars. Emir J Food Agr 25:331–341

Jain SM (2011a) Prospects of in vitro conservation of date palm genetic diversity for sustainable production. Emir J Food Agr 23:110–119

Jain SM (2011b) Date palm genetic diversity conservation for sustainable production. Acta Hort 882:785–791

Jaradat AA (2011) Biodiversity of date palm. Land use, land cover and soil sciences. Encyclopedia of life support systems (EOLSS), developed under the Auspices of the UNESCO, Eolss Publishers, Oxford, p 31. Available: http://www.eolss.net/sample-chapters/c10/E1-05A-66.pdf. Accessed 5 Aug 2016

Jaradat AA (2015) Genetic erosion of *Phoenix dactylifera* L.: perceptible, probable, or possible. In: Ahuja MR, Jain SM (eds) Genetic diversity and erosion in plants, vol 8. Springer, Netherlands, pp 131–213

Jatoi MA (2013) *In vitro* rooting and acclimatization of date palm (*Phoenix dactylifera* L.) plantlets. M. Phil thesis, Dept. of Botany, Shah Abdul Latif University, Sindh, Khairpur, Sindh, Pakistan

Jatoi MA, Markhand Z, Solangi N (2009) Dates in Sindh: facts and figures. In: Proceedings of international dates seminar, date palm Research Institute, Shah Abdul Latif University, Khairpur, Pakistan, 28 Jul 2009, pp 59–71

Johnson DV (2011) Introduction: date palm biotechnology from theory to practice. In: Jain SM, Al-Khayri JM, Johnson DV (eds) Date palm biotechnology. Springer, Netherlands, pp 1–11

Johnson DV, Al-Khayri JM, Jain SM (2013) Seedling date palms (*Phoenix dactylifera* L.) as genetic resources. Emir J Food Agr 25:809–830

Jubrael JMS (2001) Genetic characterization for some date palm cultivars in Iraq using RAPD markers. IPA J Agr Res 11:138–148 (in Arabic)

Jubrael JM, Udupa SM, Baum M (2005) Assessment of AFLP-based genetic relationships among date palm (*Phoenix dactylifera* L.) varieties of Iraq. J Americ Soc Hort Sci 130:442–447

Karim K, Chokri B, Amel S, Wafa H, Richid H, Nouredine D (2010) Genetic diversity of Tunisian date palm germplasm using ISSR markers. Int J Bot 6:182–186

Khairi MMA (2015) Date palm status and perspective in Sudan. In: Al-Khayri JM, Jain SM, Johnson DV (eds) Date palm genetic resources and utilization, vol 1. Africa and the Americas. Springer, Netherlands, pp 169–191

Khan A, Khan IA, Heinze B, Azim MK (2011) The chloroplast genome sequence of date palm (*Phoenix dactylifera* L. cv. ' Aseel '). Plant Mol Biol Rep 30:666–678

Khierallah HSM (2007) Micropropagation of two date palm (*Phoenix dactylifera* L.) cultivars using inflorescences and study the genetic stability using AFLP-PCR markers. PhD dissertation, College of Agriculture, University of Baghdad, Bagdad, Iraq

Khierallah HSM, Bader SM, Baum M, Hamwieh A (2011) Assessment of genetic diversity for some Iraqi date palms (*Phoenix dactylifera* L.) using AFLP markers. Afr J Biotech 10:9670–9676

Khierallah HSM, Al-Sammarraie SKI, Mohammed HI (2014) Molecular characterization of some Iraqi date palm cultivars using RAPD and ISSR markers. J Asia Sci Res 4:490–503

Khierallah HS, Bader SM, Ibrahim KM, Al-Jboory IJ (2015) Date palm status and perspective in Iraq. In: Al-Khayri JM, Jain SM, Johnson DV (eds) Date palm genetic resources and utilization, vol 2. Asia and Europe. Springer, Netherlands, pp 97–152

Korat date palm Thailand (2016). Available: http://koratdatepalm.com/. Accessed 24 Feb 2016

Krueger RR (2001) Date palm germplasm: overview and utilization in USA. In: Proceedings the first international conference on date palms. Al-Ain, March 1998, pp 2–37

Krueger RR (2015) Date palm status and perspective in the United States. In: Al-Khayri JM, Jain SM, Johnson DV (eds) Date palm genetic resources and utilization, vol 1. Africa and the Americas. Springer, Netherlands, pp 447–485

Kumar N, Modi AR, Singh AS, Gajera BB, Patel AR, Patel MP, Subhash N (2010) Assessment of genetic fidelity of micropropagated date palm (*Phoenix dactylifera* L.) plants by RAPD and ISSR markers assay. Phys Mol Biol Plants 16:207–213

Lédo ADS, Moura CRF, Machado CDA, Ramos SRR, Silva AVC, Lédo CAD (2014) Mannitol for coconut *ex situ* conservation by minimum growth. Pesquisa Agropecuária Bras 49:148–151

Lee DR (1963) Date cultivation in the Coachella Valley, California. Ohio J Sci 63:82–87

Mahar AQ (2007) Post-harvest studies of different cultivars of date palm (*Phoenix dactylifera* L.) fruits, their protection, identification, processing and preservation at district Khairpur, Sindh, Pakistan. Ph.D thesis, Date Palm Research Institute, Shah Abdul Latif University, Khairpur, Sindh, Pakistan

Mahmud TH (1958) Processing and marketing aspects of the date industry in the Coachella Valley, California. Master of Arts thesis in Geography. University of California Los Angeles, p 298

Maitlo WA (2015) Etiology and management of sudden decline disease of date palm (*Phoenix dactylifera* L.) in Pakistan. Ph.D. dissertation, Department of Botany, Shah Abdul Latif University, Khairpur, Sindh, Pakistan

Mancini M (2010) Desert fruit: the dates of Al Jufrah. Slow food foundation for biodiversity, Florence, Italy. Available: http://eco-library.theplanetfixer.org/docs/trees/dates/desert-fruit.pdf. Accessed 5 Aug 2016

Markhand GS, Abul-Soad AA, Mirbahar AA, Kanahr NA (2010) Fruit characterization of Pakistani dates. Pak J Bot 42:3715–3722

Marshal J (1931) Mohenjo-Daro and the Indus civilization. Reprint 2004, AES Publications Pvt. Ltd, New Delhi

Martius CFPV (1836–1853) Historia Naturalis Palmarum, 3 vols. T. O. Weigel, Munich, Germany

Maryam, Jaskani MJ, Awan FS, Ahmad S, Khan IA (2016) Development of molecular method for sex identification in date palm (*Phoenix dactylifera* L.) plantlets using novel sex-linked microsatellite markers. 3 Biotech 6:22

Mathew LS, Seidel MA, George B, Mathew S, Spannagl M, Haberer G, Torres MF, Al-Dous EK, Al-Azwani EK, Diboun I, Krueger RR, Mayer KFX, Mohamoud YA, Suhre K, Malek JA (2015) A genome-wide survey of date palm cultivars supports two major subpopulations in *Phoenix dactylifera*. G3 (Bethesda) 5:1429–1438

Maxted N, Ford-Lloyd B, Hawkes JG (1997) Complementary conservation strategies. In: Maxted N, Ford-Lloyd B, Hawkes JG (eds) Plant genetic conservation: the in situ approach. Springer, Netherlands, pp 15–39

McCubbin MJ (2007) The South African date palm industry—strengths and weaknesses. Acta Hort 736:53–57

Mikki MS (1998) Present status and future prospects of dates and dates palm industries in Saudi Arabia. In: Proceedings of the 1st international conference on the date palm, Al-Ain, Riyadh, Saudi Arabia, 8–11 Mar, pp 469–513

Miller W, Smith JG, Taylor N (1930) *Phoenix*. In: Bailey LH (ed) The standard cyclopedia of horticulture, vol 3. Macmillan, New York

Mirbahar AA, Markhand GS, Khan S, Abul-Soad AA (2013) Genetic diversity analysis among important date palm (*Phoenix dactylifera* L.) cultivars from Pakistan using ISSR markers. In: Abstracts of international conference on date palm: present status & future prospects, The Islamia University of Bahawalpur, Bahawalpur, 2–3 Sept, p 26

Mirbahar AA, Markhand GS, Khan S, Abul-Soad AA (2014) Molecular characterization of Pakistani date palm (*Phoenix dactylifera* L.) cultivars. Pak J Bot 46:619–625

Moghaieb REA, Abdel-Hadi AA, Ahmed MRA, Hassan AGM (2010) Genetic diversity and sex determination in date palms (*Phoenix dactylifera* L.) based on DNA markers. Arab J Biotech 13:143–156

Moore HE (1963) An annotated checklist of cultivated palms. Principles 7:119–183

Mortazavi SMH, Arzani K, Moieni A (2010) Optimizing storage and *in vitro* germination of date palm (*Phoenix dactylifera*) pollen. J Agric Sci Technol 12:181–189

Mowry H (1952) Native and exotic palms of Florida (revised by Dickey RD and West E). University of Florida Bulletin 152

Mueller UG, Wolfenbarger LL (1999) AFLP genotyping and fingerprinting. Trends Eco Evo 14:389–394

Munier P (ed) (1973) Le palmier-dattier—techniques agricoleset productions tropicales. Maisonneuve et Larose, Paris

Munier P (1974) Le probleme de l'origine du palmier dattier et l'Atlantide. Fruits 29:235–240

Munshi A, Osman G (2010) Investigation on molecular phylogeny of some date palm (*Phoenixdactylifra* L.) cultivars by protein, RAPD and ISSR markers in Saudi Arabia. Austr J Crop Sci 4:23–28

Nixon RW (1959) Growing dates in the United States. No. 207, U. S. Dept. Agric. Government Printing Office, Washington. P 50

Osman AMA (1984) The performance of date palms in the Sudan. Acta Hort 143:231–237

Osmanzai A, Haideri MA, Obaidi Q (2010) Performance of date palm varieties in Nangarhar, Afghanistan. The Blessed Tree—Sept 2010, pp 64–65

Panga JA (2014) Plants of AMS Garden: a garden in the Arabian Deserts of Dubai. Xlibris Corporation

Pareek S (2015) Date palm status and perspective in India. In: Al-Khayri JM, Jain SM, Johnson DV (eds) Date palm genetic resources and utilization, vol 2. Asia and Europe. Springer, Netherlands, pp 441–485

Pathak MR, Hamzah R (2008) RAPD analysis of date palm cultivars of Bahrain. FloriOrn Biotech 2:9-11Pintaud JC, Zehdi S, Couvreur TLP, Barrow S, Henderson S, Aberlenc-Bertossi F, Tregear J, Billotte N (2010) Species delimitation in the genus *Phoenix* (*Arecaceae*) based on SSR markers with emphasis on the identity of the date palm (*Phoenix dactylifera* L.). In: Sebreg O, Petersen G, Barfod AS, Davis JI (eds) Diversity, phylogeny and evolution in the monocotyledons. Aarhus University Press, Denmark, pp 267–286

Pintaud JC, Zehdi S, Couvreur TLP, Barrow S, Henderson S, Aberlenc-Berossi F, Tregear J, Billote N (2010) Species delimitation in the genus Phoenix (*Arecaceae*) based on SSR markers, with emphasis on the identity of the date palm (*Phoenix dactylifera L.*). In: Seberg O, Petersen G, Barfod AS, Davis JI (eds) Diversity, Phylogeny and Evolution in the Monocotyledons. Aarhus University Press, Denmark, pp 267–286

Popenoe P (1924) The date palm in antiquity. Sci Monthly 19:313–325

Popenoe P (1973) The date palm. Field Research Projects, Coconut Grove, Miami

Rabei S, Said WM, Rizk RM, El-Sharabasy SF (2012) Morphometric taxonomy of date palm diversity growing in Egypt. Egypt. J Bot, 2nd International conference, 29–30 Apr, Minia University, Egypt, pp 175–189

Rao VR, Sthapit BR (2012) Tropical fruit tree genetic resources: status and effect of climate change. In: Sthapit BR, Rao VR, Sthapit SR (eds) Tropical fruit tree species and climate change. Bioversity International, New Delhi, pp 97–137

Riad M (1993) The date palm sector in Egypt. CIHEAM—Options Méditerranéennes: Série A. SéminairesMéditerranéens 28:45–53

Rivera D, Obón C, Alcaraz F, Carreño E, Laguna E, Amorós A, Johnson DV, Díaz G, Morte A (2015) Date palm status and perspective in Spain. In: Al-Khayri JM, Jain SM, Jhonson DV (eds) Date palm genetic resources and utilization, vol 2. Asia and Europe. Springer, Netherlands, pp 489–526

Roberts EH (1975) Problems of long-term storage and seed and pollen of genetic resources conservation. In: Frankel OH, Hawks JG (eds) Crop genetic resources/or today and tomorrow. Cambridge University Press, Cambridge, pp 226–296

Saaidi M (1992) Comportement au champ de 32 cultivars de palmier dattier vis- ~ i-vis du Bayoud: 25 ann6es d'observation. -Agronomie 12:359–370

Sabir JSM, Abo-Aba S, Bafeel S, Zari TA, Edris S, Shokry AM, Atef A, Gadalla NO, Ramadan AM, Al-Kordy MA, El-Domyati FM, Jansen RK, Bahieldin A (2014a) Characterization of ten date palm (*Phoenix dactylifera* L.) cultivars from Saudi Arabia using AFLP and ISSR markers. C R Biol 337:6–18

Sabir JSM, Arasappan D, Bahieldin A, Abo-Aba S, Bafeel S, Zari TA, Edris S, Shokry AM, Gadalla NO, Ramadan AM, Atef A, Al-Kordy MA, El-Domyati FM, Jansen RK (2014b) Whole mitochondrial and plastid genome SNP analysis of nine date palm cultivars reveals plastid heteroplasmy and close phylogenetic relationships among cultivars. PLoS ONE 9:1–13

Saker MM, Adawy SS, Mohamed AA, El-Itriby HA (2006) Monitoring of cultivar identity in tissue culture-derived date palms using RAPD and AFLP analysis. Biol Plant 50:198–204

Sakka H, Zehdi S, Ould MSA, Rhouma A, Marrakchi M, Trifi M (2003) Tunisian date palm (*Phoenix dactylifera* L.) genotypes identification mediated by plastid PCR/RFLP based DNA. J Genet Breed 57:259–264

Sakka H, Zehdi S, Ould MSA, Rhouma A, Marrakchi M, Trifi M (2004) Genetic polymorphism of plastid DNA in Tunisian date palm germplasm (*Phoenix dactylifera* L.) detected with PCR-RFLP. Genet Resour Crop Evo 51:479–487

Salma M, Fki L, Engelmann-Sylvestre I, Niino T, Engelmann F (2014) Comparison of droplet-vitrification and D-cryoplate for cryopreservation of date palm (*Phoenix dactylifera L.*) polyembryonic masses. Sci Hortic (Amsterdam) 179:91–97

Sani LA, Aliyu MD, Hamza A, Adetunji OA, Gidado RM, Solomon BO (2010) Exploring the Nigerian date palm (*Phoenix dactylifera* L.) germplasm for *in vitro* callogenesis. Blessed Tree 177–184

Sawaya WN (2000) Proposal for the establishment of a regional network for date-palm in the near East and North Africa. A draft for discussion, FAO/RNE. Available: http://www.fao.org/docs/eims/upload/211145/Date_Palm_Proposal.pdf. Accessed 5 Aug 2016

Sedra MH (2011) Development of new Moroccan selected date palm varieties resistant to bayoud and of good fruit quality. In: JainSM, Al-KhayriJM, Johnson DV (eds), Date palm biotechnology. Springer, Netherlands, pp 513–532

Sedra MH (2013) Genetic diversity analysis of Moroccan cultivar genotyping and rapid screening for Bayoud disease resistance in date palm using molecular techniques. Acta Hort 994:271–286

Sedra MH (2015) Date palm status and perspective in Morocco. In: Al-Khayri JM, Jain SM, Johnson DV (eds) Date palm genetic resources and utilization, vol 1. Africa and the Americas. Springer, Netherlands, pp 257–323

Sedra MH, Filali H, Frira D (1993) Observation sur quelques caratéristiques phenotypique setagronomiques du fruit des variétés et clones du palmier dattier séléctionnés. Al Awamia 82:105–120

Sedra MH, El Filali H, Benzine A, Allaoui M, Nour S, Boussak Z (1996) La palmeraie dattière marrocaine: Evaluation du patrimoine phoenicicole. Fruits 1:247–259

Sedra MH, Lashermes P, Trouslot M, Hamon S (1998) Identification and genetic diversity analysis of date palm (*Phoenix dactylifera* L.) varieties of Morocco using RAPD markers. Euphytica 103:75–82

Singh BP (2009) Germplasm introduction, exchange, collection/evaluation and conservation of medicinal and aromatic plants-their export potential. In: Trivedi PC (ed) Medicinal plants: utilization and conservation. Aavishkar Publishers, Jaipur, pp 8–11

Sirjacobs M (1988) Agro-climatological criteria for selecting the most appropriate areas for protected cultivation in Egypt. Protected cultivation in the Mediterranean climate. Greenhouses in Egypt. FAO, Rome, pp 5–12

Soliman KH, Rizk RM, El-Sharabasy SF (2006) Genetic polymorphism of semi-dry date palm (Phoenix *dactylifera* L.) cultivars in Egypt. J Biotech 22:261–273

Soumaya RC, Sarra C, Salwa ZA, Khaled C, Khaled S (2014) Molecular polymorphism and phylogenetic relationships within Tunisian date palm (*Phoenix dactylifera* L.): evidence of non-coding trnL-trnF regions of chloroplast DNAs. Sci Hort 170:32–38

Srivastav VS, Kapadia CV, Mahatma MK, Jha SK, Jha S, Ahmad T (2013) Genetic diversity analysis of date palm (*Phoenix dactylifera* L.) in the Kutch region of India using RAPD and ISSR markers. Emir J Food Agr 25:907–915

Subaih WS, Shatnawi MA, Shibli RA (2007) Cryopreservation of date palm (Phoenix dactylifera) embryogenic callus by encapsulation-dehydration, vitrification and encapsulation-vitrification. Jordan J Agr Sci 3:156–171

Sudhersan C, Sudhersan J, Ashkanani J, Al-Sabah L (2015) Date palm status and perspective in Kuwait. In: Al-Khayri JM, Jain SM, Johnson DV (eds) Date palm genetic resources and utilization, vol 2. Asia and Europe. Springer, Netherlands, pp 299–321

Sulieman AME, Itimad A, Abd-Elhafise Abdelrahim AM (2012) Comparative study on five Sudanese date (*Phoenix dactylifera* L.) fruit cultivars. Food Nutr Sci 3:1245–1251

The Plant List (2012). Available: http://www.theplantlist.org/tpl1.1/search?q=Phoenix. Accessed 16 Mar 2016

Trifi M, Rhouma A, Marrakchi M (2000) Phylogenetic relationships in Tunisian date palm (*Phoenix dactylifera* L.) germplasm collection using DNA amplification fingerprinting. Agron 20:665–671

Uhl NW, Dransfield J (1987) Genera palmarum: a classification of palms based on the work of Harold E. Moore, Jr, L.H. Baily Hortorium and the International Palm Society, Lawrence, KS, pp 214–217

Wang B, Charest P, Downie B (1993) Ex situ storage of seeds, pollen and in vitro cultures of perennial woody plant species. Forestry paper 113, FAO Rome, p 85

Wasilwa LA, Esilaba AO, Moi T, Wasike VW, Kubaso S (2007) The current status of date palm production and utilization in Kenya. Afr Crop SciConf Proc 8:2209–2212

Withers LA (1991) Biotechnology and plant genetic resources conservation. In: Paroda RS, Arora RK (eds) Plant genetic resources conservation and management. International Board for Plant Genetic Resources, New Delhi, pp 273–297

Wrigley G (1995) Date palm, *Phoenix dactylifera*. In: Smartt J, Simmonds NW (eds) Evolution of crop plants, 2nd edn. Longman, London, pp 399–403

Yang M, Zhang X, Liu G, Yin Y, Chen K, Yun Q, Zhao D, Al-Mssallem I, Yu J (2010) The complete chloroplast genome sequence of date palm (*Phoenix dactylifera* L.). PLoS ONE 5:1–14

Younis RAA, Ismail OM, Soliman SS (2008) Identification of sex-specific DNA markers for date palm (*Phoenix dactylifera* L.) using RAPD and ISSR techniques. Res J Agric Biol Sci 4:278–284

Zabar AF, Borowy A (2012) Cultivation of date palm in Iraq. Ann Univ Mariae Curie Sklodowska 22:39–54

Zaid A, de Wet PF (2002) Pollination and bunch management. In: Zaid A (ed) Date palm cultivation. FAO Plant Prod. & Prot. Paper 156 Rev, vol 1, pp 227–269

Zaid ZE, El-Dawayati MM, Baker EI, Gomaa AH (2011) Studies on storage under minimal growth condition of date palm callus explants. In: Proceedings of the Arab palm conference, National Centre for Agricultural Technologies King Abdul Aziz City for Science and Technology, Riyadh, Saudi Arabia, pp 401–420

Zehdi S, Trifi M, Salem AOM, Rhouma A, Marrakchi M (2002) Survey of inter simple sequence repeat polymorphisms in Tunisian date palms (*Phoenix dactylifera L.*). J Genet Breed 56:77–83

Zehdi-Azouzi S, Rhouma S, Dkkhil-Dakhlaoui S, Salem AOM, Cherif E, Othmani A, Marrakchi M, Trifi M (2011) Polymorphism and genetic relationship in date palm using molecular markers. In: Jain SM, Al-Khayri JM, Johnson DV (eds) Date Palm Biotechnology. Springer Netherlands, Dordrecht, pp 407–425

Zhao Y, Williams R, Prakash C, He G (2013) Identification and characterization of gene-based SSR markers in date palm (*Phoenix dactylifera* L.). BMC Plant Biol 12:237

Zohary D, Hopf M (2000) Domestication of plants in the old world: the origin and spread of cultivated plants in West Asia, Europe, and the Nile Valley, 3rd edn. Oxford University Press, Oxford

Chapter 13
Approaches for In Vitro Conservation of Woody Plants Germplasm

María Teresa González-Arnao, Natalia Dolce, María Elena González-Benito, Carlos R. Castillo Martínez and Carlos A. Cruz-Cruz

Abstract The conservation of plant genetic diversity aims at preserving as much as possible extant species by using innovative and complementary approaches to guarantee the effectiveness of the safeguarding strategies and to face present problems and future threats. The development and implementation of different in vitro conservation techniques have provided improvements for the international exchange of germplasm, for the storage of different in vitro culture forms and for products generated by biotechnology. These methods are also a valuable alternative to relieve the need of large lands extensions, where reserve collections of trees belonging to many woody species are traditionally kept. This chapter provides information of several study cases, describes some useful protocols, and aims at presenting a brief overview of currently available techniques for in vitro conservation to medium- and long-term of woody plant germplasm.

M.T. González-Arnao (✉) · C.A. Cruz-Cruz
Facultad de Ciencias Químicas, Universidad Veracruzana, Prol. Oriente 6, No. 1009, C.P. 94340 Orizaba, VER, Mexico
e-mail: teregonzalez@uv.mx; mtgarnao1@hotmail.com

N. Dolce
Instituto de Botánica del Nordeste (IBONE), Universidad Nacional del Nordeste (UNNE)—Consejo Nacional de Investigaciones Científicas y Técnicas (CONICET), Facultad de Ciencias Agrarias, UNNE, Sargento Cabral No. 2131, C.C. 209, 3400 Corrientes, Argentina
e-mail: ndolce@agr.unne.edu.ar; nrdolce@gmail.com

M.E. González-Benito
Departamento de Biotecnología-Biología Vegetal, Universidad Politécnica de Madrid, Ciudad Universitaria s/n, 28040 Madrid, Spain
e-mail: me.gonzalezbenito@upm.es

C.R. Castillo Martínez
Laboratorio de Biotecnología Forestal del Centro de Investigación Disciplinaria en Conservación y Mejoramiento de Ecosistemas Forestales, INIFAP, Progreso No. 5, Delegación Coyoacán, C.P. 04010 Ciudad de México, Mexico
e-mail: castillo.carlos@inifap.gob.mx

M.R. Ahuja and S.M. Jain (eds.), *Biodiversity and Conservation of Woody Plants*, Sustainable Development and Biodiversity 17, DOI 10.1007/978-3-319-66426-2_13

Keywords Woody plant germplasm · Slow growth · Cryopreservation In vitro culture · Biodiversity conservation · Liquid nitrogen

13.1 Introduction

Many woody species are seriously threatened in different parts of the world. They are highly affected by several natural and anthropogenic factors (Augusseau et al. 2006), specifically anthropogenic factors affect forests mostly through expansion of agricultural activity, settlement, deforestation, land fragmentation, and invasive species introduction (Mebrat and Gashaw 2013). The major causes identified of woody species loss are linked to rapid human population growth rates and poverty (FAO 2009). Through the years, many woody plants have supplied products (firewood, fruits, etc.) that are very important for the subsistence of many localities and populations, who have traditionally relied on them to satisfy their needs (Shackleton et al. 2002). According to FAO, the highest rates of woody species loss are in the tropics (FAO 2009). There are some records that suggest the reduction in the forested areas to half in the last 100 years (UNDP 2007) and studies which have found that 10% of tree species are under serious threat (Williams 1998). Therefore, the increasing global need for food and fibre, and the increasing trend of extinction risks for these species, highlight the urgent need for developing and employing reliable conservation strategies to guarantee the safeguarding of woody plant germplasm.

In general, an important criterion to preserve plant genetic diversity is the implementation of different complementary strategies of conservation. In that way, it is possible to take advantage of the benefits from each method and mitigate the uncontrollable specific deficiencies. Furthermore, it is also essential to ensure the genetic stability of the stored material, and the effectiveness of the protocol in terms of reproducibility, regardless of the genotype response and of the type of biological material used.

Among the conservation approaches, there are two well-recognized categories: in situ and ex situ conservations. In situ conservation represents the use of natural habitats to maintain the gene pool of plants in ecosystems under different environmental changes (Swaminathan 1997). This method can be considered dynamic because it may provoke evolutionary modifications in plants by the effect of biotic and abiotic factors. Under these conditions, it is possible to follow the adaptation of plants which are permanently subjected to a natural and sometimes also artificial selection process. In this sense, the in situ conservation which covers biosphere reserves, national parks and other protected areas increases the amount of diversity that can be conserved, but faces several problematic situations that imply the need to develop alternative strategies to support this important type of conservation (Karp et al. 1997). The selection of the appropriate strategy should be based on several criteria including the biological nature of the species and the feasibility of applying the chosen methods (Engelmann 2012). The second conservation category which refers to ex situ approach is the preservation of plant germplasm outside its natural habitat (Heywood and Iriondo 2003). This involves different methods, some of which are also classified as dynamic, like botanic gardens and field genebanks, because both provide the opportunity of monitoring the evolutionary trajectory of samples during the storage. Other ex situ methods are classified as static, because they safeguard the genes outside of the evolutionary context (Shands 1991). They are considered safest, uninfluenced by the climate conditions, and more cost-effective than dynamic methods. That is the case of ex situ alternatives such as seed banks and the storage of plant germplasm in liquid nitrogen (−196 °C).

The development of plant biotechnology, especially the advances associated with in vitro culture and its adaptation to more than several thousand plant species (George 1996; Gonzalez-Arnao et al. 2014), has provided powerful tools to support and improve management and conservation of plant biodiversity (Withers 1995; Bunn et al. 2007).

The in vitro culture techniques have successfully supported the micropropagation of many Australian native plants, including numerous rare woody plant species (Taji and Williams 1996). At the same time, they have provided new conservation approaches, which contribute to widen the availability of options for storing plant germplasm, allowing the establishment of an additional backup for different types of materials, either derived of in vivo plants or from various in vitro biological sources. Moreover, in vitro cultures make the international exchange of plant germplasm easier from both practical and phytosanitary point of views. For instance, the entrance of fruit trees is forbidden during the growing season according to European and South American quarantine regulations; by contrast, the entrance of in vitro cultures is allowed at any time of the year (Hummer 1999). The application of in vitro techniques also provided a new valuable tool for fruit crops with the practice of micrografting. This technology consists in the placement of a maintained scion onto an in vitro grown rootstock in aseptic conditions. In vitro shoot tip grafting has often been applied for the improvement and rejuvenation of several fruit tree species, virus elimination and development of different physiological studies. Micrografting technique is used in quarantine as this method has a minimum risk for importing plants. Plant material derived from the in vitro micrografting can be further multiplied in vitro and acclimatized *ex vitro* (Rehman and Gill 2015). More recently, micrografting was also used as an efficient way for recovering cryopreserved adult shoot tips of citrus (Volk et al. 2012).

The application of in vitro culture techniques has made possible the production of a new kind of germplasm, which includes cell suspension, embryogenic callus and somatic embryos. These biotechnological products constitute a novel source of material with important genetic information useful for breeding programs (Rao 2004) and provide expanded options for preserving plant biodiversity, but require the use of in vitro methodologies for their own conservation (Gonzalez-Arnao et al. 2014). The operational use of tissue culture techniques ensures regeneration, multiplication and recovery of the biological material before and after being subjected to either conservation methods. Therefore, this is a key prerequisite to start a research work when using an in vitro approach.

The aim of this chapter is to review and provide information about the available methods for the in vitro conservation of plant germplasm. Several study cases are presenting to illustrate the development and application of these biotechnological approaches in woody plants.

13.2 In Vitro Conservation Approaches

13.2.1 Medium-Term Conservation

There are two identified approaches for in vitro conservation of plant germplasm: short- to medium-term storage and long-term storage. In the first case, conservation is achieved with the reduction in in vitro growth of tissues by altering the usual formulation of culture medium either diminishing the nutrients concentrations, as well as including the addition of osmotically active compounds, or of growth retardants (Engelmann 2011). These modifications may be combined or not with the reduction in culture temperature (Paulet and Glaszmann 1994), light intensity, storage in darkness and/or the regulation of available oxygen level (Lynch 1999). The storage at low temperature (in the range of 0–5 °C) is one of the major alternatives used for preserving cold-tolerant species of temperate plants (Orlikowska 1992; Ashmore 1997; Shibli et al. 2006).

For medium-term storage, standard in vitro culture conditions can also be used when dealing with species that have a natural slow-growing habit. Other possible options may be desiccation and/or encapsulation of the explants (Engelmann 2012). Artificial seeds, which are produced by encapsulating plant propagules (shoot buds or somatic embryos) in a synthetic matrix, enable medium-term conservation of various plant species (Devi et al. 1998).

In general, the in vitro methods used to attain medium-term conservation allow the storage of biological material from quite a few months to 2–3 years (Cruz-Cruz et al. 2013), and as was previously reviewed, this approach, which is based on the slow growth, can be induced under sub-optimal culture conditions by altering the micropropagation procedure.

13.2.2 Long-Term Conservation

The next challenge using in vitro techniques was to increase the storage duration to guarantee the long-term conservation of plant germplasm. However, this aim could only be possible with the total arrest of cellular division and of all metabolic processes. The effect produced by an ultra-low temperature, usually due to the storage of samples in liquid nitrogen (−196 °C), allowed achieving this inanimate state under a new biotechnological approach, that is, using cryopreservation.

Currently, cryopreservation represents the safest alternative for long-term preservation of plant biodiversity without requiring continuous manipulations. However, the main stages involved in the cryopreservation procedure, such as cryoprotection, cooling/warming and recovery culture, play a crucial role in attaining success, and each successive step must be optimized irrespective of the cryogenic protocol selected and depending on the biological material used (Gonzalez-Arnao et al. 2014).

Plant cryopreservation techniques have evolved significantly over the past 26 years with the development of different mixtures of cryoprotectants named as plant vitrification solutions (PVS) since 1990, the increasing of the cooling/warming rates during process since 2005 and with the partial combination of different cryogenic methods, which consequently resulted in the establishment of new alternative techniques (e.g. encapsulation-vitrification, droplet-vitrification and V- or D-Cryoplate).

All these contributions have improved the effectiveness and the adaptability of the protocols to a larger number of species, and they are closely related on the one hand to the encapsulation of tissues in calcium alginate gel, which has facilitated the manipulation and improved the tolerance of explants to drastic dehydration treatments that otherwise would have been lethal (Gonzalez-Arnao and Engelmann 2006) and, on the other hand, by the use of aluminium foil strips (Panis et al. 2005) and/or of cryoplates (Yamamoto et al. 2011, 2012) instead of cryovials, for carrying out the direct immersion of samples to liquid nitrogen during cooling and the rapid plunge into unloading solution during warming at room temperature. These innovations have increased the probability of forming a glassy state at the low temperatures, which would potentially avoid the injuries due to the formation of ice crystals when the intracellular water could not be removed sufficiently from samples prior to immersion in liquid nitrogen. At the same time, they also help to avoid the destabilization of the non-crystalline solid produced, because the warming is also performed very rapidly (Gonzalez-Arnao et al. 2008).

Several exhaustive reviews and publications have described in detail the most important characteristics of each cryopreservation techniques. They have always pointed out that the first cryogenic protocols developed, known as classical or conventional, are based on a freeze-dehydration process which induces the dehydration by the reduction in temperature usually at 0.5–1 °C min^{-1} down to around −40 °C (slow freezing regime), followed by rapid immersion in liquid nitrogen (Gonzalez Arnao et al. 2014). The next important group of cryopreservation methods comprises the vitrification-based techniques, which are supported on the transition of the liquid viscous phase to an amorphous glassy solid at the glass transition (Tg) temperature (Fahy et al. 1984). This process is induced by a severe dehydration that takes place at a non-freezing temperature, due to the exposure of samples to highly concentrated cryoprotective solutions (PVS) and/or to physical drying conditions followed by a rapid or ultra-rapid cooling in liquid nitrogen. Among the vitrification-based techniques are found: (i) vitrification method, which involves treatment of samples with cryoprotective substances (loading), dehydration with a highly concentrated plant vitrification solution (PVS), rapid cooling and rewarming, removal of cryoprotectants and recovery. This procedure has been developed for shoot tips, cell suspensions and somatic embryos of numerous different species (Sakai and Engelmann 2007); (ii) encapsulation-dehydration, where the basic protocol is encapsulation in calcium alginate beads, preculture of alginate-coated samples in liquid medium with high sucrose concentration, evaporative air or silica gel desiccation to a water content in the bead around 20% (fresh weight basis), and rapid cooling in liquid nitrogen. Rewarming of the alginate encapsulated explants usually is performed at room temperature, and for recovery, the beads are usually placed onto standard culture medium without having to extract the shoots or embryos from their alginate coating (Engelmann et al. 2008; Sherlock et al. 2005); (iii) encapsulation-vitrification, which is a combination of encapsulation-dehydration and vitrification procedures, because samples are encapsulated in alginate beads and then treated and cooled according to the vitrification technique described above (Sakai and Engelmann 2007); and (iv) droplet-vitrification, a protocol derived from the combination of the vitrification procedure with the droplet-freezing technique developed by Kartha et al. (1982). In this case, samples are treated with loading and vitrification solutions and then placed on aluminium foil strips in minute droplets of vitrification solution or just in one small drop, and the aluminium foil strip is directly immersed with the samples in liquid nitrogen (Sakai and Engelmann 2007); (v) V-Cryo-plate, which combines the encapsulation-dehydration and droplet-vitrification techniques. In this method, shoot tips are attached with a thin calcium alginate layer to an aluminium cryoplate, loaded, treated with PVS, and then cooled by direct immersion of cryoplates in liquid nitrogen (Yamamoto et al. 2011, 2012); (vi) D-Cryoplate, the only difference from V-Cryoplate is that it replaces the use of the PVS solution, by the desiccation under the laminar flow or employing silica gel, to dehydrate samples attached to the cryoplate and subjected to the loading treatment. Both cryoplate alternatives provide higher cooling and warming rates compared to other vitrification-based procedures (Engelmann 2014); (vii) pregrowth technique, which is the culture of samples in the presence of cryoprotectants, followed by rapid immersion in liquid nitrogen; (viii) desiccation, which consists of dehydrating explants or seeds usually by desiccation in the air current of a laminar airflow cabinet or with silica gel and then direct immersion in liquid nitrogen. This method is used mainly for cryopreserving seeds. Finally, Pregrowth-Desiccation is the combination of the both previously mentioned methods and it is mainly used for cryopreserving meristematic cultures, small-size seeds, polyembryonic cultures, zygotic embryos or embryonic axes extracted from seeds, respectively (Gonzalez-Arnao et al. 2008). The cryopreservation of embryonic axis is a simple and useful strategy to preserve the germplasm of plants with recalcitrant seeds, because axes can be easily dried before their rapid cooling in liquid nitrogen. In addition, according to Li and Pritchard (2009), both the recent evidence of less-than-expected longevity at conventional seed bank temperatures and the innovations in the cryopreservation of recalcitrant-seeded species are important factors which are demonstrating that ultra-cold storage should be adopted for the long-term conservation of plants (Li and Pritchard 2009).

There is no specific or unique protocol to guarantee the successful cryopreservation of plant germplasm. However, different important research groups around the world have identified empirically the best cryogenic approaches which result more effective depending on the target material (i.e. cells, callus, shoot tips, somatic embryos, etc.). Moreover, for the large-scale application of cryopreservation, sometimes the complementary use of different cryogenic techniques (e.g. classical and/or some vitrification-based procedures) is required in order to improve the response of genotypes, cultivars or species, which are less tolerant to a specific protocol, or may only require some additional modifications in the stages of cryoprotection, to achieve reproducible survival following the same technique (Gonzalez-Arnao et al. 2008; Gonzalez-Arnao and Engelmann 2013).

13.2.3 Importance of In Vitro *Conservation of Woody Plant Germplasm*

For woody plants, the most conventional conservation methods are in situ alternative and field genebanks as ex situ alternative. In situ conservation is considered the method of choice for conserving forest species and wild relatives (Brush 1995). Regarding fruit and timber trees, the field clonal collections (for vegetatively propagated species) and seed banks (for seed propagated species with orthodox seeds) are the traditional ex situ approaches for germplasm preservation. However, field collections require vast areas of lands, their management is very expensive and the system is exposed to the risks arising from different kinds of biotic and abiotic stresses. In addition, many woody species produce non-orthodox seeds, i.e. recalcitrant, which cannot be preserved under seed banks conditions (Lambardi et al. 2005). Therefore, the in vitro conservation methods play a significant complementary role.

At present, there are more reports about the development and adaptation of cryopreservation protocols for in vitro storing of woody plant germplasm, than for medium-term storage. This probably is because cryopreservation (−196 °C) currently provides more alternatives (controlled rate cooling, vitrification, encapsulation-dehydration, cryoplate techniques, desiccation, etc.) to preserve different types of materials (e.g. seeds, embryonic axis, zygotic embryos, dormant buds) and in vitro forms (e.g. cells, callus, somatic embryos) in comparison with slow growth approach, which is usually performed using shoots and/or in in vitro plantlets. In addition, long-term storage in liquid nitrogen eliminates the problem associated with the danger of maintaining big spaces required to store in vitro a backup of abundant collections, with the potential risks of contamination during the storage, and the somaclonal variation for some species (Blakesley et al. 1996).

Among the alternatives employed to induce the slow growth for different woody plant species, Shibli et al. (2006) reported in vitro medium-term storage by using high concentrations of sucrose, sorbitol or mannitol in the culture medium.

This allowed reduction in growth at room temperature of bitter almond microshoots and extended the subculture interval to four months. On the other hand, the benefits of lowering the culture temperature as a convenient strategy for medium-term storage are already known, especially when dealing with species of temperate origin. Shoot cultures of chestnut, oak and wild cherry were stored after 10 d of the last subculture at 2 °C for up to 1 year without subculturing (Janeiro et al. 1995). Shoot cultures of *Castanea sativa* (cv. "Montemarano") were stored for over 48 months at 8 °C, and the combination with a low level of lighting resulted in a positive effect on survival for such conservation period (Capuana and Di Lonardo 2013). Table 13.1 presents additional successful examples of in vitro storage for medium term.

Table 13.1 Representative examples of medium-term conservation of woody plants

Species	Tissue	Method (storage temperature)	Storage period	Reference
Acacia mangium	Seeds	4–8 °C	1.2 years	Yap and Wang (1983)
Acer saccharum	Seeds	−10 °C	5.5 years	Carl (1976)
Albizia (Paraserianthus) falcataria	Seeds	4–8 °C	1.5 years	Yap and Wang (1983)
Araucaria cunninghamii	Seeds	−15 °C	8 years	Shea and Armstrong (1978)
Casuarina equisetifolia	Seeds	−3 °C	2 years	Jones (1967)
Citrus limon	Seeds	−20 °C	9.6 years	King et al. (1981)
Cryptomera	Pollen	−196 °C	1 years	Ichikawa and Shidei (1972a)
Fagus sylvatica	Seeds	−10 °C	5 years	Suszka (1975)
Gmelina arborea	Seeds	−5 °C	2 years	Woessner and McNabb (1979)
Hop (Humulus)	Pollen	−196 °C	1 years	Haunold and Stanwood (1985)
Larix	Pollen	−196 °C	1 years	Ichikawa and Shidei (1972a)
Loblolly pine	Pollen	−18 °C	3–10 years	Bramlett and Mathews (1991)
Malus domestica cv. Antonowka	Seeds	−1 to −18 °C	3.1 years	Grzeskowiak et al. (1983)
Morus bombysis	Winter buds	−40 °C, −70 °C, −135 °C and −196 °C	1 month, 6 months, 1 year and 3.5 years	Niino et al. (1993)
Pinus spp.	Pollen	−196 °C	1 years	Ichikawa and Shidei (1972a)
P. caribea var. *hondurensis*	Seeds	8 °C	2.7 years	Yap and Wang (1983)
P. merkusii	Seeds	4–5 °C	4 years	Pausujja et al. (1986)
Populous deltoids	Seeds	−20 °C	6 years	Tauer (1979)

(continued)

Table 13.1 (continued)

Species	Tissue	Method (storage temperature)	Storage period	Reference
P. cerasifera var. *divaricate*	Seeds	−1 to −3 °C	2 years	Grzeskowiak et al. (1983)
Prunus avium	Seeds	−1 to −3 °C	2 years	Grzeskowiak et al. (1983)
P. persica	Pollen	−20 °C	9 years	Akihama et al (1980)
Pyrus spp.	Pollen	−20 °C	6 years	Akihama et al (1980)
P. bretschneideri	Shoot tips	5 °C, 8-h photoperiod; 10 °C, 16-h photoperiod	64 Weeks	Oka and Niino (1989)
P. cancasia	Seeds	−1 to −18 °C	3.1 years	Grzeskowiak et al. (1983)
P. communis	Shoot tips	4 °C, 16-h photoperiod	18 months	Wanas et al (1986)
		5 °C, 16-h photoperiod or 1 °C darkness	12 months	Moriguchi et al. (1990)
P. pyrifolia	Shoot tips	1 °C darkness	20 months	Moriguchi et al. (1990)
P. pyrifolia	Scion	−1 to −3 °C	More than 1 year	Kozaki (1975a)
		1 to −5 °C	4 years	Omura et al. (1978a)
P. pyrifolia	Pollen	Freeze drying	More than 10 years	Kozaki (1975b)
			6 years	Akihama et al. (1978)
P. pyrifolia	Seed	−3 to −5 °C	7 years	Solovieva (1966)
		Drying	3 years	Omura et al. (1978b)
Simmondsia chinensis	Pollen	−196 °C	2 years	Lee et al. (1985)

Regarding cryopreservation techniques, there are a lot of advances related to woody plants, since the first demonstrations that winter-hardy twigs survived after immersion in liquid nitrogen for one year (Sakai 1956). Related to fruit trees, pear and apple are interesting cases to mention. For these species, several cryogenic techniques have been successfully adapted and applied on large scale, such as classical protocols with controlled rate cooling (Chang and Reed 2000), encapsulation-dehydration (Niino and Sakai 1992) and vitrification (Niino et al. 1992). A significant and important fact about the impact of cryopreservation of apple germplasm is that after a loss due to diseases, over 100 unique apple accessions were re-grafted from cryopreserved dormant buds, and the cultivars were successfully restored for propagation (USDA, ARS 2013).

Cryopreservation of Japanese persimmon shoot tips has been also reported using different techniques as slow cooling (Matsumoto et al. 2004), vitrification (Matsumoto et al. 2001; Niu et al. 2012) and droplet-vitrification (Niu et al. 2012). However, the utilization of cryopreservation for persimmon germplasm is still limited. More recently, Matsumoto et al. (2015) reported the use of D-Cryoplate protocol with 10 persimmon cultivars, which resulted in high regrowth rates after cryopreservation and may facilitate the long-term conservation in the genebank context.

For non-fruit forest trees, the same three cryopreservation approaches have been applied to meristematic tissues of some important timber and pulp species such as silver birch using controlled rate cooling with in vitro meristems (Ryynänen 1996), encapsulation-dehydration with in vitro axillary buds of *Eucalyptus* species (Páques et al. 2002) and vitrification with in vitro apical segments and buds of an aspen hybrid (Jokipii et al. 2004). Embryogenic conifer cultures (Park et al. 1998), seeds or isolated embryonic axes of *Azadirachta indica* (Berjak and Durnet 1996) have been also successfully cryopreserved. Another significant advance is that several cryopreservation protocols have been already standardized for seeds of many native woody (over 30) species of Brazil, by using a rapid cooling regimen and a slow thawing at room temperature (Santos et al. 2013).

Table 13.2a, b, c, d, e presents some additional successful examples of using cryopreservation techniques for the long-term storage of woody plant germplasm.

Table 13.2 Examples of plant species cryopreserved using seeds

(**a**) Examples of plant species cryopreserved using seeds		
Acacia sp.	Direct immersion in liquid nitrogen	Salomão (2002)
Abies procera	Direct immersion in liquid nitrogen	Walters et al. (2004)
Agathis spp.	Temperatures from −20 to 42 °C	Dickie and Smith (1995)
Allocasuarina fraseriana	Direct immersion in liquid nitrogen	Touchell and Dixon (1994)
Amburana cearensis	Direct immersion in liquid nitrogen	Salomão (2002)
Azadirachta indica	Desiccation and freezing	Chaudhury and Chandel (1991)
Buchenavia tomentosa	Direct immersion in liquid nitrogen	Salomão (2002)
Camellia sinensis	Desiccation and immersion in liquid nitrogen	Kim et al. (2002)
Casuarina sumatrana	Deep freezer (–20 °C) and direct immersion in liquid nitrogen	Marzalina and Nashatul (2000)
Cedrela fissilis	Direct immersion in liquid nitrogen	Nunes et al. (2003)
Chorisia pubiflora	Direct immersion in liquid nitrogen	Salomão (2002)

(continued)

Table 13.2 (continued)

(**a**) Examples of plant species cryopreserved using seeds		
Citrus spp.	Desiccation and immersion in liquid nitrogen	Hor et al. (2005)
Coffea spp.	Desiccation and immersion in liquid nitrogen	Dussert et al. (1998)
Corylus avellane	Desication and immersion in liquid nitrogen	Normah et al. (1994)
Dialium divaricatum	Direct immersion in liquid nitrogen	Salomão (2002)
Dipterocarpus spp.	Deep freezer (–20 °C) and direct immersion in liquid nitrogen	Marzalina and Nashatul (2000)
Dyera costulata		
Enterolobium spp.	Direct immersion in liquid nitrogen	Salomão (2002)
Eriotheca gracilipes		
Eucalyptus loxophleba	Direct immersion in liquid nitrogen	Touchell and Dixon (1994)
Guettarda pohliana	Direct immersion in liquid nitrogen	Salomão (2002)
Jacaranda spp.		
Leucaena leucocephylla	Deep freezer (–20 °C) and direct immersion in liquid nitrogen	Marzalina and Nashatul (2000)
Machaerium spp.	Direct immersion in liquid nitrogen	Salomão (2002)
Pinus spp.	Direct immersion in liquid nitrogen	Pita et al. (1998)
Pistacia spp	Dehydration and one-step freezing.	Ozden-Tokatli et al. (2007)
Populus deltoids	Desiccation, deep freezer (–20 °C) and direct immersion in liquid nitrogen	Pence (1996)
Prunus spp.	Direct immersion in liquid nitrogen	Chmielarz (2009)
Spondias mombin	Direct immersion in liquid nitrogen	Salomão (2002)
Swietenia macrophylla	Deep freezer (–20 °C) and direct immersion in liquid nitrogen	Marzalina and Nashatul (2000)
Tabebuia spp.	Direct immersion in liquid nitrogen	Salomão (2002)
Tectona grandis	Deep freezer (–20 °C) and direct immersion in liquid nitrogen	Marzalina and Nashatul (2000)
Thyrsostachys siamensis		
Ulmus americana	Direct immersion in liquid nitrogen	Walters et al. (2004)
(**b**) Examples of plant species cryopreserved using cell suspensions and callus cultures		
Abies cephalonica	Cryoprotective treatments [DMSO, PGD I and II (mixture of polyethylene glycol, glucose and DMSO)], prefreezing temperature of −38 °C, and immersion in liquid nitrogen	Aronen et al. (1999)

(continued)

Table 13.2 (continued)

(**b**) Examples of plant species cryopreserved using cell suspensions and callus cultures		
Betula pendula	Slowly (0.17 °C min^{-1}) frozen down to −38 °C using a programmable freezer	Ryynänen et al. (2002)
Castanea sativa	Vitrification (PVS2)	Corredoira et al. (2007)
Citrus deliciosa	Cryoprotection (DMSO), freezing by slow cooling and storage in liquid nitrogen	Perez et al. (1999)
Citrus sinensis	PVS2 vitrification	Hao et al. (2003)
Citrus spp.	Cryoprotection (DMSO), freezing by slow cooling and storage in liquid nitrogen	Perez et al. (1997)
Elaeis guineenses	Desication and direct immersion in liquid nitrogen	Dumet et al. (2000)
Hevea brasiliensis	Programmable freezer (0.2 °C/min down to −40 °C)	Engelmann and Etienne (2000)
Mangifera indica	Encapsulation/dehydration; pregrowth/dehydration and vitrification	Wu et al. (2003)
Olea europea	Controlled rate freezing, encapsulation–dehydration, liquid nitrogen and vitrification	Lambardi et al. (2002)
Picea glauca	Long-term storage in or above liquid nitrogen	Cyr et al. (1994)
Picea mariana	PVS2 vitrification	Touchell et al. (2002)
Picea sitchensis	Two-step freezing	Find et al. (1998)
Pinus patula	Precooling tissue prior to immersion into liquid nitrogen	Ford et al. (2000b)
Pinus pinaster	Precooling tissue prior to immersion into liquid nitrogen	Marum et al. (2004)
Pinus radiate	Preculture and direct immersion in liquid nitrogen	Hargreaves et al. (2002)
Pinus roxburghii	Pretreatment, slow freezing and liquid nitrogen	Mathur et al. (2003)
Prunus avium	Pretreatments and one-step freezing	Grenier-de March et al. (2005)
Pyrus pyrifolia	Encapsulation-dehydration, vitrification, two-step frozen, encapsulation and conventional cryopreservation	Gazeau et al. (1998)
Quercus robur	Pretreatments and direct immersion in liquid nitrogen	Chmielarz et al. (2005)
(**c**) Examples of plant species cryopreserved using embryonic axes		
Aesculus hippocastanum	Desiccation and freezing	Pence (1990)
A. glabra	Desiccation and freezing	Pence (1992)
Artocarpus heterophyllus	Desiccation and freezing	Chandel et al. (1995)

(continued)

Table 13.2 (continued)

(**c**) Examples of plant species cryopreserved using embryonic axes		
Azadirachta indica	Desiccation and freezing	Berjak and Dumet (1996)
Camellia sinensis	Desiccation and immersion in liquid nitrogen	Kim et al. (2002)
Carya sp.	Desiccation and cooling	Pence (1990), Abou Taleb et al. (1992)
Castanea sativa	Desiccation and PVS vitrification	San-Jose et al. (2005)
Citrus aurantifolia	Desiccation and freezing	Cho et al. (2002c)
C. halimii	Desiccation, encapsulation–dehydration, slow freezing and vitrification	Normah and Siti Dewi Serimala (1997)
C. latipes	Desiccation–freezing, vitrification and Encapsulation–dehydration	Malik and Chaudhury (2006)
C. madurensis	Encapsulation–dehydration	Cho et al. (2003)
C. macroptera	Desiccation–freezing, vitrification and Encapsulation–dehydration	Malik and Chaudhury (2006)
C. medica	Encapsulation–dehydration	Cho et al. (2003)
C. sinensis	Desiccation and freezing, pretreated with sucrose	Santos and Stushnoff (2003)
Cocos nucifera	Pregrowth desiccation and liquid nitrogen	Sajini et al. (2006)
Coffea spp.	Desication	Dussert et al. (1999)
C. arabica	Desiccation and freezing	Martinez et al. (1996)
C. liberica	Desication	Normah and Vengadasalam (1992)
Durio zibenthinus	Dehydration and freezing	Hor et al. (1990)
Elaeis guineensis	Desiccation in silica gel and rapid freezing	Villa et al. (2007)
Euphoria longan	Desiccation and freezing	Fu et al. (1990)
Fagus	Cold-labile enzyme lactate dehydrogenase (LDH)	Pukacki et al. (2009)
Fraxinus excelsior	Desiccated and direct immersion in liquid nitrogen	Brearley et al. (1995)
Ilex brasiliensis, I. brevicuspis, I. dumosa, I. intergerrima, I. paraguariensis, I. pseudoboxus, I. taubertiana, I. theezans	Encapsulation-dehydration	Mroginski et al. (2008)
Juglans regia	Desiccation and freezing	Pence (1990)
Lansium domesticum	Desiccation, encapsulation-dehydration, vitrification and slow freezing	Normah et al. (2000)

(continued)

Table 13.2 (continued)

(**c**) Examples of plant species cryopreserved using embryonic axes		
Litchi sinensis	Desiccation followed by fast freezing	Chaudhury (2000)
Livistona chinensis	Desiccation	Wen and Song (2007)
Melia azedarach	Dehydration and freezing	Kaviani et al. (2009)
Nephelium lappaceum	Dehydration and freezing	Hor et al. (1990)
Olea europaea	Desiccation and freezing	Gonzalez-Rio et al. (1994)
Pinus radiata	One-step cooling	Hargreaves et al. (2005)
Prunus amygdalus	Desiccation and freezing	Chaudhury and Chandel (1995)
Prunus persica	Desiccation and freezing; PVS2 vitrification	De Boucaud et al. (1996)
Q. falcata	Desiccation and freezing	Pence (1992)
Q. leucotrichophora	Desiccation followed by fast freezing	Chaudhury (2000)
Q. macrocarpa	Desiccation and freezing	Pence (1992)
Q. nigra	Desiccation and freezing	Pence (1992)
Q. palustris	Desiccation and freezing	Pence (1992)
Q. robur	Dehydration protocols and encapsulation	Berjak et al. (2000)
Q. rubra	Desiccation and freezing	Pence (1992)
Swietenia macrophylla	Direct immersion in liquid nitrogen or slow cooling	Marzalina and Normah (2002)
Theobroma cacao	Desiccation–freezing in liquid nitrogen	Pence (1991)
Vigna spp	Desiccation and freezing	Normah and Vengadasalam (1992)
(**d**) Examples of plant species cryopreserved using somatic embryos		
Abies nordmanniana	Dehydration by freezing	Misson et al. (2006)
Aesculus hippocastanum	one-step freezing	Lambardi et al. (2005)
Carya illinoensis	Desiccation and freezing	Kumar and Sharma (2005)
Castanea sativa	Desiccation-based cryostorage and vitrification	Corredoira et al. (2004)
Citrus grandis	Prefreezing treatment and cryoprotectants	Oh (1997)
C. junos		
C. platymamma		
Coffea arabica	Freezing in liquid nitrogen	Bertrand-Desbrunais et al. (1998)
C. canephora	Regrowth and encapsulation	Hatanaka et al. (1994)
Juglans regia	Desiccation and freezing	Kumar and Sharma (2005)

(continued)

Table 13.2 (continued)

(**d**) Examples of plant species cryopreserved using somatic embryos		
Olea europea	Encapsulation–dehydration	Shibli and Al-Juboory (2000)
Picea. glauca	Desiccation	Percy et al. (2001)
P. glauca x engelmannii		
Pinus patula	Prefrozen and direct immersion in liquid nitrogen	Ford et al. (2000a)
Picea sitchensis	Encapsulation–dehydration	Gale et al. (2008)
Quercus robur	Desiccation and vitrification	Martinez et al. (2003)
Q. suber	Encapsulation-dehydration	Fernandes et al. (2008)
Theobroma cacao	Desiccation and freezing	Fang et al. (2009)
(**e**) Examples of plant species cryopreserved using shoot tips		
Amygdalus communis L.	Encapsulation–dehydration–vitrification	Al-Ababneh et al. (2003)
Cedrela fissilis	Encapsulation–dehydration	Nunes et al. (2003)
Citrus spp.	Encapsulation–vitrification	Wang and Deng (2004)
Cocos nucifera	Encapsulation–dehydration	Hornung et al. (2001)
Coffea racemosa	Encapsulation–dehydration	Mari et al. (1995)
Coffea sessiliflora		
Diospyros kaki	PVS2 vitrification	Matsumoto et al. (2001)
Malus spp.	Encapsulation–dehydration	Paul et al. (2000)
Phoenix dactylifera	Pregrowth and freezing	Bagniol and Engelmann (1991)
Populus alba	One-step vitrification	Lambardi et al. (2000)
Prunus domestica	One-step vitrification	De Carlo et al. (2000)
Prunus dulcis	Vitrification and encapsulation-dehydration	Shatnawi et al. (1999)
Prunus spp.	One-step vitrification	Niino et al. (1997)
Pyrus spp.	Cold acclimatation and controlled freezing	Chang and Reed (2000)
Vitis vinifera	Two-step vitrification	Matsumoto and Sakai 2003

13.3 In Vitro Medium-Term Conservation of Forest Species and Study Cases

13.3.1 Family Meliaceae

Meliaceae is a large family of tropical and subtropical woody species, comprised of 50 genera and about 575 species of trees and (rarely) shrubs (Pennington and Styles

1975; Mabberley et al. 1995; Pérez-Flores et al. 2012) occurring in a variety of habitats, from rain forests and mangrove swamps to semidesert (Pérez-Flores et al. 2012). Various species are commercially important, used for vegetable oil, soap making, insecticides and highly prized wood. Some of the best-known examples in the international timber trade are American or true mahogany (*Swietenia* spp.), African mahogany (*Khaya* spp.), sapele (*Entandrophragma* spp.), Spanish cedar or cigar box cedar (*Cedrela* spp.), toon or Australian red cedar (*Toona* spp.), besides several others from various parts of the world (Styles 1972).

Most of the timber-producing species are huge, dominant or emergent trees forming the major constituents of tropical rainforest, secondary forest or other types of woodland. Timber from members of the Meliaceae are in fact the backbone of the forest industry of many countries, but continuous exploitation of the natural forest has seriously depleted stocks of desirable specimens, notably in South America and parts of West Africa. Usually, it is the best trees which are culled during exploitation so that potential sources of elite or superior genotypes are continually decreasing with consequent depletion of the gene pool (Styles 1972). Regarding this situation, Lamb pointed out almost 50 years ago (1968) that *Cedrela* in Latin America and the Caribbean Islands had been so overcut that some species had been virtually eliminated. Since then, supplies of mature trees of good form for seed production and tree breeding have remained only in remote and inaccessible places. Similarly, in *Swietenia* only trees of a bush-like form have survived in the Caribbean Islands and are represented by the famous Cuban mahogany, *S. mahagoni* (Styles 1972).

Consequently, ex situ preservation of germplasm from members of the Meliaceae is necessary to safeguard the threatened diversity of this family, mainly due to the anthropogenic impact. In recent years, staff of the INIFAP (Jalisco, Mexico) and the IBONE (Corrientes, Argentina) has conducted research to evaluate the possibility of in vitro preservation of isolated embryos and shoot apical meristems of Meliaceae species by minimal growth.

13.3.2 Swietenia Macrophylla *King. (Mahogany)*

The genus *Swietenia* Jacq. contains three species (*S. mahagoni* Jacq., *S. macrophylla* King and *S. humilis* Zucc.) and two natural hybrids (Pennington 1981), occurring natively in the Neotropics (Figueroa 1994). They are medium-sized to large trees growing 20–50 m tall and having up to 2 m trunk diameter. The genus is famed as the supplier of mahogany, one of the most beautiful and valuable tropical timber species. At first, mahogany was yielded by *S. mahagoni*, a Caribbean species, which was so extensively used locally and exported that its trade ended by the 1950s. These days almost all mahogany is yielded by the mainland species, *S. macrophylla,* although no longer from its native locations due to the restrictions set by CITES (this species has been listed under Appendix II of the Convention on International Trade in Endangered Species of Wild Fauna and Flora, CITES 2003).

S. macrophylla, known as "caoba" in Spanish and "mahogany" or "bigleaf mahogany" in English, grows in Central America from Yucatan southwards and into South America, extending as far as Peru, Bolivia and extreme western Brazil (Lamb 1966; Biswas et al. 2002; Mishra et al. 2014). Mahogany trees may reach 50 m in height and 0.75–1.2 m in diameter (Gueilfus 1994; Standley 1946). The economic value of *S. macrophylla* timber has resulted in their overexploitation for the last centuries. However, mahogany had always been obtained from natural forests (Snook 1993) because one of the main problems for the establishment of organized planting programmes is that within its natural area of distribution in the Americas, *S. macrophylla* trees are intensely attacked by the mahogany shoot borer, *Hypsipyla grandella* Zeller (Lamb 1966; Lyhr 1992; Newton et al. 1993).

The overexploitation of natural populations and logging of the best individuals for commercial use have affected the genetic diversity of *S. macrophylla*. Consequently, it is necessary to design strategies to promote preservation and regeneration systems for this species, both in situ and ex situ. Mahogany seeds are classified as recalcitrant based on their limited desiccation tolerance and short storage life (Gómez et al. 2006). In this case, the in vitro tissue culture techniques are presented as an important tool for the establishment of ex situ germplasm conservation programs. In the last years, the INIFAP (Jalisco, México) staff has conducted studies to regenerate plants from seeds and adventitious buds (taken from field collected twigs), as well as to store apical meristems using minimal growth in vitro culture for short- and medium-term preservation of mahogany germplasm.

Regarding to the preservation process, shoot apical meristems were dissected from in vitro plants of *S. macrophylla* belonging to a working collection initiated through seeds and axillary buds from adult trees collected in natural mahogany stands, as mentioned by Martinez et al. (2013a). After dissection, apical meristems were cultured on the multiplication medium composed of MS + 1.5 mg l^{-1} BA + 0.5 mg l^{-1} ANA (control treatment) or MS lacking plant growth regulators. Cultures were kept at 24, 18 or 16 °C. After 30 and 60 d of storage at different temperatures, the number and length of developed shoots were evaluated. Data from this work demonstrated that culture on MS medium without growth regulators at 16 °C, enables maintenance of *S. macrophylla* meristems for 60 d (Martinez et al. 2013a). In a subsequent study, the effect of sucrose/mannitol concentrations and mineral oil on meristems growth was evaluated. After dissection, shoot apical meristems were cultured on the multiplication medium (control treatment) or MS supplemented with different sucrose/mannitol concentrations (30/0, 20/10, 15/15, 10/20 and 0/30 g l^{-1}). A replica of each treatment was also carried out by adding 2 ml of mineral oil (J.T. Baker®) to the media. Cultures were kept at 28, 24 or 18 °C. Every 30 d of storage at different temperatures, the number and length of developed shoots as well as their vigour and oxidation were evaluated. Results from this experiment showed that adding MS with 15 g l^{-1} sucrose and 15 g l^{-1} mannitol and keeping cultures at 18 °C was the best condition for preserving meristems during fifteen weeks without requiring subcultures of tissues. Following this procedure, the meristem growth was reduced while maintaining high survival (93%) and, after transferred to the multiplication medium, tissues showed good vigour, less oxidation and high number of shoots/explant compared with the control treatment. This methodology was also tested with meristems of *Tectona grandis* L. (teak, family Lamiaceae), which could be successfully preserved (100% survival) for fifteen weeks without requiring subcultures (Montiel-Castelán et al. 2016). Meanwhile, the addition of mineral oil to the media was not effective for in vitro preservation of mahogany or teak meristems. Researches on in vitro conservation of *Daucus carota*, *Vitis vinifera*, *Catharanthus*, *Valeriana wallichii* (Johnson et al. 2002) and *Bacopa monnieri* L. (Sharma et al. 2012) match the efficiency of mineral oil as a preservation technique. However, in this study both survival and vigour decreased significantly, and hyperhydric tissues appeared. Gaspar et al. (2002) indicate that hyperhydricity is an adaptive state induced by stress during in vitro culture.

More recently, studies have demonstrated that the previously developed storage procedure (culture of shoot apical meristems on MS + 15 g l^{-1} sucrose + 15 g l^{-1} mannitol and incubation at 18 °C) was adequate to preserve *S. macrophylla* and *T. grandis* accessions for extended periods (18 and 11 months, respectively) without requiring subcultures (unpublished data). Ex situ storage procedures are now available for the medium-term in vitro preservation of mahogany and teak. These approaches offer new opportunities for the conservation, sustainable management and utilization of these valuable timber trees.

13.3.3 Cedrela Odorata *L. (Spanish Cedar)*

The genus *Cedrela* contains seven species (Pennington 1981; Pennington and Muellner 2010) native to the tropical and subtropical New World from Mexico to northern Argentina (Verissimo et al. 1998; Cavers et al. 2003). *Cedrela* trees may reach 40 m in height and 1.2 m in diameter (Newton et al. 1995, 1999; Navarro et al. 2002). *C. odorata*, known as "Spanish cedar", "red cedar" or "cedar", is the most important and widely distributed species of the genus. Its wood is in high demand in the American tropics, as sawnwood and plywood, for moulding and cabinet work. Having an agreeable, pleasant scent, its wood is also used as packaging for Havana cigars. This species ranges from northern Mexico through Central America and the Caribbean islands to Brazil. It is found in tropical and subtropical and semihumid climates, from sea level to close to 3000 m above sea level in Bolivia (Lamb 1966). It is usually found on well-drained soils. The species tolerates prolonged periods of drought (Lamb 1966; Salas 1993); it also grows in volcanic soils (Pennington and Sarukhán 1968).

As seen above, *C. odorata* is widely distributed throughout the Neotropics and grows in mixed stands. However, the economic value of timber from this species has resulted in their overexploitation for the past two centuries and has prompted several studies concerning the sustainable production and use of *Cedrela* as a forest crop (Newton et al. 1995, 1999; Navarro et al. 2002; Cavers et al. 2003). As with *S. macrophylla*, an additional threat to the remaining trees and to organized planting programmes is that within its natural area of distribution in the Americas, *Cedrela* trees are intensely attacked by the mahogany shoot borer (*Hypsipyla grandella*), to the extent that use of these trees as a forest crop has been halted until resistant cultivars can be selected for or bred (Keay 1996; O'Neil et al. 2001). Thus, it is hoped that breeding programmes can incorporate resistance to this plague to produce resistant cultivars and/or develop hybrids which could allow resistance in another species

In the last years, the INIFAP staff has conducted researches aiming to regenerate plants from seeds and axillary buds (taken from field collected twigs), as well as to storing apical meristems using minimal growth in vitro culture for short- and medium-term preservation of Spanish cedar germplasm. Regarding the micropropagation systems, the initial step is to have a protocol for the establishment and multiplication in vitro that will generate duplicates for preservation. Significant contributions have been made recently in this sense by Sampayo et al. (2016), who suggest the establishment of apical and axillary buds on MS supplemented with 1.0 mg l^{-1} BA, 0.01 mg l^{-1} IAA, 30 g l^{-1} sucrose, 2 g l^{-1} activated charcoal (AC), 1 g l^{-1} polivinilpirrolidone (PVP) as antioxidant and 1 ml l^{-1} plant preservative mixture (PPM) as biocide. In terms of in vitro medium-term preservation of Spanish cedar, shoot apical meristems were dissected from in vitro plants of *C. odorata* belonging to a working collection initiated through seeds and axillary buds from adult trees collected in natural cedar stands, as mentioned by Martinez et al. (2013b). After dissection, apical meristems were cultured on the multiplication medium composed of MS + 1.0 mg l^{-1} BA + 0.05 mg l^{-1} ANA + 2 g l^{-1} AC (control treatment) or MS + AC lacking plant growth regulators. Cultures were kept at 24, 18 or 16 °C. After 30 and 60 d of storage at different temperatures, the number and length of developed shoots was evaluated. Data from this work demonstrated that culture on MS medium + AC without growth regulators at 16 °C, enables maintenance of *C. odorata* meristems for 60 d (Martinez et al. 2013b). Subsequent studies have demonstrated that as in *S. macrophylla* and *T. grandis*, adding MS with 15 g l^{-1} sucrose and 15 g l^{-1} mannitol and keeping cultures at 18 °C allowed to preserve *C. odorata* accessions for extended periods (14 months) without requiring subcultures (unpublished data).

Nunes et al. (2003) have developed an in vitro preservation method for germplasm from *C. fissilis*, an economically important tree of the Brazilian Atlantic Forest. This method involves the medium-term storage of artificial seeds comprising alginate encapsulated propagules (shoot tips, cotyledonary and epicotyl nodal segments) at 25 °C. Maximum post-storage (3 months) viabilities of 96–100% were achieved for encapsulated shoot tips and cotyledonary nodal segments stored on water-solidified agar (at 0.4–0.7% w/v). Encapsulated shoot tips stored on 0.4% (w/v) agar showed the highest longevity (44% survival rate after 6 months storage). Moreover, seeds of *C. fissilis* have been successfully cryopreserved (100% survival) by direct immersion in liquid nitrogen.

13.3.4 Melia Azedarach *L. (Paradise Tree)*

The genus *Melia* L. contains about five species of flowering trees from the Old World tropics. *M. azedarach* is a native of west Asia, but naturalized throughout the warm countries. The tree is well known in India and its neighbouring countries for more than 2000 years as one of the most versatile medicinal plants having a wide spectrum of biological activity. The timber is tough and durable and resembles mahogany. It is used to manufacture agricultural implements, furniture, plywood, boxes, poles, tool handles, and in cabinet making and in construction because of its resistance to termites. Aqueous and alcoholic extracts of leaves and seed reportedly control many insect, mite and nematode pests. Its oil is used for preparing the candle wax, for pest control and lice killing. Fruit stones make ideal beads and are used in making necklaces and rosaries (Katende et al. 1995). The plant is known to contain several organic molecules, i.e. terpenoids, flavonoids, steroids, acids and anthraquinones (Kumar et al. 2003).

The "paradise tree" (*M. azedarach* L. var. *gigantea*) has raised a great interest in Argentina due to the rapidity of its growth (50 cm in diameter at the height of 1.30 m and a 6-m-long trunk which can be cut after 10 years) and because its wood displays characteristics similar to Spanish cedar (*C. odorata*), which makes of it an excellent forestry species (Kunkel 1978). It shows a very good adaptability to different types of soils and climatic conditions, and it is highly resistant to insects and therefore much valued for reforestation programmes (Leonardis et al. 2000; Nardo et al. 1997). The wood is soft, easy to process manually or mechanically. It is used for making planks, furniture, coverings, frames, doors and windows (Mangieri and Tinto 1977). The paradise tree is conventionally reproduced through seeds, which results in highly heterozygous populations. Therefore, asexual propagation techniques were developed for multiplication of selected clones, including in vitro micropropagation using nodal explants (Ahmad et al. 1990; Domecq 1988; Thakur et al. 1998). However, very little work had been done to preserve in vitro germplasm of selected clones. Bernard et al. (2002) reported that embryonic axes were successfully recovered (83%) after 4 months of storage at 4 °C (using 200 μM salicylic acid). Although seed conservation at −20 °C was also tried, only 43% of the seeds germinated at the end of the experiment (Engelmann and Takagi 2000).

Great progress has been made in the last fifteen years by the IBONE (Corrientes, Argentina) staff towards the optimization of paradise tree micropropagation and preservation. A first publication documented a successful in vitro regeneration procedure through shoot tips of "paradise tree" (Vila et al. 2002), making possible the development of a preservation protocol to ensure the integrity of germplasm collections using minimal growth in vitro culture for medium-term preservation of selected genotypes (Scocchi et al. 2004).

Regarding the preservation process, shoot apical meristems (0.5–0.6 mm in length, consisting of the dome and a pair of leaf primordia) were dissected from in vitro plants of *M. azedarach* L. var. *gigantea* (clone "El dorado"), belonging to a working collection initiated through axillary buds from an adult tree (more than

10 years old) as mentioned by Domecq (1988). After dissection, apical meristems were cultured on nutritive media composed either of MS, half (1/2MS) or quarter strength (1/4MS), supplemented with different combinations and concentrations of BA and IBA, including the medium considered for optimum growth of paradise meristem (MS + 0.5 mg l^{-1} BA + 0.1 mg l^{-1} IBA) (Vila et al. 2002) as well as other 20 suboptimum media with reductions in the basal medium salts and/or reduction in the growth regulators concentrations (4–10 times less than that used in the optimum medium). In all cases, cultures were kept in a refrigerator at 4 °C in continuous darkness. After 120, 240 and 360 d of storage at 4 °C, apical meristems were transferred to shoot regeneration fresh medium (MS + 0.5 mg l^{-1} BA + 0.1 mg l^{-1} IBA) and incubated under standard culture conditions (27 ± 2 °C and 14-h photoperiod, irradiance of 116 µmol m^{-2} s^{-1}). After 60 d of culture, viability of recovered meristems was evaluated by recording the number of meristems capable of resuming growth and producing well-developed shoots (more than 25 mm long and with at least 6 expanded leaves). Subsequent root induction was obtained by culturing the regenerated shoots on MS medium supplemented with 3.5 mg l^{-1} IBA for four days and then transferring to MS medium without growth regulators.

Apical meristems remained green or pale green during storage and basically did not increase in length nor produce shoots at 4 °C. However, when apical meristems stored for 120, 240 or 360 days were transferred to shoot regeneration fresh medium and incubated in light at 27 °C, some of them remained alive and rapidly produced shoots. The percentage of meristems forming shoots was dependent on the storage duration and the medium composition. All meristems stored for 120 d were able to regenerate shoots with frequencies between 42 and 100%, independently of the storage medium. Meristems stored for 360 d only regenerated shoots when stored in certain culture media, which indicates that the reduction in the MS strength is one of the most important factors considered to obtain high survival rates of meristems after one-year storage. The culture medium composed by 1/4MS + 0.5 µM BA has allowed 67% of the explants to produce shoots (4–5 well-developed shoots/explants) after 60 d on the regeneration medium. These results are in agreement with those reported for other tropical trees where the reduction in certain components of the culture medium in conjunction with the incubation at low temperatures has allowed germplasm storage for periods of 1–4 years (Engelmann 1997). Rooting of the regenerated shoots was induced in 60% of the cultures, and the plants obtained were successfully transferred to the greenhouse. However, these results show that both shoot regeneration (67%) and rooting percentage (60%) are affected by the storage, because in the control cultures these values are around 100%. In conclusion, this investigation shows successful utilization of in vitro meristem culture for storage of *M. azedarach* during one year at 4 °C without subculture or addition of fresh medium. It is a very simple and effective technique which appears as an alternative to the one reported by Bernard et al. (2002) which is based in the culture of embryonic axes.

13.3.5 Gmelina Arborea *Roxb. (Melina)*

Gmelina arborea (melina) is an important timber-yielding tree from the family Verbenaceae, naturally distributed in the moist deciduous forests of Southeast Asia. This is a species widely used in commercial plantations for timber production according to the different uses of their wood. This medium-sized tree (30 m tall and 1.2–4.5 m trunk diameter) with clear bole of 9–15 m has received attention as a source of good-quality pulp, medium-density fibreboard (MDF) and plywood because of its beautiful white colour and stronger fibre. *G. arborea* has potential as a material for higher-grade uses such as timber for buildings and for furniture (Kojima et al. 2009). It has also been valued for its medicinal properties. Almost all parts of this tree are used in folk medicine for treating various stomach disorders, blood diseases, fevers and skin problems (Sharma et al. 2001). Roots of *Gmelina* are used in commercial Ayurvedic preparations (Tewari 1995). The plant extracts are reported to exhibit anti-inflammatory and wound healing properties (Shirwaikar et al. 2003) and are also known to inhibit platelet aggregation (Faiza and Darakhshanda 1998). Chemical constituents of *Gmelina* include lignans (Anjaneyulu et al. 1977), flavonoids (Nair and Subramanian 1975), iridoid and phenylpropanoid glycosides (Hosny and Rosazza 1998) and an isoxazole alkaloid (Barik et al. 1992).

Melina seeds are classified as recalcitrant based on their short storage life (<1 year) (Prakash 1991). In this case, the in vitro tissue culture techniques are presented as an important tool for the establishment of ex situ germplasm conservation programs. In the last years, the INIFAP (Jalisco, México) staff has conducted studies aimed at in vitro regenerating plants as well as at storing melina germplasm using minimal growth techniques for short- and medium-term preservations. Regarding the preservation process, shoot apical meristems were dissected from in vitro plants of *G. arborea* belonging to a working collection initiated through apical and axillary buds (taken from twigs collected from two-year-old plants growing in a greenhouse). In a preliminary study, apical meristems were cultured on MS lacking plant growth regulators and kept at 24, 16, 12 or 8 °C. Survival was assessed after 90 d of storage at different temperatures. Data from this work showed that incubating cultures at temperatures below 16 °C generated tissues damage. In a subsequent assay, the effect of sucrose/mannitol concentrations added to MS on meristems growth was evaluated. After dissection, shoot apical meristems were cultured on MS supplemented with different sucrose/mannitol concentrations (30/0, 25/5, 20/10, 15/15, 10/20, 5/25 and 0/30 g l^{-1}) and incubated at 24 or 18 °C. Every 30 d of storage at different temperatures, the number and length of the developed shoots as well as their vigour and oxidation were evaluated. Results from this experiment demonstrated that adding MS with 15 g l^{-1} sucrose and 15 g l^{-1} mannitol and keeping cultures at 18 °C was the best condition for preserving apical meristems during 10 months without requiring subcultures of tissues (unpublished data).

Meanwhile, Sukartiningsih et al. (2012) developed a method for synthetic seed formation and conservation by encapsulating axillary buds of *G. arborea* and storing under aseptic conditions. MS medium supplemented with 0.22 mg l^{-1} BAP, 0.02 mg l^{-1} ANA, 1.00 mg l^{-1} IBA and different concentrations of sucrose (3, 6, 12, 24 and 36%) were used as the encapsulation media in combination with 4% (w/v) sodium alginate. The synthetic seeds were stored at 4, 10, 15, 20 or 25 °C for 1, 2, 3 and 4 weeks. After 4 weeks storage, synthetic seeds were effectively preserved (90% survival) at 15 and 20 °C and they successfully sprouted shoots in a succeeding culturing at 25 °C. But those synthetic seeds stored at 4 °C presented chilling injury and no recovered were recorded.

13.4 Cryopreservation of Woody Plant Germplasm and Study Cases

13.4.1 Quercus *Species*

The genus *Quercus* (Fagaceae) comprises around 600 species and is distributed in the northern hemisphere, South of Malaysia and Colombian mountains (Mabberley 1987). Seeds of *Quercus* species are recalcitrant (Roberts and King 1980) and they cannot be stored in a dried state under conventional seed-banking conditions (Roberst 1973). Recalcitrant seeds of temperate origin species are generally less sensitive to desiccation than those of tropical (Berjak and Pammenter 2004). Recalcitrant seeds can be stored in a hydrated state, generally for short periods; however, the seeds will become more desiccation sensitive with time and, also, they will germinate, due to their ongoing metabolism (Pammenter and Berjak 2014). To widen storability, hydrated recalcitrant seeds of temperate species can be stored at low temperatures (but above zero) with aeration (Catalán Bachiller 1991). Antifungal treatments are usually also necessary. This hydrated storage can be used only for few months to a year. For long-term storage, cryopreservation could be a viable alternative (FAO 2014). Because of the large size of oak seeds, embryo axis cryopreservation is required, to allow a more uniform application of treatments and cooling/warming rates. Therefore, in vitro techniques are required for their recovery. The appropriate methodology for in vitro growth of embryonic axes should be established before attempting cryopreservation. The wounds produced by the excision of axes could enhance the production of reactive oxygen species and impair growth (Pammenter and Berjak 2014). In some instances, this problem can be overcome by the use of antioxidants or promoting antioxidant pathways (Bai et al. 2011, 2012; Naidoo et al. 2011).

Submitting the embryonic axis to low temperatures (to those of liquid nitrogen), intracellular ice crystal formation should be avoided and water should vitrified; therefore, free water should be reduced and cooling should be rapid. Due to the potential damage of low water contents, the water content window could be narrow; besides, the speed of drying is also critical. Embryo axis should be dried rapidly to avoid exposure to low water contents for long periods (Berjak and Pammenter 2008).

13.4.2 Quercus Faginea *(Gonzalez-Benito and Perez-Ruiz 1992)*

13.4.2.1 Embryo Axes

Embryo axes of *Q. faginea* Lam. had been successfully cryopreserved by previous desiccation (Gonzalez-Benito and Perez-Ruiz 1992). Green or just turning brown acorns were washed in soapy water; pericarp was removed and the embryo axis containing half of the acorn was surfaced-sterilized (0.05% NaOH for 20 min) followed by three rinses in sterile, distilled water. Axes were excised in sterile conditions and subsequently blotted dried with sterile filter paper before placing them in the airflow of a laminar flow bench. After different times (1–8 h), axes were placed inside polyethylene cryovials and plunged in liquid nitrogen. Warming took place by immersion in sterile water at 40 °C. Axes were cultured on woody plant medium (Lloyd and McCown 1981) salts plus MS-modified vitamins (1 mg L^{-1} thiamine instead of 0.1 mg L^{-1}; Murashige and Skoog 1962) supplemented with 1.5 mg L^{-1} BAP (6-benzyladenine). Incubation took placed at 25 °C in darkness for 1 week, and afterwards in a 16-h photoperiod with a photosynthetic photon flux density of 50 μmol m^{-2} s^{-1}.

Control axes (non-desiccated, not immersed in liquid nitrogen) had 64% water content (fresh weight basis), and radicle started growing two weeks after culture. Maximum recovery (development of root or/and shoot) after 4-week culture was 60% for cryopreserved axes, when they had been desiccated for 3 h (21% water content).

13.4.3 Quercus Ilex and Q. Suber *(Gonzalez-Benito et al. 2002)*

13.4.3.1 Embryonic Axis

Several factors were studied to cryopreserved embryonic axis of *Q. ilex* L. and *Q. suber* L. (Gonzalez-Benito et al. 2002). Acorns were sprayed with a fungicide mixture of 3 mL L^{-1} previcur (propamocarb 60.5% w/v) and 4 g L^{-1} benlate (methyl, 1-(butylcarbamoil)-2-benzylmidazol carbamate 50% w/w) and kept in plastic bags (not tightly closed) at 5 °C until use. After extraction, desiccation or rewarming (depending on the treatment), axes were immersed for 15 min in an ascorbic acid solution (50 mg L^{-1}) and subsequently in 0.25% NaClO for 5 min, followed by three rinses in sterile distilled water. Axes were cultured on WPM + 0.1 mg L^{-1} BA (N6-benzyladenine) for *Q. suber* or 1 mg L^{-1} BA for *Q. ilex*. In vitro incubation temperature played an important role in the appropriate development of *Q. ilex* axes, as 15 °C was superior to 25 °C: 50% shoot development *vs* 25% in control axes after twelve weeks culture in culture. On the other hand, *Q. suber* axes showed higher percentage of shoot development at 25 °C (66%) than at 15 °C (55%).

Axes were desiccated in the airflow of a laminar flow cabinet for different periods. *Q. suber* axes proved to be more sensitive to desiccation and cooling. Moderate desiccation to 34% (2 h desiccation) affected radicle elongation of *Q. suber* axes. Furthermore, desiccation to 18% (4 h) was detrimental and survival decreased to 60%. Survival after cooling (immersion in LN inside cryovials) was low in *Q. suber* axes and was present as unorganized growth. In *Q. ilex,* organized growth was decreased only after desiccation to 13% (4 h), with radicle emergence being reduced from 80% (non-desiccated) to 30% and plumule elongation from 35 to 20%.

The effect of "ultra-rapid" cooling was studied in *Q. ilex* performed axes, by immersing them directly in subcooled liquid nitrogen. Subcooling was achieved by placing a container with liquid nitrogen in a chamber where the pressure was lowered to 30 mbar. At that pressure, nitrogen had solidified. When pressure was increased to atmospheric levels, liquid and solid nitrogen coexisted (−210 °C). Axes were rapidly immersed in subcooled liquid nitrogen and kept for 1 min; cooling rates of 2700 °C min^{-1} were measured. Rewarming took place by rapidly immersing axes in WPM liquid medium at room temperature for approximately 10 min. Subsequently, axes were immersed in an ascorbic acid solution (50 mg l^{-1}) for 15 min and afterwards surface-sterilized. Although survival was high 94% for fast-cooled axes, the response as organized growth was not improved by this method compared with the standard cooling: 13% shoot elongation in embryo axes dried to 18% water content.

13.4.4 *Other Studies on Cryopreservation of* Quercus *Species*

The strategies for forest biodiversity cryopreservation have been described (Häggman et al. 2008); among them, cryopreservation is considered a viable strategy for the long-term conservation of plant cells (cell cultures, embryogenic cell cultures), seeds and embryos (somatic or zygotic).

13.4.4.1 Embryonic Axes and Plumules

Cryopreservation protocols of embryonic axes and plumules of *Quercus* species have been developed as an alternative strategy to seeds storage. As it has been mentioned before, *Quercus* seeds are recalcitrant, and, therefore, they cannot be stored under conventional seed-banking conditions.

There have been studies on the desiccation tolerance of embryo tissues of several *Quercus* species as a previous step for their cryopreservation. Black oak axes (*Q. kelloggii*, *Q. shumardii* and *Q. velutina*) survived drying to water contents of 0.20–0.25 g g^{-1} dry mass (*ca.* 17–20% water content fresh weight basis; Chmielarz and Walters 2007). Xia et al. (2014) studied the desiccation tolerance of axes of four *Quercus* different species, two from subtropical origin and two from temperate climate. They found that the species better adapted to drier environments did not produce the most desiccation-tolerant embryos, but those better adapted to freezing temperatures during winter and that plumule tissues resulted in faster drying and were more sensitive to desiccation than radicles. There was a direct relationship between survival after desiccation and after cryopreservation. The higher sensitivity of plumules could be also related to a sudden increase in extracellular superoxide related to the damage associated with cotyledon excision (Pammenter and Berjak 2014).

Chmielarz et al. (2011) approached the cryopreservation of *Quercus robur* germplasm by using plumules extracted from embryos. The procedure implied the culture of plumules in 0.5 M sucrose solution (18 h), followed by sequential immersion in 0.75 M sucrose, 1.0 M sucrose and in 1.5 M glycerol (40 min each), and subsequent desiccation to 0.5–0.6 g H_2O g^{-1} dry weight (33–38% fresh weight base) with silica gel before direct immersion to liquid nitrogen. With this protocol, survival rates of 51–76% and 8–20% plumule regrowth were obtained.

13.4.4.2 Embryogenic Cultures

Somatic embryogenesis is nowadays a useful strategy for the clonal propagation of selected genotypes of temperate hardwood trees (Pijut et al. 2011). The number of genotypes of *Quercus* species on which somatic embryogenesis has been induced has been increasing during the lasts years, and, therefore, there has been an increasing interest in maintaining those cultures. Many of the procedures developed are based on the use of vitrification solutions and fast cooling (direct immersion in liquid nitrogen).

Masses (4–6 mg) of globular–heart stage somatic embryos of *Q. robur* pre-treated resume embryogenesis after cryopreservation (Martinez et al. 2003), with a 70% recovery rate. Masses were preculture on 0.3 M sucrose medium prior to immersion in the vitrification solution (PVS2; Sakai et al 1990) for 60–90 min prior to immersion in liquid nitrogen. A similar procedure was used with six embryogenic cell culture lines obtained from selected mature *Q. robur* trees (Sánchez el at. 2008), with recovery percentages of 57–92% after 1 year storage. After that period, genetic stability (study by RAPD markers) of cryopreserved somatic embryos and seedlings was demonstrated in five out of the six lines; in the remaining line, recovered seedlings were also stable.

A similar vitrification procedure to that used for *Q. robur* was used with *Q. suber* embryogenic cultures from mature trees, obtaining high recovery (88–93%)

and germination and plant regeneration rates similar to those reached by non-cryopreserved cultures (Valladares et al. 2004).

Globular embryogenic clusters of several embryogenic lines obtained from *Q. ilex* mature trees have been successfully cryopreserved by vitrification (Barra-Jimenez et al. 2015). However, the differentiation capability was hindered by cryopreservation in one of the lines, which was related to genetic instability, detected by microsatellite markers.

Embryogenic calli has also been cryopreserved by air desiccation (Chmielarz et al. 2005). Before desiccation in the airflow of a laminar flow cabinet to 17.3% water content (fwb), calli were cultured on medium with increasing sucrose concentration (0.25, 0.5, 0.75 and 1 M).

All these previous works have contributed to the establishment of a cryopreserved collection of embryogenic cultures (globular or torpedo stages) of 51 *Q. suber* genotypes using a vitrification-based technique (Vidal et al. 2010). All lines withstood cryopreservation and were able to produce new somatic embryos by secondary embryogenesis.

13.4.5 Cryopreservation of **Ilex paraguariensis** *and Wild Relatives*

Ilex L., the only genus of the family Aquifoliaceae, is the largest genus of woody dioecious plants, with at least 600 species distributed in tropical, subtropical and temperate regions of both hemispheres (Galle 1997; Loizeau and Spichiger 2004). South America is considered one of the main areas of diversification of *Ilex*, together with East Asia (Loesener 1942; Lawrence 1951; Cuénoud et al. 2000). In southern South America, the species are mostly found in north-eastern Argentina, south-eastern Brazil and eastern Paraguay (Gottlieb et al. 2005). Most species are deciduous or evergreen shrubs or small trees, but in the tropics the genus also includes some very large trees and a few climbers (Tsang and Corlett 2005).

The genus *Ilex* comprises several species with economic importance as crops and ornamentals. Some of them, commonly named "hollies", such as "English holly" (*I. aquifolium*), "Japanese holly" (*I. crenata*) and "American holly" (*I. opaca*), have long been symbolic of Christmas and have also been cultivated by nurserymen in Europe, Asia and the USA for landscaping. Various institutions and commercial breeders are developing hybrids with improved tolerance to winter and with more foliage (Hu 1989; Walden and Wright 1995). In South America, one of the most important species of economic and pharmacological interest is *I. paraguariensis*, popularly known as "yerba mate" or "maté tree". This is obtained in the native form and is widely cultivated in north-eastern Argentina, eastern Paraguay and southern Brazil (Burris et al. 2012). The yerba mate is characterized as an important product in the socioeconomic cultural context in their original regions. It is mainly consumed as "mate", a hot beverage prepared by infusion that stimulate the central nervous system due to the presence of xanthic bases or alkaloids, such as caffeine and theobromine (Filip et al. 2001; Schinella et al. 2005). There are other products in the market made from the yerba mate leaves, as blended teas, flavoured tea, iced tea and cosmetics that use *I. paraguariensis* extracts (Mosele 2002). Some investigations are expanding the use of this species in new products, exploring its probiotic properties (Preci et al. 2011; Ril et al. 2011).

Currently, the demand for food containing biologically active substances has increased, since consumers are seeking these products for a healthier life (Melo and Guerra 2002). The yerba mate fits in this context, due to the numerous benefits to health it provides, such as hypocholesterolemic and hepatoprotective activity (Filip and Ferraro 2003; Açari et al. 2011), anti-inflammatory and anti-obesity effects (Bracesco et al. 2011), central nervous system stimulation, diuretic action (Castaldelli et al. 2011), inhibition of neoplastic cells proliferation (Mejía et al. 2010), and antioxidant activity, preventing damages caused by free radicals (Bastos et al. 2007). Studies showed that among ten plant species used as infusions, yerba-mate showed one of the highest antioxidant capacity (Asolini et al. 2006), whose extracts have higher polyphenol contents than those of green tea and similar to those of red wines (Gugliucci and Bastos 2009; Gugliucci et al. 2009).

In the natural distribution area of *I. paraguariensis*, a number of wild species of *Ilex* (like *I. brasiliensis, I. brevicuspis, I. dumosa, I. integerrima, I. pseudoboxus* and *I. theezans*) also occur which are sympatric with genuine maté (Giberti 1999). In recent years, *I. dumosa* ("yerba señorita") has received the most attention from plant breeders because this species is resistant to the attacks of certain diseases which are common for *I. paraguariensis*. Moreover, with their leaves it is possible to make "maté" with less caffeine than with the ones from the genuine yerba mate (Filip et al. 1999). Wild populations can be a valuable source of new genetic material for plant breeding. The natural diversity of plants growing in their natural habitats means that at least some individuals may carry genes of commercial importance, such as those which confer resistance to diseases and insects or are useful in stressful environments (Acquaah 2012). Therefore, it is crucial to secure as much biodiversity as is possible. However, in the maté-growing region (and their sympatric *Ilex* species) the risks of genetic erosion are high because the natural forest is gradually giving way to agroforestry and livestock production, a process accentuated by the low germinability of many species (especially that of *I. paraguariensis*). Moreover, market demands for uniformity concerning both quality and higher yields clearly restrict genetic variation of a given crop species. Consequently, ex situ preservation of *Ilex* germplasm is necessary to safeguard the threatened diversity of this genus, mainly due to the anthropogenic impact. These species are usually preserved in field collections, at the risk of disease, pest, fire, drought and human damage, besides the genetic erosion (Giberti 1999; Zhang et al. 2014). Thus, research for alternative methods to field conservation for *Ilex* genetic resources became a priority.

So far, the only reports regarding the cryopreservation of *Ilex* germplasm were carried out in the IBONE (Corrientes, Argentina). Research was aimed at the possibility of cryopreserving fruits, seeds, isolated embryos and shoot tips of diverse South American *Ilex* species, as a complementary option to field collections.

13.4.6 Cryopreservation of Ilex *Fruits, Seeds and Zygotic Embryos*

Seed storage is the most effective and efficient method for the ex situ preservation of genetic resources of plants which produce orthodox seeds, by combining low storage costs (100 times cheaper that in situ preservation of individual trees) with ease of seed distribution and regeneration of whole plants from genetically diverse material as each seed is genetically different (Linington and Pritchard 2001; Li and Pritchard 2009). *Ilex* seeds are harvested at water contents above 30%, a feature that is often associated with recalcitrant behaviour (Berjak et al. 1992). However, the levels to which *Ilex* seeds will tolerate desiccation and low temperature storage are unclear. Moreover, seeds of *Ilex* species are individually enclosed by woody endocarp and have undeveloped embryos (mostly at heart stage) when fruits reach maturity (Martin 1946; Niklas 1987; Tsang and Corlett 2005; Dolce et al. 2007), resulting in a deep dormancy and low germination rate (Hu 1975; Hu et al. 1979). For example, *I. opaca* germinates in nature after one to three years and the germination rate is about one in ten million (Ives 1923). This extremely low germination rate constitutes a serious inconvenience for breeding and conservation programs, since it leads to a loss of potentially valuable genotypes. Besides, the small embryo size (160–350 μm in length) and the high level of dormancy of *Ilex* seeds have hampered efforts to gain knowledge about their storage characteristics.

In recent years, the IBONE staff has conducted studies to evaluate the possibility to cryopreserve fruits, seeds and zygotic embryos from seven South American *Ilex* species, using the vitrification, desiccation and encapsulation-dehydration techniques, respectively. For all experiments, the source of plant material consisted in open pollinated ripened fruits (nuculanium) of *Ilex* spp., which were hand-harvested during summer ($\sim$3 months after anthesis) from trees growing in field.

13.4.6.1 Fruits

The first strategy for long-term preservation of *Ilex* germplasm involved the cryopreservation of fruits using the vitrification technique (Mroginski el al. 2006, 2011). Fruits were surface-sterilized by soaking them in 70% ethanol for 5 min, followed by immersion in an aqueous solution of 1.8% sodium hypochlorite and 0.1% Triton X-100®. Subsequently, fruits were rinsed three times with autoclaved distilled water. Superficially sterilized fruits were cold-pretreated (30 d at 4 °C).

Preconditioned fruits were placed in 5-ml cryovials and exposed to 2.5 ml of PVS3 [50% glycerol (w/v) + 50% sucrose (w/v)] (Nishizawa et al. 1993) vitrification solution for 60 min at 0 °C prior to rapid immersion in liquid nitrogen (rapid cooling), or slowly cooled at 1 °C min^{-1} from +25 to −40 °C (by using a Controller Rate freezing System, Gordiner Electronics, Inc., USA) and then immersed at −196 °C (slow cooling). After storage in liquid nitrogen for 24 h, fruits were rewarmed by immersing the cryovials in a water bath at 30 °C for 1 min and then washed three times for 15 min by replacing the PVS3 with an unloading solution composed of liquid MS medium supplemented with 1.2 M sucrose. For growth recovery, the rudimentary embryos at the heart stage were excised from seeds (Mroginski et al. 2011) and cultured in vitro on the germination medium consisting of quarter-strength MS supplemented with 0.1 mg l^{-1} zeatin (Sansberro et al. 1998, 2001). The embryos were kept in a growth room at 27 ± 2 °C, in the dark for 30 d, and then transferred to light standard culture conditions (116 μmol m^{-2} s^{-1} PPFD provided by cool white fluorescent lamps) for another 30 d. The embryo survival from fruits subjected to vitrification procedure was evaluated at 60 d after beginning the cultivation of the isolated embryos, through their germinability. Embryos from all the tested species did not tolerate cryostorage when vitrified fruits were rapidly cooled, but the ability to withstand immersion in liquid nitrogen increased when fruits were slowly cooled. In any case, the embryo germinability from slow cooled fruits was scant.

13.4.6.2 Zygotic Embryos

In another report, the IBONE staff described the successful cryopreservation of *Ilex* zygotic embryos by using the encapsulation-dehydration technique (Mroginski el al. 2008, 2011). Fruits were surface-sterilized, and rudimentary embryos were excised as above. After excision, embryos were preconditioned on semisolid germination medium supplemented with 0.3 M sucrose for 7 d, in dark conditions. Preconditioned embryos were encapsulated in 3% sodium alginate (SIGMA-ALDRICH®), polymerized with calcium chloride ($CaCl_2$) at 0.1 M, to form calcium alginate capsules approximately 5 mm in diameter. Encapsulated embryos were then pretreated in liquid MS supplemented with 0.5, 0.75 and 1.0 M sucrose, by progressively increasing the concentration from 0.5 to 1.0 M for 24 h in each condition. The pretreatment was performed by placing the samples at 27 °C on an orbital shaker at 100 rpm. After preculture, the beads were rapidly surface-dried on filter paper and dehydrated in 100 cm^3 airtight containers with 30 g silica gel (10 beads/container) for 5 h (equivalent to capsule moisture content of 25%, fresh weight basis). Samples were then placed in 5-mL cryovials and rapidly immersed in liquid nitrogen (rapid cooling), or slowly cooled at 1 °C min^{-1} from +25 to −30 °C and then immersed at −196 °C (slow cooling). After 24-h cryostorage, samples were rewarmed by immersing the cryovials in a water bath at 30 °C for 1 min, and the beads were placed on semisolid germination medium for recovery.

Table 13.3 In vitro germination (%) of *Ilex* zygotic embryos from cryopreserved fruits (subjected to the vitrification technique) and encapsulated embryos (subjected to the encapsulation-dehydration technique). Data are the mean of three replicates ± SE

Species	Fruits cryopreservation		Isolated embryos cryopreservation	
	−LN	+LN[a]	−LN	+LN[b]
I. brasiliensis	73 ± 3	3 ± 3	56 ± 8	47 ± 13
I. brevicuspis	40 ± 6	0	8 ± 3	15 ± 8
I. dumosa	40 ± 6	13 ± 3	74 ± 14	67 ± 7
I. integérrima	43 ± 7	0	27 ± 12	50 ± 12
I. paraguariensis	40 ± 6	10 ± 6	50 ± 6	27 ± 3
I. pseudoboxus	67 ± 3	23 ± 3	23 ± 9	7 ± 3
I. theezans	67 ± 3	0	67 ± 3	65 ± 15

LN liquid nitrogen
[a]Slow cooling
[b]Rapid cooling

The recovered embryos were kept in a growth room at 27 ± 2 °C, in the dark for 30 d and then transferred to light standard culture conditions for another 30 d. Survival was evaluated at 60 d after beginning the recultivation of the beads, through the germinability of the encapsulated embryos. Using the E-D technique, the ability of *Ilex* embryos to withstand cryostorage was increased (Table 13.3). In most species, the slow cooling did not improve the survival of embryos. So, the use of rapid cooling is suggested, since no sophisticated facilities are necessary.

13.4.6.3 Seeds

More recently, the possibility of cryopreserving intact seeds from seven *Ilex* species using the desiccation (D) technique was evaluated. Seeds were removed from fresh fruits, cleaned of the pulp and immediately used for assembly of experiment. Seeds were desiccated in 100 cm^3 airtight containers with 30 g silica gel (300 seeds/container) for 2-h intervals, up to 14 h. Desiccated seeds were placed in 2-ml cryovials and rapidly immersed in liquid nitrogen (rapid cooling). After 7-d cryostorage, samples were rewarmed by immersing the cryovials in a water bath at 40 °C for 2 min. Due to the deep dormancy and low germination rate of *Ilex* seeds when conventional methods are used, germinability was assessed through in vitro culture of intact/bisected seeds or isolated embryos, according to the previously optimized procedure for each species (Sansberro et al. 1998; Dolce et al. 2010, 2011, 2015). Cultures were kept in a growth room at 27 ± 2 °C, under light standard conditions for 60 d (whole and cut seeds) or in the dark for 30 d and then transferred to light conditions for another 30 d (isolated embryos). Survival and plant development from seeds subjected to desiccation procedure was evaluated at 60 d after beginning the recultivation of the seeds or embryos.

The seeds MC decreased from an initial average of 40.9 ± 0.2% to 6.2 ± 0.1% after 14-h desiccation. Seeds of all species tested tolerated desiccation down to ~6% of MC with similar germination percentages to that in their respective control (non-desiccated and non-cryostored seeds). Moreover, intact seeds of the seven species could be successfully cryopreserved when they were desiccated to 6.4–8.4% (depending on the species) prior to immersion in liquid nitrogen with no reduction in the germinability compared with the control group. It is known that the most critical factor affecting cryopreservation of seeds is MC (Pritchard 2007), so the range of MC which allows seeds to tolerate cryogenic temperatures should be determined for each species. Dehydration must be sufficient to avoid lethal intracellular freezing during cooling, but not so intense to induce extended desiccation injury. In optimal cases, no significant difference is observed in the survival rates of desiccated control and cryopreserved material (Vertucci and Farrant 1995; Pammenter and Berjak 1999; Walters et al. 2002).

Data from this study demonstrated that *Ilex* seeds of the seven South American species tested did not show high sensitivity to dehydration and cryopreservation, as might be expected for tropical species whose seeds are disseminated with high water content. Seeds tolerated desiccation to ~6% MC without a loss in viability. Furthermore, these data suggest a simple and cost-effective method for *Ilex* seed cryopreservation by using the desiccation technique. This is a very simple and cost-effective cryopreservation method because neither cryoprotectants nor any sophisticated facilities are necessary. Compared with cryopreservation of isolated embryos by encapsulation-dehydration technique, this protocol has advantages such as less consumption of time and labour and simplicity of the protocol.

13.4.6.4 Shoot Tips

Due to the allogamy of *Ilex* species, asexual reproduction would be of great value for the multiplication of select commercial lines of “mate”. Successful propagation is possible for juvenile material (Sansberro et al. 1999); however, like many other woody species, mature tissues show a low morphogenetic potential, which makes it difficult to clone mature trees by rooting cuttings or by in vitro techniques. In the same way, this makes it difficult to use cryopreservation techniques for the safe long-term storage of selected genotypes. Anyway, despite not having optimized methodologies for in vitro propagation of *Ilex* species, studies have been initiated to evaluate the possibility of cryopreserving *I. dumosa* and *I. paraguariensis* shoot tips (the two species currently cultivated for industrial purposes).

Shoot tips (2–3 mm in length, consisting of the meristematic dome and two to three leaf primordia) were dissected from in vitro regenerated shoots obtained from cuttings of plants grown in a greenhouse after 45 d of in vitro establishment (Luna et al. 2003). After dissection, shoot tips were preconditioned on semisolid multiplication medium consisting of quarter-strength MS + 0.1 mg L^{-1} BAP, supplemented with 0.3 M sucrose for 48–72 h, using standard culture conditions. Apical shoot tips were subjected to two cryopreservation techniques: encapsulation-dehydration and vitrification.

13.4.6.5 Encapsulation-Dehydration (E-D)

Preconditioned apices were encapsulated in 3% sodium alginate, polymerized with calcium chloride ($CaCl_2$) at 0.1 M, to form calcium alginate capsules approximately 5 mm in diameter. Encapsulated apices were then pretreated in liquid MS supplemented with 0.5, 0.75 and 1.0 M sucrose, by progressively increasing the concentration from 0.5 to 1.0 M for 24 h in each condition. The pretreatment was performed by placing the samples at 27 °C on an orbital shaker at 100 rpm. After preculture, the beads were rapidly surface-dried on filter paper and dehydrated in 100 cm^3 airtight containers with 30 g silica gel (10 beads/container) for 1–7 h. Samples were then placed in 5-mL cryovials and rapidly immersed in liquid nitrogen (rapid cooling), or slowly cooled at 1 °C min^{-1} from +20 to −30 °C and then immersed in liquid nitrogen (slow cooling). After 24–48-h storage in liquid nitrogen, samples were rewarmed by immersing the cryovials in a water bath at 40 °C for 2 min, and the beads were placed on semisolid MM for recovery. The recovered apices were cultured for 1 week in the dark before being transferred to standard culture conditions.

13.4.6.6 Vitrification (V)

Preconditioned apices were loaded in a 0.4 M sucrose + 2 M glycerol solution for 20–30 min at 27 °C and exposed to PVS2 [30% glycerol (w/v) + 15% ethylene glycol (w/v) + 15% (w/v) DMSO + 0.4 M sucrose] (Sakai et al. 1990) or PVS3 [50% glycerol (w/v) + 50% sucrose (w/v)] (Nishizawa et al. 1993) vitrification solutions for 0, 30, 60, 90, 120 or 150 min at 27 or 0 °C prior to rapid immersion in liquid nitrogen in cryovials with 3 mL of the respective PVS. After storage in liquid nitrogen for 24–48 h, apices were rewarmed by immersing the cryovials in a water bath at 40 °C for 2 min and then washed three times for 15 min by replacing the PVS with an unloading solution composed of liquid multiplication medium supplemented with 1.2 M sucrose. After rewarming, apices were transferred to filter papers for 1–2 min to drain off excess liquid. For growth recovery, tissues were placed on semisolid, kept one week in the dark and then transferred under standard culture conditions.

The survival from shoot tips subjected to both cryogenic procedures (E-D and V) was evaluated at 30 d after beginning the recultivation of the explants. No survival was achieved from *I. dumosa* and *I. paraguariensis* shoot tips subjected to cryostorage, whichever the freezing protocol (rapid or slow cooling) employed. With the E-D technique, survival of desiccated pretreated controls remained high (70–80%, depending on the species) until after 5-h dehydration (equivalent to capsule MC of ~25%), but it markedly decreased to 40% after 6-h dehydration and it fell to 0% after 7-h dehydration (equivalent to capsule MC of ~20%). Besides, PVS-treated controls shoot tips showed high survival rates (60–90%, depending of the PVS exposure duration) (data not published). These results suggest that the encapsulation and sucrose pretreatment or PVS exposure did not affect per se the viability of the shoot tips nor the dehydration up to MC of 25%. The negative results after cryopreservation of *Ilex* shoot tips by E-D and V techniques would not be related to the high sensitivity of this tissue to sugars or desiccation (Fig. 13.1).

Fig. 13.1 Application of vitrification and encapsulation-dehydration techniques for cryopreservation of *Ilex* spp. shoot tips. (**a**) In vitro regenerated shoots after 45 d of cuttings establishment. (**b**, **c**, **d**, **e**) *Ilex paraguariensis* shoot tips from the control treatments (−LN): untreated shoot tip after 30 (**b**) and 60 (**c**) days of culture; shoot tips treated with PVS3 for 120 min (**d**) or pretreated in liquid MS supplemented with progressively increasing sucrose concentration (**e**), after 60 days of culture. (**f–h**) *I. dumosa* encapsulated shoot tips from the control treatment (−LN) with 4-h dehydration, after 15 (**f**), 30 (**g**) and 60 (**h**) days of culture. (**i**) *I. dumosa* encapsulated shoot tips subjected to cryostorage, 10 days after recovery

Consequently, further research is still needed. Assessment of cryopreservation protocols could be supported, for example, by differential scanning calorimetry studies, manipulation of preculture treatments to enhance tissue cryotolerance (e.g. cold hardening) (Paul et al. 2000) and post-thaw culture treatments to promote tissue recovery from cryoinjury (e.g. use of nursery cultures) (Hargreaves et al 2002). Moreover, biochemical and molecular biology studies on tissue culture recalcitrance and cryoinjury could potentially be more significant (e.g. to investigate the role of oxidative stress) (Benson 2000).

Cryopreserved collections of clonally propagated woody species have already been established in several countries, including USA (apple, pear, hazelnut), France (elm), Japan (mulberry) and India (almond, citrus fruits) (Reed 2002). For other woody species, effective protocols of cryopreservation by vitrification approaches have been reported (Lambardi and De Carlo 2003; Lambardi 2002; Scocchi et al. 2004; Halmagyi et al. 2010; Volk et al. 2012; Silva et al. 2013; Soliman 2013) which would be used for cryostored germplasm collections. *Ilex* spp. cannot be ascribed to this list yet; however, efforts are continuing to overcoming the "recalcitrance" of these species to the conservation in liquid nitrogen.

13.4.7 *Family Meliaceae*

13.4.7.1 *Melia Azedarach*

As mentioned before, the "paradise tree" (*M. azedarach* L. var. *gigantea*) has raised a great interest in Argentina as an excellent forestry species. Over the last 15 years, great progress has been made towards the optimization of paradise tree micropropagation and cryopreservation, since the IBONE staff has conducted studies for regenerating plants from shoot apical meristems and zygotic embryos (Vila et al. 2002, 2003) and for cryopreserving such explants using diverse techniques (Scocchi et al. 2004, 2007).

13.4.7.2 Shoot Tips

The first strategy for long-term preservation of clonal germplasm from paradise tree involved the cryopreservation of shoot apical meristems using the encapsulation-dehydration technique (Scocchi et al. 2004). Shoot apical meristems (0.5–0.6 mm in length, consisting of the dome and a pair of leaf primordia) were dissected from in vitro plantlets of *M. azedarach* L. var. *gigantea* (clone "El dorado"). After dissection, apical meristems were preconditioned for 24 h on semisolid establishment medium consisting of MS + 2 μM BA and 0.5 μM IBA (Vila et al. 2002), using standard culture conditions. Preconditioned shoot tips were encapsulated in 3% sodium alginate (Sigma-Aldrich ®), polymerized with calcium chloride

($CaCl_2$) at 0.1 M, to form calcium alginate capsules approximately 5 mm in diameter. Encapsulated embryos were then pretreated in liquid EM supplemented with 0.5, 0.75 and 1.0 M sucrose, by progressively increasing the concentration from 0.5 to 1.0 M for 24 h in each condition. The pretreatment was performed by placing the samples at 27 °C on an orbital shaker at 100 rpm. After preculture, the beads were rapidly surface-dried on filter paper and dehydrated in 100 cm^3 airtight containers with 30 g silica gel (10 beads/container) for 5 h (equivalent to capsule MC of 25%, fresh weight basis). Samples were then placed in 5-mL cryovials and rapidly immersed in liquid nitrogen (rapid cooling) or slowly cooled at 1 °C min^{-1} from +20 to −30 °C and then immersed to −196 °C (slow cooling). After 1-h storage, samples were rewarmed by immersing the cryovials in a water bath at 30 ° C for 2 min, and the beads were transferred directly on semisolid EM for recovery. The recovered shoot apical meristems were kept in a growth room at 27 ± 2 °C, in light standard culture conditions. The survival percentage was evaluated at 7 days, shoot proliferation percentage at 28 days and number of shoots/shoot apical meristem at 60 days after beginning the recultivation of the explants subjected to encapsulation-dehydration technique.

Using the encapsulation-dehydration technique, cryopreservation of shoot apical meristems of paradise tree can be achieved with high survival rates (slow cooling = 83 ± 2%; rapid cooling = 67 ± 2%) and shoot proliferation rates (slow cooling = 60 ± 1%; rapid cooling = 43 ± 1%). Plants regenerated after cryogenic treatments preserved genetic stability, when it was evaluated using the electrophoretic patterns of nine isozyme systems and RAPD profiles. The survival rates achieved following this encapsulation-dehydration protocol were comparable to those obtained by Bernard et al. (2002), who cryopreserved embryonic axes with a salicylic acid pretreatment. However, this cryopreservation procedure using an increasing sucrose concentration as the only cryoprotectant and controlled silica gel desiccation is preferable to others previously reported. It not only resulted in high survival rates, but it also represented the first protocol by which high shoot proliferation and clonal plants regeneration could be obtained, besides confirming their genetic stability. Regarding these findings, it is known that in vitro propagation techniques based on shoot tips culture are the most desirable for long-term germplasm conservation because this is an ideal explant with few differentiated cells that tend to remain genetically stable during the regeneration process (Kartha et al. 1980; Towill 1984; Mroginski et al. 1991).

13.4.7.3 Somatic Embryos

Later a cryopreservation protocol for paradise tree somatic embryos was established by comparing three cryopreservation techniques: encapsulation-dehydration, pregrowth-dehydration and desiccation (Scocchi et al. 2007). Open pollinated immature fruits of *M. azedarach* L. var. *gigantea* were hand-harvested during late-spring (8–9 weeks after anthesis) from trees growing in field. In vitro culture of zygotic embryos and somatic embryogenesis induction were conducted according to Vila et al. (2003). Somatic embryos were subjected to three cryopreservation techniques:

Encapsulation-dehydration (E-D)
This system was adapted from that developed for shoot tips cryopreservation (Scocchi et al. 2004). The beads were dehydrated in containers with silica gel for up to 4 h, thereby reaching 21–26% of MC (fresh weight basis). Same beads were dehydrated without sucrose pretreatment. Desiccated beads were then placed in 5-mL cryovials and rapidly immersed in liquid nitrogen (rapid cooling) or slowly cooled at 1 °C min^{-1} from +20 to −30 °C and then immersed to −196 °C (slow cooling). After 1-h cryostorage, samples were rewarmed by immersing the cryovials in a water bath at 30 °C for 2 min. Somatic embryos were then transferred to recovery medium (quarter-strength MS, 3% sucrose, no growth regulators) under standard culture conditions.

Pregrowth-dehydration (P-D)
Non-encapsulated mature somatic embryos were submitted to the same pretreatment as for the encapsulation–dehydration technique, dehydrated with silica gel from 90.5% MC (0 h) to 13.5% MC (4 h), and then cooled rapidly (direct immersion of the cryotubes in liquid nitrogen) or slowly (1 °C min^{-1} from +20 to −30 °C followed by immersion of the cryotubes in liquid nitrogen). Rewarming and growth recovery took place as described previously.

Desiccation (D)
Non-pretreated and non-encapsulated mature somatic embryos were desiccated with silica gel from 83.5% MC (0 h) to 9.3% MC (4 h) and then cooled rapidly (direct immersion of the cryotubes in liquid nitrogen) or slowly (1 °C min^{-1} from +20 to −30 °C, followed by immersion of the cryotubes to −196 °C). Rewarming and growth recovery took place as described previously.

Recovery (%) was assessed after 3 weeks by counting the number of cryostored somatic embryos which developed into plantlets and/or produced adventitious embryos and gave rise to fully developed plantlets. Among the three cryopreservation techniques tested, only E-D and P-D led to successful results after slow cooling. No survival was achieved after rapid cooling, irrespective of the technique used. The highest recovery rates were achieved after 3-h dehydration, reaching 36% survival with E-D technique (21% of MC) and 30% survival with P-D technique (19% of MC). This work reported for the first time the cryopreservation of paradise tree somatic embryos. The recovery percentages achieved were intermediate, so it might be possible to further improve the results by refining the pretreatment conditions, e.g. by increasing the duration of treatment with the successive sucrose media employed, by using higher final sucrose concentrations or by testing other techniques such as vitrification, encapsulation-vitrification or droplet-vitrification (Sakai and Engelmann 2007). Finally, the protocol established would have to be validated using additional paradise tree accessions.

13.4.8 Cryopreservation of Citrus Germplasm

Citrus fruits are recognized as an important food widely consumed throughout the world, which are characterized by the distinct aroma and delicious taste, becoming an inseparable part of our diet. Citrus fruits are grown in more than 140 countries; most of the crop grows on either side of a belt around the equator covering tropical and subtropical areas of the world 35°N and 35°S latitudes with cultivation and production concentrated in major regions in the northern hemisphere (Ramana et al. 1981; UNCTAD 2004).

The origin of citrus fruit is full of interesting legends. The belief exists that citrus is native to the subtropical and tropical areas of Asia, originating in certain parts of Southeast Asia including China, India and the Malay Archipelago (Ramana et al. 1981; Gmitter and Hu 1990). According to ancient Chinese documents, in the earliest reference of citrus fruits are mentioned mandarins and pummelos as highly prized tributes only available for the imperial court. Lemon was originally grown in India, and sweet oranges and mandarins are indigenous to China. Later, some research has suggested that the true origins of citrus fruit are Australia, New Caledonia (off eastern Australia) and New Guinea (Anitei 2007). The spread of citrus also advanced to northern Africa and southern Europe, and the first introduction of citrus to America was achieved by Spanish and Portuguese explorers. Currently, the commercial production, processing and global trade of citrus are huge and have significantly increased since then. Several years ago, citrus have been considered as the most important fruit in the world (Ramana et al. 1981; UNCTAD 2004).

Citrus genetic resources are traditionally maintained as living collections in field genebanks, because most species produce seeds with recalcitrant or intermediate storage behaviour and numerous cultivars are seedless. However, as already reviewed, plant material in field conditions remains exposed to pests, diseases and other natural hazards (Gonzalez-Arnao et al. 1998; Gonzalez-Arnao and Engelmann 2013; Nagy and Attaway 1980). Greenhouses are also an advisable alternative for conserving selected materials under controlled conditions, protected from pest and insect borne disease; however, these houses demand special technical facilities, in addition to considerable expenses to maintain a limited number of specimens (Duran-Vila 1995). Therefore, cryopreservation is currently the safest and most appropriate method for long-term storage of these problem species.

So far, many studies have been conducted on cryopreservation of citrus germplasm since the first reports in the 1980s. As a result, effective cryopreservation protocols have been developed for different type of plant tissues and organs (Marin and Duran-Vila 1988; Sakai et al. 1990; Olivares-Fuster et al. 2000; Cho et al. 2001, 2002a, b, c; Gonzalez-Arnao et al. 1998, 2003; Lambardi et al. 2004; Kaya and Pillhofer 2013).

In general, cryopreservation of citrus germplasm has been performed using seeds (Mumford, and Grout 1979), ovules (Gonzalez-Arnao et al. 2003), embryogenic axes (Radhamani and Chandel 1992), somatic embryos (Marín and Duran-Vila

1988; Gonzalez-Arnao et al. 2003), embryogenic calluses and cell suspensions (Engelmann et al. 1994; Gonzalez-Arnao et al. 1997; Sakai et al. 1990), pollen (Ganeshan and Alexander 1991), and shoot tips from juvenile (Gonzalez-Arnao et al. 2003) and adult plants (Volk et al. 2012).

From the biotechnology point of view, cryopreservation of embryogenic callus and ovules has a special interest for breeding citrus programs based on protoplast fusion (Olivares-Fuster et al. 2000); additionally, cryopreservation of shoot tips from adult plants is the most desirable approach for preservation of citrus genetic resources. Plants regenerated from adult shoot tips are true-to-type and do not have juvenility characteristics. Therefore, they will not require several years to flower and produce fruits (González-Arnao et al. 2003; Volk et al. 2012).

13.4.8.1 Callus and Cell Cultures

Cryopreservation of embryogenic callus can be used for the long-term storage of totipotent lines. Embryogenic cultures have shown to be suitable as a source for protoplast isolation, somatic hybridization and plant regeneration (Olivares-Fuster et al. 2000). Therefore, cryopreserved embryogenic callus has the advantage that it can be readily used for further biotechnology applications anytime, without needing to start a process of callus formation each time, which in addition is an erratic, time-consuming, season and genotype-dependent process. The availability of frozen material contributes to solving this limitation.

13.4.8.2 Embryogenic Callus (Gonzalez-Arnao et al. 1997)

Ovules were excised under sterile conditions from immature fruits collected 2 to 8 weeks after anthesis. The fruits were surface-disinfected by soaking for 20 min in a 2% (v/v) sodium hypochlorite solution containing 0.1% (v/v) wetting agent Tween 20, and the excised ovules were cultured on semisolid MS (Murashige and Skoog 1962) medium containing 100 $mg.L^{-1}$ *i*-inositol, 0.2 $mg.L^{-1}$ thiamine HCl, 1 $mg.L^{-1}$ pyridoxine HCl, 1 $mg.L^{-1}$ nicotinic acid, 50 $g.L^{-1}$ sucrose, 500 $mg.L^{-1}$ malt extract and supplemented with 10 $g.L^{-1}$ agar. Ovules were cultured in dark for 2–3 months.

When an embryogenic mass of friable callus (white to yellow) started appearing, calluses were removed from the ovules and then subcultured every 4–6 weeks on the same medium and culture conditions until obtaining enough material. For developing cryopreservation experiments, calluses were used 3–4 weeks after the last subculture.

About one-third of sterile 2-ml cryotubes were filled with fragments of calluses which were pretreated in liquid medium with sucrose concentrations (0.15, 0.3, 0.6 or 0.9 M) or glycerol (0.1, 0.3, 0.5, or 0.7 M) combined with dimethylsulphoxide

(DMSO) for 1 h at 0 °C. DMSO was added progressively to the liquid medium over a period of 30 min until the final concentration (5 or 10%, v/v) was reached.

Freezing was performed using a programmable freezer (Bio-Cool, FTS Systems, USA) at different slow freezing rates (0.2, 0.5 or 1 °C.min^{-1}) from 0 °C to −20°, −40°, −60° or −80 °C before the rapid immersion in liquid nitrogen. Seeding was induced in samples at −10 °C. After cryopreservation, samples were kept for at least 1 h at −196 °C. Rapid thawing was carried out by plunging the cryotubes in a water bath at +40 °C, and calluses were then transferred directly (without washing) to recovery medium. Calluses, placed onto pieces of filter paper, were successively transferred to fresh medium after 1 h, 24 h and 7 days of culture. The survival rate was evaluated 60 d after freezing and corresponded to the percentage of calluses which had increased in size during the recovery period.

As result, the best cryopreservation conditions for citrus embryogenic calluses comprised the use of mixture DMSO 5% + sucrose 0.9 M or DMSO 10% + sucrose 0.6 M for 1 h at 0 °C, and freezing at 0.5 °C.min^{-1} down to −40 °C before immersion into liquid nitrogen. This protocol allowed obtaining up to 50% of viability, and recovery was evident with the formation of a new mass of calluses and then of several somatic embryos.

13.4.8.3 Cell Suspensions

The first cryopreservation report with nucellar cells of citrus (Sakai et al. 1990) and using the vitrification solution PVS2, which had been recently developed at that time, was performed with cells of navel orange (*Citrus sinensis* Osb. var. brasiliensis Tanaka). Cells were sufficiently dehydrated with the highly concentrated solution PVS2 [30% glycerol (w/v) + 15% ethylene glycol (w/v) + 15% (w/v) DMSO + 0.4 M sucrose] (Sakai et al. 1990) at 25 °C for 3 min and then transferred to plastic cryotubes and finally, directly plunged into liquid nitrogen. Citrus cells were successfully cryopreserved by vitrification, and the average rate of survival was about 90% (Sakai et al. 1990).

By using a classical protocol with slow freezing regime was performed cryopreservation of cell suspensions of *Citrus deliciosa* Tan. After a one-hour cryoprotective treatment with 0.15 M sucrose and 5% DMSO, cells were prefrozen to −40 °C at 0.5 °C.min^{-1} and immersed in liquid nitrogen. After rapid thawing, survival ranged between 50 and 60% of unfrozen controls. Proliferation in liquid medium was optimal after a 5-day recovery period on solid medium. Production of somatic embryos was higher with cryopreserved cultures than with unfrozen control cultures (Aguilar et al. 1993)

A simplified cryopreservation process (using Mr. Frosty device) was experimented with a cell suspension of willow leaf mandarin and embryogenic calluses of six varieties of Citrus. Its efficiency was comparable to that of the standard freezing protocol developed previously for these materials, which required the use of a programmable freezing apparatus. The cell suspension could be frozen without modifying the original pretreatment conditions (0.15 M sucrose + 5% DMSO). Embryogenic calluses of five varieties out of six experimented withstood cryopreservation with the simplified freezing process. Optimal results were obtained by increasing the DMSO concentration to 10 or 15% (Engelmann et al. 1994).

In general, embryogenic callus and suspension cell cultures have been cryopreserved with high level of success using different protocols based either on dehydrative slow cooling or vitrification procedures. Very high survival rates have been obtained with callus and suspension cultures originating from a wide range of genera, species and cultivars. In all instances, cryopreserved callus produced embryos and whole plants. Despite some limitations, the conservation of embryogenic callus and cell suspensions allows preservation of material in a form ready for biotechnological applications.

13.4.8.4 Cryopreservation of Ovules and Somatic Embryos (Gonzalez-Arnao et al. 2003)

Ovules are immature seeds which can be recovered in large numbers from ovaries and small fruits even from seedless cultivars. In polyembryonic species ovules cultured in vitro produce true-to-type plants through the development of somatic embryos from nucellar tissues.

Ovules of “Pineapple”, “Navel Foyos” and “Parent Navel” sweet orange (*Citrus Sinensis* Obs.), “Duncan” grapefruit (*C. paradise* Macf.), “Fino” lemon (*C. limon* (L.) Burm. f.), common mandarin (*C. deliciosa* Ten.) and sour orange (*C. aurantium* L.) were excised from immature fruits collected 2–8 weeks after anthesis collected from field grown trees of the germplasm bank maintained at Instituto Valenciano de Investigaciones Agrarias (IVIA), Valencia, Spain. The fruits were surface-disinfected by soaking for 20 min in a 2% (v/v) sodium hypochlorite solution containing 0.1% (v/v) wetting agent Tween 20. Ovules were excised from cut open fruits under sterile conditions and cultured overnight on semisolid MS (Murashige and Skoog 1962) medium containing 100 mg.L^{-1} *i*-inositol, 0.2 mg. L^{-1} thiamine HCl, 1 mg.L^{-1} pyridoxine HCl, 1 mg.L^{-1} nicotinic acid, 50 g.L^{-1} sucrose, 500 mg.L^{-1} malt extract, and supplemented with 10 g.L^{-1} agar.

Cryopreservation of ovules was performed following the basic protocol of encapsulation-dehydration technique. Ovules were encapsulated in calcium alginate (3%) beads and pregrown in liquid medium with various sucrose concentrations (0.75, 1.0 or 1.25 M) for various pregrowth durations (1–7 d.). In some cases, the sucrose concentration in the pregrowth medium was increased progressively (0.3 M + 0.5 M + 0.75 M; 0.75 M + 1 M or 0.75 M + 1 M + 1.25 M) by transferring samples after 1, 3 or 4 days in culture to media with the higher sucrose concentrations. Encapsulated ovules were then dehydrated at room temperature down to 20–25% of moisture content (MC, fresh weight basis) under the sterile air current of the laminar flow cabinet and placed in sterile 2-mL polypropylene cryotubes for freezing performed either rapidly by direct immersion of the cryotubes in liquid nitrogen or slowly, by cooling at 0.5 °C.min^{-1} from 20 to −40 °C before immersion in liquid. Survival of ovules was evaluated after 6–9 months of culture by counting the number of ovules which formed embryos.

The results obtained before cryopreservation demonstrated that ovules were able to withstand high sucrose concentrations (0.75, 1 and 1.25 M) even under prolonged treatments. However, survival was always very low and erratic after the immersion in liquid nitrogen (ranging from 1 to 16% depending on citrus specie). Other cryopreservation approaches as vitrification and encapsulation-vitrification procedures were also assayed in parallel, but both were completely unsuccessful (González-Arnao, unpublished results). It would be interesting and advisable to try the most recent cryogenic approaches concerning V- and/or D-Cryoplate techniques (Yamamoto et al. 2011; Niino et al. 2014), which would allow the manipulation of abundant number of ovules attached to cryoplates and ultra-rapid increase in the cooling rate.

13.4.8.5 Somatic Embryos (Gonzalez-Arnao et al. 2003)

The first report of cryopreservation of somatic embryos of citrus referred to embryos derived from in vitro cultured ovules of "Washington navel" sweet orange and included a conventional cryogenic protocol (Marin and Duran-Vila 1988; Marin et al. 1993). Careful selection of embryos at early developmental stages was required to achieve survival, which represented an initial drawback in the freezing procedure. The availability of sufficient material at the right physiological stage under tissue culture conditions was limited, since this is an uncontrolled biological process. Moreover, the results after cryopreservation could not be improved either by selecting uniform embryo populations or by modifying the cooling rates during freezing. Survival was always low and erratic, ranging from 3.7% to a maximum of 30.5%. Recovery of cryopreserved cultures was in the form of secondary embryogenesis originating from the surviving zones of the embryos. Therefore, this first protocol set-up was unsuitable for long-term conservation of citrus germplasm (Duran-Vila 1995).

Later, further cryopreservation studies focused on using the encapsulation-dehydration approach. Somatic embryos were induced in vitro from two biological sources: ovules isolated from immature fruits and from thin cut layer explants of stigma, style and ovaries (Carimi et al. 1999).

Ovules of the following eleven genotypes were assayed: "Washington Navel", "Navalete", "Sucreña", "Washington Foyos" and "Bonanza" sweet orange, common mandarin and "Page" [(*C. paradise* x *C. tangerina*) x *C. clementina*], "Calabacita" and "Cajel" sour orange, volkamer lemon (*C. volkameriana* Ten. & Pasq.) and alemow (*C. macrophylla* Wester). Ovules were cultured on the medium described above and under a light regime of 16 h light/8 h dark with a photosynthetic photon flux (PPF) of 43 µmol.m^{-2} s^{-1}.

Thin cut layer explants from stigma, style and ovaries of the following thirteen genotypes were assayed: "Pineapple" sweet orange, "Sevillano" sour orange, "Fino" lemon, "clemenules" clementine (*C. clementina*), "Avana Apireno" mandarin, "Fortune" (*C. clementina* x *C. tangerina*), "Kinow" (*C. nobilis* x *C. deliciosa*), satsuma (*C. unshui* (Mak.) Marc.), "Star Ruby" grapefruit, Rangpur lime (*C. limonia* Obs.), Rough lemon (*C. jambhiri* Lush), volkamer lemon and alemow. Stigma, style and ovaries were obtained from flowers collected before opening and surface-disinfected by immersion for 10 min in a 2% (v/v) sodium hypochlorite solution containing 0.1% (v/v) wetting agent Tween 20, followed by three 3-min rinses in sterile distilled water. After disinfecting, the flowers were opened under sterile conditions and whole pistils dissected with a scalpel. Stigmas, style and ovaries were cut perpendicularly to the longitudinal axis at a thickness of 0.4–0.5 mm. The thin cut layers were placed with the abaxial surface towards the semisolid culture medium supplemented with 7 $g.L^{-1}$ agar. Lighting culture conditions were the same as described above. For the cryopreservation experiments, embryos were used when they became available either at globular, torpedo or heart shape stages, without any specific selection of a physiological stage. Long-term maintenance of the embryos in the original culture media was avoided to minimize the occurrence of a secondary embryogenesis. Survival of somatic embryos was evaluated after 1 month of culture by counting those which germinated.

Thin cut layer explants from stigma, style and ovaries cultured in vitro regenerated somatic embryos only from "Fino" lemon, "Avana apireno" mandarin and "Star Ruby" grapefruit out of 13 genotypes assayed. Nevertheless, they were a useful additional source for producing citrus somatic embryos of different species. Somatic embryos obtained by in vitro culture of thin cut layer explants are illustrated in Fig. 13.2.

The best cryopreservation conditions irrespective of the source used for obtaining the somatic embryos comprised the encapsulation of tissues in calcium alginate beads (3%), preculture in liquid medium containing 0.75 M sucrose (24 h), desiccation of beads with silica gel down to around 20% (fresh weight basis), rapid immersion in liquid nitrogen followed by slow rewarming. In general, survival was high and ranged between 75 and 100% after cryopreservation. Results obtained

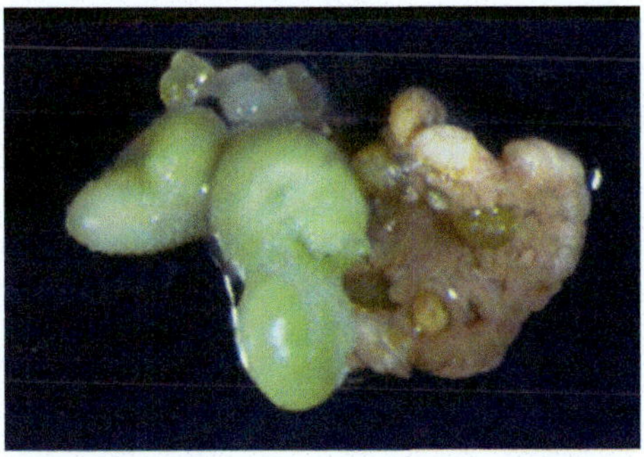

Fig. 13.2 Citrus somatic embryos derived from thin cut layer explants cultured in vitro

with explants derived from stigma, style and ovaries, respectively, are shown in Table 13.4.

In all instance, after cryopreservation the whole embryo remained green and germinated producing normal plantlets (Fig. 13.3).

Following in vitro culture of ovules or thin cut layer explants, the embryos that become available are usually in different development stages. Therefore, the selection required for the application of the classical cryogenic protocol earlier reported (Marin and Duran-Vila 1988) is cumbersome and very inconvenient. However, this difficulty was overcome using the encapsulation-dehydration method, which always allowed consistent high survival rates without any previous selection.

Somatic embryos are easily obtained from all polyembryonic species, and they regenerate true-to-type plants. Therefore, the encapsulation-dehydration approach for somatic embryos can be efficiently used for the conservation of citrus germplasm.

Table 13.4 Survival of cryopreserved somatic embryos obtained by in vitro culture of thin cut layer explants. Alginate-coated embryos were 1-d pregrown in 0.75 M sucrose and desiccated down to 20–25% moisture content prior to rapid cooling

Survival (%)[a]			
Species	Pregrowth	Desiccation	Freezing
Lemon (Fino lemon)	100 (4/4)	67 (6/9)	76 (29/38)
Mandarin (Avana Apireno)	73 (8/11)	90 (9/10)	95 (19/20)
Grapefruit (Star Ruby)	100 (3/3)	–	100 (10/10)

– not tested

[a]In brackets, number of surviving ovules out of treated ones. Reprinted from Gonzalez-Arnao et al. (2003)

Fig. 13.3 Adventitious embryos produced from a cryopreserved encapsulated somatic embryo originated from ovules

Cryopreservation of Shoot Tips

Juvenile shoot tips (Gonzalez-Arnao et al. 1998, 2000)

Seeds of *Poncirus trifoliata* (L) Raf. were devoid of testa and surface-sterilized for 10 min in 0.7% sodium hypochlorite solution with few drops of Tween 20. After 3–4 rinses with sterile distilled water, seeds were aseptically transferred in test tubes on semisolid MS medium (Murashige and Skoog 1962) supplemented with 50 g.L^{-1} sucrose, 0.5 mg.L^{-1} benzylaminopurine (BAP), 0.25 mg.L^{-1} indole butyric acid (IBA), 40 mg.L^{-1} adenine, 750 mg.L^{-1} malt extract and 10 g.L^{-1} agar. They were maintained at 26 °C under 8-h light/ 16-h dark photoperiod under a light intensity of 40 µmol.m^{-2}.s^{-1}. Shoot tips (size: 0.5–1 mm) were excised from in vitro plants 20 d after the last subculture and left overnight on standard medium for recovery.

Cryopreservation experiments were carried out following the basic protocol of encapsulation-dehydration technique. Shoot tips were encapsulated in alginate (3%) beads and pregrown in liquid medium with various sucrose concentrations (0.3 to 1 M) for various pregrowth durations (1–10 d.). In some cases, the sucrose concentration in the pregrowth medium was increased progressively by daily transfer of apices in media with higher sucrose concentration from 0.3 up to 1 M. Encapsulated shoot tips were then dehydrated at room temperature down to 20–25% of moisture content (MC, fresh weight basis) under the sterile air current of the laminar flow cabinet and transferred in sterile 2-mL polypropylene cryotubes. Samples submitted to pregrowth with progressive increase in sucrose concentrations were desiccated under the same conditions down to several MCs (36, 28, 23, 17 and 15%). After desiccation, freezing was performed either rapidly by direct immersion of the cryotubes in liquid nitrogen or slowly, by cooling at 0.5 °C.min^{-1} from 20 to −40 °C before immersion in liquid nitrogen, using a programmable freezer (Bio-Cool, FTS Systems, USA). Samples were kept for at least 1 h at −196 °C. For thawing, the cryotubes were placed under the air current of the laminar flow cabinet for 2–3 min. Beads were then transferred in Petri dishes on standard semisolid medium. Shoot tips were re-cultured for the first week in the dark and then transferred under standard lighting conditions.

The best cryopreservation conditions determined for shoot tips of *Poncirus trifoliata* rootstock comprised pregrowth in liquid medium with 0.5 M sucrose (3–4 d) or in media with increasing sucrose concentrations (0.3 M/24 h + 0.5 M/24 h + 0.75 M/24 h), desiccation down to around 20% of water content in the beads, and preferably slow freezing. These optimal conditions allowed achieving survival with shoot tips of two additional citrus rootstocks: Troyer citrange and Carrizo citrange. The protocol involved preculture with the progressive increase (from 0.3 M up to 0.75 M) in sucrose concentration, desiccation to 20–25% MC and slow freezing. Survival (%) of shoot tips from the three citrus rootstocks species after preculture, dehydration and slow freezing is shown in Table 13.5. Growth recovery of juvenile citrus shoot tips after cryopreservation occurred directly without callus formation (Fig. 13.4).

Table 13.5 Effect of encapsulation-dehydration protocol on the survival (%) of apices from three different citrus species (reprinted from Gonzalez-Arnao et al. 2000)

Survival (%)			
Species	Pregrowth	Dehydration	Slow freezing
Poncirus trifoliata	100	90 ± 5	44 ± 4
Troyer citrange	100	83 ± 4	36 ± 3
Carrizo citrange	100	95 ± 5	55 ± 5

Fig. 13.4 Growth recovery and regeneration of citrus juvenile shoot tips cryopreserved by encapsulation-dehydration

13.4.8.6 Adult Shoot Tips

Later, the next significant achievement was the cryopreservation of adult citrus shoot tips performed by Volk et al. 2012. Shoot tips were excised from actively growing vegetative flushes of protected trees, surface-disinfected, and precultured overnight on semisolid medium with 0.3 M sucrose, before the exposure to loading (2 M glycerol + 0.4 M sucrose) for 20 min and vitrification (PVS2) solutions for 30 or 60 min at 0 °C, prior to direct immersion in liquid nitrogen. Rewarmed shoot tips post-cultured overnight on survival medium were then micrografted on "Carrizo" seedling rootstocks to produce whole plants. Micrografted shoot tips recovered quickly, and rooted plants could be transferred to the greenhouse within months. Regrowth of whole plants after micrografting averaged 53 per cent for cryopreserved shoot tips of cultivars representing eight Citrus and Fortunella species (Volk et al. 2012). It was also found that freezing on foil resulted in higher levels of survival than freezing in vials. For the seven cultivars included in the experiment, the average viability of shoot tips cryopreserved using cryovials was 53%, and the viability of shoot tips cryopreserved on foil strips was 83% (Bonnaert et al. 2013).

13.4.8.7 Seeds and Embryonic Axis

King et al. (1981) determined that the longevity of various citrus seeds was improved when the storage temperature and the moisture contents of seeds were decreased.

Seeds of species tolerant to desiccation can be successfully cryopreserved by direct immersion and storage in liquid nitrogen after partial desiccation (air drying or silica gel drying), without any cryoprotective treatment. For species whose seeds are sensitive to desiccation, partially dehydrated embryonic axes excised from seeds can also be cryopreserved by direct immersion in liquid nitrogen. Marzalina and Krishnapillay (1999) reported that MC of citrus seeds should be reduced to 20% or less to achieve high germination after storage in liquid nitrogen.

Normah and Seti Dewi Serimala (1995) reported that *Citrus aurantifolia* seeds can be successfully cryopreserved at −196 °C after desiccating them to a moisture content of 12.93% (50% viability) while seeds of *C. halimii* presented only 25% viability after cryopreservation at moisture contents of 9.5%. The seeds of *C. hystrix* were highly sensitive to desiccation as they failed to germinate when the moisture contents were reduced to 27% and thus did not survive cryopreservation. On the other hand, the embryonic axes of the three Citrus species gave higher percentage of survival after cryopreservation. Survival rate was 100% in *C. aurantifolia* and *C. halimii* embryonic axes with moisture contents of 9–11% and 16.6%, respectively. With *C. hystrix* axes, the highest survival rate obtained was 60% at a moisture content of 11.04%. Mumford and Grout (1979) and Lambardi et al. (2007) have reported high tolerance of *C. limon*, *C. halimii* and *C. volkameriana* seeds to dehydration below 10% MC.

Seeds of trifoliate orange (*Poncirus trifoliata* (L.) Raf.) are sensitive to desiccation and could not withstand reduction in moisture level below 20%, whereas the excised embryonic axes could be desiccated easily to moisture levels as low as 14% without much loss in viability and could be successfully cryopreserved in liquid nitrogen (Radhamani and Chandel 1992).

Intact seeds of two citrus species were desiccated in a sterile airflow to 16% (*C. sinensis*) or 10% (*C. aurantium*) moisture content and immersed in liquid nitrogen. After cryopreservation, both citrus species showed a high (93% germination) survival (Lambardi et al. 2004).

Embryonic axes of *C. sinensis* [L.] Osb. ("Pineapple" sweet orange) of *C. reticulata* Blanco (mandarin) and *C. limon* L. (lemon) were successfully cryopreserved using the encapsulation-dehydration technique, and the best recovery rate was attained by axes with water content in the range of 0.15–0.17 g H_2O/g dry mass. Sufficiently dehydrated embryonic axes that were cooled to −196 °C developed into seedlings without any intermediary callus formation, after transfer to recovery medium (Santos and Stushnoff 2014).

13.4.8.8 Pollen

Application of cryogenic techniques to preserve pollen viability for prolonged durations has been found suitable for conservation of citrus germplasm (Kobayashi et al. 1978) For instance, pollen of four citrus cultivars (*C. limon, C. aurantifolia, C. sinensis and Poncirus trifoliata*) were collected and cryopreserved by direct immersion in liquid nitrogen (−196 °C) for 3.5 years (Ganeshan and Alexander 1991). This indicates that rapid cooling to an ultra-low temperature had no a harmful effect on the biological integrity of pollen under storing cryogenic conditions.

13.5 Conclusion

This chapter reviewed the progresses of different in vitro approaches for the conservation and safeguarding of woody plant genetic biodiversity. Several study cases were presented and described to exemplify the protocols developed for storing the germplasm to medium and long terms using different in vitro procedures.

The advances of in vitro conservation techniques have enhanced the availability of reliable strategies and provided some useful approaches, not only for woody plants, but also for improving the ex situ preservation of plant biodiversity in general. Slow growth could be more widely used to provide safer alternatives to the field genebank for species that are readily propagated in vitro, while cryopreservation is the safest option to support the storage of cell lines and of embryogenic cultures, to support the genetic transformation and the biotechnological breeding programs; in addition, it is a valuable complementary alternative for orthodox seeds in the context of seed bank. Moreover, cryopreservation is very important for the long-term conservation of biodiversity of tropical and subtropical forest species, and for many other plants with recalcitrant seeds and/or vegetatively propagated (Gonzalez-Arnao et al. 2014).

There is no single recipe for conservation, but certainly, there is a shared concern about using science to protect plant biodiversity around the world. Now, it is well recognized that the most appropriate strategy is combining approaches of ex situ and in situ methods to complement the conservation of plant germplasm.

Currently, the slow growth and cryopreservation techniques have been significantly integrated into many conservation programs, to the benefit of both germplasm management and research. Various research institutes, private companies and genebanks around the world have implemented these approaches to maintain important collections and/or backups of diverse woody species with different purposes. Such is the case of Instituto Valenciano de Investigaciones Agrarias (IVIA), Valencia, Spain, where a large cryopreserved collection of calluses is kept to support the breeding projects of citrus species (Olivares-Fuster et al. 2000). The private company Sylvagen in Vancouver, Canada, keeps conifer embryogenic cell lines (Salgotra and Gupta 2015), and NCGR (Corvallis) and NCGRP (Fort Collins) both in USA maintain collections of several temperate fruit trees (Hummer and Reed 2000).

The basic and practical knowledge and the comprehension of the physiological response of different plant species are continually improved, allowing their better adaptation and increasing the effectiveness of these in vitro methods. In many cases, the storage protocols developed are shared with other researchers and genebanks, and the expertise is used worldwide to further extend the benefits of these biotechnological approaches for the safe preservation of woody plant germplasm.

Acknowledgements Research works on cryopreservation of citrus germplasm were funded by IPGRI projects (96/098 and 99/18), INIA project (RF98/033), and a postdoctoral fellowship to the first author.

References

Abou-Taleb S, Yates I, Wood BW, Fouad MM (1992) Cryogenics and tissue culture for preserving pecan germplasm. HortScience 27:693

Açari DP, Porto VB, Rodrigues ERV, Martins F, Lima RJ, Sawaya ACHF, Ribeiro ML, Carvalho PO (2011) Effect of mate tea (*Ilex paraguariensis*) supplementation on oxidative stress biomarkers and LDL oxidisability in normo- and hyperlipidaemic humans. J Funct Foods 3(3):190–197

Acquaah G (2012) Principles of plant genetics and breeding, 2nd edn. Bowie State Uni-versity, Maryland, p 740

Aguilar ME, Engelmann F, Michaux-Ferriere N (1993) Cryopreservation of cell suspensions of *Citrus deliciosa* Tan. and histological study. Cryo-Letters 14:217–228

Ahmad Z, Zaidi N, Shah FH (1990) Micropropagation of *Melia azedarach* from mature tis-sue. Pak J Bot 22:172–178

Akihama T, Omura M, Kozaki I (1978) Further investigation of freeze-drying for deciduous fruit tree pollen. In: Akihama T, Nakajima K (eds) Long-term preservation of favourable germ plasm in arboreal crops. Fruit Tree Research Station, Ministry of Agriculture and Food, Japan, pp 1–7

Akihama T, Omura M, Kozaki I (1980) Long term storage of fruit tree pollen and its application in breeding. Farming Japan 14:53–56

Al-Ababneh SS, Shibli RA, Karam NS, Shatnawi MA (2003) Cryopreservation of bitter almond (*Amygdalus communis* L.) shoot tips by encapsulation dehydration and vitrification. Adv Hort Sci 17:15–20

Anitei S (2007) Where did citrus fruits originate from? Available from: www.news.softpedia.com/news/Where-Did-Citrus-Fruits-Originate-From-67365.htm. Accessed 2016 October

Anjaneyulu ASR, Rao AM, Rao VK, Row LR, Pelter A, Ward RS (1977) Novel hydroxy lignans from the heartwood of *Gmelina arborea*. Tetrahedron 33:133–143

Aronen T, Krajnakova J, Häggman H, Ryynänen L (1999) Genetic fidelity of cryopreserved embryogenic cultures of open-pollinated *Abies cephalonica*. Plant Sci 142:163–172

Ashmore SE (1997) Status report on the development and application of *in vitro* techniques for the conservation and use of plant genetic resources. International Plant Genetic Resources Instit., (IPGRI), Rome, Italy, p 57

Asolini FC, Tedesco AM, Carpes ST (2006) Antioxidant and antibacterial activities of phenolic compounds from extracts of plants used as tea. Braz J Food Technol 9(3):209–215

Augusseau X, Nikiéma P, Torquebiau E (2006) Tree biodiversity, land dynamics and farmers' strategies on the agricultural frontier of southwestern Burkina Faso. Biodivers Conserv 15 (2):613–630

Bagniol S, Engelmann F (1991) Effects of pregrowth and freezing conditions on the resistance of meristems of date (*Phoenix dactylifera* L. var. Bou Sthammi Noir) to freezing in liquid nitrogen. Cryo-Letters 12:279–286

Bai X, Yang L, Tian M, Chen J, Shi J, Yang Y, Hu X (2011) Nitric oxide enhances desiccation tolerance of recalcitrant *Antiaris toxicaria* seeds via protein S-nitrosylation and carbonylation. PLoS ONE 6(6): e20714; 1–11

Bai X, Chen J, Kong X, Todd CD, Yang Y, Hu X, Li D (2012) Carbon monoxide enhances the chilling tolerance of recalcitrant Baccaurea ramiflora seeds via nitric oxide-mediated glutathione homeostasis. Free Radic Biol Med 53:710–720

Barik BR, Bhowmik T, Dey AK, Patra A, Chatterjee A, Joy SS (1992) Premnazole an isoxa-zole alkaloid of *Premna integrifolia* and *Gmelina arborea* with anti-inflammatory activity. Fitoterapia 63(4):295–299

Barra-Jiménez A, Aronen TS, Alegre J, Toribio M (2015) Cryopreservation of embryogenic tissues from mature holm oak trees. Cryobiology 70:217–225

Bastos DHM, Saldanha LA, Catharino RR, Sawaya A, Cunha IB, Carvalho PO, Eberlin MN (2007) Phenolic antioxidants identified by ESI-MS from Yerba maté (*Ilex paraguariensis*) and green tea (*Camelia sinensis*) extracts. Molecules 12(3):423–432

Benson EE (2000) Do free radicals have a role in plant tissue culture recalcitrance? In Vitro Cell Dev Biol Plant 36(3):163–170

Berjak P, Dumet D (1996) Cryopreservation of seeds and isolated embryonic axes of neem (*Azadirachta indica*). Cryo-Letters 17:99–104

Berjak P, Pammenter NW (2004) Recalcitrant seeds. In: Benech-Arnold RL, Sánchez RA (eds) Handbook of seed physiology: applications to agriculture. Haworth, New York, pp 305–345

Berjak P, Pammenter NW (2008) From Avicennia to Zizania: seed recalcitrance in perspective. Ann Bot 101:213–228

Berjak P, Farrant JM, Mycock DJ, Pammenter NW (1992) Homoiohydrous (recalcitrant) seeds: developmental status, desiccation sensitivity and the state of water in axes of *Landolphia kirkii* Dyer. Planta 186(2):249–261

Berjak P, Walker M, Mycock DJ, Wesley- Smith J, Watt P, Pammenter NW (2000) Cryopreservation of recalcitrant zygotic embryos. In: Engelmann F, Takagi H (eds) Cryopreservation of tropical plant germplasm: current research progress and application. Proceedings of an international workshop, Tsukuba, Japan, October 1998. International Plant Genetic Resources Institute, Rome, pp 140–155

Bernard F, Shaker-Bazarnov H, Kaviani B (2002) Effects of salicylic acid on cold preservation and cryopreservation of encapsulated embryonic axes of Persian lilac (*Melia azedarach* L.). Euphytica 123(1):85–88

Bertrand-Desbrunais A, Fabre J, Engelmann F, Dereuddre J, Charrier A (1998) Adventitious embryogenesis recovery from coffee (*Coffea arabica* L.) somatic embryos after freezing in liquid nitrogen. C R Acad Sci Paris 307:795–801

Biswas K, Chattopadhyay I, Banerjee RK, Bandyopadhyay U (2002) Biological activities and medicinal properties of neem (*Azadirachta indica*). Curr Sci 82(11):1336–1345

Blakesley D, Pask N, Henshaw GG, Fay MF (1996) Biotechnology and the conservation of forest genetic resources: in vitro strategies and cryopreservation. Plant Growth Regul 20(1):11–16

Bonnart R, Shepherd A, Lee R, Krueger R, Volk G (2013) Cryopreservation of Citrus in the USDA-ARS national plant germplasm system collection, meeting abstract. In: 2nd international symposium on plant cryopreservation. Fort Collins, USA, p 62

Bracesco N, Sanchez AG, Contreras V, Menini T, Gugliucci A (2011) Recent advances on Ilex paraguariensis research: minireview. J Ethnopharmacol 136(3):378–384

Bramlett DL, Matthews FR (1991) Storing loblolly pine pollen. South J Appl For 15(153):157

Brearley J, Henshaw GG, Davey C, Taylor NJ, Blakesley D (1995) Cryopreservation of *Fraxinus excelsior* L. zygotic embryos. Cryo-Letters 16:215–218

Brush SB (1995) *In Situ* conservation of landraces in centers of crop diversity. Crop Sci 35 (2):346–354

Bunn E, Turner S, Panaia M, Dixon KW (2007) The contribution of in vitro technology and cryogenic storage to conservation of indigenous plants. Aust J Bot 55(3):345–355

Burris KP, Harte FM, Davidson MP, Stewart N, Zivanovic S (2012) Composition and bioactive properties of yerba mate (*Ilex paraguariensis*. St.-Hil.): a review. Chil J Agric Res 72(2):268–274

Capuana M, Di Lonardo S (2013) In vitro conservation of chestnut (*Castanea sativa*) by slow growth. Vitro Cell Dev Biol Plant 49(5):605–610

Carimi F, Tortorici MC, De Pasquale F, Crescimanno FG (1999) Somatic embryogenesis and plant regeneration from pistil thin cell layers of Citrus. Plant Cell Rep 18:935–940

Carl CM Jr (1976) Effect of separation in n-pentane on storability of sugar maple seeds. Res. Note NE-218, USDA Forest Servo Northeast. For. Exp. Sta., Broomhall, PA

Castaldelli APA, Vieira LP, Przygodda F, Martins ZN, Padoin MJ (2011) Efeito da erva mate (*Ilexparaguariensis* A. St. -Hil.) no comportamento e fisiologia de ratos Wistar. Braz J Biosci 9 (4):514–519

Catalán Bachiller G (1991) *Quercus*. In: Catalán Bachiller G (ed) Semillas de árboles y arbustos forestales. MAPA-ICONA. Colección Técnica, Madrid, pp 318–324

Cavers S, Navarro C, Lowe AJ (2003) Chloroplast DNA phylogeography reveals colonization history of a Neotropical tree, *Cedrela odorata* L., in Mesoamerica. Mol Ecol 12:1451–1460

Chandel KPS, Chaudhury R, Radhamani J, Malik SK (1995) Desiccation and freezing sensitivity in recalcitrant seeds of tea, cocoa and jackfruit. Ann Bot 76:443–450

Chang Y, Reed BM (2000) Extended alternating-temperature cold acclimation and culture duration improve pear shoot cryopreservation. Cryobiology 40:311–322

Chaudhury R (2000) Cryopreservation of seeds, embryos, embryonic axes and pollen at National Cryobank of NBPGR. In: Engelmann F, Takagi H (eds) Cryopreservation of tropical plant germplasm: current research progress and application. Proceedings of an international workshop, Tsukuba, Japan, October 1998, International Plant Genetic Resources Institute (IPGRI), Rome, pp 457–459

Chaudhury R, Chandel KPS (1991) Cryopreservation of desiccated seeds of neem (*Azadirachta indica* A. Juss.) for germplasm conservation. Indian J Plant Gen Resour 4:67–72

Chaudhury R, Chandel KPS (1995) Cryopreservation of embryonic axes of almond (*Prunus amygdalus* Batsch.) seeds. Cryo-Letters 16:51–56

Chmielarz P (2009) Cryopreservation of dormant orthodox seeds of forest trees: mazzard cherry (*Prunus avium* L.). Ann Forest Sci 66: 405

Chmielarz P, Walters C (2007) Desiccation sensitivity of white and black (red) oak embryonic axes. South Afr J Bot 73:498

Chmielarz P, Grenier-de March G, de Boucaud MT (2005) Cryopreservation of *Quercus robur* L. embryogenic callus. Cryo-Letters 26:349–355

Chmielarz P, Michalak M, Pałucka M, Wasilenczyk U (2011) Successful cryopreservation of *Quercus robur* plumules. Plant Cell Rep 30:1405–1414

Cho EG, Hor YL, Kim HH, Rao VR, Engelmann F (2001) Cryopreservation of *Citrus madurensis* zygotic embryonic axes by vitrification: importance of pregrowth and preculture conditions. Cryo-Letters 22(6):391–396

Cho EG, Hor YL, Kim HH, Rao VR, Engelmann F (2002a) Cryopreservation of *Citrus madurensis* zygotic embryonic axes by vitrification: importance of pregrowth and preculture conditions. Cryo-Letters 23:317–324

Cho EG, Hor YL, Kim HH, Rao VR, Engelmann F (2002b) Cryopreservation of *Citrus madurensis* embryonic axes by encapsulation-dehydration. Cryo-Letters 23(5):325–332

Cho EG, Noor NM, Kim HH, Rao VR, Engelmann F (2002c) Cryopreservation of *Citrus aurantifolia* seeds and embryonic axes using a desiccation protocol. Cryo-Letters 23(5):309–316

Cho EG, Hor YL, Kim HH, Gwag JG, Rao VR, Engelmann F (2003) Cryopreservation of *Citrus madurensis* zygotic embryos by encapsulation-dehydration. Korean J Breed 35:148–153

CITES, Convention on International Trade in Endangered Species of Wild Fauna and Flora (2003) Second meeting of working group on mahogany (*Swietenia macrophylla*) http://www.cites.org/common/prog/mwg/MWG2/S-MWG2-09-02-MX.pdf

Corredoira E, San-José MC, Ballester A, Vieitez AM (2004) Cryopreservation of zygotic embryo axes and somatic embryos of European chestnut. Cryo-Letters 25:33–42

Corredoira E, San-José MC, Vieitez AM, Ballester A (2007) Improving genetic transformation of European chestnut and cryopreservation of transgenic lines. Plant Cell Tiss Org Cult 91:281–288

Cruz-Cruz CA, González-Arnao MT, Engelmann F (2013) Biotechnology and conservation of plant biodiversity. Resources 2(2):73–95

Cuénoud P, Del Pero Martínez MA, Loizeau PA, Spichiger R, Andrews S, Manen JF (2000) Molecular phylogeny and biogeography of the genus Ilex L. (*Aquifoliaceae*). Ann Bot 85 (1):111–122

Cyr DR, Lazaroff WR, Grimes SMA, Quan G, Bethune TD, Dunstan DJ, Roberts DR (1994) Cryopreservation of interior spruce (*Picea glauca* engelmanni complex) embryogenic cultures. Plant Cell Rep 13:574–577

De Boucaud MT, Helliot B, Brison M (1996) Desiccation and cryopreservation of embryonic axes of peach. Cryo-Letters 17:379–390

De Carlo A, Benelli C, Lambardi M (2000) Development of a shoot-tip vitrification protocol and comparison with encapsulation-based procedures for plum (*Prunus domestica* L.) cryopreservation. Cryo-Letters 21:215–222

Devi J, Ray BK, Chetia S, Deka PC (1998) Regeneration of low temperature stored encapsulated protocorms of orchids. J Orchid Soc India 12:39–41

De-Zhu Li, Pritchard HW (2009) The science and economics of ex situ plant conservation. Trends Plant Sci 14(11):614–621

Dickie JB, Smith RD (1995) Observations on the survival of seeds of *Agathis* spp. stored at low moisture contents and temperatures. Seed Sci Res 5:5–14

Dolce NR, Gonzalo GA, Rey HY (2007) Embriología de *Ilex dumosa* e *I. brevicuspis* (Aquifoliaceae). Boletín de la Sociedad Argentina de Botánica 42(supl):40–41

Dolce NR, Mroginski LA, Rey HY (2010) Endosperm and endocarp effects on the *Ilex paraguariensis* St. Hil. (Aquifoliaceae) seed germination. Seed Sci Technol 38:441–448

Dolce NR, Mroginski LA, Rey HY (2011) Enhanced seed germination of *Ilex dumosa* R. (Aquifoliaceae) through *in vitro* culture of cut pyrenes. HortScience 46(2):278–281

Dolce NR, Medina RD, Mroginski LA, Rey HY (2015) Sowing pyrenes under aseptic conditions enhances seed germination of *Ilex brasiliensis*, *I. pseudoboxus* and *I. theezans* (Aquifoliaceae). Seed Sci Technol 43(2):1–5

Domecq CM (1988) Cultivo *in vitro* de yemas axilares de paraíso gigante (*Melia azedarach* L. var. gigantea). Phyton 48:33–42

Dumet D, Engelmann F, Chabrillange N, Dussert S, Duval Y (2000) Cryopreservation of oil palm polyembryonic cultures. In: Engelmann F, Takagi H (eds) Cryopreservation of tropical plant germplasm. Current research, progress and applications. IPGRI, International Plant Genomic Resources Institute, Rome, pp 172–177

Duran-Vila N (1995) Cryoconservation of germplasm of Citrus. In: Bajaj YPS (ed) Biotechnology in agriculture and forestry, cryopreservation of plant germplasm I, vol 32. Springer, Berlin, pp 70–86

Dussert S, Chabrillange N, Engelmann F, Anthony F, Hamon S (1998) Cryopreservation of seeds of four coffee species (*Coffea arabica*, *C. costatifructa*, *C. racemosa* and *C. sessiliflora*): importance of seed water content and of freezing rate. Seed Sci Res 8:9–15

Dussert S, Chabrillange N, Engelmann F, Hamon S (1999) Quantitative estimation of seeddesiccation sensitivity using a quantal response model: application to nine species of the genus *Coffea*. Seed Sci Res 9:135–144

Engelmann F (1997) In vitro conservation methods. In: Callow JA, Ford-Lloyd BV, Newbury HJ (eds) Biotechnology and plant genetic resources. CAB International, Oxford, UK, pp 119–161

Engelmann F (2011) Use of biotechnologies for the conservation of plant biodiversity. Vitro Cell Dev Biol Plant 47:5–16

Engelmann F (2012) Germplasm collection, storage and conservation. In: Altman A, Hasegawa PM (eds) Plant biotechnology and agriculture: prospects for the 21st Century. Academic Press, Oxford, pp 255–268

Engelmann F (2014) Cryopreservation of clonal crops: a review of key parameters. Acta Hort 1039:31–39

Engelmann F, Etienne H (2000) Cryopreservation of embryogenic callus of *Hevea brasiliensis*. In: Jain SM, Gupta PK, Newton RJ (eds) Somatic embryogenesis in woody plants, vol 6. Kluwer Academic Publishers, The Netherlands, pp 729–745

Engelmann F, Takagi H (2000) Cryopreservation of tropical plant germplasm. Current research progress and application. IPGRI, JIRCAS International Agriculture Series N° 8, Rome, p 465

Engelmann F, Dambier D, Ollitrault P (1994) Cryopreservation of cell suspensions and embryogenic calluses of *Citrus* using a simplified freezing process. Cryo-Letters 15:53–58

Engelmann F, Gonzalez-Arnao MT, Wu Y, Escobar R (2008) Development of encapsulation dehydration. In: Reed BM (ed) Plant cryopreservation: a practical guide. Springer, New York, pp 59–75

Fahy GM, MacFarlane DR, Angell CA, Meryman HT (1984) Vitrification as an approach to cryopreservation. Cryobiology 21(4):407–426

Faiza H, Darakhshanda S (1998) The inhibition of platelet aggregation and related physiological responses with crude drug extract of *Gmelina arborea*. In: Sixth international symposium on new trends in natural products chemistry, pp 279–286

Fang JY, Sacandé M, Pritchard H, Wetten A (2009) Influence of freezable/non-freezable water and sucrose on the viability of *Theobroma cacao* somatic embryos following desiccation and freezing. Plant Cell Rep 28:883–889

FAO (2009) State of the world's forests. Food and Agriculture Organization of the United Nations, Rome

FAO (2014) Genebank standards for plant genetic resources for food and agriculture. Rome

Fernandes P, Rodriguez E, Pinto G, Roldan-Ruiz I, Loose M, de Santos C (2008) Cryopreservation of *Quercus suber* somatic embryos by encapsulation-dehydration and evaluation of genetic stability. Tree Physiol 28:1841–1850

Figueroa CJC (1994) An assessment of the distribution and status of big-leaf mahogany (*Swietenia macrophylla* King). Puerto Rico Conservation Foundation and International Institute of Tropical Forestry, 25 pp, 5 maps and appendix

Filip R, Ferraro GE (2003) Researching on new species of "Mate": *Ilex brevicuspis* phytochemical and pharmacology study. Eur J Nutr 42(1):50–54

Filip R, López PG, Ferraro GE (1999) Phytochemical study of *Ilex dumosa*. Acta Hort 501:333–336

Filip R, López P, Giberti G, Coussio J, Ferraro G (2001) Phenolic compounds in seven South American *Ilex* species. Fitoterapia 72(7):774–778

Find JI, Kristensen MMH, Nørgaard JV, Krogstrup P (1998) Effect of culture period and cell density on regrowth following cryopreservation of embryogenic suspension cultures of Norway spruce and Sitka spruce. Plant Cell Tiss Org Cult 53:27–33

Ford CS, Jones NB, Van Staden J (2000a) Cryopreservation and plant regeneration from somatic embryos of *Pinus patula*. Plant Cell Rep 19:610–615

Ford CS, Jones NB, Van Staden J (2000b) Optimization of a working cryopreservation protocol for *Pinus patula* embryogenic tissue. In Vitro Cell Dev Biol Plant 36:366–369

Fu JR, Zhang BZ, Wang XP, Qiao YZ, Huang XL (1990) Physiological studies on desiccation, wet storage and cryopreservation of recalcitrant seeds of three fruit tree species and their excised embryonic axis. Seed Sci Technol 18:743–754

Gale S, John A, Harding K, Benson EE (2008) Developing cryopreservation for *Picea sitchensis* (Sitka spruce) somatic embryos: a comparison of vitrification protocols. Cryo-Letters 29:135–144

Galle FC (1997) Hollies: the genus Ilex. Timber Press, Portland, p 537

Ganeshan S, Alexander MP (1991) Cryogenic preservation of lemon (*Citrus limonia* Burm) pollen. Gartenbauwissenschaft 56:228–230

Gaspar T, Franck T, Bisbis B, Kevers C, Jouve L, Hausman JF, Dommes J (2002) Concepts in plant stress physiology. Application to plant tissue cultures. Plant Growth Regul 37(3):263–285

Gazeau C, Elleuch H, David A, Morisset C (1998) Cryopreservation of transformed *Papaver somniferum* cells. Cryo-Letters 19:147–158

Geilfus F (1994) Agricultural involution in the Dominican Republic: the dynamics of peasant farm systems in a confined environment. Ph.D. thesis, Department of Geography, Louvain University, Belgium

George EF (1996) Plant propagation by tissue culture. Part 2—in practice, 2nd edn. Exegetics, Edington

Giberti GC (1999) Recursos fitogenéticos relacionados con el cultivo y explotación de la yerba mate (*Ilex paraguariensis* St. Hil., Aquifoliáceas) en el Cono Sur de América. Acta Hort 500:137–144

Gmitter FG, Hu X (1990) The possible role of Yunnan, China, in the origin of contemporary citrus species (rutaceae). Econ Bot 44(2):267–277

Gómez TJ, Jasso-Mata JJ, Vargas-Hernández JJ, Hernández MRS (2006) Deterioro de semilla de dos procedencias de *Swietenia macrophylla* King., bajo distintos métodos de almacenamiento. Ra Ximhai 2(1):223–239

Gonzalez-Arnao MT, Engelmann F (2006) Cryopreservation of plant germplasm using the encapsulation-dehydration technique: review and case study on sugarcane. Cryo-Letters 27 (3):155–168

Gonzalez-Arnao MT, Engelmann F (2013) Consideraciones teóricas y prácticas para la crioconservación de germoplasma vegetal. In: Gonzalez-Arnao MT, Engelmann F (eds) Crioconservación de plantas en América Latina y el Caribe. IICA, San José, Costa Rica, pp 37–51

Gonzalez-Arnao MT, Cárdenas-Lara A, Urra C (1997) Crioconservación de callos embriogénicos de *Citrus sinensis* var. Pineapple. BIOTAM 8(2–3):25–32

Gonzalez-Arnao MT, Engelmann F, Urra C, Morenza M, Rios A (1998) Cryopreservation of citrus apices using the encapsulation-dehydration technique. Cryo-Letters 19(3):177–182

Gonzalez-Arnao MT, Engelmann F, Urra C, Morenza M, Rios (2000) Cryopreservation of citrus apices using the encapsulation-dehydration technique. In: Engelmann F, Takagi H (eds) Cryopreservation of tropical plant germplasm-current research progress and applications. JIRCAS, Tsukuba. IPGRI, Rome, pp 217–221

Gonzalez-Arnao MT, Panta A, Roca WM, Escobar RH, Engelmann F (2008) Development and large scale application of cryopreservation techniques for shoot and somatic embryo cultures of tropical crops. Plant Cell Tissue Organ Cult 92(1):1–13

Gonzalez-Arnao MT, Martinez-Montero M, Cruz-Cruz CA, Engelmann F (2014) Advances in cryogenic techniques for the long-term preservation of plant biodiversity. In: Ahuja MR, Ramawat KG (eds), Biotechnology and biodiversity. sustainable development and biodiversity, vol 4. Springer International Publishing, Berlin, pp 129–170

González-Arnao MT, Juárez J, Ortega C, Navarro L, Duran-Vila N (2003) Cryopreservation of ovules and somatic embryos of citrus using the encapsulation-dehydration technique. Cryo-Letters 24(2):85–94

Gonzalez-Benito ME, Perez-Ruiz C (1992) Cryopreservation of *Quercus faginea* embryonic axes. Cryobiology 29:685–690

Gonzalez-Benito ME, Prieto RM, Herradon E, Martin C (2002) Cryopreservation of *Quercus suber* and *Quercus ilex* embryonic axes: *in vitro* culture, desiccation and cooling factor. Cryo-Letters 23:283–290

Gonzalez-Rio F, Gurriaran MJ, Gonzalez-Benito E, Revilla MA (1994) Desiccation and cryopreservation of olive (*Olea europaea* L.) embryos. Cryo-Letters 15:337–342

Gottlieb AM, Giberti GC, Poggio L (2005) Molecular analyses of the genus *Ilex* (Aquifoliaceae) in southern South America, evidence from AFLP and ITS sequence data. Am J Bot 92(2):352–369

Grenier-de March G, de Boucaud M-T, Chmielarz P (2005) Cryopreservation of *Prunus avium* L. embryogenic tissues. Cryo-Letters 26:341–348

Grzeskowiak H, Miara B, Suszka B (1983) Long term storage of seeds of Rosaceae species used as rootstock for cherry, plum, apple, and pear cultivars [*Prunus avium*; *Prunus mahaleb*; *Prunus cerasifera*; *Malus domestica*]. Arbor Kornickie 28:283–320

Gugliucci A, Bastos DH (2009) Chlorogenic acid protects paraoxonase 1 activity in high-density lipoprotein from inactivation caused by physiological concentrations of hypochlorite. Fitoterapia 80:138–142

Gugliucci A, Bastos DH, Schulze J, Souza MF (2009) Caffeic and chlorogenic acids in *Ilex paraguariensis* extracts are the main inhibitors of AGE generation by methylglyoxal in model proteins. Fitoterapia 80:339–344

Häggman H, Rusanen M, Jokipii S (2008) Cryopreservation of in vitro tissues of deciduous trees. In: Reed B (ed) Plant cryopreservation: a practical guide. Springer, New York, pp 365–386

Halmagyi A, Vălimăreanu S, Coste A, Deliu C, Isac V (2010) Cryopreservation of Malus shoot tips and subsequent plant regeneration. Rom Biotechnol Lett 15(1):79–85

Hao YJ, Cheng YJ, Deng XX (2003) GUS gene remains stable in transgenic *Citrus* callus recovered from cryopreservation. Cryo-Letters 24:375–380

Hargreaves CL, Grace LJ, Holden DG (2002) Nurse culture for efficient recovery of cryopreserved *Pinus radiata* D. Don embryogenic cell lines. Plant Cell Rep 21:40–45

Hargreaves CL, Grace LJ, van der Maas SA, Menzies MI, Kumar S, Holden DG, Foggo MN, Low CB, Dibley MJ (2005) Comparative *in vitro* and early nursery performance of adventitious shoots from cryopreserved cotyledons and axillary shoots from epicotyls of the same zygotic embryo of control-pollinated *Pinus radiata*. Can J Forest Res 35:2629–2641

Hatanaka T, Yasuda T, Yamaguchi T, Sakai A (1994) Direct regrowth of encapsulated somatic embryos of coffee (*Coffea canephora*) after cooling in liquid nitrogen. Cryo-Letters 15:47–52

Haunold A, Stanwood PC (1985) Long term preservation of hop *Humulus lupulus* pollen in liquid nitrogen. Crop Sci 25:194–196

Heywood VH, Iriondo JM (2003) Plant conservation: old problems, new perspectives. Biol Cons 113(3):321–335

Hor YL, Stanwood PC, Chin HF (1990) Effects of dehydration on freezing characteristics and survival in liquid nitrogen of three recalcitrant seeds. Pertanika 13(2):309–314

Hor YL, Kim YJ, Ugap A et al (2005) Optimal hydration status for cryopreservation of intermediate oily seeds: Citrus as a case study. Ann Bot 95:1153–1161

Hornung R, Domas R, Lynch PT (2001) Cryopreservation of plumular explants of coconut (*Cocos nucifera* L.) to support programmes for mass clonal propagation through somatic embryogenesis. Cryo-Letters 22:211–220

Hosny M, Rosazza PN (1998) Gmelinosides, twelve acylated iridoid glycosides from *Gmelina arborea*. J Nat Prod 61:734–742

Hu CY (1975) *In vitro* culture of rudimentary embryos of eleven *Ilex* species. J Am Soc Hortic Sci 100:221–225

Hu CY (1989) Holly (*Ilex* spp.). In: Bajaj YPS (ed) Biotechnology in agriculture and forestry, vol 5. Springer, Berlin, pp 412–427

Hu CY, Rogalski F, Ward C (1979) Factors maintaining Ilex rudimentary embryos in the quiescent state and the ultrastructural changes during in vitro activation. Bot Gaz 140:272–279

Hummer KE (1999) Biotechnology in plant germplasm acquisition. In: Benson EE (ed) Plant conservation biotechnology. Taylor & Francis, London, Philadelphia, pp 25–40

Hummer KE, Reed BM (2000) Establishment and operation of a temperate clonal field genebank. In: Engelmann F (ed) Management of field and in vitro germplasm collections. Proceedings of a consultation meeting, CIAT, Cali, Colombia, 15–20 January, 1996. International Plant Genetic Resources Institute, Rome, pp 29–31

Ives SA (1923) Maturation and germination of seeds of *Ilex opaca*. Bot Gaz 76:60–77

Janeiro LV, Vieitez AM, Ballester A (1995) Cold storage of in vitro cultures of wild cherry, chestnut and oak. Ann For Sci 52:287–293

Johnson D, Hodgkinson MC, Joyce D (2002) Potential effects of petroleum-derived spray oils on abscission, senescence and stress physiology of citrus. In: Beattie GAC, Watson DM, Steven ML, Rae DJ, Spooner-Hart RN (eds) Spray oils beyond 2000. University of Western Sydney, pp 185–192

Jokipii SM, Ryynänen LA, Kallio PT, Aronen TS, Hägman H (2004) A cryopreservation method maintaining the genetic fidelity of a model forest tree *Populus tremula L. x P. tremuliodes* Michx. Plant Sci 166:799–806

Jones L (1967) Effect of storage at various moisture contents and temperatures on seed germination of silk oak, Australian pine, and Eucalyptus spp. USDA For. Serv., Southeast. For. Exp. Sta., Asheville, NC. Res. Note SE-83

Karp A, Kresovich S, Bhat KV, Ayad WG, Hodgkin T (1997) Molecular tools in plant genetic resources conservation: a guide to the technologies. IPGRI Tech Bull 2(2):1–47

Kartha KK, Leung L, Pahl K (1980) Cryopreservation of strawberry meristems and mass propagation of plantlets. J Am Soc Hortic Sci 105(4):481–484

Kartha KK, Leung NL, Mroginski LA (1982) In vitro growth responses and plant regeneration from cryopreserved meristems of cassava (Manihot esculenta Crantz). Zeitschrift Für Pflanzenphysiologie 107(2):133–140

Katende AB, Birnie A, Tegnas B (1995) Useful trees and shrubs for Uganda. Identification, propagation and management for agricultural and pastoral communities. Regional Soil Conservation Unit (RSCU). Technical handbook N° 10. Nairobi, RSCU, 710 pp

Kaviani B, Hashemabadi D, Mohammadi Torkashvand A, Sedaghathoor S (2009) Cryopreservation of seeds of lily [*Lilium edebourii* (Baker) Bioss.]: use of sucrose and dehydration. Afr J Biotech 8:3809–3810

Kaya D, Pillhofer JA (2013) Potential adoption of IFRS by the United States: a critical view. Acc Horiz 27(2):271–299

Keay RWJ (1996) The future of the genus Swietenia in its native forest. Bot J Linn Soc 122:3–7

Kim HH, Cha YS, Baek HJ, Cho EG, Chae YA, Engelmann F (2002) Cryopreservation of tea (*camellia sinensis* L.) seeds and embryonic axes. Cryo Lett 23(4):209–216

King MW, Soetisna U, Roberts EH (1981) The dry storage of citrus seeds. Ann Bot 48:865–872

Kobayashi S, Ikeda SI, Nakatani M (1978) Long-term storage of citrus pollen. In: Akihama T, Nakajima K (eds) Long term preservation of favourable germplasm in arboreal crop. Ministry of Agriculture and Forestry, Tsukuba-gun, Ibaraki ken, pp 8–12

Kojima M, Yamamoto H, Marsoem SN, Okuyama T, Yoshida M, Nakai T, Yamashita S, Saegusa K, Matsune K, Nakamura K, Inoue Y, Arizono T (2009) Effects of the lateral growth rate on wood quality of *Gmelina arborea* from 3.5-, 7- and 12-year-old plantations. Ann For Sci 66(5):507

Kozaki I (1975a) Storage of fruit tree scions. In: Matsuo T (ed) JIBP synthesis, gene conservation exploration, collection, preservation and utilization. University of Tokyo Press, Tokyo, pp 84–85

Kozaki I (1975b) Storage method of pollen. In: Matsuo T (ed) JIBP synthesis, gene conservation exploration, collection, preservation and utilization. University of Tokyo Press, Tokyo, pp 89–94

Kumar S, Sharma S (2005) Somatic embryogenesis and cryopreservation of walnut (*Juglans regia* L.) and pecan (*Carya illinoensis* Koch). Acta Hortic 696:143–147

Kumar R, Singh R, Meera RS, Kalidhar SB (2003) Chemical components and insecticidal properties of bakyain (Melia azedarach L.). Agric Rev 24:101–115

Kunkel G (1978) Melia azedarach Linné - Persian lilac. In: Junk W (ed) Flowering trees in subtropical gardens. Dr. W. Junk B.V. Publishers, The Hague, pp 260–261

Lamb AFA (1966) Impressions of tropical pines and hardwoods in some eastern countries. Commonwealth Forestry Institute, Oxford (Stencil.)

Lambardi M (2002) Cryopreservation of germplasm of populus (poplar) species. In: Towill L, Bajaj YPS (eds) Cryopreservation of plant germplasm II. Biotechnology in Agriculture and Forestry, vol 50. Springer, Berlin Heidelberg, pp 269–286

Lambardi M, De Carlo A (2003) Application of tissue culture to the germplasm conservation of temperate broad-leaf trees. In: Jain SM, Ishii K (eds) Micropropagation of woody trees and fruits. Kluwer Academic Publishers, Dordrecht, pp 815–840

Lambardi M, Fabbri A, Caccavale A (2000) Cryopreservation of white poplar (*Populus alba* L.) by vitrification of *in vitro* grown shoot tips. Plant Cell Rep 19:213–218

Lambardi M, Lynch PT, Benelli C, Mehra A, Siddika A (2002) Towards the cryopreservation of olive germplasm. Adv Hortic Sci 16:165–174

Lambardi M, De Carlo A, Biricolti S, Puglia AM, Lombardo G, Siragusa M, De Pasquale F (2004) Zygotic and nucellar embryo survival following dehydration/cryopreservation of citrus intact seeds. Cryo-Letters 25(2):81–90

Lambardi M, De Carlo A, Capuana M (2005) Cryopreservation of embryogenic callus of *Aesculus hippocastanum* L. by vitrification/one-step freezing. Cryo-Letters 26:185–192

Lambardi M, Halmagyi A, Benelli C, Carlo A, de Vettori C (2007) Seed cryopreservation for conservation of ancient citrus germplasm. Adv Hortic Sci 21:198–202

Lawrence GH (1951) Taxonomy of vascular plants. The Macmillan Company, Nueva York, p 576

Lee CW, Thomas JC, Buchmann SL (1985) Factors affecting *in vitro* germination and storage of Jojoba *Simmondsia-chinensis* pollen. J Am Soc Hortic Sci 110:671–676

Leonardis RFJ, Mangieri HR, Tinto JC, Alonzo A, Reuter HR (2000) In: Celulosa Argentina SA (ed) El nuevo libro del árbol. Especies exóticas de uso ornamental, tomo 3. El Ateneo, Buenos Aires, Argentina, pp 88–89

Li DZ, Pritchard HW (2009) The science and economics of ex situ plant conservation. Trends Plant Sci 14(11):614–621

Linington SH, Pritchard HW (2001) Genebanks. In: Levin S (ed) Encyclopedia of Biodiversity, vol 3. Academic Press, San Diego CA, pp 165–181

Lloyd G, McCown B (1981) Commercially fesible micropropagation of mountain laurel, Kalmia latifolia, by the use of shoot-tip culture. Comb Proc Int Plant Prop Soc 30:421–427

Loesener T (1942) Aquifoliaceae. In: Harms H, Mattfeld J (eds) Die Natürlichen Pflanzenfamilien, 2nd edn. Wilhelm Engelmann, Leipzig, pp 36–86

Loizeau PA, Spichiger R (2004) Aquifoliaceae. In: Smith N, Mori SA, Henderson A, Stevenson DW, Heald SV (eds) Flowering plants of the Neotropics. Princeton University Press, Princeton, NJ, pp 26–28

Luna C, Sansberro P, Mroginski L, Tarragó J (2003) Micropropagation of *Ilex dumosa* (Aquifoliaceae) from nodal segments in a tissue culture system. Biocell 27(2):205–212

Lyhr KP (1992) Mahogany. Silviculture and use of American Mahogany (*Swietenia* spp.). The Royal Veterinary and Agricultural University, 89 pp

Lynch P (1999) Tissue culture techniques in *in vitro* plant conservation. In: Benson EE (ed) Plant conservation biotechnology. Taylor & Francis, London, Philadelphia, pp 42–61

Mabberley DJ (1987) The plant book. Cambridge University Press, Cambridge

Mabberley DJ, Pannell CM, Sing AM (1995) Meliaceae. Flora Malesiana Series I 12:1–407

Malik SK, Chaudhury R (2006) The cryopreservation of embryonic axes of two wild and endangered citrus species. Plant Genet Res Charact Utiliz 4:204–209

Mangieri HR, Tinto JC (1977) In: Celulosa Argentina SA (ed) Libro del árbol. Esencias fo-restales no autóctonas cultivadas en la Argentina de aplicación ornamental y/o industrial, tomo 3. Impresiones Ramos Mejía S.A., Buenos Aires, Argentina, pp 61–62

Mari S, Engelmann F, Chabrillange N, Huet C, Michaux-Ferrière N (1995) Histocytological study of coffee (*Coffea racemosa* and *C. sessiliflora*) apices of *in vitro* plantlets during their cryopreservation using the encapsulation-dehydration technique. Cryo-Letters 16:289–298

Marin ML, Duran-Vila N (1988) Survival of somatic embryos and recovery of plants of sweet orange (*Citrus sinensis* L. Osb.) after immersion in liquid nitrogen. Plant Cell Tissue Organ Cult 14(1):51–57

Marín, ML, Duran-Vila, N (1988) Plant Cell Tissue Organ Culture 14: 51–57. 10. Mumford PM, Grout BWW (1979) Seed Sci Tech 7: 407–410

Marin ML, Gogorcena Y, Ortiz J, DuranVila N (1993) Recovery of whole plants of sweet orange from somatic embryos subjected to freezing thawing treatments. Plant Cell Tiss Org Cult 34:27–33

Martin AC (1946) The comparative internal morphology of seeds. Am Midl Nat 36(3):513–660

Martinez M, Gonzalez-Arnao MT, Urra C, Rojas R, Cuba M, Garcia D (1996) Preliminary studies on the cryopreservation of zygotic embryos of *Coffea arabica* variety 972. Cultivos Tropicales 17:79–81

Martinez MT, Ballester A, Vieitez AM (2003) Cryopreservation of embryogenic cultures of *Quercus robur* using desiccation and vitrification procedures. Cryobiology 46:182–189

Martínez CRC, Gutiérrez EJC, Arrazate CHA (2013a) Protocolo de multiplicación y conservación *in vitro* de cuatro especies forestales tropicales de semillas recalcitrantes CASO: Caoba (*Swietenia macrophylla*). INIFAP-SEMARNAT-CONAFOR, México, pp 19–34

Martínez CRC, Gutiérrez EJC, Arrazate CHA (2013b) Protocolo de multiplicación y conservación in vitro de cuatro especies forestales tropicales de semillas recalcitrantes CASO: Cedro Rojo (*Cedrela odorata* L.). INIFAP-SEMARNAT-CONAFOR, México, pp 2–18

Marum L, Estêvão C, Oliveira MM, Amâncio S, Rodriguez L, Miguel C (2004) Recovery of cryopreserved embryogenic cultures of maritime pine: effect of cryoprotectant and suspension density. Cryo-Letters 25:363–374

Marzalina M, Krishnapillay B (1999) Recalcitrant seed biotechnology application to rain forest conservation. In: Benson EE (ed) Plant conservation biotechnology. Taylor & Francis, London, Philadelphia, pp 265–276

Marzalina M, Nashatul ZNA (2000) Cryopreservation of some Malaysian tropical urban forestry species. In: Engelmann F, Takagi H (eds) Cryopreservation of tropical plant germplasm: current research progress and application. Proceedings of an International Workshop, Tsukuba, Japan, October 1998. International Plant Genetic Resources Institute (IPGRI), Rome, pp 465–467

Marzalina M, Normah M (2002) Cryopreservation techniques for the long-term storage of mahogany (*Swietenia macrophylla*) seeds. J Trop For Sci 14(4):525–535

Mathur G, Alkutkar VA, Nadgauda RS (2003) Cryopreservation of embryogenic culture of *Pinus roxburghii*. Biol Plant 46:205–210

Matsumoto T, Sakai A (2003) Cryopreservation of axillary shoot tips of *in vitro*-grown grape (Vitis) by a two-step vitrification protocol. Euphytica 131:299–304

Matsumoto T, Mochida K, Itamura H, Sakai A (2001) Cryopreservation of persimmon (*Diopsyros kaki* Thunb) by vitrification of dormant shoot tips. Plant Cell Rep 20:398–402

Matsumoto T, Niino T, Shirata K, Kurahashi T, Matsumoto S, Maki S, Itamura H (2004) Long-term conservation of *Diospyros* germplasm using dormant buds by a prefreezing method. Plant Biotechnol 21:229–232

Matsumoto T, Yamamoto S, Fukui K, Rafique T, Engelmann F, Niino T (2015) Cryopreservation of *Persimmon* shoot tips from dormant buds using the D cry-oplate technique. Hortic J 84 (2):106–110

Mebrat W, Gashaw T (2013) Eur J Bot Plant Sci Phytol 1(3):10–17

Mejía EG, Song YS, Heck CI, Ramírez-Mares MV (2010) Yerba mate tea (*Ilex paraguariensis*): phenolics, antioxidant capacity and in vitro inhibition of colon cancer cell proliferation. J Funct Foods 2(1):23–34

Melo EA, Guerra NB (2002) Antioxidant phenolic compounds occurring naturally in foods. Bol SBCTA 36(1):1–11

Mishra AP, Saklani S, Chandra S, Mathur A, Milella L, Tiwari P (2014) *Aphanamixis polystachya* (wall.) Parker, phytochemistry, pharmacological properties and medicinal uses: an overview. World J Pharm Pharm Sci 3(6):2242–2252

Misson JP, Druart P, Panis B, Watillon B (2006) Contribution to the study of the maintenance of somatic embryos of *Abies nordmanniana* LK: culture media and cryopreservation method. Propag Ornamental Plants 6:17–23

Montiel-Castelán P, Castillo-Martínez CR, Gómez-Reyes LA, Valle-Arizaga M, Jasso-Mata J (2016) Conservación in vitro por crecimiento mínimo de *Swietenia macrophylla* King. y *Tectona grandis* L. Agroproductividad 9(2): 20–25

Moriguchi T, Kozaki I, Yamaki S, Sanada T (1990) Low temperature storage of pear shoots in vitro. Bull Fruit Tree Res Stn 17:11–18

Mosele SH (2002) Governance in the agribusiness chain of yerba mate in the Alto Uruguai Riograndense. Master Dissertation. Universidade Federal do Rio Grande do Sul, Porto Alegre, 211 pp

Mroginski LA, Roca WM, Kartha KK (1991) Crioconservación de germoplasma. In: Roca WM, Mroginski LA (eds) Cultivo de tejidos en la agricultura: fundamentos y aplicaciones. CIAT, Cali, Colombia, pp 715–730

Mroginski LA, Sansberro PA, Scocchi AM, Luna CV, Rey HY (2006) Effect of fruit cryo-preservation on in vitro germination of zygotic embryos of several species of *Ilex*. Acta Hort 725:417–419

Mroginski LA, Sansberro PA, Scocchi AM, Luna CV, Rey HY (2008) A cryopreservation protocol for immature zygotic embryos of species of *Ilex* (Aquifoliaceae). Biocell 32(1):33–39

Mroginski L, Dolce N, Sansberro P, Luna C, Gonzalez A, Rey H (2011) Techniques and protocol on the cryopreservation of Ilex immature zygotic embryo. In: Thorpe TA, Yeung ECT (eds) Plant embryo culture: methods and protocols. Human Press, New York, EEUU, pp 215–225

Munmford PM, Grout WW (1979) Desiccation and low temperature (-196°C) tolerance of *Citrus limon* seed. Seed Sci Techno 7:407–410

Murashige T, Skoog F (1962) A revised medium for rapid growth and bioassays with tobacco tissue cultures. Physiol Plant 15:473–497

Nagy S, Attaway JA (1980) Citrus nutrition and quality: based on a symposium sponsored by the Division of Agricultural and Food Chemistry at the 179th meeting of the American Chemical Society, Houston, Texas, March 26, ACS Symposium Series (USA)

Naidoo C, Benson E, Berjak P, Goveia M, Pammenter NW (2011) Exploring the use of DMSO and ascorbic acid to promote shoot development by excised embryonic axes of recalcitrant seeds. CryoLett 32:166–174

Nair AGR, Subramanian SS (1975) Quercetagetin and other flavones from *Gmelina arborea* and *G. asiatica*. Phytochemistry 14:1135–1136

Nardo EAB, Costa AS, Lourencao AL (1997) Melia azedarach extract as an antifeedant to Bemisia tabaci (Homoptera: Aleyrodidae). Florida Entomol 80(1):92–94

Navarro C, Ward S, Hernandez M (2002) The tree *Cedrela odorata* (Meliaceae): a morphologically subdivided species in Costa Rica. Revista Biologia Tropical 50:21–29

Newton AC, Barker P, Ramnarine S, Mesen JF, Leakey RRB (1993) The mahogany shoot borer: prospects for control. For Ecol Manage 57:301–328

Newton AC, Cornelius JP, Mesen JF, Leakey RRB (1995) Genetic variation in apical dominance of *Cedrela odorata* seedlings in response to decapitation. Silvae Genetica 44:146–150

Newton AC, Allnutt TR, Gillies ACM, Lowe AJ, Ennos RA (1999) Molecular phylogeography, intraspecific variation and the conservation of tree species. Trends Evol Ecol 14:140–145

Niino T, Sakai A (1992) Cryopreservation of alginated coated in vitro grown shoot-tips of apple, pear and mulberry. Plant Sci 87:199–206

Niino T, Sakai A, Yakuwa H, Nojiri K (1992) Cryopreservation of *in vitro*-grown shoot tips of apple and pear by vitrification. Plant Cell Tiss Org Cult 28:261–266

Niino T, Koyama A, Shirata K, Ohuchi S, Suzuki M, Sakai A (1993) Long-term storage of mulberry winter buds by cryopreservation. J Serie Sci Jpn 62:431–434

Niino T, Tashiro K, Suzuki M, Ohuchi S, Magoshi J, Akihama T (1997) Cryopreservation of *in vitro*—grow shoot tips of cherry and sweet cherry by one- step vitrification. Scientia Hortic 70:155–163

Niino T, Watanabe KN, Nohara T, Rafique S, Yamamoto K, Fukui MV, Arizaga CRC, Martinez T, Matsumoto, Engelmann F (2014) Practical cryopreservation of mat rush basal stem buds by air dehydration using aluminum cry-oplate. Plant Biotechnol 31:281–287

Niklas CO (1987) Estudios embriológicos y citológicos en la yerba mate *Ilex paraguariensis* (Aquifoliaceae). Bonplandia 6:45–56

Nishizawa S, Sakai A, Amano Y, Matsuzawa T (1993) Cryopreservation of asparagus (*Asparagus officinalis* L.) embryogenic suspension cells and subsequent plant regeneration by the vitrification method. Plant Sci 91(1):67–73

Niu Y, Luo Z, Zhang Y, Zhang Q (2012) Cryopreservation of in vitro grown shoot tips of *Diospyros kaki* Thunb. using different methods. CryoLett 33:69–74

Normah MN, Seti Dewi Serimala MN (1995) Cryopreservation of seeds and embryonic axes of several citrus species. In: Ellis RH, Black M, Murdoch AJ, Hong TD (eds) Basic and applied aspects of seed biology. Kluwer Academic Publishers, Dordrecht, pp 817–823

Normah MN, Siti Dewi Serimala SD (1997) Cryopreservation of seeds and embryonic axes of several Citrus species. In: Ellis RH, Black M, Murdoch AJ, Hong TD (eds) Basic and applied aspects of seed biology. Kluwer, Dordrecht, pp 817–823

Normah MN, Vengadasalam M (1992) Effect of moisture content on cryopreservation of *Coffea* and *Vigna* seeds and embryos. Cryo-Letters 13:199–208

Normah MN, Reed BM, Yu X (1994) Seed storage and cryoexposure behavior in hazelnut (*Corylus avellana* L. cv. Barcelona). Cryo-Letters 15:315–322

Normah MN, Mainah G, Saraswathy R (2000) Cryopreservation of zygotic embryos of tropical fruit trees, a study on *Lansium domesticum* and *Baccaurea* species. In: Engelmann F, Takagi H (eds) Cryopreservation of tropical plant germplasm: current research progress and application. Proceedings of an International Workshop, Tsukuba, Japan, October 1998. International Plant Genetic Resources Institute, Rome, pp 156–160

Nunes ED, Benson EE, Oltramari AC, Araujo PS, Moser JR, Viana AM (2003) *In vitro* conservation of *Cedrela fissilis* Vellozo (Meliaceae), a native tree of the Brazilian Atlantic Forest. Biodiv Conserv 12: 837–848

O'Neill GA, Dawson I, Sotelo-Montes C, Guarino L, Guari-Guata M, Current D, Weber JC (2001) Strategies for genetic conservation of trees in the Peruvian Amazon. Biodivers Conserv 10:837–850

Oh SD (1997) The effect of prefreezing treatment and cryoprotectants on the survival of cryopreserved somatic embryos and plant regeneration in Korean native citrus species. Acta Hortic 447:499–505

Okas S, Niino T (1989) Long term-storage of pear shoots cultured in vitro by minimum growth method. Abtr Jpn See Hortic Sci Spring Meet 58(Suppl 1):120–121 (In Japanese)

Olivares-Fuster O, Asins MJ, Duran-Vila N, Navarro L (2000) Cryopreserved callus, a source of protoplasts for citrus improvement. J Hort Sci Biotech 75:635–640

Omura M, Akihama T, Kozaki L (1978a) Long term preservation of fruit tree scions. In: Akihama T, Nakajima K (eds) Long-term preservation of favourable germ plasm in arboreal crops. Friut Tree Research Station Minist Agric For Jpn, pp 26–30

Omura M, Sato Y, Seike K (1978b) Long term preservation of Japanese pear seeds under extra-low temperatures. In: Akihama T, Nikajima K (eds) Long term preservation of favourable germ plasm in arboreal crops, Fruit Tree Res Stn Minist Agric For Jpn, pp 26–30

Orlikowska T (1992) Effect of in vitro storage at 4 °C on survival and proliferation of two apple rootstocks. Plant Cell Tiss Organ Cult 31:1–7

Ozden-Tokatli Y, Ozudogru EA, Gumuse F, Lambardi M (2007) Cryopreservation of *Pistacia* spp. seeds by dehydration and one-step freezing. Cryo-Letters 28:83–94

Pammenter NW, Berjak P (1999) A review of recalcitrant seed physiology in relation to desiccation-tolerance mechanisms. Seed Sci Res 9:13–37

Pammenter NW, Berjak P (2014) Physiology of desiccation-sensitive (recalcitrant) seeds and the implications for cryopreservation. Int J Plant Sci 175:21–28

Panis B, Piette B, Swennen R (2005) Droplet vitrification of apical meristems: a cryopreservation protocol applicable to all Musaceae. Plant Sci 168(1):45–55

Páques M, Monod V, Poissonnier M, Dereudre J (2002) Cryopreservation of eucalyptus sp. shoot tips by the encapsulation-dehydration procedure. In: Towill LE, Bajaj YPS (eds) Biotechnology in agriculture and forestry, vol 50. Springer, Berlin, pp 234–245

Park YS, Bmett JD, Bonga JM (1998) Application of somatic embyogenesis in high-value clonal forestry: deployment, genetic control, and stability of cryopreserved clones. In Vitro Cell Dev Biol Plant 34:231–239

Paul H, Daigny G, Sangwan-Noreel BS (2000) Cryopreservation of apple (*Malus* x *domestica* Borkh.) shoot tips following encapsulation or encapsulation-vitrification. Plant Cell Rep 19:768–774

Paulet F, Glaszmann JC (1994) Biotechnological support for varietal extension of sugarcane. CIRAD-CA Agriculture-et-Development, Special issue, pp 47–53

Pausujja R, Granhof J, Willan RL (1986) *Pinus merkusii* Jungh. & de Vriese (including *Pinus merkusiana* Cooling & Gaussen). Danida Forest Seed Centre, Humleback, Denmark, Seed Leafl, p 7

Pence VC (1990) Cryostorage of embryo axes of several large-seeded temperate tree species. Cryobiology 27(2):212–218

Pence VC (1991) Cryopreservation of immature embryos of *Theobroma cacao*. Plant Cell Rep 10:144–147

Pence VC (1992) Desiccation and the survival of *Aesculus*, *Castanea*, and *Quercus* embryo axes through cryopreservation. Cryobiology 29:391–399

Pence VC (1996) Germination, desiccation and cryopreservation of seeds of *Populus deltoides* Bartr. Seed Sci Technol 24:151–157

Pennington TD (1981) A monograph of the neotropical Meliaceae. Flora Neotropica. The New York Botanical Gardens, New York, pp 360–390

Pennington TD, Muellner AN (2010) A monograph of Cedrela (Meliaceae). DH Books. The Manse. Chapel Lane, Milborne Port-England, 112 pp

Pennington TD, Sarukhán KJ (1968) Manual para la identificación de campo de los principales árboles tropicales de México. INIF-FAO-SAG, México, p 413

Pennington TD, Styles BT (1975) A generic monograph of the Meliaceae. Blumea 22(3):419–540

Percy REL, Livingston NJ, Moran JA, von Aderkas P (2001) Desiccation, cryopreservation and water relations parameters of white spruce (*Picea glauca*) and interior spruce (*Picea glauca* x *engelmannii complex*) somatic embryos. Tree Physiol 21:1303–1310

Perez RM, Navarro L, Duran-Vila N (1997) Cryopreservation and storage of embryogenic callus cultures of several *Citrus* species and cultivars. Plant Cell Rep 17:44–49

Perez RM, Mas O, Navarro L, Duran-Vila N (1999) Production and cryoconservation of embryogenic cultures of mandarin and mandarin hybrids. Plant Cell Tiss Org Cult 55:71–74

Pérez-Flores J, Eigenbrode SD, Hilje-Quiroz L (2012) Alkaloids, limonoids and phenols from meliaceae species decrease survival and performance of *Hypsipyla grandella* Larvae. Am J Plant Sci 2(1):988–994

Pijut PM, Lawson SS, Michler CH (2011) Biotechnological efforts for preserving and enhancing temperate hardwood tree biodiversity, health, and productivity. In Vitro Cell Dev Biol Plant 47:123–147

Pita JM, Sanz V, Escudero A (1998) Seed cryopreservation of seven Spanish native pine species. Silvae Genetica 47:220–223

Prakash R (1991) Gmelina arborea. Propagation practices of important indian trees. International Book Distributors, Dehradun, India, pp 234–236

Preci D, Cichoski AJ, Valduga AT, de Oliveira D, Valduga E, Treichel H, Toniazzo G, Cansian RL (2011) Desenvolvimento de iogurte light com extrato de erva-mate (*Ilex paraguariensis* St. Hil) e adição de probióticos. Alimentos e Nutrição 22(1):27–38

Pritchard HW (2007) Cryopreservation of desiccation-tolerant seeds. In: Day JG, Stacey GN (eds) Cryopreservation and freeze-drying protocols, 2nd edn. Humana Press Inc, Totowa, New Jersey, pp 185–201

Pukacki PM, Jarzabech M, Pukacka M (2009) Characterization of cryoprotective activity of proteins in *Acer*, *Fagus* and *Quercus* embryonic axes. In: First international symposium

"Cryopreservation in Horticultural Species". Leuven, Belgium, 5–8 April 2009, p 117 (abstract)

Radhamani J, Chandel KPS (1992) Cryopreservation of embryonic axes of trifoliate orange (*Poncirus trifoliata* L., RAF.). Plant Cell Rep 11(7):372–374

Ramana KVR, Govindarajan VS, Ranganna S (1981) Citrus fruits varieties, chemistry, technology, and quality evaluation. Part I: varieties, production, handling, and storage. Crit Rev Food Sci Nutr 15:353–431

Rao NK (2004) Plant genetic resources: advancing conservation and use through biotechnology. Afr J Biotech 3:136–145

Reed BM (2002) Implementing cryopreservation for long-term germplasm preservation in vegetatively propagated plants. In: Towill LE, Bajaj YPS (eds) Cryopreservation of plant germplasm II. Biotechnology in agriculture and forestry, vol 50. Springer, Heidelberg, pp 22–33

Rehman HU, Gill MIS (2015) Micrografting of fruit crops—a review. J Hortic 2:151

Ril FT, Loch CR, Valduga AT, Macedo SMD, Cichoski AJ (2011) Biochemical profile of rats fed yogurt containing extract of yerba mate (*Ilex paraguariensis* St. Hil). Braz J Food Technol 14 (4):332–337

Roberts EH (1973) Predicting the storage life of seeds. Seed Sci Technol 1:499–514

Roberts EH, King MW (1980) Storage of recalcitrant seeds. In: Withers LA, Williams JT (eds) Crop genetic resources. The conservation of difficult material. International Union of Biological Sciences, Series B42, pp 39–48

Ryynanen L (1996) Survival and regeneration of dormant silver birch buds stored at superlow temperatures. Can J For Res 26:617–623

Ryynänen L, Sillanpää M, Kontunen-Soppela S, Tiimonen H, Kangasjärvi J, Vapaavuori E et al (2002) Preservation of transgenic silver birch (*Betula pendula* Roth) lines by means of cryopreservation. Mol Breed 10:143–152

Sajini KK, Karun A, Kumaran PM (2006) Cryopreservation of coconut (*Cocos nucifera* L.) zygotic embryos after pre-growth desiccation. J Plantation Crops 34:576–581

Sakai A (1956) Survival of plant tissue at super-low temperature. Low Temp Sci Ser B 14:17–23

Sakai A, Engelmann F (2007) Vitrification, encapsulation-vitrification and droplet-vitrification: a review. Cryo-Letters 28(3):151–172

Sakai A, Kobayashi S, Oiyama I (1990) Cryopreservation of nucellar cells of navel orange (*Citrus sinensis* Obs. var. *brasiliensis* Tanaka) by vitrification. Plant Cell Rep 9(1):30–33

Salas EJB (1993) Árboles de Nicaragua. Instituto Nicaragüense de Recursos Naturales y del Ambiente (IRENA), Managua, Nicaragua, p 390

Salgotra RK, Gupta BB (2015) Plant genetic resources and traditional/indigenous knowledges: potential and challenges In: Salgotra RK, Gupta BB (eds) Plant genetic resources and traditional knowledge for food security. Springer Science+Business Media, Singapore, pp 1–21

Salomão AN (2002) Tropical seed species responses to liquid nitrogen exposure. Braz J Plant Physiol 14(2):133–138

Sampayo-Maldonado S, Castillo-Martínez CR, Jasso-Mata J, Jiménez-Casas M, López-Upton J, Sánchez-Monsalvo V (2016) Efecto del medio de cultivo en la propagación *in vitro* de genotipos de *Cedrela odorata* L. Agroproductividad 9(2):62–69

Sánchez C, Martínez MT, Vidal N, San-José MC, Valladares S, Vieitez AM (2008) Preservation of Quercus robur germplasm by cryostorage of embryogenic cultures derived from mature trees and RAPD analysis of genetic stability. CryoLetters 29:493–504

San-José MC, Jorquera L, Vidal N, Corredoira E, Sánchez C (2005) Cryopreservation of european chestnut germplasm. Acta Hortic 693:225–232

Sansberro PA, Rey HY, Mroginski LA, Collavino MM (1998) In vitro culture of rudimentary embryos of *Ilex paraguariensis*: responses to exogenous cytokinins. J Plant Growth Regul 17 (2):101–105

Sansberro PA, Rey HY, Mroginski LM, Collavino MM (1999) In vitro plant regeneration of *Ilex paraguariensis* (Aquifoliaceae). In Vitro Cell Dev Biol Plant 35(5):401–402

Sansberro PA, Rey HY, Mroginski LA (2001) In vitro culture of zygotic embryos of *Ilex* species. HortScience 36(2):351–352

Santos IRI, Stushnoff C (2003) Desiccation and freezing tolerance of embryonic axes from *Citrus sinensis* (L.) Osb. pretreated with sucrose. Cryo-Letters 24:281–292

Santos IRI, Stushnoff C (2014) Cryopreservation of embryonic axes of Citrus species by encapsulation-dehydration. Publi dans le 131:36–41

Santos IRI, Nassif Salomão A, Peixoto Vargas D, Da Silva PC, Ferreira Nogueira G, Carvalho M, Paiva R (2013) Situación actual y perspectivas de la investigación en crioconservación de recursos fitogenéticos en Brasil. In: Gonzalez-Arnao MT, Engelmann F (eds) Crioconservación de plantas en América Latina y el Caribe. IICA, San José, Costa Rica, pp 76–92

Schinella G, Fantinelli JC, Mosca SM (2005) Cardioprotective effect of *Ilex paraguariensis* extract: evidence for a nitric oxide-dependent mechanism. Clin Nutr 24(3):360–366

Scocchi A, Faloci M, Medina R, Olmos S, Mroginski L (2004) Plant recovery of cryopreserved apical meristem-tips of *Melia azedarach* L. using encapsulation-dehydration and assessment of their genetic stability. Euphytica 135(1):29–38

Scocchi A, Vila S, Mroginski L, Engelmann F (2007) Cryopreservation of somatic embryos of paradise tree (*Melia azedarach* L.). CryoLetters 28(4):281–290

Shackleton SE, Shackleton CM, Netshiluvhi TR, Geach BS, Ballance A, Fairbanks DHK (2002) Use patterns and value of savanna resources in three rural villages in South Africa. Econ Bot 56 (2):130–146

Shands H (1991) Complementarity of in situ and ex situ germplasm conservation from the standpoint for the future user. Israeli J Bot 40:521–528

Sharma PC, Yelne MB, Dennis TJ (2001) Database on medicinal plants used in ayurveda, vol 3. Central Council for Research in Ayurveda and Siddha, Department of ISM and H Ministry of Health and Family Welfare Government of India, pp 217–228

Sharma N, Satsangi R, Pandey R, Singh R, Kaushik N, Tyagi RK (2012) In vitro conservation of Bacopa monnieri (L.) using mineral oil. Plant Cell Tissue Organ Cult 111(3):291–301

Shatnawi MA, Engelmann F, Frattarelli A, Damiano C (1999) Cryopreservation of apices of *in vitro* plantlets of almond (*Prunus dulcis* Mill). Cryo-Letters 20:13–20

Shea GM, Armstrong PA (1978) The effect of post-harvest environmental factors on the longevity of Hoop pine seed. Queensland Dep For Res, Note, p 24

Sherlock G, Block W, Benson EE (2005) Thermal analysis of the plant encapsulation-dehydration cryopreservation protocol using silica gel as the desiccant. CryoLetters 26:45–54

Shibli RA, Al-Juboory KH (2000) Cryopreservation of 'Nabali' olive (*Olea europea* L.) somatic embryos by encapsulation-dehydration and encapsulation-vitrification. Cryo-Letters 21:357–366

Shibli RA, Shatnowi AM, Subaih SW, Ajlouni MM (2006) *In vitro* conservation and cryopreservation of plant genetic resources: a review. World J Agric Sci 2(4):372–382

Shirwaikar A, Ghosh S, Rao PGM (2003) Effect of *Gmelina arborea* Roxb. leaves on wound healing in rats. J Nat Remedies 3(1):45–48

Silva LC, Paiva R, Swennen R, Andre E, Panis B (2013) Shoot-tip cryopreservation by droplet vitrification of *Byrsonima intermedia* A. Juss.: a woody tropical and medicinal plant species from Brazilian cerrado. CryoLetters 34(4):338–348

Snook LK (1993) Stand dynamics of Mahogany (*Swietenia macrophylla* King.) and associated specie after fire and hurricane in the tropical forest of the Yucatan Peninsula, Mexico. Thesis of Forestry Ph.D. New Haven, Connecticut, Yale University, 173 pp

Soliman HIA (2013) Cryopreservation of in vitro-grown shoot tips of apricot (*Prunus armeniaca* L.) using encapsulation-dehydration. Afr J Biotech 12(12):1419–1430

Solovieva MA (1966) Long-term storage of fruit seed. In: Proceedings of the 17th International Horticulture Congress, Maryland, USA

Standley P (1946) Flora of Guatemala. Chicago Natural History Museum, Chicago, vol 24, part 5, pp. 458–459

Styles BT (1972) The flower biology of the Meliaceae and its bearing on tree breeding. Silvae Genet 21:175–182

Sukartiningsih SY, Ide Y (2012) Encapsulation of axillary buds of *Gmelina arborea* Roxb. and *Peronema canescens* Jack. Bull Univ Tokyo For 127:57–65
Suszka B (1975) Cold storage of already after-ripened beech (*Fagus silvatica* L.) seeds. Arbor Kornickie 20:299–315
Swaminathan MS (1997) Implementing the benefit sharing provisions on the convention of biological diversity: challenges and opportunities. FAO/IBPGR Plant Genetic Newsletter 112:19–27
Taji AM, Williams RR (1996) Tissue culture of Australian plants. University of New England Press, Armidale, NSW 2351, Australia
Tauer CG (1979) Seed tree, vacuum, and temperature effects on eastern cottonwood seed viability during extended storage. For Sci 25:112–114
Tewari DNA (1995) Monograph on Gamari (*Gmelina arborea* Roxb.). International Book Distributors, Dehradun, India, pp 1–84
Thakur R, Rao PS, Bapat VA (1998) In vitro plant regeneration in *Melia azedarach* L. Plant Cell Rep 18(1):127–131
Touchell DH, Dixon KW (1994) Cryopreservation for seed banking of Australian species. Ann Bot 74:541–546
Touchell DH, Chiang VL, Tsai CJ (2002) Cryopreservation of embryogenic cultures of *Picea mariana* (black spruce) using vitrification. Plant Cell Rep 21:118–124
Towill LE (1984) Survival of ultra-low temperatures of shoot-tips from *Solanum tuberosum* groups *andigena*, *phureja*, *stenotomum* and other tuber-bearing Solanum species. Cryo-Letters 5:319–326
Tsang ACW, Corlett RT (2005) Reproductive biology of the *Ilex* species (Aquifoliaceae) in Hong Kong, China. Can J Bot 83:1645–1654
UNDP (2007) Human development. In: United Nations Development Programme, Uganda Human Development Report—Rediscovering Agriculture for Human Development, p 38
United Nations Conference on Trade and Development (UNCTAD) (2004) Market information in the commodities area: information on citrus fruit. Available from: https://r0.unctad.org/infocomm/anglais/orange/sitemap.htm. Accessed 2016 October
Valladares S, Toribio M, Celestino C, Vieitez AM (2004) Cryopreservation of embryogenic cultures from mature Quercus suber trees using vitrification. CryoLetters 25:177–186
Verissimo A, Junior CS, Stone S, Uhl C (1998) Zoning of timber extraction in the Brazilian Amazon. Conserv Biol 12:128–136
Vertucci C, Farrant J (1995) Acquisition and loss of desiccation tolerance. In: Kigel J, Galilli G (eds) Seed development and germination. Marcel Dekker Inc, New York, pp 237–271
Vidal N, Vieitez AM, Fernández MR, Cuenca B, Ballester A (2010) Establishment of cryo-preserved gene banks of European chestnut and cork oak. Eur J Forest Res 129:635–643
Vila S, Scocchi A, Mroginski L (2002) Plant regeneration from shoot apical meristems of *Melia azedarach* L. (Meliaceae). Acta Physiol Plant 24(2):195–199
Vila S, Gonzalez A, Rey H, Mroginski L (2003) Somatic embryogenesis and plant regeneration from immature zygotic embryos of *Melia azedarach* (Meliaceae). In Vitro Cell Dev Biol Plant 39(3):283–289
Villa AL, Jimenez PE, Valbuena RI, Bastidas S, Nunez VM (2007) Preliminary study of the establishment of cryoconservation protocol for oil palm (*Elaeis guineensis* Jacq.). Agronomia Colombiana 25:215–223
Volk GM, Bonnart R, Krueger R, Lee R (2012) Cryopreservation of citrus shoot tips using micrografting for recovery. CryoLetters 33(6):418–426
Walden RF, Wright RD (1995) Supraoptimal root-zone temperature influences the medium solution and growth of "Helleri" holly. HortScience 30(5):1020–1025
Walters C, Farrant JM, Pammenter NW, Berjak P (2002) Desiccation and damage. In: Black M, Pritchard HW (eds) Desiccation and survival in plants: drying without dying. CAB International, London, pp 263–291
Walters C, Wheeler L, Stanwood PC (2004) Longevity of cryogenically stored seeds. Cryobiology 48:229–244

Wanas WH, Callow JA, Withers LA (1986) Growth limitation for the conservation of pear genotypes. In: Withers LA, Anderson PG (eds) Plant tissue culture and its agricultural application. Butterworthd, London, pp 285–290

Wang ZC, Deng XX (2004) Cryopreservation of shoot-tips of citrus using vitrification: effect of reduced form of glutathione. Cryo-Letters 25:43–50

Wen B, Song SQ (2007) Acquisition and loss of cryotolerance in *Livistonia chinensis* embryos during seed development. Cryo-Letters 28:291–302

Williams N (1998) Study finds 10% of tree species under threat. Science 281:1426

Withers LA (1995) Collecting *in vitro* for genetic resources conservation. In: Guarino L, Rao R, Reid R (eds) Collecting plant genetic diversity. CAB International, Wallingford, pp 511–515

Woessner RA, McNabb KL (1979) Large scale production of *Gmelina arborea* Rokb. Seed—a case study. Commonwealth For Rev 58:117–121

Wu YJ, Huang XL, Xiao JN, Li J, Zhou MD, Engelmann F (2003) Cryopreservation of mango (*Mangifera indica* L.) embryogenic cultures. Cryo-Letters 24:303–314

Xia K, Hill LM, Li D-Z, Walters C (2014) Factors affecting stress tolerance in recalcitrant embryonic axes from seeds of four *Quercus* (Fagaceae) species native to the USA or China. Ann Bot 114:1747–1759

Yamamoto SI, Rafique T, Priyantha WS, Fukui K, Matsumoto T, Niino T (2011) Development of a cryopreservation procedure using aluminium cry-oplates. Cryo-Letters 32(3):256–265

Yamamoto SI, Rafique T, Fukui K, Sekizawa K, Niino T (2012) V-cryoplate procedure as an effective protocol for cryobanks: case study of mint cryopreservation. Cryo-Letters 33(1): 12–23

Yap SK, Wang SM (1983) Seed biology of *Acacia mangium, Albizia falcataria, Eucalyptus* spp. *Gmelina arborea, Maesopsis eminii, Pinus caribaea,* and *Tectona grandls*. Malays Forester 46:26–45

Zhang N, Wen B, Ji M, Yan Q (2014) Low-temperature storage and cryopreservation of grapefruit (*Citrus paradisi* Macfad.) seeds. CryoLetters 35(5):418–426

Chapter 14
Conservation of Hardwood Forest Species

E. Corredoira, M.T. Martínez, M.C. Sanjosé and A. Ballester

Abstract Conservation of forest genetic resources is essential for meeting the demand for future wood products. Genotypes with improved characteristics are nowadays widely produced through conventional breeding programmes, by selection of mature superior trees, by genetic transformation procedures, etc., which are important for increasing the productivity of forestry clonal plantations. Strategies for forest biodiversity conservation today are well defined, among which cryopreservation is viewed as a complementary storage method, important for plant tissues with specific characteristics (vegetatively propagated species). In addition, many hardwood forest trees produce recalcitrant seeds (seeds that cannot be stored for long periods under conventional conditions) that only would be stored on a long-term basis through cryopreservation. The availability of simple, reliable and cost-effective strategies for conservation of hardwood forest species (with special attention to recalcitrant species of the Fagaceae family) will be highlighted in this review. Specifically, emphasis will be addressed to the following topics: (i) medium-term conservation through slow growth storage; (ii) cryopreservation techniques; (iii) selection of explants for cryopreservation: in vivo collection of embryonic axes and dormant buds and in vitro collection of shoot tips and embryogenic cultures; and (iv) genetic stability of cryopreserved material. The limited application of cryopreservation to the development of large cryobanks of hardwood forest species also will be mentioned.

Keyword Cold storage · Cryopreservation · Dormant buds · Micropropagation
Somatic embryogenesis · Vitrification

E. Corredoira (✉) · M.T. Martínez · M.C. Sanjosé · A. Ballester
Instituto de Investigaciones Agrobiológicas de Galicia, IIAG, CSIC,
Avenida de Vigo S/N, 15705 Santiago de Compostela, Spain
e-mail: elenac@iiag.csic.es

M.R. Ahuja and S.M. Jain (eds.), *Biodiversity and Conservation of Woody Plants*,
Sustainable Development and Biodiversity 17, DOI 10.1007/978-3-319-66426-2_14

14.1 Introduction

Forest trees cover approximately 30% of Earth's land surface, but only 3% of forested land is covered by forest plantations (Kirilenko and Sedjo 2007). An interesting strategy to increase the supply of wood to meet industrial and domestic demands is the introduction of more productive species/varieties exhibiting superior traits. Productivity in forestry plantations needs to be increased significantly, while, at the same time, native forests must be protected from further exploitation. Plantations are highly productive after improvement in genetic composition of planting stock, generally achieved through conventional breeding programmes, which require extensive resources and extended periods of time. Breeding of long-rotation hardwoods is less developed than that of short-rotation species (e.g. eucalypts, poplars, willows). Species of the genus *Quercus*, for example, usually reach seed-bearing age at 40–100 years old, and they are likely to produce good seed crops once every 3–5 years (Ducousso and Bordacs 2004). The alternative to complement the natural difficulties of conventional breeding of trees is the application of clonal forestry, if superior trees could be identified and vegetatively propagated. Clonal propagation of trees is being practised through grafting, rooting of cuttings or in vitro tissue culture techniques, including both organogenesis and somatic embryogenesis (SE) (Bonga et al. 2010). In recent years, the concept of multi-varietal forestry (MVF), defined as the deployment of a range of genetically tested varieties in plantation forestry, has been applied at the commercial level to a few conifer species. The advantages of MVF include the capture of much greater genetic gain than is possible by seed orchard breeding by exploiting both additive and non-additive genetic variations, the flexibility to rapidly deploy varieties and the ability to design diversity (Park and Bonga 2011). The identification of superior genotypes at a very early stage without phenotyping through genomic selection will provide new plant material to be used in forestry plantations (Park et al. 2016). In addition, new varieties also may be produced through genetic engineering.

According to the above considerations, the conservation of forest biodiversity should not be limited today to natural populations but also should include the great number of new and valuable genotypes identified and produced worldwide. Cryopreservation may be considered a complementary system for forest diversity conservation, and adequate procedures to carried out it are well defined (Häggman et al. 2008). In specific cases, cryopreservation is the most convenient system for long-term conservation as, for example, in the conservation of species producing recalcitrant seeds. Successful cryopreservation procedures for storage of embryo axes of chestnut (Corredoira et al. 2004; Vieitez et al. 2011, 2012) and pedunculate oak (Berjak et al. 2000) were defined, providing the possibility to develop cryobanks for long-term conservation of these species. For many angiosperm species, it is possible to combine rooting of cuttings with cryopreservation and organogenesis. Regeneration of plants from cryopreserved buds has been reported for *Populus tremuloides* (Aronen and Ryynänen 2014) and *Melia azedarach* (Yang et al. 2011). In addition, SE has become a popular system of clonal propagation

(Bonga 2016), as SE lines can be cryopreserved and retrieved in a viable state after cryopreservation. Long-term storage of these cultures and subsequent plant regeneration are an important part of clonal forestry for increasing productivity (Lambardi and De Carlo 2003). In some cases, somatic embryo germination with shoot-only elongation is the main response to germination treatments. Shoots excised from the partly germinated embryos can be multiplied, elongated and rooted via axillary shoot proliferation (Martínez et al. 2008; Ballester et al. 2016), supplementing the number of plants produced by embryo germination. Cryostorage and recovering of embryogenic cultures of different hardwood species today is well documented (see, for example, Vendrame et al. 2001; Lambardi and De Carlo 2003; Vieitez et al. 2011, 2012; Merkle 2016).

Genetic transformation experiments in forest trees are increasing, using, in most cases, somatic embryo cultures as the target material. Many transgenic lines are produced per experiment; each transgenic event must be considered as an independent embryogenic line. Cryopreservation of transgenic lines is of great importance during both the molecular analyses and field test experiments to identify the most responsive genotypes. After identification, the most promising genotypes are retrieved from cryostorage and the embryogenic lines multiplied under in vitro conditions. Successful cryopreservation of embryogenic transgenic lines has been reported for chestnut (Corredoira et al. 2016a) and oak (Mallón et al. 2013). Furthermore, Ryynänen et al. (2002) developed a cryopreservation protocol for transgenic silver birch lines using vegetative buds as material for storage. Jokipii et al. (2004) showed that it was possible to apply cryopreservation techniques to the storage of valuable transgenic lines of aspen using both in vitro shoot cultures and in vivo plants.

For many species, cryopreservation today is routine with good recovery rates, using rather simple procedures. However, the application of cryopreservation to recalcitrant hardwood forest species in the development of large gene banks will only be possible if in vitro tissue culture techniques are well defined. Only under this condition are long-term storage experiments justified. Exceptionally, in vitro conditions are not required for cryopreservation of a select number of temperate tree species (Bonnart et al. 2014). Pence (1992) applied a desiccation-based procedure to zygotic embryonic axes of different hardwood species, among which chestnut and oak were included. Although surviving axes underwent root pole elongation, no plantlets developed. One cause of the failure might be due to poorly defined in vitro culture conditions, since, at present, cryopreservation of embryonic axes and somatic embryos of both chestnut and oak is unproblematic (Vieitez et al. 2011, 2012). Some authors consider in vitro tissue culture a proper system for germplasm conservation, as it is an efficient procedure defined for the conservation of mature American elm trees (*Ulmus americana* L.) that have survived the epidemics of Dutch elm disease and are a potential source of disease resistance (Shukla et al. 2012). Alternatively, cold storage (2–4 °C) of cultures in the multiplication phase allows for extending the subculture period up to 12–16 months instead of the conventional period of 4 weeks, saving both labour and energy costs, as has been demonstrated in three different hardwood species (Janeiro et al. 1995). However, long- or medium-term maintenance under conventional conditions also involves a concomitant risk of contamination, somaclonal variation and loss of multiplication capacity. Thus, cryopreservation may be defined as the most secure method for long-term conservation.

As it will be described later, desiccation- and vitrification-based procedures have been used mainly to cryopreserve different types of explants in hardwood species. Any type of explant theoretically can be subjected to cryogenic. Cryopreservation of a core collection of 444 European elm (*Ulmus* spp.) clones in liquid nitrogen (LN) was carried out for conservation of elm genetic resources. The cryopreservation technique involved the stepwise freezing of cryotubes containing dormant buds collected from trees growing in the open (Harvengt et al. 2004). More recently, conservation of hybrid aspen germplasm also has been recorded using in vivo dormant buds as target material. The authors consider this procedure a convenient backup method for field collections (Aronen and Ryynänen 2014). As already mentioned, methods for cryopreservation of zygotic embryo axes are well defined (Berjak et al. 2000; Lambardi and De Carlo 2003; Corredoira et al. 2004; Vidal et al. 2010, 2011). In addition, shoot tips isolated from shoot proliferation cultures also have been subjected to cryopreservation. The size of the explant and the physiological conditions of the shoot are the most important conditions for successful cryopreservation (Tagaki et al. 1997; Vidal et al. 2005).

One concern related to cryopreservation methods is the genetic stability of regenerated plants due to, among other causes, the proper in vitro tissue culture technology and the mutagenic potential of the cryoprotectant dimethyl sulphoxide (DMSO). In general, there is no evidence of morphological or genetic alterations in forest trees as a result of cryopreservation (Häggman et al. 2008).

It is possible to conclude that cryopreservation is the only currently available technique that ensures safe, long-term conservation of genetic resources of plant species with recalcitrant seeds, of vegetatively propagated species, of endangered species and of biotechnology products, such as cell strains with special attributes, elite clones and genetically transformed material. In this review, we will describe the state of the art on the conservation of hardwood species, mostly related to Fagaceae trees. Topics addressed in this review include the description of the most interesting technical procedures applied to hardwoods, the selection of the most appropriate type of explant used for cryopreservation of these species, the response of the explants to the whole process and the control of the genetic stability of the post-cooled and regenerated plants.

14.2 Medium-term Conservation Through Slow Growth Storage

In vitro techniques have been used in combination with slow growth storage (also called 'minimal growth storage') for medium-term conservation of hardwoods (Table 1). Such techniques are of particular interest for conserving hardwood

Table 1 Medium-term conservation through slow growth storage of hardwood species

Species	Temperature (°C)	Light conditions	Months	Survival (%)	Reference
Alnus glutinosa	4	16h (8 μE $m^{-2}s^{-1}$)	18	75–87	San José et al. (2015a)
Betula pendula	4	Darkness	18–24	95	*
Castanea sativa	8	24h (30 μE $m^{-2}s^{-1}$)	48	82	Capuana and Di Lonardo (2013)
Castanea sativa	4	Darkness	12	85–93	*
Castanea sativa x C. crenata	2–4	16h (1.3 μE $m^{-2}s^{-1}$)	12	96	Janeiro et al. (1995)
Eucalyptus grandis	10	24h (4 μE $m^{-2}s^{-1}$)	6	70	Watt et al. (2000)
Fagus sylvatica	2–4	12h (1.3 μE $m^{-2}s^{-1}$)	18	70–90	Vieitez et al. (2003)
Fagus orientalis	2–4	12h (1.3 μE $m^{-2}s^{-1}$)	18	70–90	Vieitez et al. (2003)
Populus alba	2–4	16h (1.3 μE $m^{-2}s^{-1}$)	24	95	*
Populus alba x P. grandidentata	4	Darkness	60	25	Son et al. (1991)
Populus tremula x P tremuloides	4	Darkness	3	100	Hausman et al. (1994)
Populus tremula x P.tremuloides	2–4	16h (1.3 μE $m^{-2}s^{-1}$)	18	100	*
Prunus avium	2	16h (8 μE $m^{-2}s^{-1}$)	12	100	Janeiro et al. (1995)
Prunus serrulata	2–4	12h (8 μE $m^{-2}s^{-1}$)	12	100	*
Quercus alba	2–4	12h (8 μE $m^{-2}s^{-1}$)	12	100	*
Quercus bicolor	2–4	12h (8 μE $m^{-2}s^{-1}$)	12	100	*
Quercus petraea	2	16h (1.3 μE $m^{-2}s^{-1}$)	12	79	Janeiro et al. (1995)
Quercus robur	2–4	16h (1.3 μE $m^{-2}s^{-1}$)	12	100	Janeiro et al. (1995)
Quercus suber	5	Darkness	24	50	Romano and Martins(1999)
Ulmus minor	2–4	16h (8 μE $m^{-2}s^{-1}$)	12	100	*

*Unpublished results carried out in the authors' laboratory

species with recalcitrant seeds, for selected genotypes and for genotypes obtained after genetic transformation. This procedure is also interesting for those species for which long-term storage methods have not yet been described.

The objective of slow growth storage is to reduce the number of subcultures required, without affecting the viability of the explants nor their capacity to start growing again when normal growth conditions resume. Different strategies can be used for slow growth storage: cold storage, reduced oxygen levels (frequently achieved by covering the culture with a thin layer of oil), desiccation of the plant material and modification of the culture medium. Possible modifications include reduction in mineral elements or carbon source, addition of osmotically active compounds (e.g. chlorocholine chloride, abscisic acid) or growth retardants (e.g. mannitol), and/or changes in the nature and/or concentration of growth regulators (Kavani 2011; Cruz-Cruz et al. 2013). Cold storage is by far the most widely used approach for medium-term conservation of in vitro cultures because it is a simple method that does not require special equipment (only refrigerators). The reduction in temperature leads to decreased metabolic activity (i.e. reduction in respiration, water loss, wilting) and, therefore, reduced growth of the explants. The optimal storage temperature mainly depends on the species considered (Lambardi and De Carlo 2003). Cultures of hardwood species are usually stored at 2–5 °C, and cold storage is generally combined with low light intensity or even darkness (Table 1).

Other factors affecting the efficiency of slow storage include the type and physiological stage of the explants. Although several explant types have been used for cold storage, shoot tips and nodal segments are generally preferred for in vitro preservation, owing to the high survival and regrowth rates and high level of genetic stability (San José et al. 2015a). Survival of the stored material is greatly affected by the time between the last subculture under standard conditions and the placement of the material in cold storage (Engelmann 1991, 2011a).We have found that for optimal storage conditions of different oak, chestnut, alder, elm and cherry species, preculture of explants for 10–12 days under standard conditions is required prior to storage in low temperature, while, for poplar, birch and beech, at least 2 weeks of preculture are needed.

We are using a routine protocol consisting of the preculture of nodal segments (for 10–15 days) on the same medium as that used in the multiplication stage followed by storage in cool cabinets (340 L Sanyo Medicool MPR 311D) at 2–4 °Cwith 12 h per day of dim light (8 $\mu E\ m^{-2}\ s^{-1}$) or in darkness. Explants of all of the above-mentioned species can be maintained for at least 1 year without transfer to fresh medium (Table 1 and Fig. 1a,b). For beech, poplar, elm, alder, birch and some oak genotypes, the storage period can be extended to 18 months or more. After optimisations of the preculture time and cold storage conditions, survival rates between 85 and 100% have been recorded for transfer of explants to standard growth conditions (Janeiro et al. 1995; Vieitez et al. 2003, 2012; San José et al. 2015a).

We also are maintaining transgenic lines of poplar, oak and chestnut obtained in our laboratory, where evaluation is carried out. Transgenic poplar lines carrying genes that inhibit pollen formation, confer resistance to 2,4,6-trinitrotoluene (TNT) or exhibit increased tolerance to heavy metals are currently under cold storage, and a survival rate of 100% has been observed after at least 1 year in storage. In transgenic chestnut and oak lines transformed with genes that encode pathogenesis-related proteins, 100% survival has been observed after 6–9 months in cold storage.

Fig. 1 In vitro conservation of hardwood forest species. **a** Axillary shoot cultures of *Quercus robur* after 12 months of cold storage. **b** Axillary shoot cultures recovered after 12 months of cold storage and regrowth under standard conditions for 4 weeks. **c** Shoots developing from successfully cryopreserved shoot apices of *Alnus glutinosa*, after exposure to PVS2 solution for 30 min. **d** New somatic embryos developed from cryopreserved proembryogenic masses of *Quercus ilex*, following exposure to PVS2 solution for 30 min

14.3 Long-term Conservation Through Cryopreservation

Cryopreservation is defined as the storage of living material at extremely low temperatures in LN (−196 °C) or in seldom its vapour phase (−150 °C). The technique involves immersing plant cells, tissues or organs in LN. At this temperature, the cells enter a state of 'absolute quiescence', as all cell divisions and metabolic processes are almost totally halted, enabling conservation for a theoretically unlimited period of time (Engelmann 2004). Most vegetative explants subjected to cryopreservation contain high levels of free water in their cells and are therefore highly sensitive to damage caused by the ice that forms during the freezing process. The plant material is usually dehydrated to protect it from the damage caused by conversion of the intracellular water to ice. Classic and modern methods can be used for this purpose. These methods are differentiated by the techniques used and by the physical mechanisms on which they are based. In classic procedures, the material is dehydrated by freezing, while today, cellular dehydration is induced before freezing by exposure of the samples to concentrated cryoprotectants and/or desiccation.

14.3.1 Cryopreservation Techniques

14.3.1.1 Slow Cooling (Classic Method)

The first method to be widely used for the cryopreservation of plant tissue was the classic or conventional 'slow cooling', 'slow freezing' or 'two-step' method. This procedure involves the use of a programmable freezer to achieve a slow decrease in the temperature (usually 0.1–1 °C min^{-1}) to −35 or −40 °C before the samples are immersed in LN (Pence 2014). The slow decrease in temperature causes the formation of extracellular ice; a differential water gradient is created across the membrane and intracellular water moves to the outside (Benson 2008). This process reduces the amount of water available to form ice and is referred to as freeze-induced desiccation. The speed of cooling and freezing determines the amount of water that moves to the extracellular space before the cell contents solidify. However, excessive dehydration may cause damage due to an increase in the concentration of salts and changes in the cell membrane. In addition to strict control of the rate of freezing, a cryoprotectant such as DMSO can be added. The use of cryoprotectants has improved the results of the slow cooling method in many species (Reed and Uchendu 2008). Finally, the defrosting process must be carried out as quickly as possible to prevent recrystallisation of the water, as the crystals that form during this stage are larger and more damaging.

The procedure is much more complex and expensive than non-precooling methods, as it requires programmable freezers and is technologically more demanding. A more economical alternative is the use of Mr. Frosty™ freezing containers or other similar domestic appliances containing methanol or isopropanol. In this case, the container or device containing isopropanol is placed in a freezer (−70 °C), the temperature within the container decreases by approximately 1 °C min^{-1}, and the samples are then immersed in LN. The slow cooling procedure has been used successfully with cell suspensions, calluses and apices excised from in vivo and in vitro cultures of cold-tolerant species, among them hardwood species (Engelmann 2011a).

14.3.1.2 Vitrification-Based Procedures

The new freezing techniques developed are based on the vitrification process, defined as the transition of water directly from the liquid phase to an amorphous or metastable phase, thus preventing the formation of ice crystals (Fahy et al. 1984). In this case, the solid that forms is actually a highly viscous supersaturated solution with the mechanical properties of a solid, although no crystalline structure is formed. Transition to the vitreous state does not imply chemical but rather physical changes in the viscosity of the liquid. As mentioned above, to achieve this, techniques have been developed in which the plant material is exposed to concentrated cryoprotectant solutions and/or desiccation before direct freezing in LN. The development of vitrification-based techniques has enabled the cryopreservation of a large number of plant species, including many hardwood species. These techniques are generally more suitable for differentiated explants (such as apices and somatic embryos) that are formed by various types of specialised cells, and each has different temperature requirements, which hampers application of the classic method.

Of the different vitrification-based methods available, vitrification, desiccation, encapsulation–dehydration and droplet–vitrification have been applied to hardwood forest species.

Vitrification Vitrification is easy to use and highly reproducible, and a wide range of species (including hardwood species) already have been cryopreserved successfully using this procedure (Engelmann 1997, 2011a; Kaczmarczyk et al. 2012; Pence 2014). As such, the vitrification procedure probably has been one of the most commonly used in recent years. In general, this technique involves treatment of samples with cryoprotective substances (usually called loading solutions), dehydration in a highly concentrated vitrification solution and ultra-rapid cooling in LN. After storage in LN, the plant material is rapidly defrosted, the vitrification solution is removed, and the material is finally cultured in a suitable recovery medium.

Various compounds can be used as cryoprotectants, including glycerol, mannitol, sorbitol, DMSO, ethylene glycol, polyethylene glycol (PEG) and sugars (i.e. glucose, sucrose). These compounds are responsible for protecting plant tissues from the potential damage produced by cryopreservation. They can be used either alone or in combination, forming parts of loading and vitrification solutions. Loading solutions contain cryoprotectants (usually sucrose, glycerol and ethylene glycol), although at lower concentrations than in vitrification solutions, and loading solutions are applied for short periods of time (5–90 min, depending on the type of material), with the aim of increasing the tolerance to dehydration subsequently induced by the vitrification solution. The most commonly used loading solution is a mixture of 2 M glycerol and 0.4 M sucrose, as described by Matsumoto et al. (1994). Vitrification solutions are highly concentrated and with more complex mixtures of cryoprotectants. The most commonly used is the vitrification solution 'plant vitrification solution 2' (PVS2), which consists of glycerol (30% w/v), ethylene glycol (15% w/v) and DMSO (15% w/v) in liquid medium with 0.4 M sucrose (Sakai et al. 1990).

In vitrification protocols, optimisation of the time and temperature of exposure to PVS2 is essential in order to induce vitrification without invoking osmotic stress or toxicity in the cultures. The time of application of PVS2 varies depending on the type of explant and species, although periods of 15–120 min are usually used. The use of low temperatures (0–4 °C) has been found to be optimal, as this enables the cryoprotectant to be incorporated gradually, thus minimising damage. In addition to loading solutions, the samples also can be preconditioned in order to increase their tolerance to PVS2. The most popular preconditioning treatments are cold acclimation/cold hardening (mainly for temperate species) and preculture on a medium with high concentrations of sugar; these treatments can be applied simultaneously, successively or individually (Sakai and Engelmann 2007). Pretreatment by cold hardening has been described for excised cultures but also for stock plants and is sometimes combined with a reduction in light intensity and/or inclusion of abscisic acid in the culture medium. Although different sugars (glucose, sorbitol, sucrose or fructose) are included in pretreatment media, sucrose (usually 0.3–0.7 M for 1–3 days) is the most commonly used. The aim of these stepwise treatments is to increase the accumulation of sugars and protective proteins in order to increase the tolerance to PVS2 (Kavani 2011).

Droplet vitrification is a cryopreservation procedure derived from vitrification. Although this procedure has been used increasingly in recent years, to date it has been applied scarcely to forest hardwoods. In the droplet vitrification procedure, samples are loaded, treated with the vitrification solution, frozen in a microdroplet (1–2 μL) of vitrification solution and placed on aluminium foil strips, which are immediately immersed in LN (Sakai and Engelman 2007). The high thermal conductivity of aluminium foil together with the small volume of the vitrification solution yields higher cooling and warming rates (Cruz-Cruz et al. 2013).

Desiccation (air drying) Desiccation is one of the simplest and easiest procedures, as it only involves the dehydration of plant material and direct immersion in LN. Desiccation can be carried out in a laminar airflow cabinet or over silica gel or by using a flow of sterile compressed air (flash drying). Although desiccation over silica gel or with sterile compressed air is highly recommended, as these methods are more precise and reproducible than air desiccation, the latter is the most commonly used method. Desiccation is, above all, used with pollen, orthodox seeds, zygotic embryos and embryogenic axes excised from recalcitrant seeds. In the case of embryos, the main parameters to consider are the stage of development, size and water content. The optimal moisture content, which usually ranges from 20–30%, is achieved by 4–5 h of desiccation in a laminar airflow cabinet at room temperature.

Encapsulation–dehydration The encapsulation–dehydration technique is a variant of the air desiccation technique. The procedure is based on the production of artificial or synthetic seeds. For encapsulation, explants first are loaded into sodium alginate (usually 3% w/v). The plant material and alginate solution are then aspirated with a pipette and added dropwise to a solution containing 100 mM calcium. The calcium causes polymerisation of the alginate around the explant and the formation of beads, each containing one sample. The beads are then precultured in medium enriched with sucrose (at concentrations between 0.75 and 1.25 M) for 24–72 h to increase the intracellular solute concentrations and promote desiccation. After the preculture period, the beads are partially desiccated by drying, generally, in a laminar airflow cabinet or over silica gel, to approximately 20–30% of water content (WC), before being immersed in LN. Although the idea of encapsulation is to protect the explants from handling stress, the technique also has the advantage in that growth regulators (e.g. abscisic acid) and/or nutritional compounds can be added to the alginate matrix to help improve the recovery rate and to overcome the toxicity of the cryoprotectants.

Encapsulation–dehydration methods are continuously being improved and are, along with vitrification procedures, frequently used for cryopreservation of vegetatively propagated plants; however, they have not been widely applied to hardwood forest species.

14.3.2 Selection of Explants for Cryopreservation

In developing a cryopreservation system to deal with genetic diversity, an initial question is what propagule to cryopreserve. In principle, all parts of hardwoods may be stored in LN: callus cultures, dormant buds, apical meristems, embryonic axes, seeds, somatic embryos and pollen. Meristematic tissues are the preferred explants for cryopreservation of vegetatively propagated species, as somaclonal variation is lower in organised tissues. The propagules may be collected under either in vivo or in vitro conditions.

14.3.2.1 In Vivo Propagule Collection

In vivo-collected embryonic axes or dormant buds are currently amenable to cryopreservation and can be useful for conservation of hardwood species. Plant material often must be adapted to, or exist in, a desirable physiological state before the propagule is harvested, especially for the cryostorage of clonal lines. Seeds of recalcitrant hardwoods need to be collected at an appropriate maturity, which may represent a rather narrow time window in their development. Furthermore, those species that possess considerable cold hardiness are easier to cryopreserve. Isolated buds, nodal sections or whole scions can be cooled; twigs can be rooted, only if this ability exists, as in the model trees poplar and salix; or cryostored buds can be recovered by grafting or under in vitro culture conditions (Towill 2002). Only for specific accessions of model trees is in vitro tissue culture not required, but, for most hardwood species, this technology is essential.

Embryonic axes Seeds of many hardwood species have recalcitrant or intermediate storage behaviour and therefore cannot be stored dry at low temperature as can orthodox seeds. Others, such as silver birch or poplar, have orthodox seeds, but their viability decreases during cold storage. The most promising method for long-term conservation of intermediate and recalcitrant-seeded species is cryopreservation (Ballesteros et al. 2014). However, research in this field is still at a very preliminary stage (Engelmann 2011a), with this fact being especially true for forest hardwoods (Table 2). To date, few successful studies have been reported.

Table 2 Conservation of hardwood recalcitrant seeds by cryopreservation

Species	Explant type	Pretreatment	Desiccation procedure	Storage in LN	Thawing procedure	Recovery medium	Survival (%)	Plant recovery (%)	Reference
Acer saccharinum	Embryonic axes	20 µM ABA and tetcyclacis	Desiccation with sales	Precooled in a programable freezer from 0–40 °C at rate of −0.33 °C min^{-1}	Water bath at 40 °C	WPM	nm	50–54	Beardmore and Whittle (2005)
Acer pseudoplatanus	Embryonic axes with 0.5 mm cotyledon	Seed stratification at 3 °C 8–11 w	Fast desiccation (15–20%) with air drying and over activated silica gel	Precooled in a programable freezer from 0–40 °C at rate of −0.25 °C min^{-1}	Water bath at 10 °C for 30 s, washed in 0.1 M suc, and stored at 0 °C for 20 h	WPM	70	50	Pukacki and Juszczyk (2015)
Castanea sativa	Embryonic axes	–	Desiccation in the laminar airflow cabinet (20–24%)	Direct immersion in LN	Water bath at 40–42 °C for 2 min	MS (1/2 Nitrates) + BA 0.2 mg/l	93–100	63	Corredoira et al. (2004)
Juglans regia	Embryonic axes	Propanediol and suc	Desiccation in the laminar airflow cabinet (20%)	Direct immersion in LN	Water bath at 40–45 °C	DKW + BA 1 mg/l	85	75	de Boucaud et al. (1991)

(continued)

Table 2 (continued)

Species	Explant type	Pretreatment	Desiccation procedure	Storage in LN	Thawing procedure	Recovery medium	Survival (%)	Plant recovery (%)	Reference
Juglans cinerea	Embryonic axe with 3 mm cotyledon	–	Desiccation in the laminar airflow cabinet (4.8%)	Direct immersion in LN	Water bath at 40–40 °C 5 min	WPM	nm	36	Beardmore and Vong (1998)
Quercus faginea	Embryonic axes	–	Desiccation in the laminar airflow cabinet (21%)	Direct immersion in LN	Water bath at 40 °C for 1 min	WPM + BA 0.1 mg/l	nm	60	González-Benito and Pérez-Ruiz (1992)
Quercus robur	Plumules (0.5–1.0 mm)[1]	0.75 M suc + 1 M suc + 1.5 M glyc[2]	Desiccation over activated silica gel 10 min	Direct immersion in LN	1.2 M suc solution bath at 42 °C 40 min and 0.5 M suc solution 40 min at 23 °C	WPM + BA 0.8 mg/l	5–71	8–20[3]	Chmielarz et al. (2011)

[1] Shoot apical meristems of embryos; [2] Each solution 40 min at 23 °C; [3] Only shoot development. Abbreviations: *BA* benzylaminopurine; *DKW* Driver and Kuniyuki (1984); *glyc* glycerol; *LN* liquid nitrogen; *MS* Murashige and Skoog (1962); *nm* not mentioned; *suc* sucrose; *WPM* Lloyd and McCown (1980); – not applied

Protocols must be developed for disinfection, in vitro inoculation, germination and development of plantlets as stages prior to storage in LN. The moisture content of the seeds is the most critical factor for a successful cryopreservation protocol (Pammenter and Berjak 1999; Pritchard 2007). Desiccation technique is the most commonly used procedure prior to cryopreservation in LN of seeds. In hardwoods, prior to cryopreservation, explants have been dehydrated by air desiccation, either in a laminar airflow cabinet or over activated silica gel. The optimum moisture content for cryopreservation ranges from 4.8 to 24%, depending on the species. In all reports, cryopreservation has been performed using embryonic axes (Table 2).

It is widely accepted that embryonic axes are the most appropriate explant because, in recalcitrant seeds, axes tolerate much more severe dehydration than does the entire seed (Berjak et al. 2000). Isolation of the embryonic axis in this type of seed occurs because the seeds are relatively large and mainly consist of the cotyledons. The embryonic axis represents a small proportion of the volume and is sufficiently small to be rapidly dehydrated and theoretically successfully cryopreserved. In the majority of papers, embryonic axes are directly immersed in LN into polypropylene cryovials, and they are warmed using the fast warming method—immersing cryovial envelopes in a water bath at 40–42 °C for 1–5 min. 'Woody Plant Medium' (Lloyd and McCown 1980) with or without benzyl-amino-purine (BA) is the medium most commonly used to germinate embryonic axes, with the exception of embryogenic axes of European chestnut and walnut, where MS medium (Murashige and Skoog 1962) with half the nitrates and DKW medium (Driver and Kuniyuki 1984) are utilised, respectively.

In Fagaceae species, successful cryostorage of European chestnut zygotic embryonic axes has been reported (Corredoira et al. 2004). Briefly, chestnut seeds without their external coat are sterilised by successive immersion in 70% ethanol for 2 min and a 5% solution for 30 min. The embryonic axes are aseptically dissected from the surrounding cotyledons and transferred to empty Petri dishes (20–25 axes per dish). Then, the embryonic axes are desiccated in the open Petri dishes in a laminar airflow cabinet until their moisture content is reduced to 20–24% of fresh weight. The desiccated axes are transferred to 2-mL cryovials and rapidly plunged into LN. For thawing, the vials are immersed for 2 min in a water bath at 40–42 °C. The axes are cultured in tubes containing recovery medium consisting of MS medium with half the nitrates. After 6 weeks of culture, between 93 and 100% of excised embryonic axes survived storage in LN, and some 63% subsequently developed as whole plants.

Zygotic embryo axes of pedunculate oak have been cryopreserved using the simplest desiccation procedure, using flash drying and direct immersion in LN, which yielded appreciable germination recovery rates (Berjak et al. 2000). These rates were not improved when more complex dehydration procedures were applied (Chmielarz et al. 2011). In this report, plumules consisting of shoot meristem surrounded by leaf primordia were desiccated over activated silica gel and were cryopreserved by being placed intocryovials and immersed directly into LN. Using this procedure, only 8–20% of cryopreserved plumules developed shoots, which should be rooted to obtain whole plants. Better results were reported with embryonic axes of *Quercus faginea* (González-Benito and Pérez-Ruiz 1992). The embryogenic axes are desiccated either under sterile conditions in a laminar airflow cabinet for different periods of time or by being soaked in 15% DMSO for 1 h and subsequently immersed in LN. The best germination response after freezing (60%) was observed when embryonic axes were desiccated from 53 to 21% moisture content (on a fresh weight basis).

Unorganised post-cooled growth of axes was reported for cork oak, holm oak and beech species (González-Benito et al. 2002), while only a slight elongation of the hypocotyls has been recorded for American beech embryonic axes (Pence 1990).

Dormant buds Procedures for cryostorage of dormant buds for which in vitro technologies are not required have been proposed for *Fraxinus* (Volk et al. 2009), *Populus* and *Salix* (Bonnart et al. 2014). In the three cases, the methodology is similar. Briefly, scions (between 5 and 15 mm in diameter) containing vegetative buds from field-grown trees are collected during winter. The bud wood is cut into 4–6-cm nodal sections with 2–3 buds per section. Ten sections are stacked and sealed in polyolefin tubing. The tubes are cooled to −35 °C for 24 h and then placed into LN. Cryopreserved sections are warmed to 5 °C for 24 h and either grafted in appropriate rootstocks or rooted after sections are dipped in rooting hormone solution. Recovery percentages ranged from 34–100% in *Fraxinus* and 42–100% in *Populus trichocarpa* and were 40% in *Salix*. The authors claim that dormant buds are often the most economical method for cryopreservation when adequate quantities of appropriate source materials are available. In our opinion, however, the procedure only will be valid for a small number of species/accessions/genotypes within hardwood trees.

According to Jokipii et al. (2004) and Aronen and Ryynänen (2014), cryostorage of in vivo dormant buds of hybrid aspen (*Populus tremula* x *P. tremuloides*) is the most suitable method as a backup for field collections. However, contrary to the preceding examples, these authors propose the use of in vitro tissue culture techniques as the most convenient for plant regeneration. The procedure is based on previous experiments using dormant silver birch buds as target material (Ryynänen 1996). Similar protocols also have been applied to cryostorage of transgenic lines of birch (Ryynänen et al. 2002) and hybrid aspen (Aronen and Ryynänen 2014). Briefly, axillary and apical buds with scales and a short piece of attached twig (1 cm) are collected between August and February and sealed in 2-mL cryotubes, 17–20 buds per tube. The cryotubes are kept at 0 °C for 24 h and then frozen to the prefreezing temperature of −38 °C using a controlled-rate freezer. The tubes are then immersed in LN for 1–7 months. The buds are thawed and sterilised, and the scales, young leaves and attached twigs are removed from the buds. The innermost leaves with the remaining meristems are then cultured on appropriate culture media. At least 75% of the buds regenerated through micropropagation, and there was no difference compared to non-frozen controls (Aronen and Ryynänen 2014).

Regardless of the method used for plant regeneration after cryopreservation, the procedures already described were applied to a small number of genotypes per species in order to demonstrate the validity of the technology. However, development of large gene banks of cryopreserved hardwoods is limited, to our knowledge, to the example of European elms (Harvengt et al. 2004). Cryopreservation of a core collection of 444 elm (*Ulmus* spp.) clones in LN was carried out by two laboratories participating (to duplicate the collection) in a European project of conservation of elm genetic resources. Plant material, collected from nine European countries, represented a large sample of the genetic diversity within three European elm species (*Ulmus minor*, *U. laevis* and *U. glabra*) and their hybrids. The technology applied is well defined in Harvengt et al. (2004). From a random sample of cryopreserved dormant buds of the three major species, in vitro regrowth was assayed. For *U. minor*, *U. minor* x *U. glabra* and *U. laevis*, all clones were easily regenerated with standard techniques, but this was not the case for *U. glabra*. Here, only a few explants started to show signs of development, and, in all cases, they failed to grow. To solve this problem, the authors utilised the micrografting technique, using vigorous *U. minor* rooted shoots as rootstocks. Plants regenerated from material cryopreserved for up to 5 years did not exhibit any loss in regrowth capacity over time, as compared with fresh samples. These results suggest that cryopreservation may be suitable for long-term conservation of elm buds. The authors concluded that dormant bud cryopreservation of elms is a very economically competitive solution with respect to the classic technique of field clonal conservation archives (Harvengt et al. 2004).

Unfortunately, these operative methods could not be applied to all hardwood species. In our experience (unpublished results), the cryopreservation of in vivo dormant buds of recalcitrant species, such those of the genera *Castanea* or *Quercus*, is very difficult, as, in general, only a few explants started to show signs of development, and they fail to growth. This situation is critical if the dormant buds are collected from mature trees. Consequently, other alternatives should be applied.

14.3.2.2 In Vitro Propagule Collection

The approach of cryopreserving tissues from in vitro shoot multiplication cultures can be an alternative when the response of zygotic embryos or dormant buds to cryostorage is unsatisfactory or not feasible (Pence 2014). Shoot tips and/or buds (excised from apical or axillary buds of in vitro shoot cultures or from shoots of in vitro-germinated zygotic embryos) and embryogenic cultures (i.e. embryogenic callus, somatic embryos and cell suspensions) are the specimens commonly used for cryopreservation of vegetatively propagated species, including hardwood species (Fig. 1c,d). These types of explants are available throughout the year and are easy to manipulate; they may be produced in large amounts, and, consequently, they provide many clonal replicates for the different experiments necessary to optimise the cryogenic protocols. An important point for successful cryopreservation is closely related to the definition of effective protocols for proliferation of donor cultures and for recovery of shoots/somatic embryos after LN storage. Another critical point is the physiological state of the propagules: it is important to use donor shoots and somatic embryos in an optimal physiological stage, sampling only high-quality in vitro plants and somatic embryos to ensure successful recovery post-cryopreservation (Benson and Harding 2012).

Shoot apices During recent years, cryopreservation protocols using shoot apices have been established for many woody species, including fruit trees and several hardwood forest species (Table 3). In addition, shoot apices derived from genetically transformed shoot cultures of the hardwoods *Betula pendula* (Ryynänen et al. 2002) and *P. tremula* x *P. tremuloides* (Jokipii et al. 2004) also have been cryopreserved.

The size of the shoot and its physiological stage are critical components for successful regrowth of cryopreserved shoots. Cold hardening of in vitro donor plants often shows a positive effect, especially for cold-tolerant species. The exposure to low temperatures increases the intrinsic tolerance to ultra-low temperatures by triggering genes responsible for cold stress (Takagi 2000). In hardwood species, hardening is achieved when cultures are grown at 5 °C in darkness or under a short photoperiod for 14–21 days (Table 3), although in some species, such as cherry, a cultivation period of 45 days in cold conditions is required (Niino et al. 1997). Another option is to induce cold tolerance of excised shoots by cultivation at 4 °C for 2 weeks (Vidal et al. 2005, 2010).

The most used specimen for cryopreservation is the apical/terminal meristem plus a couple of leaf primordia measuring 0.5–2 mm in length. Also, axillary buds of birch showed a good recovery rate (Ryynänen and Aronen 2005a). Shoot tips are cryopreserved by applying slow cooling procedures, encapsulation–dehydration and vitrification-based cryopreservation techniques (Table 3). Slow cooling techniques, involving slow prefreezing followed by immersion in LN, are applied exclusively to shoot tips of birch, following a similar procedure to that described for dormant buds. The encapsulation–dehydration procedure has, however, been applied scarcely to cryopreservation of shoot tips derived from hardwood species (Table 3). In these reports, shoot tips are encapsulated in sodium alginate (3% w/v), and then the beads are cultured in increased sucrose solutions, dehydrated in an air flow or over silica gel (20–25% WC) and rapidly cooled in LN.

Shoot tips of hardwood species have been cryopreserved mainly using the vitrification-based techniques. Shoot tips, excised from axillary shoot cultures treated with or without cold hardening, are precultured on media containing high sucrose concentrations (0.09–7 M), generally in combination with cold storage at 5 °C, as shown in Table 3. Pretreatment with osmotic agents such as sucrose desiccates tissue prior to cryopreservation, and pretreatment also can stimulate some cold acclimation responses (Benson and Harding 2012). After being precultured, shoot tips are exposed to cryoprotectants. Samples are usually treated with a loading solution with an intermediate concentration of cryoprotectants (frequently 2 M glycerol and 0.4 M sucrose, Matsumoto et al. 1994), followed by dehydration with highly concentrated vitrification solutions. Then, the shoot tips are rapidly immersed in LN. The vitrification solution PVS2 is largely used with these explants of hardwood species, and the exposure time varied from 25–120 min,

Table 3 Cryopreservation of shoot-tips isolated from in vitro shoot multiplication cultures of hardwood species

Species	Origin (number of clones)[1]	Preculture shoot donor plants	Explant type (lengh mm)	Preculture of dissected shoot tips	Loading treatment	Precryogenic treatment	Cryogenic treatment	Shoot recovery (%)	Reference
Alnusglutinosa	Adult (3)	–	Apical shoot tips (0.5–1 mm)	0.2 M suc at 4 °C 2d	2 M glyc + 0.4 M suc at 25 °C 20 min	PVS2 (half strength) at 0 °C 30 min + PVS2 (full strength) at 0 °C 30 min	Direct immersion in LN	50	San José et al. (2014)
Betula pendula	Adult (5)	Cold hardening + 100 μM ABA (at 5 °C 21–28d)	Apical shoot tips (0.3–0.4 mm)	5% DMSO + 100 μM ABA (at 5 °C 3d)	–	10% PEG + 10% glucose + 10% DMSO at 0 °C 30 min	Slow cooling	6.4–87	Ryynänen (1996, 1998), Ryynänen and Häggman (2001), Ryynänen and Aronen (2005b)
Betula pendula	One-two year old (2nonGM and 4GM)	–	Stem buds, apical and axillary buds	–	–	–	Slow cooling	72–100	Ryynänen et al. (2002)
Betula pendula	Adult	Cold hardening (at 5 °C 28d)	Axillary buds	0.7 M suc (at 5 °C 1d)	2 M glyc + 0.4 M suc at 25 °C 20 min	PVS2 at 0 °C 120 min	Direct immersion in LN	71	de Ryynänen and Aronen (2005a)

(continued)

Table 3 (continued)

Species	Origin (number of clones)[1]	Preculture shoot donor plants	Explant type (lengh mm)	Preculture of dissected shoot tips	Loading treatment	Precryogenic treatment	Cryogenic treatment	Shoot recovery (%)	Reference
Castanea sativa	Juvenile (3)/ adult (2)	Cold hardening terminal shoot buds (10 mm) (at 4 °C 14d)	Apical shoot tips (0.5–1 mm)	Excised shoot apices in 0.2 M suc (at 4 °C 2d)	2 M glyc + 0.4 M suc at 25 °C 20 min	PVS2 at 0 °C 120 min	Direct immersion in LN	38–54	Vidal et al. (2005)
Castanea sativa	Adult (46)	Cold hardening terminal shoot buds (10 mm) (at 4 °C 14d)	Apical shoot tips (1–2 mm)	Excised shoot apices in 0.2 M suc (at 4 °C 2d)	2 M glyc + 0.4 M suc at 25 °C 20 min	PVS2 at 0 °C 120 min	Direct immersion in LN	2–53	Vidal et al. (2010)
Eucalyptus sps.	Juvenile (13)	–	Apical shoot tips (1–1.5 mm)	0.25 M suc (1d)→ 0.625 M suc (1d)	–	Dropletvitrification PVS2 at 0 °C 60 min	Direct immersion inLN	43.4–84.8[2] 68.6[3] 37.9[4] 46.7–58[5] 43.3–82.5[6]	Kaya et al. (2013)
Eucalyptus grandis x E. camaldulensis	(1)	–	Axillary bud	Encapsulation 3% alginate→4 M suc + 0.4Mglyc (3d)→ 0.7M suc + 0.7M glyc (3d)	–	Desiccation in laminar air flow 3–5 h ($\leq$ 25% WC)	Slow cooling	44–96	Blakesley and Kiernan (2001)

(continued)

Table 3 (continued)

Species	Origin (number of clones)[1]	Preculture shoot donor plants	Explant type (lengh mm)	Preculture of dissected shoot tips	Loading treatment	Precryogenic treatment	Cryogenic treatment	Shoot recovery (%)	Reference
Fraxinus excelsior	Adult/Juvenile	Cold hardening (at 5 °C 10d)	Apical shoot tips	0.8 M glyc (2d)	2 M glyc at 22 °C 20 min	PVS2 at 0 °C 25 min	Direct immersion in LN	67–73	Schoenweiss et al. (2005)
Populus alba	(1)	Cold hardening (at 5 °C 21d)	Apical shoot tips and buds (1–2 mm)	0.09 M suc (at 5 °C 2d)	2 M glyc + 0.4 M suc at 25 °C 20 min	PVS2 at 0 °C 60 min	Direct immersion in LN	25–93	Lambardi et al. (2000)
Populus tremula x tremuloides	Adult (1 nonGM and 3GM)	Cold hardening (at 5 °C 21d)	Apical segments and buds (2–4 mm)	0.09 M suc (at 5 °C 3d)	2 M glyc + 0.4 M suc at 0 °C 20 min	PVS2 at 0 °C 30 min	Directimmersion in LN	2.5–75	Jokipii et al. (2004)
Prunus avium	Adult (5)	Cold hardening (at 5 °C 45 d)	Apical shoot tips (3 mm)	0.7 M suc (at 5 °C 1d)	–	PVS2 at 5 °C 120 min	Direct immersion in LN	70–80	Niino et al. (1997)
Prunus avium	Adult (1)	–	Apical shoot tips (2–3 mm)	0.3 M suc (at 24 °C 1d)	5% DMSO + 5% glyc + 5% suc at 0 °C 20 min	PVS2 at 0 °C 60 min	Direct immersion in LN	77.8	Shatnawi et al. (2007)

(continued)

Table 3 (continued)

Species	Origin (number of clones)[1]	Preculture shoot donor plants	Explant type (lengh mm)	Preculture of dissected shoot tips	Loading treatment	Precryogenic treatment	Cryogenic treatment	Shoot recovery (%)	Reference
Prunus avium	Adult (1)	–	Apical shoot tips (2–3 mm)	Preculture of shoot tips in 0.3 M suc (25 °C 1d)→ Encapsulation in 3% sodium alginate→beads cultured 0.75 M suc (1d)	–	Desiccation beads over silica gel (20–22% WC)	Direct immersion in LN	76.3	Shatnawi et al. (2007)
Salix hybrid	(1)	–	Apical shoot tips (0.7–1 mm)	Encapsulation in alginate gel→preculture bead in 0.4 M suc (15h)	–	Desiccation beads in laminar airflow 3–5h	Direct immersion in LN	8–83	Blakesley et al. (1996)

[1] Number of clones evaluated is given in parentheses, when reported in cited reference; [2] *E. grandis x camaldulensis*; [3] *E. camaldulensis* [4] *E. grandis*; [5] *E. grandis x urophylla*; [6] *E. urophylla x grandis*. Abbreviations: *ABA* abscisic acid; *DMSO* dimethylsulphoxide; *GM* genetically modified; *glyc* glycerol; *LN* liquid nitrogen; *nonGM* not genetically modified; *PEG* polyethylene glycol; *PVS2* plant vitrification solution 2; *suc* sucrose; *WC* water content; – not applied

always at 0 °C (Table 3). For hardwood species, the use of the droplet–vitrification technique is still applied sporadically and limited, to our knowledge, to five *Eucalyptus* species (Table 3), obtaining better results than with the encapsulation–dehydration procedure (Kaya et al. 2013).

In Fagaceae species, cryopreservation using shoot tips as target explants has been reported only in chestnut. The most efficient method for chestnut shoot regrowth employed 0.5–1 mm shoot tips isolated from 1-cm-long terminal buds excised from 3–5-week-old shoot cultures and cold hardened at 4 °C for 2 weeks. After preculture of isolated shoot tips at 4ºfor 2 days, they are treated for 20 min with a loading solution (2 M glycerol + 0.4 M sucrose) and 120 min at 0 °C with PVS2 solution prior to immersion in LN. To recover the cryostored shoot tips, the cryotubes are rapidly rewarmed in a water bath at 40 °C for 2 min, and the PVS2 solution is drained off and replaced with liquid Gresshoff and Doy medium (1972) containing 1.2 M sucrose for 20 min. Shoot apices are then transferred to sterilised filter paper discs placed in Petri dishes over recovery medium consisting of basal medium supplemented with 2.2 μM BA, 2.9 μM 3-indoleacetic acid and 0.9 μM zeatin. Using this protocol, 38–54% shoot recovery rates were achieved, and, in all cases, plant regeneration also was obtained (Vidal et al. 2005).

This procedure was of great interest for the establishment of a cryopreserved gene bank of European chestnut (Vidal et al. 2010). The vitrification-based procedure previously defined was applied to 46 chestnut genotypes selected for their resistance to ink disease. Forty-three out of 46 genotypes evaluated survived the freezing process, and 63% recovered their capacity to produce new shoots. The results achieved demonstrated the feasibility of cryopreservation for long-term conservation of European chestnut germplasm, including genetically transformed lines using in vitro-grown shoot tips.

Embryogenic cultures As mentioned for shoot apices, cryopreservation has been performed with many embryogenic cultures of woody plants (Engelmann 2011b). Nevertheless, most of these protocols have been defined for fruit trees, and only a few papers report long-term cryostorage of somatic embryos in hardwoods (Table 4). In our experience, the developmental stage of the somatic embryos is a key factor for obtaining high embryo recovery rates after cryopreservation. A careful selection of somatic embryos at early developmental stages is necessary to achieve survival after LN immersion. For example, clumps of 3–4 globular- or heart-stage cork oak somatic embryos showed better embryo recovery rates after LN storage than more differentiated, cotyledonary-stage embryos, in which cells exhibit higher levels of vacuolisation and differentiation than do the active embryogenic cells at globular/early torpedo stages (Valladares et al. 2004). In holm oak, when globular embryos were used as a source of explants, embryo recovery was only observed after 24 h in LN, and their differentiation ability was completely lost after 4 weeks in LN (Barra-Jiménez et al. 2015). However, working with the same embryogenic holm oak lines and the same protocol, we demonstrated that long-term cryostorage was feasible if pro-embryogenic masses instead of globular embryos were utilised as the target material (unpublished results). In most reports, somatic embryos are precultured on a conditioning medium with high sucrose concentrations as a means to induce freeze tolerance prior to cryogenic storage. As shown in Table 4, the precryogenic treatments are mainly limited to the vitrification procedure or desiccation of somatic embryos in a laminar airflow cabinet. There is only one case of application of the encapsulation–dehydration technique (Fernandes et al. 2008). In almost all reports in which vitrification and desiccation techniques were compared, the best results of embryo recovery were obtained with the vitrification procedure (Table 4). Among vitrification solutions, DMSO at different concentrations or, especially, PVS2, was the solutions most frequently used. The temperature as well as the duration of exposure to the vitrification solution PVS2 has been evaluated in several experiments. In most cases, the optimal exposure time to PVS2 was 60–90 min at 25 or 0 °C (Table 4). Freezing of embryogenic explants was usually rapid, but some positive results also have been obtained using slow cooling.

Significant progress has been carried out during recent years on the cryopreservation of somatic embryos in Fagaceae species. Embryogenic lines were induced previously from selected elite trees. In addition, improved lines also were obtained through genetic transformation of somatic embryos. The pool of embryogenic cultures was subjected for cryopreservation experiments as a proper system of long-term conservation (Corredoira et al. 2014, 2016a). Cryopreservation was successfully applied to somatic embryos of *Castanea sativa*, *Quercus robur*, *Q. suber* and *Q. alba* following vitrification-based procedures. The basic procedure is a combination of sucrose pretreatment, dehydration with PVS2 solution and rapid freezing. In brief, embryo clumps of globular- and/or heart-stage embryos isolated from stock embryogenic cultures (3–4 weeks after thc last subculture) are incubated for 3 days on a 0.3-M sucrose medium. After preculture, the embryo clumps are placed in 2 mL cryovials with a 1.8-mL PVS2 vitrification solution and left to sit for 60 min at 25 °C (*Q. robur* and *Q. alba*) or at 0 °C (*C. sativa* and *Q. suber*). After this period, the embryo clumps are resuspended in 0.6 mL of fresh PVS2, and the cryovials are placed in a cryostorage cane or cryobox, which is then immersed rapidly in LN. For thawing, the vials are immersed for 2 min in a water bath at 40–42 °C, and the embryo clumps are cultured on proliferation media for 6–8 weeks. After rewarming, efficient resumption rates have been achieved in the four species studied, with values ranging 57–92% for *Q. robur* (Martínez et al. 2003, Sánchez et al. 2008), 88–93% for *Q. suber* (Valladares et al. 2004), 68% for *C. sativa* and about 54% for *Q. alba* (Corredoira et al. 2014). In the case of chestnut, cork oak and pedunculate oak, these embryo recovery rates were maintained after 10 years of cryostorage of somatic embryo clumps (unpublished results), demonstrating the validity of the vitrification-based procedure defined. An operational application of the vitrification-based protocol defined by Valladares et al. (2004) was the establishment of a cryopreserved cork oak gene bank with embryogenic lines derived from selected genotypes (Vidal et al. 2010). Currently, all 51 cork oak genotypes withstood freezing and were able to produce new somatic embryos after LN storage. Furthermore, three experimental plots have been established with the selected materials to determine their productivity and heritable characters by investigation of

Table 4 Cryopreservation of somatic embryos in hardwood species

Species	Explant type[1]	Preconditioning treatment	Precryogenic treatment	Cryogenic treatment	Embryo recovery (%)	Reference
Alnus glutinosa	Clumps of globular heart SE (2–3 mg)	0.3 M suc (3d)	LS: 2 M glyc + 0.4 M suc at 25 ºC 20 min; PVS2 at 0 ºC 60 min	Direct immersion in LN	90	San José et al. (2015b)
Aesculus hippocastanum	Globular SE (1.5–3.0 mm)	0.74 µM ABA (4d)	0.5 DMSO + 0.5 M glyc + 1 M suc at 0 ºC 60 min	Slow cooling-35 ºC at 1 ºC (30 min)→LN-196 ºC	43	Jekkel et al. (1998)
Aesculus hippocastanum	Globular SE (1.5–3.0 mm)	0.74 µM ABA (4d)	Desiccation in the laminar airflow cabinet (4 h; 13% WC)	Slow cooling-35 ºC at 1 ºC (30 min)→LN-196 ºC	46	Jekkel et al. (1998)
Aesculus hippocastanum	Torpedo SE	Cold store 4 ºC (5d)	LS: 2 M glyc + 0.4 M suc at 25 ºC 20 min; PVS2 at 0 ºC 90 min	Direct immersion in LN	70	Lambardi et al. (2005)
Castanea sativa	Clumps of 2–3 globular early heart SE	0.3 M suc (3d)	PVS2 at 0 ºC 60 min	Direct immersion in LN	68–92	Corredoira et al. (2004,2007, 2012)
Castanea sativa	Clumps of 2–3 globular early heart SE	0.3 M suc (3d)→0.7 M suc (4d)	Desiccation in the laminar airflow cabinet (2 h; 25% WC)	Direct immersion in LN	33	Corredoira et al. (2004)
Fagus grandifolia	Globular SE	–	10% DMSO + 5% suc	Direct immersion in LN	90[1]	Jörgensen (1990)
*Fraxinus excelsior*sis	Embryogenic callus	0.6 M suc	0.6 M suc + 7.5% DMSO 60 min	Slow cooling Mr. Frosty-40 ºC	83[1]	Ozudogru et al. (2010)
Juglans nigra x J. regia	Early globular SE	Cold hardening (4 ºC 1w), 5% DMSO + 0.5% proline, suc (from 0.25–1 M)	Desiccation in the laminar airflow cabinet	Direct immersion in LN	43	de Boucaud et al. (1994)

(continued)

Table 4 (continued)

Species	Explant type[1]	Preconditioning treatment	Precryogenic treatment	Cryogenic treatment	Embryo recovery (%)	Reference
Liriodendron tulipifera	Cell suspensions	Sorbitol 0.4 M (24 h)	5% DMSO (0 °C)	Slow cooling Mr. Frosty-70 °C	100	Vendrame et al. (2001)
Liquidambar sp.	Cell suspensions	Sorbitol 0.4 M (24 h)	5% DMSO (0 °C)	Slow cooling Mr. Frosty-70 °C	100	Vendrame et al. (2001)
Prunus avium	Embryogenic callus	Suc (0.25 M 1d→0.5 M 1d→0.75 M 2d→1 M 3d)	Desiccation in the laminar airflow cabinet (20% WC)	Direct immersion in LN	89	Grenier-de March et al. (2005)
Quercus ilex	Globular clusters	0.3 M suc (3d)	PVS2 at 25 °C 30 min	Direct immersion in LN	80[2]	Barra-Jiménez et al. (2015)
Quercus robur	Clumps of 2–3 globular early heart SE	0.3 M suc (3d)	PVS2 at 25 °C 60 min	Direct immersion in LN	70–92	Martínez et al. (2003) and Sánchez et al. (2008)
Quercus robur	Clumps of 2–3 globular early heart SE	0.3 M suc (3d)→0.7 M suc (4d)	Desiccation in the laminar airflow cabinet (2–3 h; 24–34% WC)	Direct immersion in LN	56	Martínez et al. (2003)
Quercus suber	2–3 globular SE (2–4 mg)	0.3 M suc (3d)	PVS2 at 0 °C 60 min	Direct immersion in LN	88–93	Valladares et al. (2004)
Quercus suber	SE clusters	Encapsulation 3% alginate + 0.1 M CaCl2→beads precultured in 0.7 M suc (3d)	Desiccation in the laminar airflow cabinet (25–35% WC)	Direct immersion in LN	90	Fernandes et al. (2008)

[1] Survival percentage, no mention of embryo recovery is given; [2] Embryo differentiation was only observed after 24 h LN; their differentiation ability was completely lost after 4 weeks in LN. Abbreviations: *ABA* abscisic acid; *DMSO* dimethylsulphoxide; *glyc* glycerol; *LN* liquid nitrogen; *LS* loading solution; *PVS2* plant vitrification solution 2; *SE* somatic embryos; *suc* sucrose; *WC* water content; – not applied

progenies. During field testing, copies of the material were preserved under LN (Vidal et al. 2010). In chestnut, the embryogenic resumption rates of three lines transformed with marker genes, ranging from 52 to 65%, were found to be similar to those achieved with cryopreserved somatic embryos derived from the wild-type line (66%) (Corredoira et al. 2007). Moreover, this procedure was successfully used in the cryopreservation of chestnut embryogenic lines transformed with a thaumatin-like protein gene (Corredoira et al. 2012) and a chitinase gene (Corredoira et al. 2016b). The conservation of embryogenic transgenic lines under LN is of great importance during the field evaluation of embryogenic lines transformed with genes of interest.

The alternative method to the vitrification-based procedure studied in hardwood species is the desiccation-based technique, which, in general terms, has produced worse results compared to those of the former protocol. Comparative studies of vitrification- and desiccation-based methods clearly showed the higher recovery rates of the vitrification procedure. In pedunculate oak, clumps of globular-/heart-stage somatic embryos were pretreated by successive preculture on 0.3- and 0.7-M sucrose-supplemented media followed by desiccation in the air flow of a laminar airflow cabinet to water contents of 24–34% and then immersed in LN. Regardless of the embryogenic line used and the desiccation period applied, up to 56% of the cryopreserved somatic embryos resumed their embryogenic capacity. As has been described for oak, when globular- or heart-shaped chestnut somatic embryos were subjected to the same desiccation protocol prior to storage in LN, only 33% of embryos resumed the embryogenic capacity after thawing in comparison with 68–92% when vitrification is applied (Corredoira et al. 2004, 2012).

Another alternative proposal is the use of the encapsulation–dehydration methodology, as it has been reported in cork oak by Fernandes et al. (2008) with a recovery rate of 90%. However, this figure is also achieved in the same species using the least laborious and complex technique of vitrification (Valladares et al. 2004).

In conclusion, for hardwood species the vitrification-based methodology is highly recommended because it provides significantly better results than the desiccation-based method and is simpler and considerably less time-consuming than techniques involving encapsulation.

14.3.3 Genetic Stability of Cryopreserved Material

As it has been mentioned in the Introduction section, the genetic stability of regenerated plants after cryopreservation should be taken into consideration. The in vitro technology as well as the mutagenic potential of DMSO may be sources of instability although there is no evidence of morphological or genetic alterations in forest trees as a result of cryopreservation (Häggman et al. 2008). It is necessary to confirm whether the cooled material is 'true to type' after cryopreservation

(Harding 2004). This fact is especially important in the case of long-living hardwood trees because in these species, genetic alterations may not be observed in young plants but may appear in mature trees (Häggman et al. 2008; Vieitez et al. 2012). It is usually accepted that the different treatments applied before immersion of explants in LN can induce genetic and epigenetic changes. However, in the majority of reports published until now, there is no evidence of morphological, cytological or molecular modifications after LN storage. In hardwood species, there are few reports in which genetic stability after cryopreservation was evaluated, but, in all cases, genetic or morphological alterations were not detected. In these papers, the assessment of genetic stability was performed using PCR-based markers, such as random amplified polymorphic DNAs (RAPDs) and microsatellites (single sequence repeats (SSRs)), and cytological studies, such as flow cytometry and chromosome counting.

In *Betula pendula*, no morphological and genetic changes using RAPD analysis and chromosome counting were found in plants recovered from cryopreserved shoot tips previously treated with slow cooling (Ryynänen and Aronen 2005b) and vitrification procedures (Ryynänen and Aronen 2005a). In the same species, Ryynänen et al. (2002) confirmed by Southern blot and northern blot that the transcript levels of the neomycin phosphotransferase (*nptII*) gene did not change due to cryopreservation of shoots isolated from transgenic plants. Also, Jokipii et al. (2004) confirmed by RAPD analysis that slow cooling and vitrification procedures did not affect the genetic stability of in vivo- and in vitro-collected buds of *P. tremula* x *P. tremuloides*. In this study, the presence of the transgenic *nptII* gene, analysed by Southern blot, also was found to be stable in shoots of cryopreserved transgenic lines. On the other hand, no changes in DNA content were found in somatic embryos and plantlets of *Alnus glutinosa* as determined by flow cytometry (San José et al. 2015b).

In Fagaceae species, RAPD, SSR and flow cytometry analyses have been applied to assess the genetic fidelity of cryogenic somatic embryos and plantlets derived from them. In *Q. robur*, no differences were detected between cryopreserved plants and mother plants of five embryogenic lines derived from mature trees treated with PVS2 solution and immersed in LN for 1 year using RAPD analysis (Sánchez et al. 2008). In *Q. suber*, the genetic stability of somatic embryos recovered from cryopreservation was confirmed by SSR and flow cytometry analyses (Fernandes et al. 2008). Lastly, chestnut and pedunculate oak transgenic lines also were successfully cryopreserved following the vitrification-based procedure, and the stable integration of the *uidA* gene into the transgenic plants that were regenerated subsequent to cryopreservation was demonstrated (Corredoira et al. 2007; Mallón et al. 2013).

In conclusion, there are no references of genetic alterations in cryopreserved material of hardwood species according to the genetic tools used for screening. More advanced technologies should be applied to confirm these results.

14.4 Concluding Remarks

The rapid development of biotechnological techniques, including in vitro tissue culture and cryopreservation, provides an opportunity for long-term conservation of hardwood tree genetic resources. Cryopreservation techniques are well defined, and the use of the most appropriate method is related to the species under study and the type of explant to be stored. For conservation of recalcitrant seeds, the simplest and recommended method consists of the isolation of embryonic axes, reduction in water content up to 20–25% of their initial level and direct immersion in LN. A more laborious procedure is required for cryostorage of both shoot tips and embryogenic cultures. In the first case, the steps to be covered prior to direct immersion of the explants in LN are cold hardening of donor shoots (2–4 weeks), preculture of isolated shoot tips on sucrose solution for 2–3 days and precryogenic treatment with vitrification solution for 60–90 min. A similar manner must be followed for the cryopreservation of somatic embryos and embryogenic cultures. The most important factor to be taken into consideration in this case is the type of explants selected for cryostorage. Clumps of 2–3 somatic embryos at globular/early or heart stage are highly recommended instead other materials.

An interesting approach that, in our opinion, should be explored deeply in the coming years is the storage under LN of dormant buds collected directly from selected hardwood trees growing in the open. The procedure is simple and even in very specific cases in vitro tissue culture technology is not required: cryopreserved twigs bearing 1–2 buds may be rooted directly in appropriate substrates. However, in most cases, the typical steps of in vitro establishment of dormant buds are needed. In any case, the great potential of this approach for cryopreservation of genetic diversity in hardwoods seems clear.

From the data available, genetic stability of cryopreserved hardwoods is unproblematic. More efforts should be addressed, however, to diminish the mutagenic potential of DMSO by either decreasing its concentration within the vitrification solutions or by searching for new chemicals with no mutagenic properties.

Assuming that cryopreservation technology is today efficient and well defined for hardwood forest species, the practical application in the development of cryobanks is still very limited. Efforts should be made in this direction in the coming years for long-term preservation of hardwood biodiversity.

Acknowledgements We would like to thank Dr. Ana Mª Vieitez for sharing her enthusiasm and knowledge on somatic embryogenesis and cryopreservation. This research was partially funded by the Ministerio de Economía y Competitividad (Spain) through project AGL2016-73143-C4-4-R.

References

Aronen T, Ryynänen L (2014) Cryopreservation of dormant *in vivo*-buds of hybrid aspen: timing as critical factor. CryoLetters 35:385–394

Ballester A, Corredoira E, Vieitez AM (2016) Limitations of somatic embryogenesis in hardwood trees. In: Park YS, Bonga JM, Moon HK (eds) Vegetative propagation of forest trees. National Institute of Forest Science (NiFos), Seoul, Korea, pp 56–74

Ballesteros D, Sershen Varghese B, Berjak P, Pammenter NW (2014) Uneven drying of zygotic embryos and embryonic axes of recalcitrant seeds: challenges and considerations for cryopreservation. Cryobiology 69:100–109

Barra-Jiménez A, Aronen TS, Alegre J, Toribio M (2015) Cryopreservation of embryogenic tissues from mature holm oak trees. Cryobiology 70:217–225

Beardmore T, Vong W (1998) Role of the cotyledonary tissue in improving low and ultralow temperature tolerance of butternut (*Junglans cinerea*) embryonic axes. Can J For Res 28:903–910

Beardmore T, Whittle CA (2005) Induction of tolerance to desiccation and cryopreservation in silver maple (*Acer saccharinum*) embryonic axes. Tree Physiol 25:965–972

Benson EE (2008) Cryopreservation Theory. In: Reed BM (ed) Plant cryopreservation: a practical guide. Springer, New York, pp 15–32

Benson EE, Harding K (2012) Cryopreservation of shoot tips and meristems: an overview of contemporary methodologies. In: Loyola-Vargas VM, Ochoa-Alejo N (eds) Plant cell culture protocols. Methods in molecular biology, vol 877. Springer, New York, p 191–226

Berjak P, Walker M, Mycock DJ, Wesley-Smith J, Watt P, Pammenter NW (2000) Cryopreservation of recalcitrant zygotic embryos. In: Engelman F, Takagi H (eds) Cryopreservation of tropical plant germplasm. Current Research Progress and Application, IPGRI, Rome, pp 140–155

Blakesley D, Kiernan RJ (2001) Cryopreservation of axillary buds of a *Eucalyptus grandis* x *eucalyptus camaldulensis* hybrid. CryoLetters 22:13–18

Blakesley D, Pask N, Henshaw GG, Fay MF (1996) Biotechnology and the conservation of forest genetic resources: *in vitro* strategies and cryopreservation. Plant Growth Regul 20:11–16

Bonga JM (2016) Conifer clonal propagation in tree improvement programs. In: Park YS, Bonga JM, Moon HK (eds) Vegetative propagation of forest trees. National Institute of Forest Science (NiFos), Seoul, Korea, pp 3–31

Bonga JM, Klimaszewska K, von Aderkas P (2010) Recalcitrance in clonal propagation, in particular of conifers. Plant Cell, Tissue Organ Cult 100:241–254

Bonnart R, Waddell J, Haiby K, Widrlechner MP, Volk GM (2014) Cryopreservation of *Populus trichocarpa* and *Salix* dormant buds with recovery by grafting or direct rooting. CryoLetters 35:507–515

Capuana M, Di Lonardo S (2013) *In vitro* conservation of chestnut (*Castanea sativa*) by slow growth. In Vitro Cell Dev Biol Plant 49:605–610

Chmielarz P, Michalak M, Palucka M, Wasilenczyk U (2011) Successful cryopreservation of *Quercus robur* plumules. Plant Cell Rep 30:1405–1414

Corredoira E, San-José MC, Ballester A, Vieitez AM (2004) Cryopreservation of zygotic embryo axes and somatic embryos of European chestnut. CryoLetters 25:33–42

Corredoira E, San José MC, Vieitez AM, Ballester A (2007) Improving genetic transformation of European chestnut and cryopreservation of transgenic lines. Plant Cell, Tissue Organ Cult 91:281–288

Corredoira E, Valladares S, Allona I, Aragoncillo C, Vieitez AM, Ballester A (2012) Genetic transformation of European chestnut somatic embryos with a native thaumatin-like protein (*CsTL1*) gene isolated from *Castanea sativa* seeds. Tree Physiol 32:1389–1402

Corredoira E, Toribio M, Vieitez AM (2014) Clonal propagation via somatic embryogenesis in *Quercus* spp. In: Ramawhat KG, Mérillon J-M, Ahuja MR (eds) Tree Biotechnology. CRC Press, Boca Raton, Florida, USA, pp 262–302

Corredoira E, Vieitez AM, San José MC, Vieitez FJ, Ballester A (2016a) Advances in somatic embryogenesis and genetic transformation of European chestnut: Development of transgenic resistance to ink and blight disease. In: Park YS, Bonga JM, Moon HK (eds) Vegetative propagation of forest trees. National Institute of Forest Science (NIFoS), Seoul, Korea, pp 279–301

Corredoira E, San José MC, Vieitez AM, Allona I, Cipriano A, Ballester A (2016b) *Agrobacterium*-mediated transformation of European chestnut somatic embryos with a *Castanea sativa* (Mill.) endochitinase gene. New Forest doi: 10.1007/s11056-016-9537-5

Cruz-Cruz CA, González-Arnao MT, Engelmann F (2013) Biotechnology and conservation of plant biodiversity. Resources 2:73–95

de Boucaud MT, Brison M, Ledoux C, Germain E, Lutz A (1991) Cryopreservation of embryogenic axes of recalcitrant seed: *Juglans regia* L. cv Franquette. Cryoletters 12:163–166

de Boucaud MT, Brison M, Negrier P (1994) Cryopreservation of walnut somatic embryos. CryoLetters 15:151–160

Driver JA, Kuniyuki AH (1984) In vitro propagation of paradox walnut rootstock. HortScience 19:507–509

Ducousso A, Bordacs S (2004) EUFORGEN technical guidelines for genetic conservation and use for pedunculate and sessile oaks (*Quercus robur* and *Q. petraea*). International Plant Genetic Resources Institute, Rome, Italy, 6 pages

Engelmann F (1991) *In vitro* conservation of tropical plant germplasm-a review. Euphytica 57:227–243

Engelmann F (1997) *In vitro* conservation methods. In: Callow JA, Ford-Lloyd BV, Newburry HJ (eds) Biotechnology and plant genetic resources. Conservation and use, Wallingford, UK, pp 119–161

Engelmann F (2004) Plant cryopreservation: progress and prospects. In Vitro Cell Dev Biol—Plant 5:427–433

Engelmann F (2011a) Use of biotechnologies for the conservation of plant biodiversity. In Vitro Cell Dev Biol—Plant 47:5–16

Engelmann F (2011b) Cryopreservation of embryos: an overwiew. In: Thorpe TA, Yeung EC (eds) Plant embryo cultures: methods and protocols, methods in molecular biology, vol 710. Springer, New York, pp 155–184

Fahy GM, MacFarlane DR, Angell CA, Meryman HT (1984) Vitrification as an approach to cryopreservation. Cryobiology 21:407–426

Fernandes P, Rodríguez E, Pinto G, Roldán-Ruíz I, De Loose M, Santos C (2008) Cryopreservation of *Quercus suber* somatic embryos by encapsulation-dehydration and evaluation of genetic stability. Tree Physiol 28:1841–1850

González-Benito ME, Pérez-Ruiz C (1992) Cryopreservation of *Quercus faginea* embryonic axes. Cryobiology 29:685–690

González-Benito ME, Prieto RM, Herradon E, Martin C (2002) Cryopreservation of *Quercus suber* and *Quercus ilex* embryonic axes: *In vitro* culture, desiccation and cooling factors. CryoLetters 23:283–290

Grenier-de March G, de Boucaud MT, Chmielarz P (2005) Cryopreservation of *Prunus avium* L. embryogenic tissues. CryoLetters 26:341–348

Gresshoff PM, Doy CH (1972) Development and differentiation of haploid *Lycopersicon esculentum*. Planta 107:161–170

Häggman H, Rusanen M, Jokipii S (2008) Cryopreservation of *in vitro* tissues of deciduous trees. In: Reed B (ed) Plant cryopreservation: a practical guide. Springer, NY, pp 365–386

Harding K (2004) Genetic integrity of cryopreserved plant cells: a review. CryoLetters 25:3–22

Harvengt L, Meier-Dinkel A, Dumas E, Collin E (2004) Establishment of a cryopreserved gene bank of European elms. Can J For Res 34:43–55

Hausman JF, Neys O, Kevers C, Gaspar T (1994) Effects of *in vitro* storage at 4 °C on survival and proliferation of poplar shoots. Plant Cell, Tissue Organ Cult 38:65–67

Janeiro LV, Vieitez AM, Ballester A (1995) Cold storage of *in vitro* cultures of wild cherry, chestnut and oak. Ann Sci For 52:287–293

Jekkel Zs, Gyulai G, Kiss J, Kiss E, Heszky LE (1998) Cryopreservation of horse-chestnut (*Aesculus hippocastanum* L.) somatic embryos using three different freezing methods. Plant Cell Tissue Organ Cult 52:193–197

Jokipii S, Ryynänen L, Kallio PT, Aronen T, Häggman H (2004) A cryopreservation method maintaining the genetic fidelity of a model forest tree, *Populus tremula x Populus tremuloides* Michx. Plant Sci 166:799–806

Jörgensen J (1990) Conservation of valuable gene resources by cryopreservation in some forest tree species. Plant Physiol 136:373–376

Kaczmarczyk A, Funnekotter B, Menon A, Phang PY, Al-Hanbali A, Bunn E, Mancera RL (2012) Current issues in plant conservation. In: Katkov I (ed) Current frontiers in Cryobiology. Available from: htpp://www.intechopen.com/books/current-frontiers-in-cryobiology/current-issues-in plant-cryoconservation

Kavani B (2011) Conservation of plant genetic resources by cryopreservation. Aust J Crop Sci 5:778–800

Kaya E, Alves A, Rodrigues L, Jenderek M, Hernandez-Ellis M, Ozudogru A, Ellis D (2013) Cryopreservation of *Eucalyptus* genetic resources. CryoLetters 34:608–618

Kirilenko AP, Sedjo RA (2007) Climate change impacts on forestry. PNAS 104:19697–19702

Lambardi M, De Carlo A (2003) Application of tissue culture to germplasm conservation of temperate broad-leaf trees. In: Jain SM, Ishii K (eds) Micropropagation of woody trees and fruits. Kluwer Academic Publihers, Netherlands, pp 815–840

Lambardi M, Fabbri A, Caccavale A (2000) Cryopreservation of white (*Populus alba* L.) by vitrification of *in vitro*-grown shoot tips. Plant Cell Rep 19:213–218

Lambardi M, De Carlo A, Capuana M (2005) Cryopreservation of embryogenic callus of *Aesculus hippocastanum* L. by vitrification/one-step freezing. CryoLetters 26:185–192

Lloyd G, McCown B (1980) Commercially feasible micropropagation of mountain laurel, *Kalmia latifolia*, by use of shoot tip cultures. Comb Proc Int Soc 30:421–427

Mallón R, Vieitez AM, Vidal N (2013) High-efficiency *Agrobacterium*-mediated transformation in *Quercus robur*: selection by use of a temporary immersion system and assessment by quantitative PCR. Plant Cell, Tissue Organ Cult 114:171–185

Martínez MT, Ballester A, Vieitez AM (2003) Cryopreservation of embryogenic cultures of *Quercus robur* using desiccation and vitrification procedures. Cryobiology 46:182–189

Martínez MT, Corredoira E, Valladares S, Jorquera L, Vieitez AM (2008) Germination and conversion of somatic embryos derived from mature *Quercus robur* trees: the effects of cold storage and thidiazuron. Plant Cell, Tissue Organ Cult 95:341–351

Matsumoto T, Sakai A, Yamada K (1994) Cryopreservation of *in vitro* grown apical meristems of wasabi (*Wasabia japonica*) by vitrification and subsequent high plant regeneration. Plant Cell Rep 13:442–446

Merkle SA (2016) Application of somatic embryogenesis and transgenic technology to conserve and restore threatened forest tree species. In: Park YS, Bonga JM, Moon HK (eds) Vegetative propagation of forest trees. National Institute of Forest Science (NIFoS), Seoul, Korea, pp 261–278

Murashige T, Skoog F (1962) A revised medium for rapid growth and bioassays with tobacco tissue cultures. Physiol Plant 15:473–497

Niino T, Tashiro K, Suzuki M, Susumu O, Magoshi J, Akihama T (1997) Cryopreservation of *in vitro* grown shoot tips of cherry and sweet cherry by one-step vitrification. Sci Hortic 70:155–163

Ozudogru EA, Capuana M, Kaya E, Panis B, Lambardi M (2010) Cryopreservation of *Fraxinus excelsior* L. embryogenic callus by one-step freezing and slow cooling techniques. CryoLetters 31:63–75

Pammenter NW, Berjak P (1999) A review of recalcitrant seed physiology in relation to desiccation-tolerance mechanisms. Seed Sci Res 10:13–37

Park YS and Bonga JM (2011) Application of somatic embryogenesis in forest management and research. In: Park YS, Bonga JM, Moon HK (eds) Proceedings of the IUFRO Working Party 2.09.02: somatic embryogenesis of trees/ conference proceedings/advances in somatic embryogenesis in trees and its application for the future forests and plantations. August 19–21, 2010, Suwon, Republic of Korea, p 3–8

Park YS, Beaulieu J, Busquet J (2016) Multi-varietal forestry integrating genomic selection and somatic embryogenesis. In: Park YS, Bonga JM, Moon HK (eds) Vegetative propagation of forest trees. National Institute of Forest Science, Seoul, Korea, pp 355–378

Pence VC (1990) Cryostorage of embryo axes of several large-seeded temperate tree species. Cryobiology 27:212–218

Pence VC (1992) Desiccation and the survival of *Aesculus, Castanea* and *Quercus* embryo axes through cryopreservation. Cryobiology 29:391–399

Pence VC (2014) Tissue cryopreservation for plant conservation: potential and challenges. Int J Plant Sci 175:40–45

Pritchard HW (2007) Cryopreservation of desiccation-tolerant seeds. Methods Mol Biol 368:185–201

Pukacki PM, Juszczyk K (2015) Desiccation sensitivity and cryopreservation of the embryonic axes of the seeds of two *Acer* species. Trees 29:385–396

Reed BM, Uchendu E (2008) Controlled rate cooling. In: Reed BM (ed) Plant cryopreservation: a practical guide. Springer, New York, pp 77–92

Romano A, Martins-Loução MA (1999) *In vitro* cold storage of cork oak cultures. Plant Cell, Tissue Organ Cult 59:155–157

Ryynänen L (1996) Survival and regeneration of dormant silver birch buds stored at super-low temperatures. Can J Forest Res 26:617–623

Ryynänen L (1998) Effect of abscisic acid, cold hardening, and photoperiod on recovery of cryopreserved *in vitro* tips of silver birch. Cryobiology 36:32–39

Ryynänen L, Aronen T (2005a) Vitrification, a complementary cryopreservation method for *Betula pendula* Roth. Cryobiology 51:208–219

Ryynänen L, Aronen T (2005b) Genome fidelity during short- and long-term tissue culture and differentially cryostored meristems of silver birch (*Betula pendula*). Plant Cell, Tissue Organ Cult 83:21–32

Ryynänen L, Häggman H (2001) Recovery of cryopreserved silver birch shoot tips is affected by the pre-freezing age of the cultures and ammonium substitution. Plant Cell Rep 20:354–360

Ryynänen L, Sillanpää M, Kontunen-Soppela S, Timonen H, Kangasjärvi J, Vapaavuori E, Häggman H (2002) Preservation of transgenic silver birch (*Betula pendula* Roth) lines by means of cryopreservation. Mol Breeding 10:143–152

Sakai A, Engelmann F (2007) Vitrification, encapsulation-vitrification and droplet-vitrification: a review. CryoLetters 28:151–172

Sakai A, Kobayashi S, Oiyama I (1990) Cryopreservation of nucellar cells of navel orange (*Citrus sinensis* Osb. var. *brasiliensis* Tnaka) by vitrification. Plant Cell Rep 9:30–33

San José MC, Valladares S, Vázquez-Janeiro L, Corredoira E (2014) Cryopreservation of *in vitro*-grown shoot tips of *Alnus glutinosa* (L.) Gaertn. Acta Physiol Plant 36:109–116

San José MC, Corredoira E, Oliveira H, Santos C (2015a) Cryopreservation of somatic embryos of *Alnus glutinosa* (L.) Gaertn. and confirmation of ploidy stability by flow cytometry. Plant Cell, Tissue Organ Cult 123:489–499

San José MC, Janeiro LV, Corredoira E (2015b) Simple strategy for the *in vitro* conservation of *Alnus glutinosa* (L.) Gaertn. germplasm. Trees 27:1597–1608

Sánchez MC, Martínez MT, Vidal N, San José MC, Valladares S, Vieitez AM (2008) Preservation of *Quercus robur* germplasm by cryostorage of embryogenic cultures derived from mature trees and RAPD analysis of genetic stability. CryoLetters 29:493–504

Schoenweiss K, Meier-Dinkel A, Grotha R (2005) Comparison of cryopreservation techniques for long-term storage of ash (*Fraxinus excelsior* L.). CryoLetters 26:201–212

Shatnawi MA, Shibli R, Qrunfleh I, Bataeineh K, Obeidat M (2007) *In vitro* propagation and cryopreservation of *Prunus avium* using vitrification and encapsulation dehydration methods. J Food Agric Environ 5:204–208

Shukla MR, Jones AMP, Sullivan JA, Liu C, Gosling S, Saxena PK (2012) *In vitro* conservation of American elm (*Ulmus americana*): potential role of auxin metabolism in sustained plant proliferation. Can J For Res 42:686–697

Son SH, Chun YW, Hall RB (1991) Cold storage of *in vitro* cultures of hybrid poplar shoots (*Populus alba* L. x *P. grandidentata* Mich). Plant Cell, Tissue Organ Cult 27:161–168

Tagaki H, Thinh NT, Islam OM, Senboku T, Sakai A (1997) Cryopreservation of *in vitro* grown shoot tips of taro (*Colocasia esculenta* (L.) Schott) by vitrification. I. Investigation of basic conditions of the vitrification procedure. Plant Cell Rep 16:594–599

Takagi H (2000) Recent developments in cryopreservation of shoot apices of tropical species. In: Engelman F, Takagi H (eds) Cryopreservation of tropical plant germplasm. Current Research Progress and Application, IPGRI, Rome, pp 178–193

Towill LE (2002) Cryopreservation of plant germoplasm: Introduction and some observations. In: Towill LE, Bajaj YPS (eds) Biotechnology in agriculture and forestry 50. Cryopreservation of plant germoplasm II. Springer, Berlin, p 3–21

Valladares S, Toribio M, Celestino C, Vieitez AM (2004) Cryopreservation of embryogenic cultures from mature *Quercus robur* trees using vitrification. CryoLetters 25:177–186

Vendrame WA, Holliday CP, Montello PM, Smith DR, Merkle SA (2001) Cryopreservation of yellow-poplar and sweetgum embryogenic cultures. New Forest 21:283–292

Vidal N, Sánchez C, Jorquera L, Ballester A, Vieitez AM (2005) Cryopreservation of chestnut by vitrification of *in vitro*-grown shoot tips. In Vitro Cell Dev Biol Plant 41:63–68

Vidal N, Vieitez AM, Fernández MR, Cuenca B, Ballester A (2010) Establishment of cryopreserved gene banks of European chestnut and cork oak. Eur J For Res 129:635–643

Vieitez AM, San José MC, Sánchez MC, Ballester A (2003) Micropropagation of *Fagus* spp. In: Jain SM, Ishii K (eds) Micropropagation of woody trees and fruits. Kluwer Academic Publishers, Netherlands, pp 181–215

Vieitez AM, San José MC, Corredoira E (2011) Cryopreservation of zygotic embryonic axes and somatic embryos of European chestnut. In: Thorpe TA, Yeung EC (eds) Plant embryo cultures: methods and protocols, methods in molecular biology, vol 710. Springer, New York, pp 201–213

Vieitez AM, Corredoira E, Martínez MT, San José MC, Sánchez C, Valladares S, Vidal N, Ballester A (2012) Application of biotechnological tools to *Quercus* improvement. Eur J Forest Res 131:519–539

Volk GM, Bonnart R, Waddell J, Widrlechner MP (2009) Cryopreservation of dormant buds from diverse *Fraxinus* species. CryoLetters 30:262–267

Watt MP, Thokoane NL, Mycock D, Blackeway F (2000) *In vitro* storage of *Eucalyptus grandis* germplasm under minimal growth conditions. Plant Cell, Tissue Organ Cult 61:161–164

Yang BH, Hur SD, Hong YP, Hong KN (2011) Mass propagation and Cryopreservation of In Vitro -cultured Axillary Bud Meristems of *Melia azedarach*. In: Park YS, Bonga JM, Moon HK (eds) Vegetative propagation of forest trees. National Institute of Forest Science, Seoul, Korea, p 132

Chapter 15
Diversity of Trees in the Mesoamerican Agroforestry System

Noel A. González-Valdivia, William Cetzal-Ix, Saikat Kumar Basu, Fernando Casanova-Lugo and Jesús F. Martínez-Puc

Abstract Biodiversity conservation and production of foods imply a hard trade-off with no simple solutions routes. However, changing conventional agrarian and animal husbandry models to an agroforestry one is an important approach for achieving a better equilibrium between both economical and conservative goals. Agroforestry systems (AFS) for tropical lands provide a renewed old approach to deal with the need for feeding a growing population while avoiding damage to ecosystems on which food production is achieved. Neotropical studies on the subject have increased since the 1980s, accumulating evidences that it is possible to consider the potential of agroforestry to improve the status of biodiversity without hampering regular agricultural production. Three countries stand out for their contributions in the Mesoamerican region: Costa Rica, Mexico and Nicaragua. Notably, the researchers of biological diversity associated with Mesoamerican agroforestry systems have turned their attention primarily on nine biological groups: ants, bats, birds, butterflies, dung beetles, mammals, soil macrofauna, terrestrial mollusks and plants. Most research in Mesoamerica, including that in Colombia and Venezuela, dealing with biodiversity in the production systems has been abandoned. But there is an increasing trend of studies on biodiversity conservation in areas under active cultivation or livestock since both include trees. It is clear that the AFS can only help in reducing the negative impacts of agriculture and livestock grazing systems on the natural biodiversity. In conjunction with the

N.A. González-Valdivia (✉) · W. Cetzal-Ix · J.F. Martínez-Puc
Instituto Tecnológico de Chiná, Calle 11 entre 22 y 28, Colonia Centro Chiná
24050 Campeche, Mexico
e-mail: rolito22@hotmail.com

S.K. Basu
Department of Biological Sciences, University of Lethbridge, Lethbridge
AB T1K 3M4, Canada

F. Casanova-Lugo
Instituto Tecnológico de la Zona Maya, Othón P. Blanco, Quintana Roo, Mexico

M.R. Ahuja and S.M. Jain (eds.), *Biodiversity and Conservation of Woody Plants*,
Sustainable Development and Biodiversity 17, DOI 10.1007/978-3-319-66426-2_15

network of protected areas in the region, this synergistic effect may increase the ability of biological conservation of the territory, alongside increasing economic benefits to the local rural society.

Keywords Agroforestry · Biodiversity · Conservation · Plants Mesoamerica · Neotropical

15.1 Introduction

Mesoamerica typically represents a unique American sociocultural region that geographically extends between Mexico and Central American countries like Belize, Guatemala, El Salvador, Honduras, Nicaragua and Costa Rica. However, various scholars over time have further expanded the region of Mesoamerica spreading between the two continents of North America and South America connected through the Central American bridge, thereby including in broader definition countries even in the northern reaches of the South American continent up to present-day Venezuela and Colombia. Although the definition of the Mesoamerican region will continue to be debated and discussed among various historians, archeologists and sociologists; however, none can ignore the distinct sociocultural and socioeconomic history of this unique region of the globe (Andrade et al. 2008a, b). Mesoamerica is interesting from the perspectives of human history being an active hot bed of the evolution of major human civilizations in the continents of the Americas. Mesoamerica represents some of the major pre-Columbian indigenous societies of the continents of North America and South America. The boundaries of various nation states within this region has changed and redrawn several times with the current nations emerging on the strong foundation of their rich history, traditions, societies and cultures.

Another important aspect of the Mesoamerican region has been its unique environment being located at the heart of the Neotropics. Mesoamerican region is blessed with great biodiversity of life forms in the form of both flora and fauna with particular reference to tress (Montagnini 2012). The region has been geographically and geologically characterized by conspicuous rain forests that has stood the test of time for several centuries, and shaped as well as culturally impacted the development of various Mesoamerican civilizations and societies across the pages of history (Harvey et al. 2008). The rich biodiversity of the region has been amply recorded in the archeological as well as historical accounts and artifacts of the past civilizations and has continued to be one of the most important factors shaping the life and culture of numerous indigenous communities distributed across the present-day Mesoamerica. Even the European settlers who later integrated with these indigenous societies and finally settled down in this regions also adapted similar sociocultural values strongly associated with the rich biodiversity of the Neotropics (Perfecto and Vandermeer 1994).

Trees are absolutely important for the Neotropical ecosystems as they constitute distinct ecosystem of their own and support both the soil ecosystems and the vast forest ecosystems in several ways. Without the trees, the basic parameters of the sustenance of the local ecosystem of the Mesoamerican region will get drastically altered and modified in an extremely detrimental and negative manner (Harvey et al. 2008). The tree constitutes the canopy feature of the Neotropical forest systems and is extremely important for the survival of several micro-niches and dependent ecosystems of such complex tropical forests and also plays a vital role in both the mineral and nutrient cycling as well has vital roles with respect to global warming and climate change impacting our planet (Maffi 2005). Our review research has tried to capture how the rich tree biodiversity of the region has been ecologically shaping Mesoamerican AFS over the decades.

15.1.1 Agroforestry Systems or AFS in Mesoamerica

Protected natural areas is one of the most important regional strategies implemented by law in the Neotropics for the purpose of successful biodiversity conservation (Halffter 2011). In most cases, the conservation laws are imposed without any consideration for the affected human population. All these efforts are important, but insufficient by not including strategies that incorporate the social component from the beginning, since conscious appropriation of a culture of conservation often exists in a population and is often more important than the formal laws or regulations. In the Neotropics, the heterogeneity is evident along with these conservation practices by local, traditional and indigenous human societies for millennia (Gómez-Pompa and Kaus 1992; Gómez-Pompa et al. 2003).

An example of human adaptation of farming practices of semipermanent, rotational nature is commonly known as slash-and-burn agriculture; which still remains a common practice in various parts of Mesoamerica. This system, more appropriately called maize field system (in Spanish "milpa"), is the result of the implementation of comprehensive knowledge of natural, biological, ecological and geochemical cycles, as well as subsequent improvement and adaptation of crop varieties (Hernández-X et al. 2011) that should be able to sustain an average population density 142 persons/km^2 (Ford et al. 2011). This adaptive knowledge has been summarized in an expression, biocultural diversity (Toledo 1996; Maffi 2005), and can help us deal with the uncertainty of global warming and climate change (Montagnini 2012).

Agroforestry system (AFS) is as an expression of cultural diversity of Mesoamerican people. It has been an ancient and integral part of the agricultural traditions of the Neotropics for several centuries. When introduced with livestock and other new crops after the arrival of European settlers, a modification of these traditional processes resulted in the implementation of new production strategies. It gradually shifted to the rotational agroforestry system or "corn field" system and its variants known as "conuco" (Petit-Aldana and Uribe-Valle 2006), for monoculture of crops and open pastures, which are set for long periods in soil, making it difficult for the latter to recover their natural fertility and ecological balance. By removing forest cover extensively, local biodiversity is negatively impacted, and the system becomes unbalanced, causing, among other detrimental impacts, soil degradation and soil erosion as well as promoting the preponderance of unwanted pests.

Several studies have demonstrated that the incorporation of the tree as an element of management of the agricultural systems allows many species of plants to successfully survive and coexist within these productive areas (Harvey et al. 2008; Vandermeer and Perfecto 2007). However, it is clear from the review of the available primary literature that the effect of removing original forests, in most cases, involves species that are either generalist or specialist types in both open and/or disturbed habitats. There are species in the forest that are present in the early ecological succession, taking advantage of the fall of dominant trees and the dynamics of availability of light in the forest floor seriously challenged by the predominance of high canopy growth trees competing with one another for space, light and other necessary nutrients. Species inside the dense forest represent wooded specialists that tend to disappear in those territories that are being extensively used for grazing and for growing conventional crops (González-Valdivia et al. 2014).

The strategy that supports an increase in the agricultural and livestock production includes different arboreal strata, with multiple purposes for the human users. Trees should belong to a broad and representative diversity of native functional groups, including pioneer trees to persistent dominant upper canopy members of the original undisturbed forest. The arrangements should include different forms and spatial patterns, incorporating them from the set of scattered trees, through scattered woods, connected by lines of trees with multiple species and vegetation strata of variable amplitude reaching the edges of the permanent strips of vegetation adjacent to cultivated fields or open pastures.

Even when there is available local knowledge of significant conservation value, the plant diversity associated with different AFS in tropical America (Neotropics) needs to be studied carefully. Very little is known about the agroforestry systems designed to improve efficiency simultaneously in agricultural production and wildlife conservation in the Mesoamerican region. The studies conducted till date frequently describe the AFS host species richness such as the preserved habitats (Tobar-López and Ibrahim 2008; González-Valdivia et al. 2011), which may even represent a relatively high percentage of flora and fauna of a state of a country (González-Valdivia 2003). In this review, some research trends on the biodiversity of agroforestry systems (AFS) in the Mesoamerican region have been discussed and analyzed.

15.1.2 Information of AFS for the Mesoamerican Region

Information on the botanical diversity studied in AFS for the Mesoamerican region (Fig. 15.1 and Table 15.1) has been obtained.

The cutoff point is detected in the search when the titles and authors cited begin to repeat themselves and the number of new entrants to the list declines significantly. This method of sampling is characteristic of qualitative research techniques because it reaches the generalization of information following the scientific communication networks (Bernard 2006).

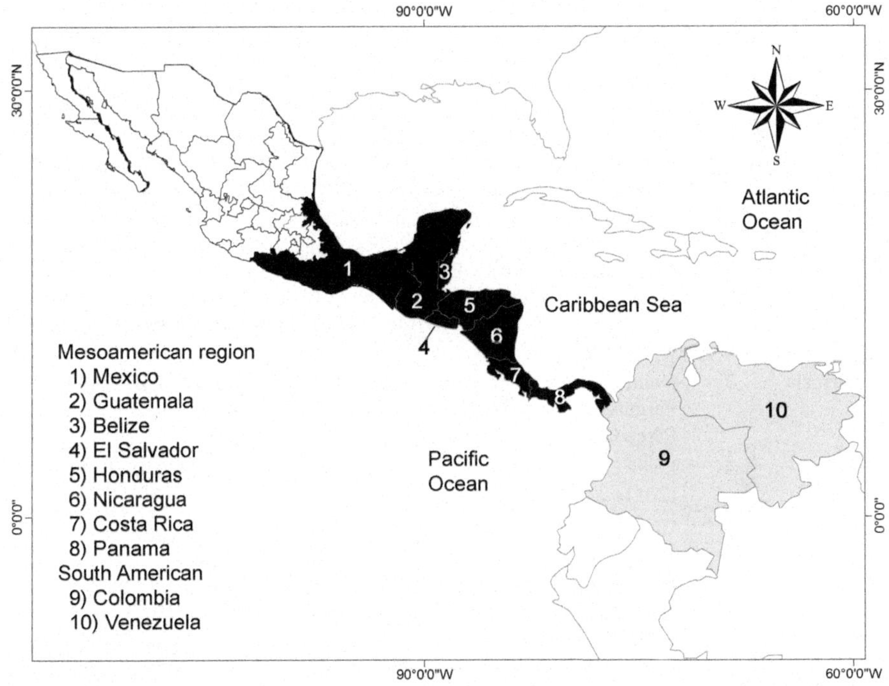

Fig. 15.1 Countries included in this review of publications for AFS for the Mesoamerican region and South America area (Colombia and Venezuela)

Table 15.1 Selected publications available and accessible online on AFS and associated biodiversity country in Mesoamerica (including Colombia and Venezuela)

Country	NP	References
México	53	Guevara et al. (1986, 1992); Nestel et al. (1992); (Guevara and Laborde 1993); Greenberg et al. (1997aa, b), Ibarra-Núñez and García-Ballinas (1998); Moguel and Toledo (1999); Otero-Arnaiz et al. (1999); Ibarra et al. (2001); Montes de Oca (2001); Perfecto and Vandermeer (2002); Mas and Dietsch (2003); Brown et al. (2004); Armbrecht et al. (2004); Perfecto et al. (2004); Tejeda-Cruz and Sutherland (2004); Mas and Dietsch (2004); Díaz Valdiviezo (2005); Perfecto et al. (2005); Cruz-Angón and Greenberg (2005); Larellano et al. (2005), Pineda et al. (2005), Pascacio-Damián (2006); Petit-Aldana and Uribe-Valle (2006); García-Estrada et al. (2006); Philpott et al. (2006a, b); Dietsch et al. (2007); Soto-Pinto et al. (2008); Williams-Guillén et al. (2008); Macip-Ríos and Casas-Andreu. (2008); Macip-Ríos and Muñoz-Alonzo (2008); Cruz-Angón et al. (2009); González-Valdivia, et al (2011); Salgado-Mora et al. (2007); Philpott et al. (2008a); Picone (2000); Casasola et al. (2001); Nieto (2001); González-Valdivia (2003); Hernández et al. (2003); Harvey et al. (2004a, b); Vílchez et al. (2004); Gómez et al. (2004); Pérez et al. (2005); Harvey et al. (2005); Pérez et al. (2006); Sáenz et al. (2006); Harvey et al. (2006a, b); Medina et al. (2007); Méndez et al. (2010); Siles et al. (2013); Goodall et al. (2015); Philpott et al. (2008b); Toledo et al. (2008); Jackson et al. (2009); Jha and Vandermeer (2009); Ramírez-Albores (2010); Mariaca-Méndez (2012); Van Der Wal et al. (2012); González-Valdivia et al. (2012); Guevara et al. (2012); Basto-Estrella et al. (2012); González-Valdivia et al. (2013); Martínez-Encino et al. (2013); Toledo-Aceves et al. (2013); González-Valdivia et al. (2014); De la Mora-Rodríguez (2014); Arellano et al. (2013)
Guatemala	7	Vaninni (1994); Greenberg et al. (1997b); Calvo and Blake (1998); Ferguson (2001); Ferguson et al. (2003); Kraker-Castañeda and Pérez-Consuegra (2011); Cerezo et al. (2013)
Honduras	4	Pérez et al. (2005); Alemán-Mejía (2008); Fonte et al. (2010); Pérez et al. (2011)
El Salvador	2	García et al. (2001); Méndez et al. (2010)
Nicaragua	18	Picone (2000); Casasola et al. (2001); González-Valdivia (2003); Hernández et al. (2003); Harvey et al. (2004a, b); Vílchez et al. (2004); Gómez et al. (2004); Pérez et al. (2005a, b); Harvey et al. (2005); Pérez et al. (2006); Sáenz et al. (2006); Harvey et al. (2006a, b); Medina et al. (2007); Méndez et al. (2010); Siles et al. (2013); Goodall et al. (2015)
Costa Rica	48	Janzen (1982a, b); Somarriba (1985); Perfecto and Vandermeer (1994); Perfecto and Snelling (1995); Somarriba (1995); Perfecto et al. (1996, 1997); Vandermeer and Perfecto (1997); Holl (1998); Harvey et al. (1999); Harvey and Haber (1999); Harvey (2000); Nielsen and DeRosier (2000); Peterson and Haines (2000); Slocum and Horvitz (2000); Ibrahim et al. (2001); Daily et al. (2001); Guiracocha et al. (2001); Zamora et al. (2001); Rojas et al. (2001); Cárdenas et al. (2003); Cleveland et al. (2003); Esquivel et al. (2003); Munroe et al. (2004); Harvey et al. (2004a, b); Pérez et al. (2005a); Harvey et al. (2005); Sáenz et al. (2006);

(continued)

Table 15.1 (continued)

Country	NP	References
		Estrada et al. (2006); Tobar-López et al. (2006); Sánchez-De León et al. (2006); Rossi et al. (2011); Harvey et al. (2007); Villanueva et al. (2007); Andrade et al. (2008a, b); Harvey et al. (2008); Tobar-López and Ibrahim (2008); Nadkarni and Haber (2009); Peters and Greenberg (2013); Sánchez et al. (2013); Stephen and Sánchez (2014); Vásquez-Vela (2014); Durango et al. (2015)
Panama	2	Van Bael et al. (2007); Córdoba et al. (2013)
Colombia	18	Decaëns et al. (1994); Botero and Russo (1998); Jiménez et al. (1998); Posada et al. (2000); Cajas and Sinclair (2001); Esquivel-Sheik and Calle Díaz (2002); Decaëns et al. (2004); Sáenz et al. (2006); Armbrecht et al. (2006); Fajardo et al. (2009); Calonge et al. (2010); Giraldo et al. (2011); Botero and De la Ossa (2011); Neita and Escobar (2012); De la Ossa et al. (2012); Rojas-Sánchez et al. (2012); López et al. (2013); De la Ossa-Lacayo (2013); Díaz-Bohórquez et al. (2014)
Venezuela	2	Araujo and López-Hernández (1999); Hernández et al. (2012)

NP Number of publications

15.2 AFS and Plant Diversity by Countries in Mesoamerica

The majority of the published studies on diversity in AFS are concentrated in Costa Rica, Nicaragua and Mexico, with the more recent inclusion of other South American countries like Colombia and Venezuela (near region), which has increased the number of publications derived from research on the subject. In Table 15.1, the authors and the number of publications per country are presented.

15.3 Publications on Plant Diversity in Mesoamerican AFS

The result of the literature review studies on plant diversity in AFS initiated during the 1980s, with relatively few articles and other scientific papers appearing in media prior to this period (Fig. 15.2). In the 1990s, there was a marked increase in the number of publications on the subject of biological diversity associated with production systems. Then, studies of fragmentation of habitats and its effect on diversity predominated, as well as those that study plant successions and ecosystem recoveries after the abandonment; the focus has been to understand how these processes occur and how they can be used in the restoration of the disturbed ecosystems.

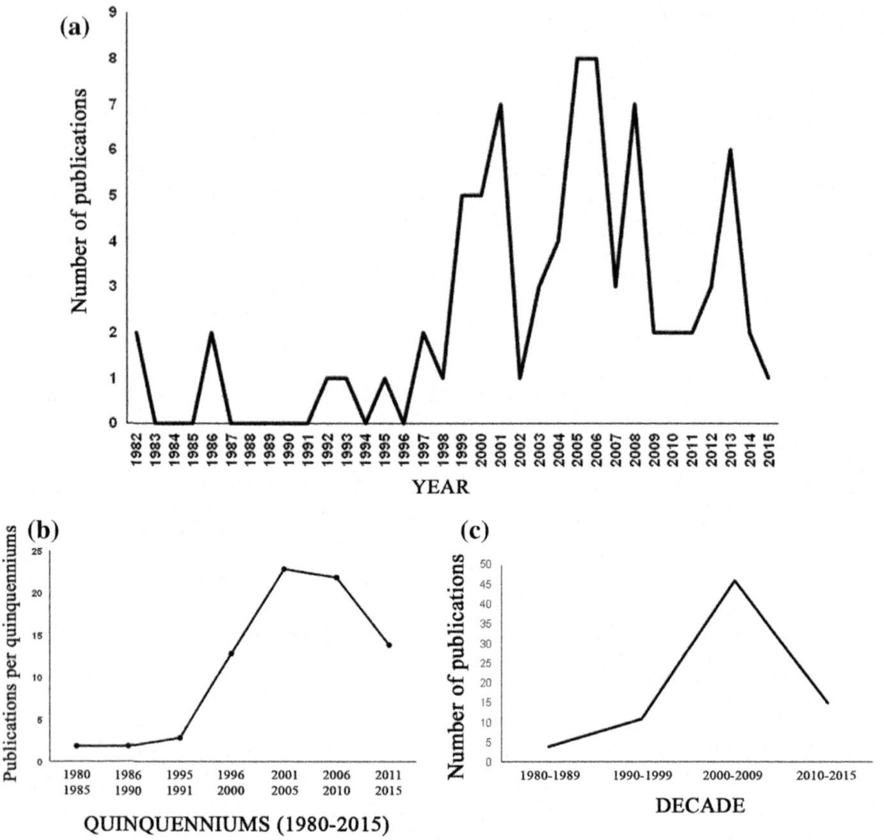

Fig. 15.2 Publications on biodiversity in Mesoamerican AFS (including Colombia and Venezuela) according to reviewed scientific literature between 1980 and 2015. **a** Publications per year. **b** Publications by five-year period. **c** Publications by ten-year period

15.4 Trends in the Diversity of AFS Flora

In the 2000s, the research has been predominantly focused on agricultural management systems, particularly those categorized as AFS. In this case, most of the researches have been focused on one or a few species and handful studies with multiple taxa are available (Table 15.2). The trend remains high in this type of approach during the current decade, appearing more frequently in AFS investigations assessing biodiversity based on various biological groups, not only vegetation (trees), but rather observed as a whole. Figure 15.1 shows the pattern of publications on plant diversity in wooded farming systems, AFS type. The greatest dynamism of publications on Mesoamerican AFS and diversity of plants were initiated in the early 1990s, peaking in 2000, from which it has decreased in general, however, maintaining a superior rate to what it demonstrated before 1990.

Table 15.2 Growth forms diversity and origin taxa of AFS per countries

Country	Taxa	Herb	Shrub	Tree	Climber	Vine	NT	NAP	N
Mexico	94	2	1	90	1	–	1	12	81
Belize	10	–	–	10	–	–	1	–	9
Guatemala	38	4	8	26	–	–	3	–	35
El Salvador	26	–	1	25	–	–	2	4	20
Honduras	10	1	–	9	–	–	–	–	10
Nicaragua	165	59	13	75	13	5	9	11	145
Costa Rica	43	3	–	40	–	–	9	6	28
Panama	20	–	–	20	–	–	2	5	13
Colombia	5	1	1	3	–	–	1	–	4
Venezuela	8	–	–	8	–	–	2	1	5

A total of 45 of the 77 studies of vegetation in AFS distributed in different countries including some type of flora as the only object of study, predominating in this case the tree layer (Fig. 15.3). Of the total, 31 documents include a multi-taxonomic perspective, plus vegetation and other biological groups. Of these, up to 16 documents consider the relationships of vegetation with two or more distinct biological groups. On the other hand, vegetation studies directly related to ornithology (14 reports), arthropodology (3 reports) and herpetology (2 reports) have also been documented (Fig. 15.4).

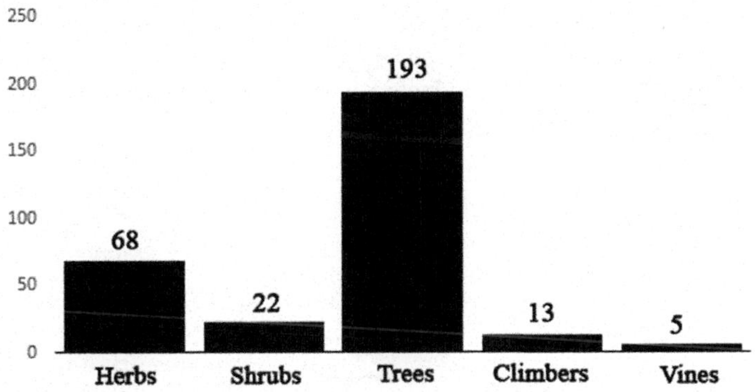

Fig. 15.3 Growth form diversity in AFS for the Mesoamerican region

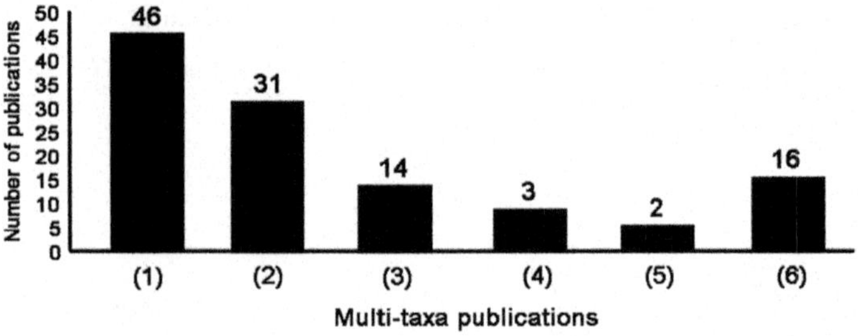

Fig. 15.4 Publications with multi-taxonomic perspective of the reported total. Number in parentheses: **1** Plants. **2** Multi-taxa. **3** Plants and birds. **4** Plants and arthropods. **5** Plants and herpetofauna. **6** Plants and other groups

15.5 Diversity of Flora in AFS

The flora recorded for AFS includes about 301 taxa. Of these, 286 have been identified to the species level and 15 at the genus level. Families with more species diversity are Fabaceae (73), Poaceae (23), Malvaceae (14), Bignoniaceae (11), Rutaceae (11), Euphorbiaceae (10), Moraceae (9) and Rubiaceae (9) which together accounts for 53% of the floral assemblage (Fig. 15.5). These plant families with more species in the Mesoamerican area coincides with the best represented families in the flora of AFS. The flora recorded in the AFS are mainly native species (244), followed by non-native species from other continents (33) and non-natives that do not grow naturally in those countries and are cultivated by humans (24).

The growth forms recorded in the studies reviewed is dominated by the arboreal habit (193 species), followed by the herbs (69 species), shrubs (22 species), climbers (12 species) and vines (5 species) (Fig. 15.6). The predominance of trees would increase if two studies conducted in Nicaragua (González-Valdivia 2003;

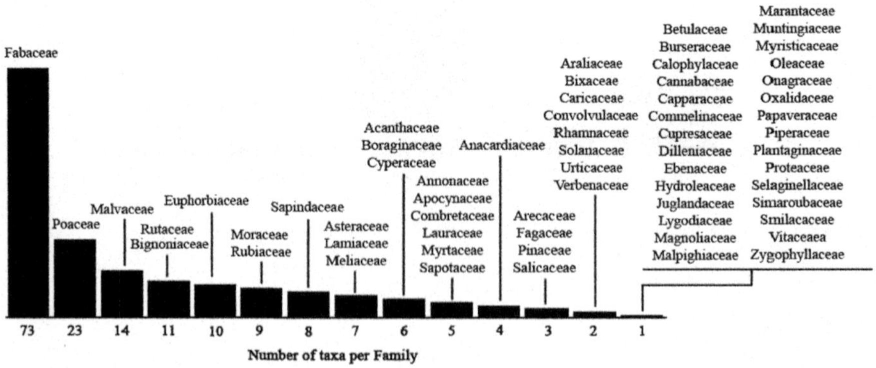

Fig. 15.5 Botanical families registered in the flora of AFS for the Mesoamerican region (including all growth forms)

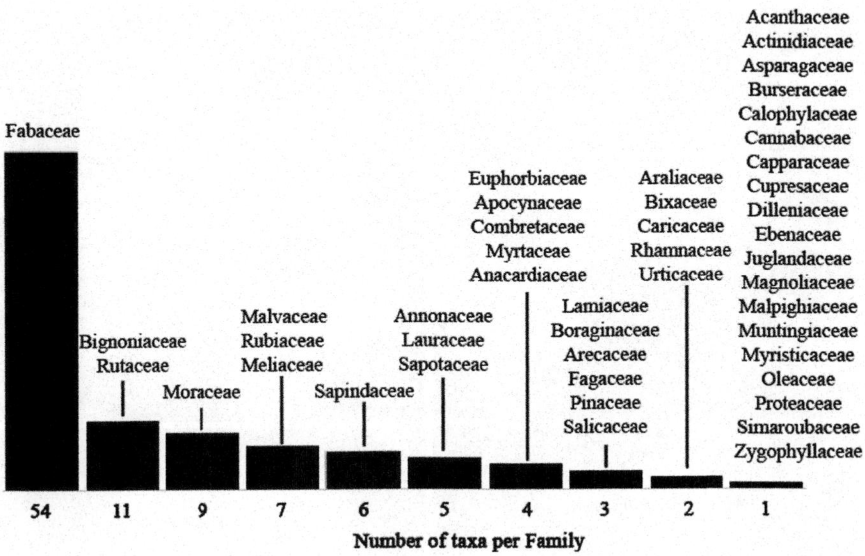

Fig. 15.6 Botanical families registered in the flora of AFS for the Mesoamerican region (including only trees)

Gómez et al. 2004) were not included. In the previous two works, herbs that grow spontaneously in the land dedicated to livestock are included, a situation quite widespread in several local studies.

15.6 Diversity of Trees in AFS

For the purpose of providing equality to the results between countries, we proceeded to the restricted comparison of trees, which by definition are the essential component of the agroforestry system. When considering only the trees (including palms), the number of species demonstrating less extreme, the order of importance of trees has been presented in the AFS by changes according to the countries. Of the 193 registered taxa of trees in the AFS, Mexico is the country with highest richness (90 species), followed by Nicaragua (75 species), Costa Rica (40) and other countries with less than 20 species (Table 15.2).

Studying in dctail thc botanical familics (including natives, non natives and non-American plant families), we found that legume (Fabaceae) members predominate over other plant families in the region (Fig. 15.6). The families such as Bignoniaceae, Moraceae, Rubiaceae and Meliaceae are also frequently represented in the AFS in the region. Within the families represented by non-native American species, Rutaceae provides an important component to tree flora in the AFS in Mesoamerica. This family is widely cultivated in Mesoamerica for its edible (*Citrus*) as well as medicinal (*Zanthoxylum*) importance (Fig. 15.7).

Fig. 15.7 Representative species of the AFS for the Mesoamerican region

Table 15.3 Plant species in agroforestry systems in Mesoamerica (based on the studies cited here)

Family	Taxa	HG	DO	M	B	G	E	H	N	C	P	Co	V
Acanthaceae	*Blechum pyramidatum* (Lam.) Urb.	Hb	N						×				
Acanthaceae	*Bravaisia integerrima* (Spreng.) Standl.	T	N						×				
Acanthaceae	*Dyschoriste quadrangularis* (Oertst.) Kuntze	Hb	N						×				
Acanthaceae	*Elytraria imbricata* (Vahl) Pers.	Hb	N						×				
Acanthaceae	*Nelsonia canescens* (Lam.) Spreng.	Hb	NT*						×				
Acanthaceae	*Trichanthera gigantean* (Humb. & Bonpl.) Nees	S	N									×	
Actinidiaceae	*Saurauia villosa* DC.	T	N	×									
Amaranthaceae	*Achyranthes aspera* L.	Hb	N						×				
Anacardiaceae	*Astronium graveolens* Jacq.	T	N	×	×				×	×			
Anacardiaceae	*Mangifera indica* L.	T	NT*	×			×						
Anacardiaceae	*Spondias mombin* L.	T	N	×		×			×				
Anacardiaceae	*Spondias purpurea* L.	T	N	×			×		×				
Annonaceae	*Annona cherimola* Mill.	T	NT						×				
Annonaceae	*Annona diversifolia* Saff.	T	N	×									
Annonaceae	*Annona muricata* L.	T	N	×							×		
Annonaceae	*Annona purpurea* Moc. & Sessé ex Dunal	T	N	×									
Annonaceae	*Annona reticulata* L.	T	N	×					×				
Apocynaceae	*Aspidosperma cruentum* Woodson	T	N		×								
Apocynaceae	*Aspidosperma megalocarpon* Müll. Arg.	T	N	×		×							
Apocynaceae	*Gonolobus barbatus* Kunth.	Cl	N						×				
Apocynaceae	*Stemmadenia donnell-smithii* (Rose) Woodson	T	N	×		×							
Apocynaceae	*Thevettia* sp.	S	N			×							
Araliaceae	*Dendropanax arboreus* (L.) Decne. & Planch.	T	N	×									
Araliaceae	*Sciadodendron excelsum* Griseb.	T	N						×				
Arecaceae		T	N						×	×			

(continued)

Table 15.3 (continued)

Family	Taxa	HG	DO	M	B	G	E	H	N	C	P	Co	V
	Acrocomia aculeata (Jacq.) Lodd. ex Mart.												
Arecaceae	*Attalea cohune* Mart.	T	N	×		×							
Arecaceae	*Cocos nucifera* L.	T	NT*	×									
Asparagaceae	*Dracaena fragrans* (L.) Ker Gawl	T	NT*							×			
Asteraceae	*Ageratum conyzoides* L.	Hb	N						×				
Asteraceae	*Baltimora recta* L.	Hb	N						×				
Asteraceae	*Calea urticifolia* (Mill.) DC.	Hb	N						×				
Asteraceae	*Elephantopus spicatus* B. Juss ex Aubl.	Hb	N						×				
Asteraceae	*Liabum* sp.	Hb	N			×							
Asteraceae	*Melampodium divaricatum* (Rich. ex Rich.) DC.	Hb	N						×				
Asteraceae	*Tridax procumbens* L.	Hb	N						×				
Betulaceae	*Alnus acuminata* Kunth.	T	N							×			
Bignoniaceae	*Crescentia alata* Kunth	T	N	×					×	×			
Bignoniaceae	*Crescentia cujete* L.	T	N	×							×		
Bignoniaceae	*Jacaranda caucana* Pittier	T	NT								×		
Bignoniaceae	*Parmentiera aculeata* (Kunth.) Seem.	T	N	×									
Bignoniaceae	*Tabebuia donnell-smithii* Rose	T	N	×		×							
Bignoniaceae	*Tabebuia guayacan* (Seem.) Hemsl.	T	N								×		
Bignoniaceae	*Tabebuia pentaphylla* (L.) Hemsl.	T	N	×									
Bignoniaceae	*Tabebuia ochraceae* A. H. Gentry	T	N						×	×			
Bignoniaceae	*Tabebuia rosea* (Bertol.) DC.	T	N	×		×	×		×	×	×		
Bignoniaceae	*Tabebuia serratifolia* (Vahl) G.Nicholson	T	N										×
Bignoniaceae	*Tabebuia* sp.	T	N	×									
Bixaceae	*Bixa orellana* L.	S	N	×									
Bixaceae	*Cochlospermum vitifolium* (Willd.) Spreng.	T	N						×				
Boraginaceae	*Bourreria mollis* Standl.	S	N			×							

(continued)

Table 15.3 (continued)

Family	Taxa	HG	DO	M	B	G	E	H	N	C	P	Co	V
Boraginaceae	*Cordia alliodora* (Ruiz & Pav.) Oken	T	N	×		×	×	×	×	×	×	×	
Boraginaceae	*Cordia bicolor* A. DC.	S	N						×				
Boraginaceae	*Cordia collococca* L.	T	N						×				
Boraginaceae	*Cordia dentata* Poir.	T	N						×				
Boraginaceae	*Cordia panamensis* L. Riley	S	N						×				
Burseraceae	*Bursera simaruba* (L.) Sarg.	T	N	×	×	×	×		×	×	×		
Calophyllaceae	*Calophyllum brasiliensis* (Cambess.)	T	N	×		×							
Cannabaceae	*Aphanathe monoica* (Hemsl.) J.F. Leroy	T	N	×									
Capparaceae	*Capparidastrum frondosum* (Jacq.) Cornejo & Iltis	T	N						×				
Caricaceae	*Carica papaya* L.	T	N	×		×							
Caricaceae	*Jacaratia mexicana* A. DC.	T	N			×							
Combretaceae	*Combretum fruticosum* (Loefl.) Stuntz	V	N						×				
Combretaceae	*Laguncularia racemosa* (L.) C.F. Gaertn.	T	N				×						
Combretaceae	*Terminalia amazonia* (J.F. Gmel.) Exell	T	N	×						×			
Combretaceae	*Terminalia ivorensis* A. Chev.	T	NT*							×	×		
Combretaceae	*Terminalia oblonga* (Ruiz & Pav.) Steud.	T	NT			×							
Commelinaceae	*Commelina diffusa* Burm. f.	Hb	N						×				
Convolvulaceae	*Ipomoea batatas* (L.) Lam.	Cl	N						×				
Convolvulaceae	*Ipomoea hederifolia* L.	Cl	N						×				
Cupresaceae	*Cupressus lusitanica* Mill.	T	NT*							×			
Cyperaceae	*Cyperus articulatus* L.	Hb	N						×				
Cyperaceae	*Cyperus luzulae* (L.) Retz.	Hb	N						×				
Cyperaceae	*Cyperus odoratus* L.	Hb	NT*						×				
Cyperaceae	*Cyperus rotundus* L.	Hb	NT*						×				
Cyperaceae	*Dichromena ciliata* Pers.	Hb	NT						×				
Cyperaceae		Hb	N						×				

(continued)

Table 15.3 (continued)

Family	Taxa	HG	DO	M	B	G	E	H	N	C	P	Co	V
	Scleria melaleuca Rchb. ex Schltdl. & Cham.												
Dilleniaceae	*Curatella americana* L.	T	N						×				
Ebenaceae	*Diospyros salicifolia* Humb. & Bonpl. ex Willd.	T	N						×				
Euphorbiaceae	*Acalypha alopecuroides* Jacq.	Hb	N						×				
Euphorbiaceae	*Acalypha* sp.	Hb	N			×							
Euphorbiaceae	*Caperonia palustris* (L.) A. St.-Hil.	Hb	N						×				
Euphorbiaceae	*Croton draco* Schltdl.	T	N						×				
Euphorbiaceae	*Hevea brasiliensis* (Willd. ex A. Juss.) Müll. Arg.	T	NT									×	
Euphorbiaceae	*Hura crepitans* L.	T	NT										×
Euphorbiaceae	*Manihot esculenta* Crantz.	S	NT						×				
Euphorbiaceae	*Phyllanthus amarus* Schumach. & Thonn.	Hb	N						×				
Euphorbiaceae	*Phyllanthus brasiliensis* (Aubl.) Poir.	S	N			×							
Euphorbiaceae	*Sapium macrocarpum* Müll. Arg.	T	N						×				
Fabaceae	*Acacia collinsii* Saff.	T	N						×				
Fabaceae	*Acacia cornigera* (L.) Willd.	S	N						×				
Fabaceae	*Acacia farnesiana* (L.) Willd.	S	N						×				
Fabaceae	*Acacia mangium* Willd.	T	NT							×			
Fabaceae	*Acacia pennatula* (Schltdl. & Cham.) Benth.	T	N	×					×				
Fabaceae	*Acrocarpus fraxinifolius* Arn.	T	NT			×							
Fabaceae	*Andira inermis* (Wright.) DC.	T	N							×			
Fabaceae	*Bauhinia divaricata* L.	S	N				×						
Fabaceae	*Bauhinia monandra* Kurz	S	NT*						×				
Fabaceae	*Bauhinia ungulata* L.	T	N				×						
Fabaceae		T	N									×	

(continued)

Table 15.3 (continued)

Family	Taxa	HG	DO	M	B	G	E	H	N	C	P	Co	V
	Caesalpinia pinnata (Griseb.) C. Wright												
Fabaceae	*Calliandra calothyrsus* Meisn.	T	N						×				
Fabaceae	*Calliandra houstoniana* (mil.) Standl.	T	N			×							
Fabaceae	*Calopogonium mucunoides* Desv.	Cl	N						×				
Fabaceae	*Cassia fistula* L.	T	NT*								×		
Fabaceae	*Cassia grandis* L. f.	T	NT				×		×				
Fabaceae	*Centrosema pubescens* Benth.	Cl	N						×				
Fabaceae	*Centrosema plumieri* (Pers.) Benth.	Cl	N						×				
Fabaceae	*Chloroleucon eurycyclum* Barneby & J.W. Grimes	T	NT							×			
Fabaceae	*Crudia choussyana* (Standl.) Standl.	T	NT				×						
Fabaceae	*Dalbergia retusa* Hemsl.	T	N					×	×	×			
Fabaceae	*Dalbergia stevensonii* Standl.	T	N	×									
Fabaceae	*Desmodium distortum* (Aubl.) J.F. Macbr.	Hb	N						×				
Fabaceae	*Desmodium procumbens* (Mill.) Hitchc.	Hb	N						×				
Fabaceae	*Diphysa americana* (Mill.) M. Sousa	T	N				×		×				
Fabaceae	*Diphysa robinioides* Benth.	T	N	×				×		×	×		
Fabaceae	*Enterolobium cyclocarpon* (Jacq.) Griseb.	T	N	×	×		×		×	×			
Fabaceae	*Erythrina americana* Mill.	T	N	×									
Fabaceae	*Erythrina berteroana* Urb.	T	N						×		×		
Fabaceae	*Erythrina costaricensis* Micheli	T	N							×			
Fabaceae	*Erythrina poepiggiana* (Walp.) O.F. Cook	T	N							×			
Fabaceae	*Erythrina* sp.	T	N	×									
Fabaceae		T	N	×			×		×		×		

(continued)

Table 15.3 (continued)

Family	Taxa	HG	DO	M	B	G	E	H	N	C	P	Co	V
	Gliricidia sepium (Jacq.) Walp.												
Fabaceae	*Haematoxylum campechianum* L.	T	N	×									
Fabaceae	*Hymenaea courbaril* L.	T	N				×		×				
Fabaceae	*Indigofera jamaicensis* Spreng.	Cl	N						×				
Fabaceae	*Inga densiflora* Benth.	T	N							×			
Fabaceae	*Inga jinicuil* Schltdl.	T	N	×									
Fabaceae	*Inga micheliana* Harms	T	N	×									
Fabaceae	*Inga punctata* Willd.	T	N						×				
Fabaceae	*Inga ruiziana* G. Don.	T	N				×						
Fabaceae	*Inga vera* Willd.	T	N	×					×				
Fabaceae	*Leucaena shannonii* Donn. Sm.	T	N						×				
Fabaceae	*Lysiloma aurita* (Schltdl.) Benth.	T	N				×		×				
Fabaceae	*Lysiloma divaricatum* (Jacq.) J.F.Macbr.	T	N				×						
Fabaceae	*Lonchocarpus guatemalensis* Benth.	T	N		×	×							
Fabaceae	*Lonchocarpus hondurensis* Benth.	T	N	×									
Fabaceae	*Lonchocarpus parviflorus* Benth.	T	N						×				
Fabaceae	*Lonchocarpus rugosus* Benth.	T	N	×									
Fabaceae	*Macroptilium atropurpureum* (DC.) Urb.	Cl	N	×					×				
Fabaceae	*Mimosa albida* Willd.	S	N						×				
Fabaceae	*Mimosa pigra* L.	S	N						×				
Fabaceae	*Mucuna pruriens* (L.) DC.	Cl	N						×				
Fabaceae	*Myrospermum frutescens* Jacq.	T	N						×				
Fabaceae	*Peltogyne purpurea* Pittier	T	NT		×								
Fabaceae	*Piscidia grandifolia* (Donn. Sm.) I.M. Johnst.	T	N						×				
Fabaceae	*Piscidia piscipula* (L.) Sarg.	T	N	×	×								
Fabaceae	*Pithecellobium dulce* (Roxb.) Benth.	T	N	×					×				

(continued)

Table 15.3 (continued)

Family	Taxa	HG	DO	M	B	G	E	H	N	C	P	Co	V
Fabaceae	*Pithecellobium oblongum* Benth.	T	N						×				
Fabaceae	*Platymiscium dimorphandrum* Donn. Sm.	T	N	×									
Fabaceae	*Platymiscium parviflorum* Benth.	T	N						×				
Fabaceae	*Platymiscium yucatanum* Standl.	T	N	×									
Fabaceae	*Platypodium elegans* Vogel	T	N								×		
Fabaceae	*Pterocarpus officinalis* Jacq.	T	N						×				
Fabaceae	*Rhynchosia minima* (L.) DC.	Cl	N						×				
Fabaceae	*Samanea saman* (Jacq.) Merr.	T	NT	×					×	×	×		×
Fabaceae	*Schyzolobium parahyba* (Vell.) S.F. Blake	T	N	×						×			
Fabaceae	*Senna atomaria* (L.) H. S.Irwin & Barneby	T	N	×					×				
Fabaceae	*Senna obtusifolia* (L.) H.S.Irwin & Barneby	Hb	N						×				
Fabaceae	*Senna* sp.	Hb	N			×							
Fabaceae	*Tamarindus indica* L.	T	NT*	×									
Fabaceae	*Teramnus labialis* (L. f.) Spreng.	Cl	N						×				
Fabaceae	*Teramnus uncinatus* (L.) Sw.	Cl	N						×				
Fagaceae	*Quercus oleoides* Schltdl. & Cham.	T	N	×									
Fagaceae	*Quercus sapotifolia* Liebm.	T	N						×				
Fagaceae	*Quercus segoviensis* Liebm.	T	N						×				
Hydroleaceae	*Hydrolea spinosa* L.	S	N						×				
Juglandaceae	*Juglans neotropica* Diels	T	NT							×			
Lamiaceae	*Cornutia pyramidata* L.	S	N			×			×				
Lamiaceae	*Gmelina arborea* Roxb.	T	NT*	×									
Lamiaceae	*Hyptis capitata* Jacq.	Hb	N						×				
Lamiaceae	*Hyptis pectinata* (L.) Poit.	Hb	N						×				

(continued)

Table 15.3 (continued)

Family	Taxa	HG	DO	M	B	G	E	H	N	C	P	Co	V
Lamiaceae	*Hyptis suaveolens* (L.) Poit.	Hb	N						×				
Lamiaceae	*Tectona grandis* L. f.	T	NT*	×			×						×
Lamiaceae	*Vitex gaumeri* Greenm.	T	N						×				
Lauraceae	*Cinnamomum triplinerve* (Ruiz & Pav.) Kosterm.	T	N						×				
Lauraceae	*Nectandra hihua* (Ruiz & Pav.) Rohwer	T	N			×							
Lauraceae	*Ocotea* sp.	T	N							×			
Lauraceae	*Persea americana* Mill.	T	N	×			×						
Lauraceae	*Persea schiedeana* Nees	T	N	×									
Lygodiaceae	*Lygodium venustum* Sw.	Cl	N						×				
Magnoliaceae	*Magnolia yoroconte* Dandy	T	N					×					
Malpighiaceae	*Byrsonima crassifolia* (L.) Kunth	T	N	×				×	×	×			
Malvaceae	*Byttneria aculeata* Jacq.	Hb	N						×				
Malvaceae	*Ceiba aesculifolia* (Kunth) Britten & Baker f.	T	N						×				
Malvaceae	*Ceiba pentandra* (L.) Gaertn.	T	N	×	×				×				×
Malvaceae	*Guazuma ulmifolia* Lam.	T	N	×		×	×		×	×			
Malvaceae	*Luehea candida* (Moc. & Sessé ex DC.) Mart.	T	N	×									
Malvaceae	*Luehea seemannii* Triana & Planch	T	N						×				
Malvaceae	*Melochia tomentosa* L.	Hb	N						×				
Malvaceae	*Pachira quinata* (Jacq.) W.S. Alverson	T	N	×					×	×			
Malvaceae	*Robinsoniella lindeniana*	S	N						×				
Malvaceae	*Sida acuta* Burm. f.	Hb	N						×				
Malvaceae	*Sida jussiaeana* DC.	Hb	N						×				
Malvaceae	*Theobroma bicolor* Humb. & Bonpl.	T	N	×									
Malvaceae	*Triumfetta lappula* L.	Hb	N						×				
Malvaceae	*Waltheria indica* L.	Hb	N						×				
Marantaceae	*Maranta arundinaceae* L.	Hb	N						×				

(continued)

Table 15.3 (continued)

Family	Taxa	HG	DO	M	B	G	E	H	N	C	P	Co	V
Meliaceae	*Cedrela odorata* L.	T	N	×	×	×	×	×	×	×	×		×
Meliaceae	*Guarea* sp.	T	N								×		
Meliaceae	*Melia azedarach* L.	T	N		×								
Meliaceae	*Swietenia humilis* Zucc.	T	N			×							
Meliaceae	*Swietenia macrophylla* King	T	N	×				×		×			×
Meliaceae	*Trichilia americana* (Sessé & Moc.) T.D. Penn.	T	N						×				
Meliaceae	*Trichilia hirta* L.	T	N								×		
Moraceae	*Artocarpus altilis* (Parkinson ex F.A. Zorn) Fosberg	T	NT*	×									
Moraceae	*Artocarpus communis* J.R. Forst. & G. Forst.	T	NT*	×									
Moraceae	*Brosimum alicastrum* Sw.	T	N	×		×							
Moraceae	*Ficus insipida* Willd.	T	N						×				
Moraceae	*Ficus aurea* Nutt.	T	N						×				
Moraceae	*Ficus pertusa* L. f.	T	N	×					×	×			
Moraceae	*Ficus* sp.	T	N	×									
Moraceae	*Poulsenia armata* (Miq.) Standl.	T	N	×									
Moraceae	*Trophis racemosa* (L.) Urb.	T	N			×							
Muntingiaceae	*Muntingia calabura* L.	T	N	×					×				
Myristicaceae	*Virola koschnyi* Warb.	T	N							×			
Myrtaceae	*Eucalyptus camaldulensis* Dehnh.	T	NT*				×						
Myrtaceae	*Eucalyptus deglupta* Blume	T	NT*							×			
Myrtaceae	*Eucalyptus globulus* Labill.	T	NT*							×			
Myrtaceae	*Myrciaria floribunda* (H.West ex Willd.) O. Berg	S	N						×				
Myrtaceae	*Psidium guajava* L.	T	N	×			×	×	×				
Oleaceae	*Fraxinus uhdei* (Wenz.) Lingelsh.	T	NT							×			
Onagraceae	*Ludwigia decurrens* Walter	Hb	N						×				
Oxalidaceae	*Oxalis corniculata* L.	Hb	N						×				
Papaveraceae	*Argemone mexicana* L.	Hb	N						×				
Pinaceae		T	NT							×			

(continued)

Table 15.3 (continued)

Family	Taxa	HG	DO	M	B	G	E	H	N	C	P	Co	V
	Pinus caribaea Morelet												
Pinaceae	*Pinus oocarpa* Schiede	T	N					×	×	×			
Pinaceae	*Pinus maximinoi* H.E. Moore	T	N						×				
Piperaceae	*Piper* sp.	S	N			×							
Plantaginaceae	*Scoparia dulcis* L.	Hb	N	×									
Poaceae	*Andropogon gayanus* Kunth	Hb	NT						×				
Poaceae	*Axonopus compressus* (Sw.) P. Beauv.	Hb	NT						×				
Poaceae	*Bambusa vulgaris* Schrad.	Hb	NT*	×									
Poaceae	*Brachiaria brizantha* (A. Rich.) Stapf	Hb	NT							×			
Poaceae	*Brachiaria decumbens* Stapf	Hb	NT							×			
Poaceae	*Chloris radiata* (L.) Sw.	Hb	N						×				
Poaceae	*Cynodon nlemfuensis* Vanderyst	Hb	NT*						×				
Poaceae	*Cynodon plectostachyus* (K. Schum.) Pilg.	Hb	NT*						×				
Poaceae	*Dichanthium aristatum* (Poir.) C.E. Hubb.	Hb	NT*						×				
Poaceae	*Echinochloa colona* (L.) Link	Hb	NT*						×				
Poaceae	*Guadua angustifolia* Kunth	Hb	N									×	
Poaceae	*Hyparrhenia rufa* (Nees) Stapf	Hb	NT*						×	×			
Poaceae	*Ixophorus unisetus* (J. Presl) Schltdl.	Hb	N						×				
Poaceae	*Melinis minutiflora* P. Beauv.	Hb	NT*						×				
Poaceae	*Oplismenus burmannii* (Retz.) P. Beauv.	Hb	N						×				
Poaceae	*Panicum maximun* Jacq.	Hb	NT*						×				
Poaceae	*Panicum* sp.	Hb	NT						×				
Poaceae	*Paspalum centrale* Chase	Hb	N						×				
Poaceae	*Paspalum conjugatum* P.J. Bergius	Hb	N						×				
Poaceae		Hb	N						×				

(continued)

Table 15.3 (continued)

Family	Taxa	HG	DO	M	B	G	E	H	N	C	P	Co	V
	Paspalum notatum Flüggé												
Poaceae	*Paspalum plicatulum* Michx.	Hb	N						×				
Poaceae	*Paspalum virgatum* L.	Hb	N						×				
Poaceae	*Setaria parviflora* (Poir.) M. Kerguélen	Hb	N						×				
Proteaceae	*Grevillea robusta* A. Cunn. Ex R. Br.	T	NT			×				×			
Rhamnaceae	*Karwinskia calderonii* Standl.	T	N						×				
Rhamnaceae	*Ziziphus mexicana* Rose	T	N	×									
Rubiaceae	*Alibertia edulis* (Rich.) A. Rich. ex DC.	T	N	×									
Rubiaceae	*Blepharidium mexicanum* Standl.	T	N	×									
Rubiaceae	*Calycophyllum candidissimum* (Vahl) DC.	T	N						×				
Rubiaceae	*Genipa americana* L.	T	N	×			×		×				
Rubiaceae	*Genipa* sp.	T	N	×									
Rubiaceae	*Hamelia patens* Jacq.	S	N			×			×				
Rubiaceae	*Randia armata* (Sw.) DC.	T	N						×				
Rubiaceae	*Simira salvadorensis* (Standl.) Steyerm.	T	N	×									
Rubiaceae	*Spermacoce remota* Lam.	Hb	N					×					
Rutaceae	*Citrus aurantiifolia* (Christm.) Swingle	T	NT*								×		
Rutaceae	*Citrus x aurantium* L.	T	NT*				×						
Rutaceae	*Citrus limon* (L.) Osbeck	T	NT*	×									
Rutaceae	*Citrus mitis* Blanco	T	NT*								×		
Rutaceae	*Citrus nobilis* Lour.	T	NT*	×									
Rutaceae	*Citrus reticulata* Blanco	T	NT*	×									
Rutaceae	*Citrus sinensis* (L.) Osbeck	T	NT*	×							×		
Rutaceae	*Zanthoxylum* sp.	T	N	×									
Rutaceae	*Zanthoxylum elephantiasis* Macfad.	T	NT						×				
Rutaceae	*Zanthoxylum kellermanii* P. Wilson	T	N	×									

(continued)

Table 15.3 (continued)

Family	Taxa	HG	DO	M	B	G	E	H	N	C	P	Co	V
Rutaceae	*Zanthoxylum juniperinum* Poepp.	T	N			×							
Salicaceae	*Casearia corymbosa* Kunth	T	N						×				
Salicaceae	*Casearia sylvestris* Sw.	T	N						×				
Salicaceae	*Zuelania guidonia* (Sw.) Britton & Millsp.	T	N	×					×				
Sapindaceae	*Cupania belizensis* Standl.	T	N			×							
Sapindaceae	*Cupania dentata* Moc. & Sessé ex DC.	T	N	×					×				
Sapindaceae	*Cupania glabra* Sw.	T	N	×									
Sapindaceae	*Cupania guatemalensis* (Turcz.) Radlk.	T	N							×			
Sapindaceae	*Sapindus saponaria* L.	T	N	×					×				
Sapindaceae	*Serjania atrolineata* C. Wright	V	N						×				
Sapindaceae	*Serjania* sp.	V	N						×				
Sapindaceae	*Thouinidium decandrumc* (Humb. & Bonpl.) Radkl.	T	N						×				
Sapotaceae	*Chrysophyllum cainito* L.	T	N	×									
Sapotaceae	*Chrysophyllum mexicanum* Brandegee	T	N			×							
Sapotaceae	*Manilkara sapota* Van Royen	T	N	×									
Sapotaceae	*Pouteria amygdalina* (Standl.) Baehni	T	N		×								
Sapotaceae	*Pouteria sapota* (Jacq.) H.E. Moore & Stearn	T	N	×									
Selaginellaceae	*Selaginella sertata* Spring	Hb	N						×				
Simaroubaceae	*Simarouba amara* Aubl.	T	N				×		×	×			
Smilacaceae	*Smilax spinosa* Mill.	V	N						×				
Solanaceae	*Cestrum racemosum* Ruiz & Pav.	S	N			×							
Solanaceae	*Solanum* sp.	S	N			×							
Urticaceae	*Cecropia obtusifolia* Bertol.	T	N	×		×							
Urticaceae	*Cecropia peltata* L.	T	N	×									
Verbenaceae	*Lantana camara* L.	Hb	N			×			×				
Verbenaceae	*Stachytarpheta frantzii* Pol.	Hb	N						×				

(continued)

Table 15.3 (continued)

Family	Taxa	HG	DO	M	B	G	E	H	N	C	P	Co	V
Vitaceae	*Cissus microcarpa* Vahl	V	N						×				
Zygophyllaceae	*Bulnesia arborea* (Jacq.) Engl.	T	N										×

Abbreviations **Habit growth (HG):** *Cl* climber. *Hb* herb. *T* tree. *S* shrub. *V* vines. **Distribution origin (DO):** *N* native. *NT* non-native for the country. * Non-American plant. **Countries:** *M* Mexico. *B* Belize. *G* Guatemala. *E* El Salvador. *H* Honduras. *C* Costa Rica (Mesoamerica). *P* Panama. *Co* Colombia. *V* Venezuela (South America)

Most of the trees that have been identified in the AFS have low frequency of occurrence and are not represented in all the countries, with the exception of a few species (Table 15.3). The common species recorded in the studies on AFS are *Cedrela odorata* (eight countries), *Bursera simarouba* and *Cordia alliodora* (both in seven countries). This is followed by *Tabebuia rosea* (six countries) and finally *Enterolobium cyclocarpon* and *Guazuma ulmifolia* (in five countries each one) (Table 15.3).

The most common species in the Mesoamerican AFS are part of a set of trees that are considered to be useful for multiple purposes (Roman-Miranda 2001; Zamora et al. 2001; Ochoa-Gaona et al. 2012). For example, the common species providing fuel, fodder, fiber, timber, fixing atmospheric nitrogen/carbon, providing shade and/or posts for local producers and/or farmers, as well as being able to resist floods, droughts, fires, grazing or frequent pruning, sprouting successfully and quickly recovering the coverage of their foliage are more widely cultivated (Lombo-Ortiz 2012). The case of *Cedrela odorata* may be influcnced by reforestation programs that are being promoted in the region for the past several decades.

15.7 Conclusion

It is interesting to note that in this region, a research process has been initially established on a particular species or a specific group of plants (trees) within the AFS, but more recent studies tend to include a perspective that extends to the flora as a whole. It is also important to note that there has been a recent tendency to also include other biological groups, highlighting the birds. The number of publications has been gradually increased, and these have also increased the localized efforts to establish the role of the AFS, as adequate sites to conserve species for Mesoamerican flora, mostly include species with at least some degrees of threat. The countries that have conducted more efforts to document the associated AFS in the Mesoamerica plants predominantly include Mexico, Costa Rica and Nicaragua. Other countries in Mesoamerica have fewer studies available and traceable following the sampling method used in this particular study; however, they participated as a whole throughout the period involved in this specific review.

References

Alemán-Mejía JB (2008) Caracterización de reptiles y percepción local hacia las serpientes en fincas ganaderas de la subcuenca del Río Copán, Honduras. MC Thesis, CATIE. Turrialba

Andrade HJ, Segura M, Somarriba E, Villalobos M (2008a) Valoración biofísica y financiera de la fijación de carbono por uso del suelo en fincas cacaoteras indígenas de Talamanca, Costa Rica. Agroforestería en las Américas 46:45–50

Andrade HJ, Esquivel H, Ibrahim M (2008b) Disponibilidad de forrajes en sistemas silvopastoriles con especies arbóreas nativas en el trópico seco de Costa Rica. ZT 26: 289–292

Araujo Y, López-Hernández D (1999) Caracterización de las poblaciones de lombrices de tierra en un sistema de agricultura orgánica ubicado en una sabana en el Amazonas venezolano. Ecotropicos 12:49–55

Arellano L, Favila ME, Huerta C (2005) Diversity of dung and carrion beetles in a disturbed Mexican tropical montane cloud forest and on shade coffee plantations. Biodivers Conserv 14:601–615

Arellano L, León-Cortés JL, Halffter G, Montero J (2013) Acacia woodlots, cattle and dung beetles (Coleoptera: Scarabaeinae) in a Mexican silvopastoral landscape. Rev Mex Biodivers 84:650–660

Armbrecht I, Perfecto I, Vandermeer J (2004) Enigmatic biodiversity correlations: Ant diversity responds to diverse resources. Science 304:284–286

Armbrecht I, Perfecto I, Silverman E (2006) Limitation of nesting resources for ants in Colombian forests and coffee plantations. Ecol Entomol 31:403–410

Basto-Estrella G, Rodríguez-Vivas R, Delfín-González H, Reyes-Novelo E (2012) Escarabajos estercoleros (Coleoptera: Scarabaeidae: Scarabaeinae) de ranchos ganaderos de Yucatán, México. Rev Mex Biodivers 83:380–386

Bernard HR (2006) Research methods in anthropology: qualitative and quantitative approaches, 4th edn. Library of Congress, Washington DC

Botero R, Russo RO (1998) Utilización de árboles y arbustos fijadores de nitrógeno en sistemas sostenibles de producción animal en suelos ácidos tropicales. Conferencia electrónica de la FAO sobre "Agroforestería para la producción animal en Latinoamérica", San José

Brown G, Moreno AG, Barois I, Fragoso C, Rojas P, Hernández B, Patrón JC (2004) Soil macrofauna in SE Mexican pastures and the effect of conversion from native to introduced pastures. Agric Ecosyst Environ 103:313–327

Cajas YS, Sinclair F (2001) Characterization of multistrata silvopastoral systems on seasonally dry pastures in the Caribbean Region of Colombia. Agrofor Syst 53:215–225

Calvo L, Blake J (1998) Bird diversity and abundance on two different shade coffee plantations in Guatemala. Bird Conserv Int 8:297–308

Cárdenas G, Harvey C, Ibrahim M, Finegan B (2003) Diversidad y riqueza de aves en diferentes hábitats en un paisaje fragmentado en Cañas, Costa Rica. Agroforestería en las Américas 10:78–85

Casasola F, Ibrahim M, Harvey C, Kleinn C (2001) Caracterización y productividad de sistemas silvopastoriles tradicionales en Moropotente, Estelí. Nicaragua. Agroforestería en las Américas 8(30):17–20

Cerezo A, Robbins CS, Dowell B (2013) Uso de hábitats modificados por aves dependientes de bosque tropical en la región caribeña de Guatemala. Rev Biol Trop 57:401–419

Cleveland CC, Towsend AR, Schmidt SK, Constance BC (2003) Soil microbial dynamics and biogeochemistry in tropical forests and pastures, southwestern costa rica. Ecol Appl 13:314–326

Córdoba C, Cerda R, Deheuvels O, Hidalgo E, Declerck F (2013) Polinizadores, polinización y producción potencial de cacao en sistemas agroforestales de Bocas del Toro, Panamá. Agroforestería en las Américas 49:26–32

Cruz-Angón A, Greenberg R (2005) Are epiphytes important for birds in coffee plantations? An experimental assessment. J Appl Ecol 42:150–159

Cruz-Angón A, Baena ML, Greenberg R (2009) The contribution of epiphytes to the abundance and species richness of canopy insects in a Mexican coffee plantation. J Trop Ecol 25:453–463
Daily GC, Sanchez-Azofeifa GA, Ehrlich PR (2001) Countryside biogeography: use of human-dominated habitats by the avifauna of southern Costa Rica. Ecol Appl 11:1–13
De la Mora-Rodríguez A (2014) Hormigas en cafetales: Estudio sobre la calidad de la matriz en fincas cafetaleras en el Soconusco, Chiapas. PhD Thesis, Colegio de la Frontera Sur, San Cristóbal de las Casas
De la Ossa-Lacayo A (2013) Cercas vivas y su importancia en la conservación de avifauna nativa. Rev Colombiana Cienc Anim 5:171–193
De la Ossa J, Galván-Guevara S, De la Ossa-Lacayo A (2012) Importancia del jagüey ganadero en la conservación local de aves silvestres en el Caribe colombiano. Revista U.D.C.A Actualidad & Divulgación Científica 15:181–186
Decaëns T, Jiménez JJ, Barros SE, Chauvel A, Blanchart E, Lavelle P (2004) Soil macrofaunal communities in permanent pastures derived from tropical forest or savanna. Agric Ecosyst Environ 103:301–312
Decaëns T, Lavelle P, Jiménez JJ, Escobar G, Rippenstein G (1994) Impact of land management on soil macrofauna in the Oriental Llanos of Colombia. Eur J Soil Biol 30:157–168
Díaz-Bohórquez AM, Bayley NJ, Botero JE, Gómez C (2014) Aves migratorias en agroecosistemas del norte de Latinoamérica, con énfasis en Colombia. Ornitología Colombiana 14:3–27
Dietsch TV, Perfecto I, Greenberg R (2007) Avian foraging behavior in two different types of coffee agroecosystem in Chiapas, Mexico. Biotropica 39:232–240
Durango W, Uribe L, Henríquez C, Mata R (2015) Respiración, biomasa microbiana, y actividad fosfatasa del suelo en dos agroecosistemas y un bosque en Turrialba, Costa Rica. Agron Costarric 39:37–46
Estrada A, Sáenz J, Harvey C, Naranjo E, Muñoz D, Rosales-Medina M (2006) Primates in agroecosystems: conservation value of some agricultural practices in Mesoamerican landscapes. In: Estrada A, Garber P, Pavelka M (eds) Study of Mesoamerican Primates: distribution, ecology, behavior and conservation. Kluwer Academic Publisher, New York, pp 437–470
Esquivel-Sheik MJ, Calle Díaz Z (2002) Árboles aislados en potreros como catalizadores de la sucesión en la Cordillera Occidental Colombiana. Agroforestería en las Américas 9:43–47
Esquivel H, Ibrahim M, Harvey C, Villanueva C, Benjamín T, Sinclair F (2003) Árboles dispersos en potreros de fincas ganaderas de un ecosistema seco de Costa Rica. Agroforestería en las Américas 10(39–40):24–29
Fajardo D, Johnston-González R, Neira L, Chará J, Murgueitio R (2009) Influencia de los sistemas silvopastoriles en la diversidad de aves en la Cuenca del río La Vieja, Colombia. Revista Recursos Naturales y Ambiente 58:9–16
Ferguson BG, Vandermeer J, Morales H, Griffith DM (2003) Post agricultural succession in El Petén. Guatemala. Conserv Biol 17(3):818–828
Ferguson BG (2001) Seed dispersal by livestock accelerates succession in a degraded Neotropical field (Guatemala). Ecol Restor 19:117
Fonte SJ, Barrios E, Six J (2010) Earthworms, soil fertility and aggregate-associated soil organic matter dynamics in the Quesungual agroforestry system. Geoderma 155:320–328
Ford A, Clarke KC, Morlet S (2011) Calculating late classic lowland Maya population for The Upper Belize River area. Res Rep Belizean Archaeol 8:75–87
García-Estrada C, Damon A, Sánchez-Hernández C, Soto-Pinto L, Ibarra-Núñez G (2006) Bat diversity in montane rainforest and shaded coffee under different management practices in southeastern Chiapas, Mexico. Biol Conserv 132:351–361
García E, Jaime M, Mejía B, Guillén L, Harvey CA (2001) Árboles disperses dentro de cultivos anuales en el municipio de Ilobasco. El Salvador. Agroforestería en las Américas 8(31):39–44
Gómez-Pompa A, Kaus A (1992) Taming the wilderness myth. BioScience 42:271–279
González-Valdivia NA, González-Escolástico G, Barba-Macías E, Hernández-Daumás S, Ochoa-Gaona S (2013) Mirmecofauna asociada con sistemas agroforestales en el Corredor Biológico Mesoamericano en Tabasco, México. Rev Mex Biodivers 84:306–317

González-Valdivia NA, Barba-Macías E, Hernández-Daumás S, Ochoa-Gaona S (2014) Avifauna en sistemas silvopastoriles en el Corredor Biológico Mesoamerícano, Tabasco, México. Rev Biol Trop 62:1031–1052

Gómez R, López M, Harvey CA, Villanueva C (2004) Caracterización de las fincas ganaderas y relaciones con la cobertura arbórea en potreros en el municipio de Belén, Rivas, Nicaragua. Revista Encuentros (UCA, Nicaragua) 36:94–113

Gómez-Pompa A, Allen M, Fedick SL, Jiménez-Osornio JJ (2003) The lowland Maya area. Three millennia at the human-wildland interface. Food Product Press, New York

González-Valdivia NA (2003) Dos sistemas silvopastoriles como refugios de vida silvestre en el municipio de Estelí. MC Thesis, Universidad Nacional Autónoma de Nicaragua-Léon, León

González-Valdivia NA, Ochoa-Gaona S, Ferguson BG, Pozo C, Rangel-Ruiz LJ, Arriaga-Weiss SL, Kampichler C (2011) Indicadores ecológicos de hábitat y biodiversidad en un paisaje neotropical: perspectiva multitaxonómica. Rev Biol Trop 59:1433–1451

Goodall KE, Bacon CM, Méndez VE (2015) Shade tree diversity, carbon sequestration, and epiphyte presence in coffee agroecosystems: A decade of smallholder management in San Ramón, Nicaragua. Agric Ecosyst Environ 199:200–206

Greenberg R, Bichier P, Sterling J (1997a) Acacia, cattle and migratory birds in Southeastern México. Biol Cons 80:235–247

Greenberg R, Bichier P, Sterling J (1997b) Bird populations in rustic and shade coffee plantations of eastern Chiapas, Mexico. Biotropica 29:501–514

Guevara S, Laborde J (1993) Monitoring seed dispersal at isolated standing trees in tropical pastures: consequences for local species availability. Vegetatio 107(108):319–338

Guevara S, Purata S, van der Maarel E (1986) The role of remnant forest trees in tropical secondary succession. Vegetatio 66:77–84

Guevara S, Meave J, Moreno-Casasola P, Laborde J (1992) Floristic composition and structure of vegetation under isolated trees in neotropical pastures. J Veg Sci 3:655–664

Guiracocha G, Harvey C, Somarriba E, Krauss U, Carrillo E (2001) Conservación de la biodiversidad en sistemas agroforestales con cacao y banano en Talamanca, Costa Rica. Agroforestería en las Américas 8:7–11

Halffter G (2011) Reservas de la biósfera: problemas y oportunidades en México. Acta Zool Mex 27:177–189

Harvey C, Saénz J, Montero J, Medina A, Sánchez D, Vílchez S, Hernández B, Maes JM, Sinclair FL (2004a) Abundance and species richness of trees, birds, bats, butterflies and dung beetles in a sylvopastoral systems in the agricultural landscapes of Costa Rica and Rivas, Nicaragua. In: ′t. Manetje L, Ramírez L, Ibrahim M, Sandoval C, Ojeda N, Ku J (eds) The importance of silvopastoral system in rural livelihoods to provide ecosystem services, Memorias del Segundo Simposio Internacional de Sistemas Silvopastoriles, Universidad Autónoma de Yucatán, Mérida,73–76

Harvey CA, Tucker N, Estrada A (2004b) Live fences, isolated trees and windbreaks: tools for conserving biodiversity in fragmented tropical landscapes? In: Schroth G, Fonseca GAB, Harvey CA, Gascon C, Vasconcelos HL, Izac AMN (eds) Agroforestry and biodiversity conservation in tropical landscapes, Island Press, Washington, pp. 261–289

Harvey C (2000) Evaluación y documentación de proyectos agroforestales en pequeña escala en América Central. Agroforestería en las Américas 7:72–73

Harvey C, Haber W (1999) Remnant trees and the conservation of biodiversity in Costa Rica pastures. Agrofor Syst 44:37–68

Harvey C, Haber WA, Solano R, Mejías F (1999) Árboles remanentes en potreros de Costa Rica: Herramientas para la conservación. Agroforestería de las Américas 6:19–22

Harvey CA, Villanueva C, Villacís J, Chacón M, Muñoz D, López M, Ibrahim M, Gómez R, Taylor R, Martínez J, Navas A, Sáenz J, Sánchez D, Medina A, Vílchez S, Hernández B, Pérez A, Ruiz F, López F, Lang I, Sinclair FL (2005) Contribution of live fences to the ecological integrity of agricultural landscapes. Agric Ecosyst Environ 111:200–230

Harvey CA, Medina A, Merlo-Sánchez D, Vílchez S, Hernández B, Sáenz J, Maes JM, Casanoves F, Sinclair FL (2006a) Patterns of animal diversity in different forms of tree cover in agricultural landscapes. Ecol Appl 16(5):1986–1999

Harvey CA, Medina A, Sánchez DM, Vílchez S, Hernández B, Sáenz JC, Maes JM, Casanoves F, Sinclair FL (2006b) Patterns of animal diversity in different forms of tree cover in agricultural landscapes. Ecol Appl 16:1986–1999

Harvey CA, Komar O, Chazdon R, Ferguson BG, Finegan B, Griffith DM, Martínez-Ramos M, Morales H, Nigh R, Soto-Pinto L, Van Breugel M, Wishnie M (2008) Integrating agricultural landscapes with biodiversity conservation in the Mesoamerican Hotspot. Conserv Biol 22:8–15

Hernández B, Maes JM, Harvey CA, Vílchez S, Medina A, Sánchez D (2003) Abundancia y diversidad de escarabajos coprófagos y mariposas diurnas en un paisaje ganadero en el departamento de Rivas, Nicaragua. Agroforestería en las Américas 10:93–102

Hernández-X E, Inzunza FR, Solano CB, Arias LM, Parra MR (2011) La tecnología del cultivo. Revista de Geografía Agrícola 46–47:91–96

Hernández L, Ojeda A, López-Hernández D (2012) Características bio-ecológicas en poblaciones de lombrices (Oligochaeta: Glossoscolecidae) de una sabana natural y una protegida en los llanos centrales de Venezuela. Rev Biol Trop 60:1217–1229

Holl KD (1998) Do bird perching structures elevate seed rain and seedling establishment in abandoned tropical pasture? Restor Ecol 6:253–261

Ibarra-Núñez G, García-Ballinas JA (1998) Diversidad de tres familias de arañas tejedoras (Araneae: Araneidae, Tetragnathidae, Theridiiae) en cafetales del Soconusco, Chiapas, Mexico. Folia Entomol Mex 102:11–20

Ibarra A, Estrada A, Arriaga S (2001) Avifauna asociada a dos cacaotales tradicionales en la región de la Chontalpa, Tabasco, México. Universidad y Ciencia 17:101–112

Ibrahim M, Schlonvoig A, Camargo JC, Souza M (2001) Multi-strata silvipastoral systems for increasing productivy and conservation of natural resources in Central America. In Gomide JA, Soares Mattos WE, Carneiro da Silva S (eds) Proceedings of the XIX International Grassland Congress, Piracicaba, pp 645–649

Jackson D, Vandermeer J, Perfecto I (2009) Spatial and temporal dynamics of a fungal pathogen promote pattern formation in a tropical agroecosystem. TOECOLJ 2:62–73

Janzen DH (1982a) Natural history of guacimo fruits (Sterculiaceae: *Guazuma ulmifolia*) with respect to consumption by large mammals. Am J Bot 69:1240–1250

Janzen DH (1982b) Horse response to *Enterolobium cyclocarpum* (Leguminosae) fruit crop size in a Costa Rican deciduous forest pasture. Brenesia 19–20:209–219

Jha S, Vandermeer JH (2009) Contrasting bee foraging in response to resource scale and local habitat management. Oikos 118:1174–1180

Jiménez JJ, Moreno AG, Decaëns T, Lavelle P, Fisher MJ, Thomas RJ (1998) Earthworm communities in native savannas and man-made pastures of the Eastern Plains of Colombia. Biol Fertil Soils 28:101–110

Kraker-Castañeda C, Pérez-Consuegra SG (2011) Contribución de los cafetales bajo sombra en la conservación de murciélagos en la antigua Guatemala. Guatemala. Acta Zool Mex 27:291–303

Lombo-Ortiz DF (2012) Evaluación de la disponibilidad de biomasa y capacidad de rebrote de leñosas forrajeras en potreros del trópico seco de Nicaragua. MC Thesis, CATIE, Turrialba

López L, Armbrecht I, Montoya-Lerma J, Molina EJ (2013) Diversidad de avispas parasitoides en un sistema silvopastoril orgánico de producción ganadera de Colombia. Avances en Investigación Agropecuaria 17:65–78

Macip-Ríos R, Casas-Andreu G (2008) Los cafetales en México y su importancia para la conservación de los anfibios y reptiles. Rev Mex Biodivers 24:143–159

Macip-Ríos R, Muñoz-Alonzo A (2008) Diversidad de lagartijas en cafetales y bosque primario en el Soconusco chiapaneco. Rev Mex Biodivers 79:185–195

Maffi L (2005) Linguistic, cultural, and biological diversity. Ann Rev Anthropol 29:599–617

Mariaca-Méndez R (2012) La complejidad del huerto familiar Maya del sureste de México. El huerto familiar del sureste mexicano, SERNAPAN, Villahermosa

Martínez-Encino C, Villanueva-López G, Casanova-Lugo F (2013) Densidad y composición de árboles dispersos en potreros en la sierra de Tabasco, México. Agrociencia 47:483–496

Mas AH, Dietsch TV (2003) An index of management intensity for coffee agroecosystems to evaluate butterfly species richness. Ecol Appl 13:1491–1501

Mas AH, Dietsch TV (2004) Linking shade coffee certification to biodiversity conservation: butterflies and birds in Chiapas, Mexico. Ecol Appl 14:642–654

Medina A, Harvey C, Sánchez D, Vilchez S, Hernández S (2007) Bat diversity and movement in a neotropical agricultural landscape. Biotropica 39:120–128

Méndez VE, Bacon CM, Olson M, Morris KS, Shattuck A (2010) Agrobiodiversity and shade coffee smallholder livelihoods: a review and synthesis of ten years of research in Central America. Prof Geogr 62:357–376

Moguel P, Toledo VM (1999) Biodiversity conservation in traditional coffee systems of Mexico. Conserv Biol 13:1–12

Montagnini F (2012) Los sistemas agroforestales y su contribución para la mitigación y adaptación al cambio Climático. Alcance- Revista de la Facultad de Agronomía UCVMaracay 71:1–24

Montes de Oca E (2001) Escarabajos coprófagos de un escenario ganadero típico de la región de los Tuxtlas, Veracruz, México: importancia del paisaje en la composición de un gremio funcional. Acta Zool Mex 82:111–132

Munroe T, Pettersson B, Ibrahim M, Harvey C (2004) A Biodiversity inventory and evaluation of forest and silvopastoral systems in Costa Rica. In: ′t. Manetje L, Ramírez L, Ibrahim M, Sandoval C, Ojeda N, Ku J (eds) The importance of silvopastoral system in rural livelihoods to provide ecosystem services, Memorias del Segundo Simposio Internacional de Sistemas Silvopastoriles, Universidad Autónoma de Yucatán, Mérida, pp. 117–120

Nadkarni NM, Haber WA (2009) Canopy seed banks as time capsules of biodiversity in pasture-remnant tree crowns. Conserv Biol 23:1117–1126

Neita JC, Escobar F (2012) The potential value of agroforestry to dung beetle diversity in the wet tropical forests of the Pacific lowlands of Colombia. Agroforest Syst 85:121–131

Nestel D, Dicksche F, Altieri M (1992) Diversity patterns of soil macro-Coleoptera in Mexican shaded and unshaded coffee agroecosystems: an indication of habitat perturbation. Biodivers Conserv 2:70–78

Nielsen K, DeRosier D (2000) Windbreaks as corridors for birds. In: Nadkarni MN, Wheelwright NT (eds) Monteverde: ecology and conservation of a tropical cloud forest. Oxford University Press, New York, pp 351–376

Nieto H, Gómez M, Somarriba E (2001) Contribución de *Acacia pennatula* (carbón) a la productividad agroforestal sostenible de la Reserva Natural Miraflor-Moropotente, Estelí, Nicaragua. Agroforestería en las Américas 8:21–23

Ochoa-Gaona S, Zamora-Cornelio LF, Cabrera-Pérez S, González-Valdivia NA, Pérez-Hernández I, López-Moreno V (2012) Flora leñosa útil de la sierra de Tenosique, Tabasco. El Colegio de la Frontera Sur, Villahermosa

Otero-Arnaiz A, Castillo S, Meave J, Ibarra-Manríquez G (1999) Isolated pasture and the vegetation under their canopies in the Chiapas coastal plain, Mexico. Biotropica 31:243–254

Pascacio-Damián MG (2006) Influencia del paso por el tracto digestivo y defecación por ganado bovino sobre el destino de semillas de tres fabaceas de bosque tropical subcaducifolia, en el Ejido La Florida Municipio de Cintalapa, Chiapas. BCs Thesis, Universidad de Ciencias y Artes de Chiapas, Tuxtla Gutiérrez

Pérez AM, Bornemann G, Campo L, Sotelo M, Ramírez F, Arana I (2005) Relaciones entre biodiversidad y producción en sistemas silvopastoriles de América Central. Ecosistemas 14:132–141

Pérez E, Richers B, DeClerck F, Casanoves F, Gobbi J, Benjamín T (2011) Uso y manejo de la cobertura arbórea en sistemas silvopastoriles en la subcuenca del río Copán, Honduras. Agroforestería en las Américas 48:26–35

Perfecto I, Vandermeer JH (1994) Understanding biodiversity loss in agroecosystems: reduction of ant diversity resulting from transformation of the coffee ecosystem in Costa Rica. Entomology (Trends Agric. Sci.) 2:7–13

Perfecto I, Snelling R (1995) Biodiversity and tropical ecosystem transformation: ant diversity in the coffee agroecosystem in Costa Rica. Ecol Appl 5:1084–1097

Perfecto I, Vandermeer J (2002) Quality of agroecological matrix in a tropical montane landscape: ants in coffee plantations in southern Mexico. Conserv Biol 16:174–182

Perfecto I, Ricc R, Greenberg R, van der Voolt M (1996) Shade coffee as refuge of biodiversity. Bioscience 46:598–608

Perfecto I, Vandermeer J, Hansen P, Cartín V (1997) Arthropod biodiversity loss and the transformation of a tropical agro-ecosystem. Biodivers Conserv 6:935–945

Perfecto I, Vandermeer J, López-Bautista G, Ibarra-Núñez G, Greenberg R, Bichier P, Langridge S (2004) Greater predation in shaded coffee farms: the role of resident neotropical birds. Ecology 85:2677–2681

Perfecto I, Vandermeer J, Mas A, Soto-Pinto L (2005) Biodiversity, yield, and shade coffee certification. Ecol Econ 54:435–446

Peterson CJ, Haines BL (2000) Early successional patterns and potential facilitation of woody plant colonization by rotting logs in premontane Costa Rican pastures. Restor Ecol 8:361–369

Peters VE, Greenberg R (2013) Fruit supplementation affects birds but not arthropod predation by birds in Costa Rican agroforestry systems. Biotropica 45:102–110

Petit-Aldana J, Uribe-Valle G (2006) Unidad modelo de enseñanza y transferencia de tecnología en conuco (agricultura migratoria): una propuesta. Revista Forestal Venezolana 50:85–91

Philpott SM, Perfecto I, Vandermeer J (2006a) Effects of management intensity and season on arboreal ant diversity and abundance in coffee agroecosystems. Biodivers Conserv 15:139–155

Philpott SM, Uno S, Maldonado J (2006b) The Importance of ants and high-shade management to coffee pollination and fruit weight in Chiapas, Mexico. Biodivers Conserv 15:487–501

Philpott SM, Perfecto I, Vandermeer J (2008a) Behavioral diversity of predatory arboreal ants in coffee agroecosystems. Environ Entomol 37:181–191

Philpott SM, Perfecto I, Vandermeer J (2008b) Effects of predatory ants on lower trophic levels across a gradient of coffee management complexity. J Anim Ecol 77:505–511

Picone C (2000) Diversity and abundance of arbuscular-mycorrhizal fungus spores in tropical forest and pasture. Biotropica 32:734–750

Pineda E, Moreno C, Escobar F, Halffter G (2005) Frog, bat, and dung beetle diversity in the cloud forest and coffee agroecosystems of Veracruz, Mexico. Conserv Biol 19:400–410

Posada JM, Aide TM, Cavelier J (2000) Cattle and weedy shrubs as restoration tools of tropical montane rainforest. Restoration Ecol 8:370–379

Ramírez-Albores JE (2010) Diversidad de aves de hábitats naturales y modificados en un paisaje de la Depresión Central de Chiapas, México. Rev Biol Trop 58:511–528

Román-Miranda ML (2001) Evaluación de cinco especies arbóreas nativas como alimentación de rumiantes en el trópico seco. PhD Thesis, Universidad de Colima, Colima

Rojas L, Godoy C, Hanson P, Hilje L (2001) A survey of homopteran species (Auchenorrhyncha) from coffee shrubs and poró and laurel trees in shaded coffee plantations, in Turrialba, Costa Rica. Rev Biol Trop 49:1057–1065

Rojas-Sánchez A, Hartman-Ulloa K, Almonacid-Márquez R (2012) El impacto de la producción de café sobrela biodiversidad, la transformación del paisaje y las especies exóticas invasoras. Ambiente y Desarrollo 16:93–104

Rossi E, Montagnini F, Virginio-Filho EM (2011) Effects of management practices on coffea productivity and herbaceous species diversity in agroforestry systems in Costa Rica. In: Montagnini F, Francesconi W, Rossi E (eds) Agroforestry as a tool for landscape restoration. Nova Science Publishers, New York, pp 115–132

Sáenz JC, Villatoro F, Ibrahim M, Fajardo D, Pérez M (2006) Relación entre las comunidades de aves y la vegetación en agropaisajes dominados por la ganadería en Costa Rica, Nicaragua y Colombia. Agroforestería en las Américas 45:37–48

Salgado-Mora MG, Ibarra-Nuñez G, Macías-Sámano JE, López-Báez O (2007) Diversidad arbórea en cacaotales del Soconusco, Chiapas, México. Interciencia 32:763–768

Sánchez NV, Vargas-Castro LE, Sánchez A, Amador M (2013) Riqueza y abundancia de mariposas diurnas, escarabajos coprófagos y plantas en cultivos orgánicos y convencionales de tres regiones de Costa Rica. Cuadernos de Investigación UNED 5:249–259

Sánchez-De León Y, De Melo E, Soto G, Johnson J, Lugo-Pérez J (2006) Earthworm populations, microbial biomass and coffee production in different experimental agroforestry management systems in Costa Rica. Caribbean J Sci 42:397–409

Siles P, Martínez-Rayo J, Andino-Rugama F, Molina L (2013) Diversidad arbórea en cercas vivas y dos fragmentos de bosque en la comunidad de Santa Adelaida, Estelí. Encuentro 96:60–76

Slocum MG, Horvitz CC (2000) Seed arrival under different genera of trees in a neotropical pasture. Plant Ecol 149:51–62

Somarriba E (1985) Arboles de guayaba (*Psidium guajava* L.) en pastizales. II. Consumo de fruta y dispersión de semillas. Turrialba 35:329–332

Somarriba E (1995) Guayaba en potreros: establecimiento de cercas vivas y recuperación de pasturas degradadas. Agroforestería en las Américas 6:27–29

Soto-Pinto L, Armijo-Florentino C, Anzueto-Martínez M (2008) La milpa con árboles ixim ´te o taunguya, un prototipo agroforestal. El Colegio de la Frontera Sur, San Cristóbal de las Casas

Stephen C, Sánchez R (2014) Species richness and relative species abundance of Nymphalidae (Lepidoptera) in three forests with different perturbations in the North-Central Caribbean of Costa Rica. Rev Biol Trop 62:919–928

Tejeda-Cruz C, Sutherland WJ (2004) Bird responses to shade coffee production. Anim Conserv 7:169–179

Tobar-López D, Ibrahim M (2008) Valor de los sistemas silvopastoriles para conservar la biodiversidad en fincas y paisajes ganaderos en América Central. CATIE, Turrialba

Tobar-López D, Ibrahim M, Casasola F (2006) Diversidad de mariposas en un paisaje agropecuario del Pacífico Central de Costa Rica. Agricultura en las Américas 45:58–65

Toledo VM (1996) México: diversidad de culturas. CEMEX Agrupación Sierra Madre, México

Toledo VM, Barrera-Bassols N, García-Frapolli E, Alarcón-Chairés P (2008) Uso múltiple y biodiversidad entre los mayas de la península de Yucatán, México. Interciencia 33:345–352

Toledo-Aceves T, Mehltreter K, García-Franco JG, Hernández-Rojas A, Sosa VJ (2013) Benefits and costs of epiphyte management in shade coffee plantations. Agric Ecosyst Environ 181:149–156

Vandermeer J, Perfecto I (1997) The agroecosystem: a need for the conservation biologist´s lens. Conserv Biol 11:591–592

Vandermeer J, Perfecto I (2007) The agricultural matrix and a future paradigm for conservation. Conserv Biol 21:274–277

Van Bael SA, Bichier P, Ochoa I, Greenberg R (2007) Bird diversity in cacao farms and forest fragments of western Panama. Biodivers Conserv 16:2245–2256

Van Der Wal H, Peña-Álvarez B, Arriaga-Weiss SL, Hernández-Daumás S (2012) Species, Functional Groups, and Habitat Preferences of Birds in Five Agroforestry Classes in Tabasco, Mexico. Wilson J Ornithol 124:558–571

Vaninni J (1994) Nearctic avian migrants in coffee plantations and forest fragments of south-western Guatemala. Bird Conserv Int 4:209–232

Vásquez-Vela ALM (2014) Valoración comparativa de la macrofauna de lombrices en sistemas agroforestales de café orgánico y convencional en contraste con cultivos en pleno sol y bosque, durante la época lluviosa y seca en Turrialba, Costa Rica. MC Thesis, CATIE, Turrialba

Vílchez S, Harvey CA, Sánchez D, Medina A, Hernández B (2004) Diversidad de aves en un paisaje fragmentado de bosque seco en Rivas, Nicaragua. Managua. Nicaragua. Encuentros 68:60–75

Villanueva C, Tobar D, Ibrahim M, Casasola F, Barrantes-Arguedas R (2007) Árboles dispersos en potreros de fincas ganaderas del pacifico central de Costa Rica. Agroforestería en las américas 45:12–20

Williams-Guillen K, Perfecto I, Vandermeer J (2008) Bats limit insects in a netropical agroforestry system. Science 320:70

Zamora S, García J, Bonilla G, Aguilar H, Harvey CA, Muhammad I (2001) ¿Cómo utilizar los frutos del Guanacaste (*Enterolobium cyclocarpum*), guácimo (*Guazuma ulmifolia*), genizaro (*Pithecellobium saman*) y jícaro (*Crescentia alata*) en la alimentación animal? Agroforestería en las Américas 8:45–49

Chapter 16
Native Trees and Shrubs for Ecosystems Services and the Redesign of Resilient Livestock Production Systems in the Mexican Neotropics

Francisco Solorio, Luis Ramírez, Saikat Basu, Liz Trenchard, Lucero Sarabia, Julia Wright, Carlos Aguilar, Baldomero Solorio, Armin Ayala and Juan Ku

Abstract Animal production systems in the tropics are currently based mainly on the grazing of animals in the fallow lands and pastures. Conventional livestock systems such as these are, however, strongly affected by limited precipitation. During dry periods, biomass production is reduced; this in turn affects the quality of animal diet and ultimately negatively affects livestock productivity. Often natural woody vegetation is removed to provide grazing areas. In regions where there are

---

F. Solorio (✉) · L. Ramírez · L. Sarabia · C. Aguilar · B. Solorio · A. Ayala · J. Ku
Campus de Ciencias Biológicas Y Agropecuarias, Universidad Autónoma de Yucatán, Carretera Mérida-Xmatkuil Km. 15.5. C.P, 97100 Mérida, Yucatán, Mexico
e-mail: ssolorio@correo.uady.mx

L. Ramírez
e-mail: raviles@correo.uady.mx

L. Sarabia
e-mail: lucy_34_88_sarsal@hotmail.com

C. Aguilar
e-mail: caperez@correo.uady.mx

A. Ayala
e-mail: aayala@correo.uady.mx

J. Ku
e-mail: kvera@correo.uady.mx

S. Basu
Department of Biological Sciences, University of Lethbridge, AB T1K 3M4 Lethbridge, Canada
e-mail: saikat.basu@alumni.uleth.ca

L. Trenchard · J. Wright
Centre for Agroecology, Water and Resilience, Coventry University, Coventry, UK
e-mail: apy157@covnetry.ac.uk

J. Wright
e-mail: aa9469@coventry.ac.uk

M.R. Ahuja and S.M. Jain (eds.), *Biodiversity and Conservation of Woody Plants*, Sustainable Development and Biodiversity 17, DOI 10.1007/978-3-319-66426-2_16

competing demands for natural resources, alternative approaches to livestock production which include a tree or shrub component can have many benefits both for the conservation of natural resources and for improved food security. Integration of native trees and shrubs along with grasses in the fallow pastures could improve plant productivity, nutritional quality of animal diet and help in sustaining natural biodiversity of the local ecosystems. A significant contribution to the future livestock systems could be made with the reforestation of the degraded land pastures. Reforesting would lead to increased water use efficiency, thereby reducing the impact of pests and diseases. Through sequestration of carbon they contribute towards climate mitigation in the form of methane reduction from the ruminants by improving the quality of animal diet with foliage trees and shrubs. Integration of different trees and shrub species in the livestock systems will integrate agroecosystems services for long-term ecological sustainability. Furthermore, incorporating trees and shrubs into the animal systems production will also improve habitat for the local wildlife.

Keywords Ecosystems services · Greenhouse gasses · Livestock systems
Native tree species · Silvopastoral systems

16.1 Introduction

It has been estimated that over 50% of the Earth's surface lies in the tropics and that 75% of the world's inhabitants live in these regions (Thompson 2000). Moreover, more than half of the bovine population is also located in the tropics (Wolfenson et al. 2000). Livestock production in the tropics, especially beef and milk, is performed mainly under extensive systems which are based on grazing. In this sense, the climate conditions greatly influence production of both the animal and the pasture.

In most of the tropics, milk production and beef production show a pattern of seasonality; this is influenced mainly by the rains which normally occur during the summer, but also by the drought season in which the grass is scarce and of low quality. Furthermore, the animals are exposed to extreme heat which in turn leads to low dry matter intake (Rhoads et al. 2009) and poor animal performance (Nienaber and Hahn 2007). Fibrous forages and high environmental temperature not only reduce voluntary feed intake in ruminants, but also increase fermentative heat and need for drinking water (Shibata and Mukai 1979).

Some model climate projections report that precipitation will increase at high latitudes and will decrease in the tropical and subtropical regions (IPCC 2007). Due to a combination of warming and drying, crop yields in the tropics and subtropics are expected to fall 10–20% by 2050 which will be a serious constraint for the existing livestock systems (Jones and Thornton 2003). Moreover, up to 50% loss of available biomasses in pastoral systems has been forecast for Central America (Nardone et al. 2010). Apart from changes in the productivity of rangelands and yields of feed crops, climate change will directly impact livestock production. Heat stress is a major constraint for livestock production (meat, milk, egg yield and quality), reproductive performance, animal health and welfare (Thornton et al. 2009; Nardone et al. 2010; Gaughan 2012; Lara and Rostagno 2013); significant economic losses have been projected due to global warming both in intensive and in extensive livestock production systems (IPCC 2007). According to St-Pierre et al. (2003), North American farms suffered economic losses around 1.69 and 2.36 billion dollars due to heat stress in livestock. This study also reported that economic losses about 58% occurred in the dairy industry, 20% in the beef industry, 15% in pigs and 7% in the poultry industry (Nardone et al. 2010).

As a consequence of the changing climate, animals kept under extensive systems, especially during summer months, will be exposed to an increased risk of health problems, due to increasing populations of internal and external parasites as well as vector-borne diseases (Patz et al. 2000a, b; Wittmann et al. 2001). The predicted modifications of precipitation regimes and the concurrent increase of aridity will have a great impact on the availability of feedstuff for animals as grazing and mixed rain-fed systems will be the most damaged by climate change (Thornton et al. 2009; Nardone et al. 2010).

In the Mexican neotropics, a considerable section of previously forested land has been converted into agricultural lands, especially to grasslands for the livestock. This has led to the degradation of the local ecosystems and loss of biodiversity compared to native forests. There are ways in which the biodiversity and ecosystem service value of pastures can be improved through the inclusion of native trees and shrubs in the current farming system. Not only do agroforestry systems such as these improve the environment; they can also improve crop and livestock yield, reduce reliance on agrochemicals and improve agroecosystem resilience. Agroforestry systems also provide additional benefits for human well-being, directly supplying food and medicines which can improve livelihoods and enhance food security. There are several studies demonstrating native tree species that are adapted to the local conditions, have greater ecosystem service value than many fast-growing exotic species (Garen et al. 2009). However, systems need to be managed well to maximise their benefits; although farmers prefer traditional multipurpose native species, current government policies and extension services in the region favour exotic timber species (Harrison et al. 2005).

16.2 Importance of Reforestation

This section deals with the use of trees, especially native trees, to maintain biodiversity, to improve the landscape connectivity and to provide environmental services, such as carbon sequestration, favour insect biodiversity, etc. In many cases, felling of trees has been justified by replacement planting with seedlings, but this compensation is not adequate. It will take several years before a seedling reaches maturity, and it cannot immediately replace the role a mature trees has in an ecosystem, for example in mitigating climate change. Trees can provide multiple services to human kind. Unfortunately, in many cases these services are not recognised until it is too late. Trees are the translators and moderators of the energy that reach the earth. At crown and canopy levels, trees convert all the energy coming from the wind, sunlight and rainfall into life and growth. Without trees we cannot live on the earth.

Reforestation can be defined as the process of planting trees (either with seedlings or seed) in areas where they have been previously removed. It also includes the rehabilitation of land areas that have been seriously degraded. Reforestation could be carried out with foreign woody species (mainly commercial species) or with native species; these two approaches have strong consequences, which will be discussed in this chapter. Reforestation can be achieved by either sowing seeds or planting seedlings. This could be achieved either as timber plantations of foreign species or by trying to emulate the original forests by planting native trees in a similar structure as the native forests, which may take 50 years or more under tropical conditions to mature. Currently, the objectives of reforestation have been expanded to include socio-economic and environmental benefits, in addition to traditional focus of wood production, soil erosion prevention and water flow.

16.2.1 Fragmented Agricultural Systems

According to Gibson et al. (2011), fragmentation is a process that occurs when habitats are broken down into smaller areas, which are then separated by different land cover types. Anthropogenic activities have resulted, in several cases, in the fragmentation of natural habitats, for example, the deforestation to enable sowing of large areas of grain crops or grasses for supporting the local livestock. The total number of fragmented areas has increased. It has been proposed that the pattern and intensity of fragmentation are associated with land tenure, environmental conditions and production activities (Chakravarty et al. 2012). According to Ochoa-Gaona (2001), the prevalence of rather poor soil conditions, small holdings, growing human population densities, increasing poverty and the absence of alternative employment opportunities will promote high rates of deforestation and forest fragmentation. Therefore, a holistic strategy should be undertaken to mitigate this deforestation and fragmentation process, since it is an extremely challenging and complex task.

16.2.2 Importance for Insect Biodiversity

Trees are important refuge for many organisms, including a great diversity of insect species. Forest provides refugees for insects for pollination. Therefore, deforestation also results in loss of habitats for several insect species, including those that are beneficial as biological pests used in plant disease control. It has been shown that insects belonging to the families *Hemiptera* and *Lepidoptera* have the tendency to increase the number of species in forested areas compared with the monocrop-cover lands (e.g. grasslands, grain crops), which suggests the positive effects of vegetation diversity (which provide a diversity of habitats) on the richness of insect species. However, it has been found that species richness is higher in natural forests than in the reforested areas.

The reforestation to recover the canopy cover could benefit the local insect biodiversity, but the use of local tress species has been more preferable than the exotic genera. Integration of native trees in reforestation increases the habitat value and promotes better and stable growth performances since they are better adapted to the local growing conditions (Piotto 2007). However, the lack of knowledge in the establishment and management of these native species warranties their broad incorporation into reforestation programs (Stimm et al. 2008). It has been reported that the microclimate buffering effect (provided mainly by the canopy) of the surrounding vegetation due to reforestation is more important than the biodiversity of plant species for insect species richness (Cunningham et al. 2015). In addition, early canopy cover is important in the reforestation to stimulate early association with high insect diversity.

16.3 Importance of Trees for Climate Change Mitigation

One of the main roles of trees is the capture of large amounts of carbon from the atmosphere. However, the difficulty to achieve this potential role arises in the complexity for determining the amount of the forest carbon sink that can be managed to reduce the impact of CO_2 emissions. Near to 200 pg of CO_2 is emitted by the change on land use (Canadell and Raupach 2008). According to Canadell and Raupach (2008), climate policies and the inclusion of the tropical regions have key role to promote forestry for climate change protection through carbon sequestration and socio-economic benefits. Deforestation in developing countries, especially in the tropical regions, remain high, where almost 13 Mha/year is deforested (Canadell and Raupach 2008). Nevertheless, these regions could contribute with about 65% of the total carbon mitigation. It is important to indicate that the reforestation, although it could level off large proportion of the carbon emitted to the atmosphere, it has the risk of returning this carbon through disturbances (e.g. uncontrolled fire, which is increased in frequency and intensity due to hotter and longer draughts). The positive effect of trees on climate mitigation is increased in the tropical regions by the cloud formations that help to improve sunlight reflection. Reforestation is a major strategy to mitigate forest emissions (Corbera et al. 2010). Currently, it is estimated that the land transformation (into forest) required is large to cover the CO_2 emissions. For example, China has used about 24 Mha of new forest plantation to level off the CO_2 emissions of 100 years (Canadell and Raupach 2008).

In contrast, reforestation of larger areas aimed to increase carbon sequestration and therefore climate change mitigation and could lead to unwanted negative impacts, e.g. decrease food security, reduce stream flows, loss of biodiversity and local income of farmers. Therefore, it is important that reforestation plans should be undertaken in a sustainable way that considers, in addition to the biophysical benefits, the socio-economic impacts such as the food production and cash flow (additional income) for rural development.

16.4 The Value of Native Trees in Ecosystem Services in Mesoamerica

Trees in the landscape provide a wide range of ecosystem services that are of benefit to the environment and human well-being (Fig. 16.1). A wide range of ecosystem services are recognised, and these services are categorised as provisioning, regulating, supporting and cultural services according to the Millennium Ecosystem Assessment (MEA 2005).

Native trees, adapted over time for their particular biotic and abiotic conditions, arguably provide greater benefits in terms of their ecosystem service value than introduced trees. In agroforestry systems (AFS), native trees can provide additional

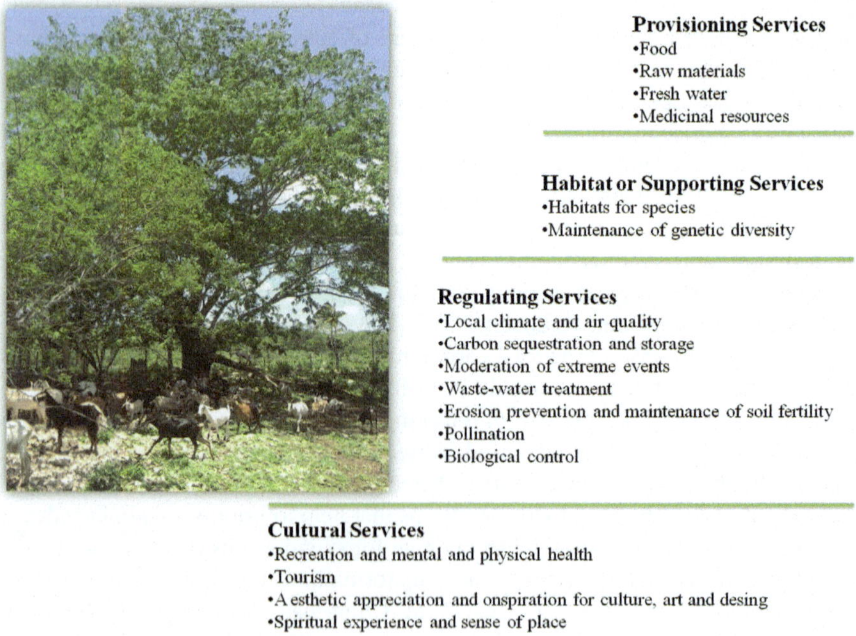

Fig. 16.1 Ecosystem services (modified from Garbach et al. 2014) by native trees

sources of income and food for farmers, as well as other ecosystem functions. In Mesoamerica, there is a long tradition of using trees in home gardens and in mixed farming systems; best examples from Mexico include the '*milpa*' home garden system. These traditional systems pre-date the arrival of European settlers and are still practised in some places as effective and productive small-scale low input agriculture. The '*milpa*' is commonly thought to be a simple three-crop system of maize, beans and squash, but in fact it comprises a succession spanning many years that progresses from 'slash and burn' of the primary or secondary forest, through annuals, to orchards and then forest gardens (Ford and Nigh 2015).

More conventional farmers also report ecosystem service benefits from trees in their landscapes (Fig. 16.2). These trees might be individuals which are self-sown and have been left in pasture for shade or timber, or remnants of primary forest adjacent to their farmland which provide a source of pollinators or beneficial insects

Fig. 16.2 Native trees provide ecosystem services within agricultural landscapes

for the growing crops. Nevertheless, there has been a huge reduction in all types of woodland cover throughout Central America; FAO estimates a loss of 40% of forest cover in the region over the last 40 years (FAO 2006). This need to address the loss of woody cover is widely recognised and there are both international and national reforestation schemes which support the planting of trees for various ecosystem services including carbon sequestration and water management (Hall et al. 2011).

Trees used in reforestation schemes tend often to be exotic timber-producing trees rather than the native multipurpose trees preferred by the local farmers. While all trees, including exotic and invasive species, can provide ecosystem service benefits, invasive trees can have serious negative consequences on both the local environment and farming communities. Exotic trees can have a limited range of benefits compared to multipurpose native trees which are considered to offer a great range of positive ecosystem benefits. Despite this general consensus, the research available to support this aspect is rather limited. The following section discusses the research available supporting the value of native trees and shrubs for ecosystem services. It focuses on the agroforestry practices in Mexico where evidence is available, but also includes primary literature sources from elsewhere in Mesoamerica and from non-native trees where there are gaps.

16.4.1 Provisioning Services and Native Trees

Provisioning services are those which provide food for humans and animals, such as wild fruit trees, or natural resources such as wood for construction and tools; and sources of medicines both for traditional medicine and for pharmaceuticals (MEA 2005). On agricultural land, trees may be planted specifically to provide one or more of these services or may be self-seeded trees which are tolerated and/or managed to maximise resources. Additionally, provisioning services may also be provided by remnants of natural forests or secondary forests close to or on farmland.

Native trees can provide a wide range of direct food products for humans including fruit, pods, nuts, syrups, gums, as well as processed foods or beverages that use various similar or other tree parts such as pods, leaves or bark. Fruit, pods and leaves can also provide an additional source of forage for livestock which is especially useful in the dry season when grass forage may be unavailable (Nair 2012). The type and value of food products provided by native trees on the agricultural landscape vary widely and depend on the species used, type of farming system and environment as well as farmer knowledge. The number and range of species used to provide food in agroforestry also vary widely. Suarez et al. (2012) reported on a study in Veracruz that a third of the useful species identified were used to provide food. Twenty fruit species were reported, and non-fruit species mentioned were *Erythrina* spp., *Yucca* spp. and *Gliricidia sepium* (Jacquin) Kunth ex Walpers (flowers), *Nopalea* spp. (stems), *Leucaena lanceolata* S. Watson ssp. lanceolata and *L. leucocephala* (Lam.) de Wit subsp. glabrata (seeds). In this study, farmers also used the dead wood of trees to cultivate edible mushrooms (Suarez et al. 2012). In Quiotepec, the most important trees and shrubs identified by participants were mezquites (*Prosopis laevigata* (Humb. et Bonpl. ex Willd)), 'guajes' (*L. esculenta* (Moc. Et Sessé ex Dc.) Benth.), lemon ((*Citrus limon* L.) Burm), 'manteco' (*Parkinsonia praecox* (Ruiz et Pavón) Harms) and 'nopales' (*Opuntia* spp.) (Murgueitio et al. 2011). Native trees in AFS may also provide habitat for edible insects such as 'cuchamá' the larvae of the butterfly *Paradirphia fumosa*, and 'cocopache' (Hemiptera). Both have economic value. The larvae cuchamá are normally associated with 'manteco' *P. praecox*. Another important food source is 'pitahaya' *Hylocereus undatus* (Haw.) (Britton & Rose), which is normally associated with mezquite (*P. laevigata*), a legume which also produces edible fruits (Vallejo et al. 2015).

Traditionally, the species grown in agroforestry systems tend to be of multi-purpose. Key ecosystem services provided by woody species include food, forage, timber, fuel and shade (Fig. 16.3 and Table 16.1). While it might be expected that food would be one of the most important ecosystem services, in a number of studies in Mexico other ecosystem services were considered to be of greater importance. For example, in a study carried out in Veracruz, by Moreno-Casasola and Paradowska (2009) reported that of the 18 uses reported-wood used for construction purposes, fences and firewood were all considered by respondents to be more important than food. Similarly, Murgueitio et al. (2011) reported that the most important ecosystem service of the 20 given by respondents in Zapotitlán, was '*beauty of the crop fields*'.

Fig. 16.3 *Brosimun alicastrum* Swarts. (Ramon), a native neotropical tree species, important for livestock production

Table 16.1 List of native of native trees that may be appropriate for reforestation and restoration of the livestock systems production redesign

Common name	Latin name	Common uses	Potential uses	References
Capomo, Ramón, Ojoche	*Brosimum alicastrum*	Fodder	Timber, food, medicine and reforestation	Niembro et al. (2010); Olivares-Pérez et al. (2011)
Cedro, Calicedro, Cedro rojo, Culché	*Cedrela odorata* L.	Timber	Medicine	Cintron (1990)
Ceiba, Pochote, Ochoe, Árbol de algodón.	*Ceiba pentandra* A. DC.	Timber	Medicine	Rodriguez et al. (2003)
Bonete, cuaguayote	*Jacaratia mexicana*	Food and medicine	Honey production and timber	Arias et al. (2010)
Palo mulato o Chaca	*Bursera simaruba* a (L.) Sarg.	Living Fences	Timber, medicine and honey production	Pennington y Sarukhán (2005); Niembro et al. (2010)
Cueramo, bocote, borcino	*Cordia elaeagnoides* A. DC.	Living fences	Timber and medicine	González-Gómez et al. (2006)
Parota, Guanacaste, Orejon, Pich	*Enterolobium Cyclocarpum* (Jacq.) Griseb.	Timber	Fodder and food	Castañeda et al. (2003); Zamora et al (2001); Olivares-Pérez et al. (2011)
Guácimo, guásima, cuaulote, pixoy	*Guazuma ulmifolia* (Roxb.) Benth	Living fences	Fodder and food	Zamora et al. (2001); Hernández et al. (2014)
Pinzan, guamúchil	*Pithecellobium dulce* (Roxb.) Benth	Living fences	Fodder and food	Olivares-Pérez et al. (2011); Hernández et al. (2014)
Mezquite	*Prosopis laevigata*	Living fences	Fodder and food	
Matarratón, Cocoite	*Gliricidia sepium*	Living fences	Fodder	Olivares-Pérez et al. (2011)
Huaje o huaxin	*Leucaena* ssp.	Fodder	Food and timber	Olivares-Pérez et al. (2011)

Trees and shrubs can also provide ingredients for traditional medicine and sources for the local pharmaceutical industry. Several reports describe in general terms the use of native trees in traditional medicine (Ngarivhume et al. 2015); often little detail is supplied other than to state that there are medicinal uses for a particular species, e.g. the Ramon tree in Yucatan (Fig. 16.3), Chiapas and Tabasco states (Hall et al. 2011).

In studies which evaluated the relative importance of various ecosystem services in Mexico, medicine was more often than not a relatively low priority. For example, in a study based in the Tehuacán-Cuicatlán Biosphere Reserve, Vallejo et al. (2015) found that of the 20 reasons given for maintaining trees in their agroforestry system, respondents listed provision of fodder for livestock, fuel, wood and fruit as all more important than medicine. However, farmers are not traditional healers who tend to forage for wild medicines rather than trying to domesticate them.

A wide range of native species provide timber for a multitude of uses including construction, fence posts, tools and furniture (Fuentealba and Martinez-Ramos 2014). Many native species can provide valuable hardwood timber (Aluja et al. 2014). However, in many restoration and reforestation schemes, the preference is often for fast-growing exotic timber species such as eucalyptus, and the use of these is supported by government policy. In Veracruz, government support has been provided for both plantations and restoration forestry, but this support normally includes commercial timber species, such as *Cedrela odorata* L., *Tabebuia rosea* (Bertol.), *T. donnellsmithii* Rose, *Gmelina arborea* Roxb. ex Sm. and *Tectona grandis* L.f., of these only the first two are native (Suarez et al. 2012). However, monocultures of exotic species have a lower ecosystem service value when compared to native woody species. In agroforestry systems, timber is only one desirable ecosystem service and multipurpose native species are preferred by rural communities, some researchers suggesting that this disparity is the reason why many reforestation schemes fail (Suarez et al. 2012). In pastures, trees may principally be left as shade for cattle and people, but by cultivating trees that also have economic value such as timber, they may also act as an insurance policy or as a bank account for difficult times (Fuentealba and Martinez-Ramos 2014).

In agroforestry systems, a wide variety of timber species are used; for example in Chiapas, Orantes-Garcia et al. (2013) identified 35, mostly multipurpose species being used. The main species grown were bojón (*Cordia alliodora* (R. & P.) Oken), palo amarillo (*Terminalia obovata* (R. & P.) Steud), mahogany (*Swietenia macrophylla* G. King), the jolocín (*Heliocarpus donnellsmithii* Rose), cedar (*Cedrela odorata* M. Roem. King) and copalchi (*Croton guatemalensis* Lotsy). There are a number of reasons why exotic timber species are promoted; key reasons include availability of seed and local understanding of species requirements and knowledge of management techniques (Suarez et al. 2012). However, in AFS farmers may prefer slower-growing timber species both because their value is greater when harvested and also because these native species provide additional benefits, for the same cost (Fuentealba and Martinez-Ramos 2014).

16.4.2 Regulating Services and Native Trees

Trees and shrubs can provide a wide range of regulating ecosystem services; these are indirect or direct benefits which enhance, for example, local and global air quality, soil fertility and structure, water availability, quality and storage and pest control. In intensive agricultural and forestry systems, farmers rely on the systematic application of broad spectrum agrochemicals to deal with pest and disease outbreaks, but these have many negative impacts on the environment and human health. An alternative strategy to deal with pests and diseases in more sustainable farming systems is to work with natural pest cycles and ecosystems to understand and prevent pest outbreaks. Incorporating trees and shrubs either in or adjacent to farmland can provide a 'critical reservoir' for beneficial insects, birds and mammals, both pollinators and predators (Chandler et al. 2013). The authors found that there were only very few examples of farmers manipulating woody vegetation to enhance natural enemies and improve pest control in the literature reviewed. In an assessment of the value of native woodlands as a source of beneficial insects, Aluja et al. (2014) reported that fragments of native woodland adjacent to farms provided a range of parasitoids which prey on fruit fly pests and suggested that this may be useful in integrated pest control. However, the authors found also that farmers were not aware of the value of these woodland fragments for pest control, and some farmers were removing native plant hosts such *Spondias radlkoferi* J.D. Smith. Aluja et al. (2014) suggested that there is much scope to develop more sustainable agriculture systems based on a more integrated system of pest control such as that in Asian rice systems. This would be more sustainable and less susceptible to pest outbreaks (Ruiz-Guerra et al. 2014).

The incorporation of trees into the farming landscape can also affect local climate and hydrological cycles, as well as gas exchange and energy balance; these impacts can be demonstrated at various scales. Therefore, careful choice of appropriate woody species and management can ensure that ecosystem services are maximised in the existing AFS. Vegetation structure affects the energy balance in an ecosystem; the main input is solar radiation which drives photosynthesis and provides heat for the environment. Incorporating trees and shrubs can reduce the amount of solar radiation received at the soil surface, altering soil heat fluxes and protecting understory plants and soil biota from overheating. Soil absorbs heat from the sun during the day and releases this heat at night; the structure of vegetation can positively affect this process and impact the thermal movement of air at different scales (Morales-Romero and Molina-Freaner 2015). The structure, density and type of vegetation also change the water balance and water quality in the landscape. Trees and shrubs have a number of benefits in tropical systems where water is limited; vegetation intercepts rainfall, which not only alters the infiltration rate, but also reduces surface run off and evaporation rates (Murgueitio et al. 2011).

Water availability and storage can be improved by the increase in organic matter provided by leaf litter and root systems of trees and shrubs. Also, woody vegetation can also improve water quality, as active processes within the rhizosphere can remove pollutants. In addition to this, some management practices in agroforestry can enhance water availability. A number of tree species are very draught tolerant and can be grown in dry regions to provide additional forage during the dry season (Table 16.1). Some of these trees have deep tap roots as well as a shallow lateral root system; therefore, when water is scarce, they can access groundwater at some depth. *Prosopis* spp., for example, can reach water some 15 m below the soil surface (McCulley et al. 2004; Landmeyer 2011). Trees can also improve the physical conditions in the root zone; root systems improve soil structure, counteract compaction and reduce soil erosion (Murgueitio et al. 2011). Many traditional farming systems practised in Central America that use terraces in their AFS, also conserve water and prevent soil and water erosion.

In the Mezquital Valley of Central Mexico, the Ñañhú people have constructed terraces and borders to manage water and sediments to improve soil and humidity for crops (Vallejo et al. 2015). In surveys carried out in various parts of Mexico, farmers frequently report that water retention is an important reason for their choice of AFS (Murgueitio et al. 2011; Vallejo et al. 2015). For example, in a study by Moreno-Calles et al. (2012) interviewees commented that "*trees call water*" and that this was a very important consideration. Government schemes such as Payment for Ecosystem Services (PES) aim to enhance watershed protection, biodiversity conservation and protect carbon reservoirs and sinks. In Mexico, PES for forest conservation and protection of watersheds, the use of native trees could reduce the need for inputs of chemical pesticides and synthetic chemical fertilisers in agricultural intensification which contaminate water supplies (Chandler et al. 2013).

Trees in the landscape can also provide physical and structural benefits, such as shade, windbreaks, erosion prevention and wind protection (Hernández-Mucino et al. 2015). Trees in pasture systems can provide shade for cattle in the dry season, reducing heat stress and increasing dairy production (Griscom and Ashton 2011). Shade can also provide a favourable microclimate beneath the canopy of the tree; at ground level this can increase humidity, reduce temperature and solar radiation compared to open fields and can enhance crop production. There are many examples of positive benefits in comparison with open pasture; in Arizona, Schade et al. (2009) found that *Prosopis velutina* Woot. improved both soil moisture and soil nitrogen. Other species such as *Schinus molle* L., *Jacaranda mimosifolia* D. Don and *Polaskia chichipe* Rol.-Goss. provide shade for both people and livestock (Moreno-Calles et al. 2012). The main shade trees in the Tehuacán Valley for livestock were *Lysiloma acapulcense* (Kunth) Benth., *Ehretia tinifolia* M. Martens & Galeotti, *Maclura tinctoria* (L.) D. Don ex Steud. and *Ficus cotinifolia* Kunth; of these *L. acapulcense* was preferred because it produces lighter shade that allows grass to grow underneath the canopy (Vallejo et al. 2014).

The benefits of shade in AFS are widely recognised by farmers, and this is often given as one of the important benefits of AFS. In the Tehuacán-Cuicatlán Biosphere Reserve, traditional and intensive practices are often combined, where terraces and borders help to provide shade and modify growing conditions (Moreno-Calles et al. 2012; Muro-Perez et al. 2012; Vallejo et al. 2015). Trees and shrubs also serve as living fences that can provide protection for animal or crops against harsh and desiccating winds, therefore improving growth and yield (Fig. 16.4). In research carried out by Vallejo et al. (2015), *Prosopis laevigata* (H. B. ex Willd.) Johnst. M. C. is noted for its use as a barrier against strong winds as well as its high number of uses.

Fig. 16.4 *Gliricidia sepium* living fences provide a wide range of ecosystems services in the Mexican neotropics

16.4.3 Trees and Shrubs in Living Fences

Living fences can provide a number of positive benefits, including shade, fruit and forage as well as continuity between fragments of woody vegetation (Fig. 16.4); they can also stabilise soil, improve water retention and provide beneficial microclimates for seed germination and for invertebrates (Love et al. 2009). In Mexico, studies have indicated that living fences can support a considerable range of plant species, including species which are typical of mature forest systems. A study of living fences in pasture systems in Los Tuxtlas identified more than 200 higher plant species (Guevara et al. 1994). Living fences may also support a greater range of bird and mammal species and provide links between forest fragments that improve the movement of species in the landscape.

On the other hand, trees and shrubs can enhance soil fertility. The two key macronutrients required for plant growth, N and P, can be deficient in agricultural systems. In intensive systems, these are supplemented by the application of chemical fertilisers; however, these have a wide range of detrimental environmental effects (De Haan et al. 1997). A number of tree species used in AFS are nitrogen fixing, such as *Inga* spp., and can therefore improve nitrogen availability in the agroecosystem (Diemont et al. 2011). Others form symbiotic associations with mycorrhizae, and these can improve P availability in the soil. The addition of trees to the farmed landscape can also enhance carbon sequestration, a valuable ecosystem service.

16.4.4 Cultural Ecosystem Services and Native Trees

A range of services can be included in this category, including recreation and physical health, tourism, aesthetics and appreciation for culture, heritage, art and design, spiritual experience and sense of place (Infield and Mugisha 2013). There was little discussion in the literature of the value of AFS for recreation, health or tourism; the most frequently cited cultural benefits of trees in the landscape by practitioners are aesthetic, spiritual and social cohesion (Bacon et al. 2012). Traditional agroforestry systems involve a degree of community organisation and therefore support positive interaction and co-operation (Vallejo et al. 2015). Surveys of the value ES of native trees also report that respondents gave ethical and spiritual reasons for valuing trees, including aesthetic reasons for maintaining trees and shrubs, referring to plants as '*wild luxury*', e.g. '*sotolín*' (*Beaucarnea gracilis* Lem), 'viejito' (*Cephalocereus columna-trajani* (Karw.) K.Schum). Or stated that '*if there is no reason to remove them, why to do it*' or commented that certain plants provide protections against '*bad spirits*', e.g. the '*pirul*' *Schinus molle* (Vallejo et al. 2014).

16.5 Supporting Services and Native Trees: Habitat and Genetic Diversity

The provision of genetic diversity, which is the diverse range of genes both within and between species, is crucial in order that cultivars and breeds may adapt over time and also serve as a gene pool for the breeding of domesticated and commercial plants and animals. Habitats that have a very large number of diverse species are termed biodiversity hot spots and are protected (Dıaz et al. 2005). Pastures and crop systems tend to be simplified agricultural landscapes with reduced diversity (Harvey et al. 2011). By retaining trees in a pasture system or by incorporating trees and shrubs in an AFS, the biodiversity of available habitat is significantly increased. A mosaic of native vegetation with a range of species and varied structure helps to support an increased number and diversity of local fauna, providing cover for beneficial insects as well as predators (Muro-Perez et al. 2012). Some studies in the Central American region found that AFS support more than 90% of the genetic variation in arid zones (Otero-Arnaiz et al. 2005; Casas et al. 2006; Parra et al. 2010; Cruse-Sanders et al. 2013). In Central America and Mexico, a programme call CAMCORE is orientated to promote ex situ conservation of threatened native forest species from these regions, including the test of native species over a range of environments in the neotropics (FAO 2006; Koskela et al. 2014).

Incorporating trees in the landscape provides a greater diversity of habitats for the pollinators and also improves connectivity between fragments of woodland to encourage seed and gene dispersal. More than three quarters of the most popular food crops require animal pollination; an ecosystem service provided mostly by insects, but also to a lesser extent birds and mammals (Cruse-Sanders et al. 2013).

16.6 Trees and Shrubs for the Redesign of Resilient Livestock Production Systems in the Tropics

A model based on diverse native tree species is presented; the aim is to restore degraded land caused by tropical livestock production (Fig. 16.5). Reconversion of pasture monocrops will be more efficient if the areas dedicated to livestock are reforested with multipurpose trees and shrubs that are arranged in an ecological way that can exploit the positive and complementary interactions (Celis and Sibu 2011). Nitrogen-fixing trees and shrubs should be planted in association with non-leguminous species for timber or fruit production forming alleys (Fig. 16.5); soil root–interactions can maintain high soil biological activity. A significant portion of nitrogen fixed can be released through leaves and root nodules and be transferred to the associated non-leguminous trees. Livestock paddocks reforested with trees such as *B. alicastrum* and *G. sepium* (medium size trees) associated with *E. cyclocarpum* in alley arrange in order to avoid grass shade; these species are suitable for cultivation on rocky and shallow soils, as well as dry conditions like those of the Yucatan Peninsula. *E. cyclocarpum* and *C. pentandra* also improve microclimate conditions by reducing solar radiation (Baligar et al. 2008; Georgi and Zafiriadis 2006).

Intercropping *B. alicastrum* and *Leucaena* can enhance ecological restoration and improve farm productivity. Planting trees in a linear order or in isolated paddocks enables seed dispersal in the agriculture landscape. Enhancing the diversity of tree and shrub species within animal production systems further improves resilience (Harvey and Haber 1998; Dumont et al. 2014). The researchers

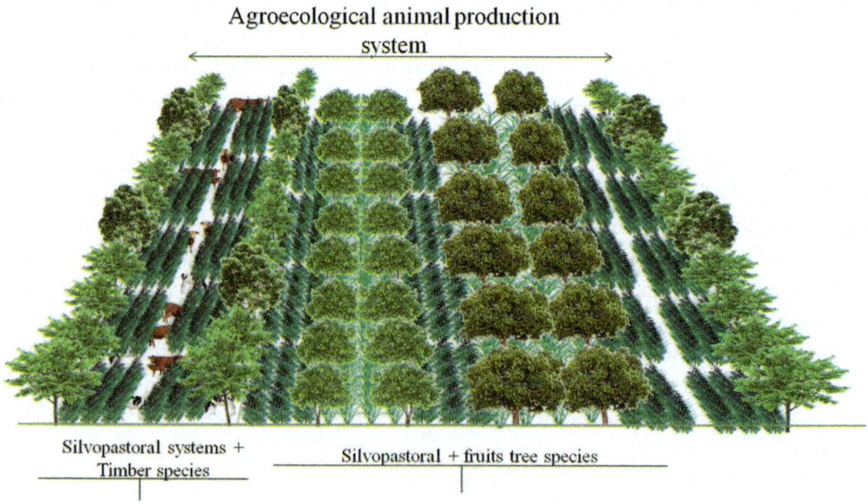

Fig. 16.5 Conceptual agroecological livestock production model based on native tree and shrub species associated with grasses to enhance diversity within animal production systems

recommended five ecological principles for the redesign of animal production systems-introducing leguminous species improving foliage production and quality; hence, reducing input needed for production and decreasing pollution by eliminating the need of nitrogen fertilisers. Combination of valuable native timber species with fodder trees such as *C. odorata* with *B. alicastrum* encourages the successful conservation of local biodiversity.

Livestock systems with diverse tree species are less vulnerable to climate change and produce better quality forage than grasses in monocrops even under adverse conditions. Multipurpose tree species allow diversification of the production system. Additionally, animals can achieve greater weight gains and/or increase milk production including more efficient recycling of nutrients. The livestock redesign with multiple strata formed by trees, shrubs and grasses will encourage positive relationships between the different components to improve resistance to draught by reducing water evaporation and increasing soil water retention during the harsh summer. Forage availability (inter-annual and season) and quality are the main constraints to animal production in the tropics; hence, multiple layers formed by trees also improve animal behaviour allowing animal to acquire plasticity in foraging choices and reducing environmental stress.

16.6.1 The Effect of Trees and Shrubs on Animal Welfare and Productivity

The weather in tropical regions has a direct impact on animal health, growth and reproduction (Rovira 2014). In extensive livestock production systems, livestock is often subject to heat stress as animals are typically left in treeless paddocks without any artificial shade. Without trees to provide shade, animals in open pasture have to endure high ambient temperatures as well as high humidity; animals may also be negatively affected by increases in wind velocity and intensity of solar radiation. The resulting heat stress has a number of adverse effects on the animal and reduces its productivity (Tucker et al. 2008; Aggarwal and Upadhyay 2013). In all animals, heat stress in livestock can negatively affect both fodder intake and conversion efficiency. In lactating animals, a lack of shade can also reduce an animal's resting time and this leads to a decrease in milk production (Bach et al. 2008) as well as lower conception rates (Aggarwal and Upadhyay 2013). While there are several methods available that can be used to alleviate an animal's heat stress (Tucker et al. 2008; Renaudeau et al. 2012; Das et al. 2016), one of the most sustainable ways is to introduce trees into livestock systems (Amendola et al. 2016). Shade provided by trees and shrubs in silvopastoral systems can provide comfort and improve animal performance (Paciullo et al 2011). The positive effects of tree shade on livestock production are summarised in Fig. 16.6.

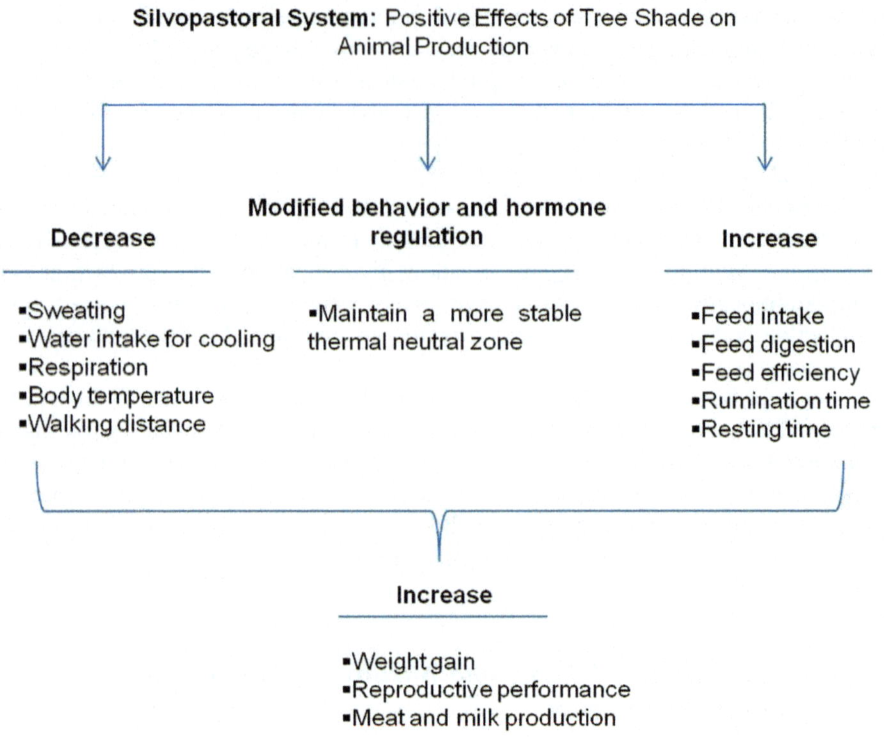

Fig. 16.6 Schematic of the general effects of tree shade in silvopastoral systems for livestock animal production

16.7 Conclusions

Producing livestock using well-designed silvopastoral systems can provide a range of environmental and socio-economic benefits compared to intensive livestock production. In the neotropics, livestock is mainly produced using rain-fed grass open pasture. Compared to the natural woodland habitats they have replaced these simple agricultural ecosystems are much less diverse, less sustainable and less resilient. By including trees and shrubs in pasture, especially carefully chosen native species, the ecological value of farmed landscapes can be vastly improved. Native trees provide additional habitat for local fauna, including beneficial insects, and improve connectivity between remnant woodland habitats. Key benefits for the farmer include increased animal productivity as trees provide shade to reduce heat stress in livestock and additional forage during the dry season. The inclusion of native trees can also improve rural food security when trees also provide fruit or other foodstuffs, as well as producing an additional source of income from non-food products such as timber. However, despite the clear benefits of native tree species for land regeneration and livestock improvement, the potential in Mexico continues to be limited. Research suggests that livestock farmers are interested in using multipurpose native trees on their land and so perhaps further research is required to determine the constraints to their wider use and to develop the policy interventions that may be necessary to achieve the wider adoption of sustainable resilient livestock systems such as these.

Acknowledgements This publication would not have been possible without the generosity supported by the SEVENTH FRAMEWORK PROGRAMME, THE PEOPLE PROGRAMME: International Research Staff Exchange Scheme throughout the project 'Sharing Best Agroecological Practice for Resilient Production Systems in Dryland and Drought Conditions', Grant No. 612686, funded by the European Commission.

References

Aggarwal A, Upadhyay R (2013) Shelter management for alleviation of heat stress in cows and buffaloes. In: Heat stress and animal productivity. Springer, pp 169–183

Aluja M, Sivinski J, Van Driesche R, Anzures-Dadda A, Guillen L (2014) Pest management through tropical tree conservation. Biodivers Conserv 23:831–853

Améndola L, Solorio FJ, Ku-Vera JC, Améndola-Massiotti RD, Zarza H, Galindo F (2016) Social behaviour of cattle in tropical silvopastoral and monoculture systems. Animal 10(05):863–867

Arias DJ, Peñaloza-Ramírez O, Dorado P, Cuevas-Reyes E, Leyva AU, Albarrán-Lara and G. Rangel-Altamirano (2010) Phylogeographic patterns and possible incipient domestication of *Jacaratia mexicana* A. DC. (Caricaceae) in Mexico. Genet Resour Crop Evol 57:1227–1238

Bacon CM, Getz C, Kraus S, Montenegro M, Holland K (2012) The social dimensions of sustainability and change in diversified farming systems. Ecol Soc 17(4):41

Bach A, Valls N, Solans A, Torrent T (2008) Associations between nondietary factors and dairy herd performance. J Dairy Sci 91(8):3259–3267

Baligar VC, Fageria NK, Paiva A, Silveira A, de Souza JO Jr, Lucena E, Jorda J Jr (2008) Light intensity effects on growth and nutrient-use efficiency of tropical legume cover crops. In: Toward agroforestry design. Springer, Netherlands, pp. 67–79

Canadell JG, Raupach MR (2008) Managing forests for climate change mitigation. Science 320 (5882):1456–1457

Casas A, Cruse-Sanders J, Morales E, Otero-Arnaiz A, Valiente-Banuet A (2006) Maintenance of phenotypic and genotypic diversity in managed populations of *Stenocereus stellatus* (Cactaceae) by indigenous peoples in Central Mexico. Biodivers Conserv 15(3):879–898

Castañeda Nieto Y, Álvarez Morales G, Melgarejo Velásquez L (2003) Ganancia de peso, conversión y eficiencia alimentaria en ovinos alimentados con fruto (semilla con vaina) de parota (*Enterolobium cyclocarpum*) y pollinaza. Veterinaria México 34(1):39–46

Celis G, Shibu J (2011) Restoring abandoned pasture land with native tree species in Costa Rica: effects of exotic grass competition and light. For Ecol Manag 261(10):1598–1604

Chakravarty S, Ghosh SK, Suresh CP, Dey AN, Shukla G (2012) Deforestation: causes, effects and control strategies. Glob Perspect Sustain For Manag 1:1–26

Chandler RB, King DI, Raudales R, Trubey R, Chandler C, Chavez VJA (2013) A small-scale land-sparing approach to conserving biological diversity in tropical agricultural landscapes. Conserv Biol 27:785–795

Cintron BB (1990) *Cedrela odorata* L. Cedro hembra, Spanish cedar. Silvics N Am 2(654):250–257

Corbera E, Estrada M, Brown K (2010) Reducing greenhouse gas emissions from deforestation and forest degradation in developing countries: revisiting the assumptions. Clim Change 100 (3–4):355–388

Cruse-Sanders JM, Parker KC, Friar EA, Huang DI, Mashayekhi S, Prince LM, Casas A (2013) Managing diversity: domestication and gene flow in *Stenocereus stellatus* Riccob.(Cactaceae) in Mexico. Ecol Evol 3(5):1340–1355

Cunningham SC, Mac Nally R, Baker PJ, Cavagnaro TR, Beringer J, Thomson JR, Thompson RM (2015) Balancing the environmental benefits of reforestation in agricultural regions. Perspect Plant Ecol Evol Syst 17(4):301–317

Das R, Sailo L, Verma N, Bharti P, Saikia J (2016) Impact of heat stress on health and performance of dairy animals: a review. Vet World 9(3):260–268

De Haan C, Steinfeld H, Blackburn H (1997) Livestock & the environment: finding a balance. European Commission Directorate-General for Development, Development Policy Sustainable Development and Natural Resources, Rome, Italy

Dıaz S, Tilman D, Fargione J, Chapin III FS, Dirzo R, Ktzberber T (2005) Biodiversity regulation of ecosystem services. Trends Conditions: 279–329

Diemont SAW, Bohn JL, Rayome DD, Kelsen SJ, Cheng K (2011) Comparisons of Mayan forest management, restoration and conservation. For Ecol Manage 261:1696–1705

Dumont B, González-García E, Thomas M, Fortun-Lamothe L, Ducrot C, Dourmad JY, Tichit M (2014) Forty research issues for the redesign of animal production systems in the 21st century. Animal 8(8):1382–1393

Dvorak WS, Donahue JK, Hodge GR (1997) Fifteen years of ex situ gene conservation of Mexican and Central American forest species by the CAMCORE cooperative. Rome, Italy, Recursos Geneticos Forestales (FAO)

FAO (2006) Livestock Policy Brief No. 3: Cattle ranching and deforestation. Online available at www.lead.virtualcentre.org or www.fao.org/ag/aga.html

Ford A, Nigh R (2015) The Maya forest garden: eight millennia of sustainable cultivation of the tropical woodlands. Left Coast Press

Fuentealba BD, Martinez-Ramos M (2014) Transplanting native tree seedlings to enrich tropical live fences: an ecological and socio-economic analysis. Agrofor Syst 88:221–236

Garbach K, Milder JC, Montenegro M, Karp DS, DeClerck FAJ (2014) Biodiversity and ecosystem services in agroecosystems. Encycl Agric Food Syst 2:21–40

Garen EJ, Saltonstall K, Slusser JL, Mathias S, Ashton MS, Hall JS (2009) An evaluation of farmers' experiences planting native trees in rural Panama: implications for reforestation with native species in agricultural landscapes. Agrofor Syst 76(1):219–236

Gaughan JB (2012) Basic principles involved in adaption of livestock to climate change. In: Sejian V, Naqvi SMK, Ezeji T, Lakritz J, Lal R (eds) Environmental stress and amelioration in livestock production. Springer, Berlin, pp 245–261

Georgi NJ, Zafiriadis K (2006) The impact of park trees on microclimate in urban areas. Urban Ecosystems 9(3):195–209

Gibson L, Lee TM, Koh LP, Brook BW, Gardner TA, Barlow J, Sodhi NS (2011) Primary forests are irreplaceable for sustaining tropical biodiversity. Nature 478(7369):378–381

González-Gómez JC, Madrigal-Sánchez X, Ayala Burgos A, Juárez-Caratachea A, Gutiérrez Vázquez E (2006) Especies arbóreas de uso múltiple para la ganadería en la región de Tierra Caliente del Estado de Michoacán, México. Livestock Research for Rural Development 18(8)

Griscom HP, Ashton MS (2011) Restoration of dry tropical forests in Central America: a review of pattern and process. For Ecol Manag 261:1564–1579

Guevara S, Meave J, Moreno-Casasola P, Laborde J, Castillo S (1994) Vegetación y flora de potreros en la Sierra de los Tuxtlas, México. Acta Bot Mex 28:1–27

Hall JS, Ashton MS, Garen EJ, Shibu J (2011) The ecology and ecosystem services of native trees: implications for reforestation and land restoration in Mesoamerica. For Ecol Manag 261:1553–1557

Harrison SR, Venn TJ, Sales R, Mangaoang EO, Herbohn JL (2005) Estimated financial performance of exotic and indigenous tree species in smallholder plantations in Leyte Province. Ann Trop Res 27(1):67–80

Harvey CA, Haber WA (1998) Remnant trees and the conservation of biodiversity in Costa Rican pastures. Agrofor Syst 44(1):37–68

Harvey CA, Chacon M, Donatti CI, Garen E, Hannah L, Andrade A, Clement C (2014) Climate-smart landscapes: Opportunities and challenges for integrating adaptation and mitigation in tropical agriculture. Conserv Lett 7(2):77–90

Harvey CA, Villanueva C, Esquivel H, Gómez R, Ibrahim M, López M, Villacís J (2011) Conservation value of dispersed tree cover threatened by pasture management. For Ecol Manage 261(10):1664–1674

Hernández D, Gómez H, Cobos MA, Quiroga R, Pezo D (2014) Árboles Forrajeros de Tres Regiones Ganaderas de Chiapas, México: Usos y Características Nutricionales. Ecosistemas y Recursos Agropecuarios 6(1)

Hernández-Mucino D, Sosa-Montes E, Ceccon E (2015) *Leucaena macrophylla*: an ecosystem services provider? Agrofor Syst 89(1):163–174

Infield M, Mugisha A (2013) Culture, values and conservation: a review of perspectives, policies and practices. Fauna & Flora International, Cambridge UK, p 2

IPCC (Intergovernmental Panel on Climate Change: AR4) 2007 The intergovernmental panel on climate change 4th assessment report. Jackson Institute, University College, London. www.ipcc.ch/

Jones PG, Thornton PK (2003) The potential impacts of climate change on maize production in Africa and Latin America in 2055. Glob Environ Change 13(1):51–59

Koskela J, Vinceti B, Dvorak W, Bush D, Dawson IK, Loo J, Jamnadass R (2014) Utilization and transfer of forest genetic resources: a global review. For Ecol Manag 333:22–34

Landmeyer JE (2011) Introduction to phytoremediation of contaminated groundwater: historical foundation, hydrologic control, and contaminant remediation. Springer Science & Business Media

Lara LJ, Rostagno MH (2013) Impact of heat stress on poultry production. Animals 3:356–369

Love BE, Bork EW, Spaner D (2009) Tree seedling establishment in living fences: a low-cost agroforestry management practice for the tropics. Agrofor Syst 77(1):1–8

McCulley RL, Jobbagy EG, Pockman WT, Jackson RB (2004) Nutrient uptake as a contributing explanation for deep rooting in arid and semi-arid ecosystems. Oecologia 141(4):620–628

MEA (2005) Ecosystems and human well-being: current state and trends. World Resources Institute, Washington, DC

Morales-Romero D, Campo J, Godinez-Alvarez H, Molina-Freaner F (2015) Soil carbon, nitrogen and phosphorus changes from conversion of thornscrub to buffelgrass pasture in northwestern Mexico. Agric Ecosyst Environ 199:231–237

Moreno-Calles AI, Casas A, García-Frapolli E, Torres-García I (2012) Traditional agroforestry systems of multi-crop "milpa" and "chichipera" cactus forest in the arid Tehuacán Valley, Mexico: their management and role in people's subsistence. Agrofor Syst 84(2):207–226

Moreno-Casasola P, Paradowska K (2009) Useful plants of tropical dry forest on the coastal dunes of the center of Veracruz State. Madera Y Bosques 15:21–44

Murgueitio E, Calle Z, Uribe F, Calle A, Solorio B (2011) Native trees and shrubs for the productive rehabilitation of tropical cattle ranching lands. For Ecol Manag 261:1654–1663

Muro-Perez G, Jurado E, Flores J, Sanchez-Salas J, Garcia-Perez J, Estrada E (2012) Positive effects of native shrubs on three specially protected cacti species in Durango, Mexico. Plant Species Biol 27:53–58

Nair PKR (2012) Climate change mitigation: a low-hanging fruit of agroforestry. In: Agroforestry-The future of global land use. Springer, Netherlands, pp 31–67

Nardone A, Ronchi B, Lacetera N, Ranieri MS, Bernabucci U (2010) Effects of climate changes on animal production and sustainability of livestock systems. Livest Sci 130:57–69

Ngarivhume T, van't Klooster CI, de Jong JT, Van der Westhuizen JH (2015) Medicinal plants used by traditional healers for the treatment of malaria in the Chipinge district in Zimbabwe. J Ethnopharmacol 159:224–237

Niembro Rocas A, Vázquez Torres M, Sánchez Sánchez YO (2010) Árboles de Veracruz: 100 especies para la reforestación estratégica. Comision del Estado de Veracruz para la Conmemoración del Bicentenario de la Independencia Nacional y el Centenario de la Revolución Mexicana, Xalapa, México, 253 pp

Nienaber JA, Hahn GL (2007) Livestock production system management responses to thermal challenges. Int J Biometeorol 52:149–157

Ochoa-Gaona S (2001) Traditional land-use systems and patterns of forest fragmentation in the highlands of Chiapas, Mexico. Environ Manag 27(4):571–586

Olivares-Pérez J, Avilés-Nova F, Albarrán-Portillo B, Rojas-Hernández S, Castelán-Ortega OA (2011) Identificación, usos y medición de leguminosas arbóreas forrajeras en ranchos ganaderos del sur del estado de México. Trop Subtrop Agroecosyst 14(2):739–748

Orantes-Garcia C, Angel Perez-Farrera M, Uriel Del Carpio-Penagos C, Tejeda-Cruz C (2013) Native tropical timber resource use in the community of Emilio Rabasa, at the Selva El Ocote Biosphere Reserve in Chiapas, Mexico. Madera Y Bosques 19:7–21

Otero-Arnaiz A, Casas A, Hamrick JL, Cruse-Sanders J (2005) Genetic variation and evolution of Polaskia chichipe (Cactaceae) under domestication in the Tehuacán Valley, central Mexico. Mol Ecol 14(6):1603–1611

Paciullo DSC, de Castro CRT, de Miranda Gomide CA, Maurício RM, Pires MDFÁ, Müller MD, Xavier DF (2011) Performance of dairy heifers in a silvopastoral system. Livest Sci 141 (2):166–172

Parra F, Casas A, Peñaloza-Ramírez JM, Cortés-Palomec AC, Rocha-Ramírez V, González-Rodríguez A (2010) Evolution under domestication: ongoing artificial selection and divergence of wild and managed *Stenocereus pruinosus* (Cactaceae) populations in the Tehuacán Valley, Mexico. Ann Bot 106(3):483–496

Patz JA, Graczyk TK, Geller N, Vittor AY (2000a) Effects of environmental change on emerging parasitic diseases. Int J Parasitol 30(12):1395–1405

Patz JA, Graczyk TK, Geller N, Vittor AY (2000b) Effects of environmental change on emerging parasitic diseases. Int J Parasitol 30:1395–1405

Pennington T, Sarukhán J (2005) Árboles tropicales de México: manual para la identificación de las principales especies (tercera edición). México D.F., México, Universidad Nacional Autónoma de México y Fondo de Cultura Económica, México D.F., México, 523 pp

Piotto D (2007) Growth of native tree species planted in open pasture, young secondary forest and mature forest in humid tropical Costa Rica. J Trop For Sci:92–102

Renaudeau D, Collin A, Yahav S, De Basilio V, Gourdine JL, Collier RJ (2012) Adaptation to hot climate and strategies to alleviate heat stress in livestock production. Animal 6(05):707–728

Rhoads ML, Rhoads RP, Vanbaale MJ, Collier RJ, Sanders SR, Weber WJ, Crooker BA, Baumgard LH (2009) Effects of heat stress and plane of nutrition on lactating Holstein cows: I. Production, metabolism, and aspects of circulating somatotropin. J Dairy Sci 92(5):1986–1997

Rodríguez JAA, Guido JSF, Garrido JT, Bojórquez MMC (2003) Nomenclatura, Forma de Vida, Uso, Manejo Y Distribución de Las Especies Vegetales de la Península de Yucatán. Etnaflora Yucatanense Fasiculo 20 (No. 20). UADY

Rovira P (2014) The effect of type of shade on physiology, behaviour and performance of grazing steers. Animal 8(03):470–476

Ruiz-Guerra B, Velazquez Rosas N, Lopez-Acosta JC (2014) Plant diversity in live fences and pastures, two examples from the mexican humid tropics. Environ Manag 54:656–667

Schade JD, Sponseller R, Collins SL, Stiles A (2009) The influence of *Prosopis* canopies on understorey vegetation: effects of landscape position. J Veg Sci 14(5):743–750

Shibata M, Mukai A (1979) Effect of heat stress and hay concentrate rations on milk production, heat production and some physiological responses to lactating cows. Jpn J Zoot Sci 50:630–644

Stimm B, Beck E, Günter S, Aguirre N, Cueva E, Mosandl R, Weber M (2008) Reforestation of abandoned pastures: seed ecology of native species and production of indigenous plant material. In: Gradients in a tropical mountain ecosystem of Ecuador. Springer, Berlin, pp 417–429

St-Pierre NR, Cobanov B, Schnitkey G (2003) Economic losses from heat stress by US livestock industries. J Dairy Sci 86:52–77

Suarez A, Williams-Linera G, Trejo C, Valdez-Hernandez JI, Manuel Cetina-Alcala V, Vibrans H (2012) Local knowledge helps select species for forest restoration in a tropical dry forest of central Veracruz, Mexico. Agrofor Syst 85:35–55

Thompson LG (2000) Ice core evidence for climate change in the Tropics: implications for our future. Quat Sci Rev 19:19–35

Thornton PK, Van De Steeg J, And Notenbaert A, Herrero M (2009) The impacts of climate change on livestock and livestock systems in developing countries: a review of what we know and what we need to know. Agric Syst 101:113–127

Tucker CB, Rogers AR, Schütz KE (2008) Effect of solar radiation on dairy cattle behaviour, use of shade and body temperature in a pasture-based system. Appl Anim Behav Sci 109(2):141–154

Vallejo M, Casas A, Blancas J, Moreno-Calles AI, Solis L, Rangel-Landa S, Davila P, Tellez O (2014) Agroforestry systems in the highlands of the Tehuacan Valley, Mexico: indigenous cultures and biodiversity conservation. Agrofor Syst 88:125–140

Vallejo M, Casas A, Perez-Negron E, Moreno-Calles AI, Hernandez-Ordonez O, Tellez O, Davila P (2015) Agroforestry systems of the lowland alluvial valleys of the Tehuacan-Cuicatlan biosphere reserve: an evaluation of their biocultural capacity. J Ethnobiol Ethnomed 11(8):1–18

Wittmann EJ, Mellor PS, Baylis M (2001) Using climate data to map the potential distribution of Culicoides imicola (Diptera: Ceratopogonidae) in Europe. Rev Sci Tech 20:731–740

Wolfenson D, Roth Z, Meidan R (2000) Impaired reproduction in heat stressed cattle: basic and applied aspects. Anim Rep Sci 60–61:535–547

Zamora S, García J, Bonilla G, Aguilar H, Harvey CA, Ibrahim M (2001) ¿ Cómo utilizar los frutos de guanacaste (*Enterolobium cyclocarpum*), guácimo (*Guazuma ulmifolia*), genízaro (*Pithecellobium saman*) y jícaro (*Crescentia alata*) en alimentación animal? Agroforesteria en las Americas 8(31):45–49

The manufacturer's authorised representative in the EU is Springer Nature Customer Service Centre GmbH, Europaplatz 3, 69115 Heidelberg, Germany. If you have any concerns regarding our products, please contact ProductSafety@springernature.com

Printed and bound by CPI Group (UK) Ltd, Croydon, CR0 4YY

30/09/2026

02225563-0014